Statistical Computing
in C++ and R

Chapman & Hall/CRC
The R Series

Series Editors

John M. Chambers
Department of Statistics
Stanford University
Stanford, California, USA

Torsten Hothorn
Institut für Statistik
Ludwig-Maximilians-Universität
München, Germany

Duncan Temple Lang
Department of Statistics
University of California, Davis
Davis, California, USA

Hadley Wickham
Department of Statistics
Rice University
Houston, Texas, USA

Aims and Scope

This book series reflects the recent rapid growth in the development and application of R, the programming language and software environment for statistical computing and graphics. R is now widely used in academic research, education, and industry. It is constantly growing, with new versions of the core software released regularly and more than 2,600 packages available. It is difficult for the documentation to keep pace with the expansion of the software, and this vital book series provides a forum for the publication of books covering many aspects of the development and application of R.

The scope of the series is wide, covering three main threads:
- Applications of R to specific disciplines such as biology, epidemiology, genetics, engineering, finance, and the social sciences.
- Using R for the study of topics of statistical methodology, such as linear and mixed modeling, time series, Bayesian methods, and missing data.
- The development of R, including programming, building packages, and graphics.

The books will appeal to programmers and developers of R software, as well as applied statisticians and data analysts in many fields. The books will feature detailed worked examples and R code fully integrated into the text, ensuring their usefulness to researchers, practitioners and students.

The R Series

Statistical Computing
in C++ and R

Randall L. Eubank

Arizona State University
Tempe, California, USA

Ana Kupresanin

Lawrence Livermore National Laboratory

Livermore, California, USA

CRC Press
Taylor & Francis Group
Boca Raton London New York

CRC Press is an imprint of the
Taylor & Francis Group an **informa** business

A CHAPMAN & HALL BOOK

Cover Image: Goran Konjevod/Organic Origami/organicorigami.com

CRC Press
Taylor & Francis Group
6000 Broken Sound Parkway NW, Suite 300
Boca Raton, FL 33487-2742

First issued in paperback 2022

Version Date: 2011928

ISBN 13: 978-1-03-247761-9 (pbk)
ISBN 13: 978-1-4200-6650-0 (hbk)

DOI: 10.1201/b11538

Visit the Taylor & Francis Web site at
http://www.taylorandfrancis.com

and the CRC Press Web site at
http://www.crcpress.com

For Lisa and Goran

Preface

When one looks at a book with "statistical computing" in the title, the expectation is most likely for a treatment of the topic that has close ties to numerical analysis. There are many texts written from this perspective that provide valuable resources for those who are actively involved in the solution of computing problems that arise in statistics. The presentation in the present text represents a departure from this classical emphasis in that it concentrates on the writing of code rather than the development and study of numerical algorithms, per se. The goal is to provide a treatment of statistical computing that lays a foundation for original code development in a research environment.

The advancement of statistical methodology is now inextricably linked to the use of computers. New methodological ideas must be translated into usable code and then numerically evaluated relative to competing procedures. As a result, many statisticians expend significant amounts of their creative energy while sitting in front of a computer monitor. The end products from the vast majority of their efforts are unlikely to be reflected in changes to core aspects of numerical methods or computer hardware. Nonetheless, they are modern statisticians that are (often very) involved in computing. This book is written with that particular audience in mind.

What does a modern statistician need to know about computing? Our belief is that they need to understand at least the basic principles of algorithmic thinking. The translation of a mathematical problem into its computational analog (or analogs) is a skill that must be learned, like any other, by actively solving relevant problems. It is also important to have some comprehension of how computers work in order to avoid or even recognize the more common pitfalls that arise from, e.g., finite precision arithmetic.

Perhaps the most fundamental skill is one that provides a means to communicate with a computer—usually through a computer language. However, many statistical computing texts focus on general developments that are not language specific. It is our belief that there is much to be learned from translating, e.g., pseudo-code into something that actually carries out computations and generates output. The downside of taking this path is that specific language choices must be made thereby running the risk of not meeting the needs of every reader. Our specific choices of C++ and R have been made with an eye toward minimizing this risk.

The R language is, arguably, the de facto standard for statistical research purposes. There are now many books that detail its use (along with that of add-on packages) for the solution of data analysis problems. In a broader sense, R is a very powerful functional language that merely happens to have built-in (and add-on) tools that perform some of the standard (and not so standard) statistical calculations with data. We take this latter viewpoint here and focus primarily on describing the use of the functional and object-oriented nature of the R language.

We feel that a good working knowledge of R is a must for any statistician whose job description involves methodological development. We also firmly believe that R is not (and was never meant to be) suited for the solution of all computing problems that arise in the practice of statistics. Our own experience has shown that having familiarity with another compiled (rather than interpreted like R) language is an essential ingredient for computer based problem solving. Our choice for this "second language" is C++. There are various reasons for this that include its object-oriented structure that has some similarity with features that are present in R. It is also easy to use C++ code in R through the creation of shared libraries. Thus, the use of C++ does not preclude the use of R and conversely. Indeed, we view the relationship between our use of C++ and R as symbiotic rather than competitive. For example, the two languages provide environments that differ sufficiently to where simultaneous (either totally or in part) creation of code in both languages can be invaluable for the detection of coding/logical/mathematical errors and we routinely use them in this manner. But, more importantly, both languages offer unique features that, when used in tandem, can take code development beyond what can be obtained from either language alone.

After a brief discussion of object-oriented programming concepts in Chapter 1, the book proceeds in the following manner. Chapter 2 discusses floating-point representation of numbers and the consequences of round-off error while beginning the introduction to C++. The Chapter 2 treatment is expanded in Chapter 3 with a sketchy, but sufficient for our purposes, development of the C++ language. A somewhat detailed discussion of random number generation is provided in Chapter 4. Then, Chapters 5 and 6 deal with programming in R. The Chapter 5 material is directed toward writing new functions that rely on existing classes and methods. Chapter 6 then explores the R class structure and the idea of generic functions.

Chapter 7 represents a change point in the emphasis of the text relative to Chapters 1–6. The focus shifts from learning languages to code development for the solution of specific mathematical problems. For this reason, code listings become more complex and some programs are not shown explicitly. This is partially for space considerations. But, the presumption at this point is that the reader has emerged from the R/C++ "boot-camp" of Chapters 3, 5 and 6 with the ability to write their own code to fill in any gaps that are left in the text. Indeed, many of the missing functions and programs are posed as exercises at the end of each chapter. There are also two appendices that give complete listings of the matrix, vector and random number generation classes that occupy important roles in Chapters 7–8.

Statistics is about the manipulation of data. Thus, it is somewhat surprising that the concept of abstract data structures (ADTs) has not received more interest in the statistics community. As data becomes more and more complex, we believe it will become increasingly important for statisticians to have a working knowledge of this area. Chapter 9 gives a broad overview of ADTs with applications in both C++ and R. Then, Chapter 10 follows up with a treatment of ADT implementations that are available in C++ through the language's container classes.

Parallel computing is no longer a concept accessible by the elite few. Our desktop machines are perfectly capable of carrying out computations in parallel and, by doing so, speeding up the computations by a factor that can be as large as the number of processors. Moreover, the application programming interfaces (APIs) that provide the key to creating parallel code are relatively easy to learn and use. Chapter 11 gives an introduction to parallel computing that includes the OpenMP and MPI APIs for shared and distributed memory systems. Parallel computing in R is also covered through consideration of the Rmpi and multicore packages.

The subject matter in this book has been taught at two levels: first year, first semester statistics graduate students (at Texas A&M) and third (or higher) year statistics Ph.D. students (at ASU). The present treatment is most suited and aimed toward students with the latter skill set. As a result, there is the presumption of a pre-existing working knowledge of both R and Unix shell programming. Appendices are provided to aid in filling any knowledge gaps the reader might have in these two areas.

The serial and OpenMP programs in the book were tested on both Linux and Mac OS X operating systems. The MPI code was tested on the ASU Saguaro cluster. No attempt has been made to test our programs in a Windows environment.

A rough timetable for teaching the topics in the book over a fifteen week semester might allocate one week to each of Chapters 2, 4, 9 and 10, two weeks to Chapters 3, 7, 8 and 11 and three weeks to Chapters 5–6. The material in Chapter 11 could be augmented by access to and use of local high performance computing resources. The class that has been taught at ASU ended with students learning how to run code on the Saguaro cluster at the Fulton High Performance Computing Center. The ASU teaching environment has always been a Linux computer lab wherein both students and instructor wrote and ran programs during the course of each lecture.

Many of the programs presented in the book can be downloaded from `https://math.asu.edu/~eubank/CandR`. While we have paid some attention to writing efficient code, the resulting programs are (very) unlikely to provide optimal performance in any sense. Our goal is the illustration of concepts and the most direct (even brute force) formulations that follow this mandate are what we have used. We encourage readers to rework our programs to obtain alternative approaches that improve on our attempts.

Every chapter, except the first, includes a number of exercises. Their level of difficulty ranges from elementary to challenging. In this latter instance we have developed prototype solutions. But, complicated problems allow for different solution options and we would hope that student solutions would be both varied and novel in many of these instance.

This book has gradually evolved over a period of about 10 years. There are many people that have contributed significantly to its progress and final form. From Texas A&M, the fingerprints of James Hardin, Joe Newton and Shane Reese (now at BYU) can be found at each turn of the page. Phil Smith at Texas Tech University and Andrew Karl, Steven Spiriti, Dan Stanzione and Guoyi Zhang at ASU have aided immeasurably in helping us to learn some of the ins and outs of parallel computing and random number generation. Of course, none of this would have been possible without Henrik Schmiediche (at A&M) and Renate Mittleman and Vishnu Chintamaneni (at ASU) that helped us with all our hardware and software issues over the years. We gratefully acknowledge the input from a diligent reviewer that caught so many errors and inconsistencies while pointing out new (at least to us) ideas to consider and implement. A special word of thanks is also due to NSF who provided the funding for the Council on Statistics server that has played a pivotal role in both the development and teaching of the material in the text.

Last, but nonetheless first, is recognition of our spouses Lisa and Goran that have provided both technical and emotional support throughout the writing process. More than that of any others, their friendship and understanding has made it possible to undertake and finish this project.

RANDALL L. EUBANK

ANA MARIA KUPRESANIN

Contents

List of Algorithms

Chapter 1

Introduction

It is not uncommon to find that a book on computational statistics will be absent of an explicit definition of what comprises its subject matter area. This may be due, in part, to the amorphous nature of the field that stems from its broad scope. Here we will venture to give a definition of the topic that will at least be applicable to the subsequent pages of this text. Specifically,

Definition 1.1. *Computational statistics is the development and application of computational methods for problems in statistics.*

A somewhat more refined notion of computational statistics might view it as the translation of intractable, as well as tractable in many cases, mathematical problems that arise from the statistics genre into a form where they are amenable to (mostly approximate) solution using computers. This translation generally takes the form of an *algorithm* that represents a step-by-step description of the calculations that must be undertaken to provide the desired solution. But, there is more than just algorithmic development involved in computational statistics in that "exact" solutions cannot be expected from computer calculations. This is by design in many instances where the intent is actually to compute an approximation to an "exact" solution that agrees with the target value only to within some specified tolerance. There is also the issue of round-off error that arises from the fact that irrational numbers cannot be stored in their entirety in the finite amount of memory available in a computer. The fact that approximations are employed means that some type of error analysis is needed and, as a result, this is an aspect of numerical analysis as well as computational statistics.

To carry out scientific computing one needs to be "conversant" in an effective programming language. For statistics, a popular language is provided by R. This furnishes an open source computing environment that has strong ties to the S language developed at Bell Labs. A commercial implementation of S is available under the name of S-Plus. R is not to be confused with computer packages that provide computational tools for calculating various statistical quantities of interest for specific data sets. Certainly R is capable of working in that capacity. However, R is a powerful language in its own right that is ideal for addressing many computational statistics problems in the sense of our definition.

R is very good at what it has been designed to do. But, not every statistical computing problem is amenable to solution using R and even some that can be solved with R are more efficiently handled using another approach. There are many compiled languages that can be used as a supplement to R and conversely. Standard choices include C, C++, Fortran and Java. Much of the R language relies on underlying C code and, not surprisingly, it is relatively straightforward to import C programs into R. The same is true for C++ and this latter option has the additional feature of, like R itself, having an object-oriented structure. As a result, it seems natural to use R and C++ individually or in tandem as languages to provide the computer interface component for computational statistics. That is the path that will be pursued here.

1.1 Programming paradigms

The purpose of a computer language is to provide an avenue of "communication" between a (typically human) user and a computer. There are many ways such avenues can be created and each language will differ in terms of the specific way it performs this task. Rather than focus on implementational details, the differences and similarities between computer languages can be more readily appreciated by taking the broader view of this section that examines languages in the context of their relation to common programming paradigms.

Computer programming languages are sometimes characterized as being *imperative* or *declarative*. The distinction is based on whether the program specifies a sequence of statements to be executed or a sequence of definitions and requirements without directly specifying how to achieve them. Imperative languages are usually also *procedural*, allowing one to build procedures as blocks of executable statements that can be executed repeatedly by calls from various points within a program. Examples of procedural languages are C, C++, Java, Algol, COBOL and Fortran.

Historically, computer programming began with the use of *machine code* that communicated instructions to the computer in its native binary language. This was followed by *assembly language* that paired specific binary instructions with shortcut symbols that were then translated to machine code through an assembler. Procedural languages provided the next big advance by allowing code to be developed using standard words, phrases and mathematical expressions that are related to a desired action by the processor. Languages of this nature (notably Fortran) used *compilers* to translate commands into binary sequences that could be understood by the computer.

Procedural programming decomposes a problem into component parts that can be solved using one or more subprograms, usually called *functions* or *subroutines*, that act on variables that are defined in the program. Generally a subroutine has arguments but no return value. Functions, on the other hand, typically take arguments and return information after they have completed their task. In C/C++ the `void` return type allows functions to bypass the return step which enables them to also serve in a subroutine capacity.

Procedural programming is algorithmic by nature with each subprogram or procedure in a set of code often corresponding to a single step in a global algorithm that has been designed to solve the problem at hand. This approach lends itself to code reuse in that subroutines and functions can be employed again when similar subproblems arise. A disadvantage is that, unless safeguards are put in place, data can move freely across subprograms in a manner that allows for possibly inappropriate access or use.

Declarative languages are sometimes further divided into *functional* and *goal-oriented*. Generally, in functional languages one expresses the computation as the evaluation of a function. This evaluation, of course, may be divided into smaller pieces, each again involving the evaluation of a single function. Common Lisp, Mathematica and R are functional languages. In a goal-oriented language, the programmer specifies definitions and rules and lets the system find a "solution" satisfying the definitions by using a built-in general search loop. An example of a goal-oriented language is Prolog.

A different categorization may be based on the basis of how a programming language deals with data. In this context a frequently used paradigm is *object-oriented* programming. Many modern programming languages are object-oriented, regardless of whether they are otherwise procedural or functional. For example, both C++ and Java are object-oriented, procedural languages while Common LISP and R are functional languages that also have extensive object-oriented features.

The basic notion underlying the object-oriented approach is that both data and the methods that manipulate it are packaged into a single object. This creates a programming philosophy wherein objects, rather than procedures, are the fundamental building blocks that work individually or in combination to carry out a specific task. These objects are designed by first creating an abstract conceptualization of some aspect of a problem and then formalizing the notion in a computer code construct called a *class*. An instance of a class that is created inside a program is called an *object*. It encapsulates its own internal data and the methods/functions/subroutines that are appropriate for use with the data it contains. This "encapsulation" feature has the benefit of providing a mechanism for data protection. The object-oriented framework is also amenable to code design and reuse in that the creation of objects tends to enforce subtask modularity.

As one might expect, problem decomposition in functional programming involves the creation of a set of functions that work in tandem to find a solution. Under strict adherence to the functional programming philosophy, functions can only take inputs and produce outputs; they are not allowed to make changes in the data provided to the program. Instead, each function operates on its input with its output then providing the source of input for another function or functions along the solution path. By disallowing changes in input data functional programming represents a kind of counterpoint to the object-oriented approach where the creation and alteration of objects is the focus of a program.

1.2 Object-oriented programming

This section gives a high-level introduction to the object-oriented programming (OOP) concept. These ideas recur in later chapters and it will be helpful to have a common technical dialect with which to discuss them.

The focus of OOP is creation of classes. A class provides an abstract description of a concept through specification of its attributes and its behavior. For scientific computing the "concept" in question will usually represent a mathematical operation such as solving a system of equations, generating a random number or finding the minimum of a given function. In this respect a class should be designed with a single meta-task in mind. For example, a class for numerical linear algebra should probably not contain methods for checking that an integer is a prime number. The construction of focused classes has the benefit of simplifying code development as well as enhancing the opportunities for code reuse.

The attributes of a class are its data components or *members* and its member functions or *methods*. Data members often belong to *primitive data types* in the sense of being integer, floating-point or other types of variables that are inherent to the programming language. However, they can be more abstract in the sense of deriving from other user-created data types or classes. As noted in the previous section, an instance of a class for which data members have been given specific values is called an *object*.

The behavior of a class object depends on the class methods that determine the type of actions the object can perform as well as the ways the object can be altered. In combination the data and function members provide answers to the essential design questions

- What are objects of the class supposed to do?
- What data or information will an object require to carry out its task?

The members of a class can be provided with different levels of protection. Specifically, class members can be classified as *private, public* or *protected*. The private designation has the consequence that such members and methods can be used only within the class itself. Members and methods that are public are open to use outside the class by other functions and objects. Protected class member and methods are only available to derived class objects that inherit from the class in a sense to be described shortly.

The OOP approach provides us with various types of *polymorphism*. The word "polymorphism" means that a single thing can take different forms. In the context of OOP this translates to functions and classes that adapt or behave in different ways depending on the data types furnished as function arguments or during object construction. In C++ polymorphism is obtained through 1) function overloading, 2) template classes and functions and 3) virtual functions.

Function overloading happens when there are multiple functions with the same name that take different argument types. In C++ we can use this facility to create methods for matrix multiplication, scalar multiplication of a matrix, matrix-vector multiplication, etc., that are all invoked by the ∗ operator normally used for scalar multiplication.

The word "template" suggests a formatted structure that has been created for repeated use in recurring, similar applications. For example, a template form for a job application letter would likely contain preset information about an applicant's qualifications, contact addresses, etc., that would be pertinent for all its possible recipients. Other information such as the salutation and address of the recipients would be left blank so the document could be adapted to specific cases. Class and function templates in object-oriented programming behave analogously to the letter template idea. The role of "blank spaces" is played by parameters that specify data types. For example, using this approach an input/output operation can be tailored to work differently and appropriately for the input/output of integers, floating-point values, characters or strings of characters. In this sense a template provides a family of classes or functions whose elements are indexed by the possible values for their parameters.

Inheritance is a feature of object-oriented languages that allows new classes to be constructed from existing or *base* classes. The new or *derived* classes inherit the public members and methods of the base class and can modify them if necessary to produce a polymorphic behavior. For example, a banded matrix is a matrix whose special structure can be exploited to realize savings in terms of both storage and speed of various matrix operations including inversion. Thus, we might envision a general matrix class from which a banded matrix class is derived. While some of the methods from the base matrix class may work well with banded matrices, methods such as matrix-vector multiplication should be specialized to allow for proper use of the band limited structure. The approach is to again create another version of the ∗ operator, except that in this instance the solution lies in making ∗ a *virtual method*; this allows ∗ to coexist as a method in both the base and derived class but behave differently depending on whether a full or banded matrix object uses the operator. Note that this is not the same as function overloading where the function's arguments determine its behavior.

At this point a few (possibly) new words have been learned and some (hopefully) descriptive, albeit somewhat vague, explanations have been provided of what these words might mean. The material in succeeding chapters is aimed at providing some depth to the general OOP concept as it has been described here. However, the focal theme of this text is scientific computing which carries the consequence that a thorough treatment of OOP is beyond its scope. A comprehensive development of OOP can be found in, e.g., Booch, et al. (2007).

1.3 What lies ahead

The next chapter begins with a discussion of how computers represent numbers. The fact that computers have only finite amounts of memory poses problems for dealing with irrational numbers. Exact storage would require infinitely many memory locations which makes it is necessary to compromise and settle for an approximate representation. This leads to the discussion of round-off error and other consequences of working with floating-point arithmetic. Some treatment of this topic is almost obligatory in a book on numerical methods. This is more likely due to the importance of the subject than maintenance of a convention.

A bad bit of advice on round-off error is "use double precision and don't worry about it". The bad part of this recommendation is not the use of "double precision", which is certainly

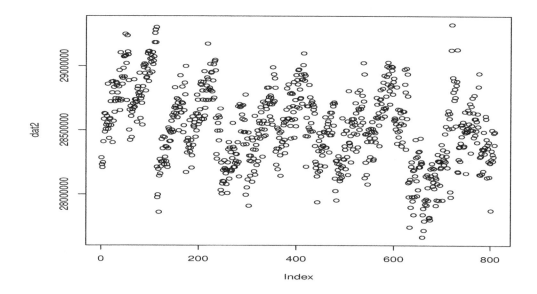

Figure 1.1 *Time series data*

a good practice in most settings. Instead the problem is in believing that increasing the precision solves all numerical problems and allows one to take a cavalier view of the whole round-off issue.

The data in Figure 1.1 represents values of a computer system performance variable that were taken at one-second intervals for a total of 810 seconds. The data will be analyzed using an autoregressive model of order 2; that is, the observed responses $x_1, \ldots, x_{810}$ will be treated as realizations from

$$X_t = \beta_0 + \beta_1 X_{t-1} + \beta_2 X_{t-2} + e_t, \ t = 3, \ldots, 810,$$

with the e_t being uncorrelated random errors with zero means and common variance. The goal then is to obtain estimators of β_0, β_1 and β_2. One approach to estimation is through a conditional least squares criterion where we minimize

$$\sum_{t=3}^{810} (x_t - b_0 - b_1 x_{t-1} - b_2 x_{t-2})^2 \tag{1.1}$$

as a function of b_0, b_1, b_2 conditional on the first two observations that are used to initialize the recursion (e.g., Shumway and Stoffer 2010).

One can obtain the minimizers of (1.1) directly. However, if observations are arriving in real time it may be of interest to compute the estimators at each time point and then update them when a new observation arrives. The first step in this updating algorithm is to compute coefficient estimators corresponding to a fit of the fifth response based on the first four observations. This leads us to consideration of the normal equations $X^T X b = X^T x$ for $b = (b_0, b_1, b_2)^T, x = (x_3, x_4, x_5)^T$ and X given explicitly by

```
> X
      [,1]        [,2]        [,3]
[1,]      1  28232697  28285807
[2,]      1  28209724  28232697
[3,]      1  28248568  28209724
```

Now, this X matrix is nonsingular and the coefficients can be determined directly using its inverse. But, the computational scheme evolves through updating $(X^T X)^{-1}$ which requires it to be computed at this initial phase of the algorithm. An attempt to invert $X^T X$ in R produces the response

```
> solve(t(X)%*%X)
Error in solve.default(t(X) %*% X) :
  system is computationally singular:
  reciprocal condition number = 1.33467e-22
```

But, X is nonsingular which means that the same is true for $X^T X$. The problem is one of numerical singularity as indicated by the R error message. This can be viewed as meaning that in terms of the precision that the machine can maintain for the size of numbers involved in the computation, the matrix in question cannot be distinguished from one that is singular. Another interpretation is that the numbers in the calculations have exceeded the ability of the computer to represent them in any meaningful sense.

The cause of the problem is actually quite easy to diagnose. It stems from the disparity between the size of values in the first unit column of X and those in the two columns that contain the responses. As we will see in Chapter 2, manipulation, particularly addition, of numbers of very different magnitudes causes problems for computer arithmetic. The solution is to take a standard piece of advice from courses on linear regression: standardize variables by, e.g., subtracting their means and dividing by their standard deviations. Once this transformation is performed the computations involving the data in Figure 1.1 proceed without difficulty. Although the solution is not novel, this example does illustrate that a little knowledge of a computer's limitations can remove the mystery from computational outcomes that might otherwise seem puzzling.

Chapter 2 is also where some of the elementary aspects of C++ will be introduced. A detailed exploration of the language then begins in Chapter 3 with discussions of the C/C++ function concept and of execution control using conditional `if/else` blocks and `for/while` loops. With this as background, the next step is into the OOP aspects of the language that allow for creation of new data types with classes and their object instances. We also introduce the C++ approach to generic programming using template classes and functions.

One of the powerful aspects of C++ is the ability it provides for dynamic memory allocation. This allows one to delay specification of the size of objects, such as arrays, until run-time. In particular, this makes it possible for the memory needs of a program to be determined by input. Dynamic memory management is accomplished through a special type of variable called a *pointer* whose value is a location in the computer's physical memory. The use of pointers to create array structures is a key to effective use of C/C++ in scientific computing. Accordingly, we spend some time in Chapter 3 on developing this idea and then illustrate its utility via the construction of C++ classes that can represent matrices and vectors.

Although statistics is a mathematical science, it has a commonality with the lab sciences as well. In this respect the computer represents the statistician's lab and she/he uses it, just like a lab scientist, to experiment with new ideas related to statistical methodology. These experiments are usually of the Monte Carlo variety that rely on artificial data to assess the performance of new (and old) procedures. This makes random number generation an obvious candidate for Chapter 4 that marks the beginning of our topical treatment of statistical computing. The basic issue is how to produce a random sample from a uniform distribution on the interval $[0, 1]$. This particular problem is examined in some detail before moving on to generation of random numbers for general distribution types.

Chapters 5–6 are devoted to the use of R. Some of the elementary aspects of R are presented in Appendix B. Chapter 5 starts from this foundation and then focuses on writing new functions to carry out tasks in the R environment. R and C++ are quite compatible in the sense that it is possible to use C++ functions in R and conversely. This means that the two languages can be used in tandem, rather than exclusively, to solve statistical computing problems. Chapter 5 contains a discussion of how one can tap into this powerful interface with shared C++ libraries that can be loaded into R.

R possesses a class structure that has certain similarities to the one available in C++. On the other hand, the functional nature of R creates a number of important differences. Basically, C++ and R classes are similar in terms of how they deal with data members. The difference is that classes in R have no member functions. Instead, generic functions are created where the class of an argument determines a function's behavior. The result is that the effective use of the OOP features of R requires developments that go beyond and in different directions than those that were presented in Chapter 3 for C++. These new ideas are explored in Chapter 6 where the techniques for creating new classes and new generic functions (and their associated method functions) in the R language are demonstrated.

Numerical linear algebra is a dominant area of scientific computing. While many problems are inherently linear, even nonlinear problems are often linearized with the consequence that linear equations, eigenvalue problems, etc., arise in a myriad of contexts. In statistics, linear equations are frequently encountered that arise from fitting linear models and, more generally, linear least-squares problems. Eigenvalues and singular values play a crucial role in multivariate analysis, for example.

Chapter 7 treats some of the topics from numerical linear algebra that have particular relevance for statistical computing. The Cholesky decomposition features prominently in this development as a useful tool for solving both full and band-limited linear systems. Banded matrices and the Cholesky method also provide our first foray into the use of the C++ inheritance mechanism. Chapter 7 represents a departure from its predecessors in that C++ and R are used in concert with ideas being presented in both languages.

Statisticians often encounter numerical optimization methods through nonlinear least-squares problems. Numerical optimization is also the key to practical implementation of the uniquely statistical maximum likelihood methodology. Chapter 8 introduces the standard golden section and Newton optimization algorithms from numerical analysis. Random number generation also comes into play through stochastic search algorithms that provide an effective tool for avoiding issues with functions possessing multiple extrema.

Chapter 9 is dedicated to a high-level introduction to theoretical aspects of data storage. The treatment is in terms of abstract data structures that provide models for standard storage paradigms that are encountered in practice.

Perhaps the most familiar data structure is an array of numbers. An abstract view of this concept leads to the idea of a dynamic array that can hold user-defined (abstract) data types and expand or contract its size to accommodate different storage needs. Of course, arrays are not suited for every storage problem and there are many other options that will be more efficient for access and manipulation of data in some instances. Examples of such structures include the binary search trees, hash tables, lists and queues that are defined and illustrated in Chapter 9. This is also where we introduce an abstraction of the pointer concept called an iterator that plays an important role in Chapter 10.

Chapter 10 is the practical companion of Chapter 9. Here the focus is on the container classes and algorithms that reside in the C++ Standard Template Library. Specific applications for this set of tools include the creation of flexible code for data import, methodology for analyzing streaming data and efficient techniques for implementation of discrete event simulations.

Parallel processing has often been viewed as an esoteric side of computing. However, now that multiple-core processors are the industry standard, the operating systems of desktop machines and laptops are all actively engaged in parallel processing. The problem for the typical code developer lies in obtaining access to the multithreading capabilities of the machines at their disposal. This resource becomes available through an *application programming interface* or *API* that contains a set of commands (usually referred to as *bindings*) that link a language such as C++ with a code library that manages communication to and between individual processors. OpenMP is a simple yet sophisticated API for shared memory settings that is amenable to use even in a desktop environment. The industry standard for parallel programming in distributed memory environments is the *Message Passing Interface* or MPI. The OpenMP and MPI interfaces are not competitive in that they are designed to solve very different communication problems. For MPI the task is sending messages between processors that occupy different physical locations where they do not have access to a common block of memory. In contrast, with OpenMP processors communicate directly through memory that they share; part of the function of the API in this instance is to control memory usage in a way that will prevent data corruption.

Chapter 11 provides an introduction to parallel programming in OpenMP and MPI. The essential bindings that are needed for writing programs that use both APIs are presented and their use is illustrated in the examples and exercises for the chapter. There are several approaches that now exist for parallel computing with R. Two of these are considered: the Rmpi package that provides an interface to MPI and the multicore package that furnishes a means to effectively use R in shared memory settings.

Simulation studies are one obvious statistical application for parallel processing. In many ways they are an ideal application in that a large Monte Carlo experiment can be portioned out to different processors that conduct their individual simulations without the need for inter-processor communication. Near optimal speedups are possible in such instances that make it feasible to perform experiments with both scope and replication levels that could not be contemplated with serial code. Of course, this all rests on the ability to obtain high quality, independent random number streams for the different processors. It turns out this is not so easy to accomplish using standard serial generators and steps must be taken to guard against inter-processor stream dependence. This can be easily accomplished in C++ with the RngStreams package and we conclude the text by illustrating how this can be used in OpenMP, MPI and R.

Chapter 2

Computer representation of numbers

2.1 Introduction

Before delving into code development, we first need to think about the basic ingredient that lies at the core of all scientific computations: numbers. Computer languages generally provide specific ways of representing fundamental or *primitive* data types that correspond to numbers in some direct manner. For C++ these primitive data types include

- *boolean* (or `bool`) for variables that take on the two values `true` and `false` or, equivalently, the values 1 (= `true`) and 0 (= `false`),

- *integer*, which translates into specific types such as `short`, `unsigned short`, `int`, `long int` and `unsigned long int` that provide different amounts of storage for integer values,

- *character*, indicated by the modifier `char` in source code, that is used for variables that have character values such as "A" and

- *floating point* which encompasses basically all non-integer numbers with the three storage types `float`, `double` and `long double`.

Ultimately, all computer operations, no matter how sophisticated, reduce to working with representations of numbers that must be stored appropriately for use by a computer's central processing units or CPUs. This process is accomplished through the manipulation of transistors (in CPUs and random access memory or RAM) that have two possible states: off and on or, equivalently, 0 (= off) and 1 (= on). By combining transistors it becomes possible to create a dyadic representation for a number that allows it to be stored and used in arithmetic operations. Of course, the number of transistors is finite which affects both how and how much information can actually be stored. To manage overall memory (i.e., storage) levels effectively it is necessary to restrict the amount of memory that can be allocated to different kinds of numbers with the consequence that there are limits on how big or small a number can be and still be stored in some meaningful fashion. To appreciate the implications of this we first need to think a bit about computer arithmetic.

In the familiar decimal (or base 10) system, numerical values are represented in units or powers of 10. For simplicity, let us work with only nonnegative integers for the moment. Then, the basis representation theorem from number theory (e.g., Andrews 1971) has the consequence that any such integer, k, can be written as

$$k = \sum_{j=0}^{m} a_j (10)^j \tag{2.1}$$

for some unique (if we require $a_m \neq 0$) integer m and some unique set of integer coefficients $\{a_0, \ldots, a_m\}$. As an example,

$$193 = 1 * (10)^2 + 9 * (10)^1 + 3 * (10)^0$$

with $*$ indicating multiplication. So, in this case $m = 2$, $a_0 = 3$, $a_1 = 9$ and $a_2 = 1$. Note that for the coefficients in (2.1) to be unique their range must be restricted to, e.g., $\{0, \ldots, 9\}$.

Since transistors have only two states it is not surprising that the base of choice for computer arithmetic is binary or base 2. The base 2 version of (2.1) is the *binary expansion*

$$k = \sum_{j=0}^{m} b_j (2)^j \tag{2.2}$$

with m a nonnegative integer for which $b_m \neq 0$. The b_j's are coefficients that can take on only the values 0 or 1. Returning to our example, it can be seen that

$$193 = 1 * (2)^7 + 1 * (2)^6 + 1 * (2)^0.$$

This corresponds to (2.2) with $m = 7$, $b_7 = b_6 = b_0 = 1$ and $b_5 = \cdots = b_1 = 0$.

The fact that the coefficients of a number in binary form are all zeros and ones means that it can be represented as a sequence of zeros and ones with the zeros and ones corresponding to the b_j in (2.2) and their location in the sequence indicating the associated power of two for the coefficient. The slots that hold the coefficients in a sequence are called *bits* with a sequence of eight bits being termed a *byte*. For the number 193 this translates into a tabular representation such as

power of two	2^7	2^6	2^5	2^4	2^3	2^2	2^1	2^0
coefficient	1	1	0	0	0	0	0	1

That is, in binary 193 can be represented exactly in one byte as the value 11000001.

The connection between machine memory and binary arithmetic becomes complete once bits and bytes are identified with individual transistors and blocks of eight transistors. To represent a number in memory its binary representation is physically created by allocating it a block of memory (e.g., a group of contiguous transistors), identifying the individual transistors in the block with a power of 2 from its binary representation and then turning on those transistors that correspond to powers of 2 that have unit coefficients. To belabor the point a bit, a block of eight transistors would be needed to hold the integer 193; the first, seventh and eighth transistor in the block would be turned on and the other five would be turned off. Note that it will take all eight bits/transistors to store 193 and any fewer would not be capable of doing the job.

Just as 193 cannot be stored in anything less than eight bits there is a limit to the size of numbers that can be stored in any finite number of bits. For example, the largest number that can be stored in one byte is

$$\sum_{j=0}^{7} (2)^j = 2^7 \sum_{j=0}^{7} 2^{-j} = 2^7 \frac{1 - (1/2)^8}{1 - (1/2)} = 2^8 - 1 = 255. \tag{2.3}$$

More generally, the largest number that can be stored in m bits is $2^m - 1$ (Exercise 2.1).

2.2 Storage in C++

Let us now consider how the ideas about number representation from the previous section are implemented in C++. While investigating this issue we will also introduce some of the language features that will allow us to begin to write, compile and execute C++ programs.

The amount of storage that is allocated to different data types can be determined using the `sizeof` operator that C++ inherits from C. Listing 2.1 below demonstrates the use of this operator to determine information about storage space, in bytes, that is provided for some of the primitive data types discussed in the previous section.

Listing 2.1 *storageSize.cpp*

```cpp
//storageSize.cpp
#include <iostream>

using namespace std;

int main()
{
  cout << "The storage allocated for a char is " << sizeof(char)
       << " bytes" << endl;
  cout << "The storage allocated for an unsigned short integer is "
       << sizeof(unsigned short int) << " bytes" << endl;
  cout << "The storage allocated for an integer is " << sizeof(int)
       << " bytes" << endl;
  cout << "The storage allocated for a long integer is "
       << sizeof(long int)  << " bytes" << endl;
  cout << "The storage allocated for an unsigned long integer is "
       << sizeof(unsigned long int) << " bytes" << endl;
  cout << "The storage allocated for a float is " << sizeof(float)
       << " bytes" << endl;
  cout << "The storage allocated for a double is " << sizeof(double)
       << " bytes" << endl;
  cout << "The storage allocated for a long double is "
       << sizeof(long double) << " bytes" << endl;
  return 0;
}
```

Now this is a very simple program. But, it illustrates some of the basic syntax that will be seen again and again. At the outset there is a *comment statement* (signified by //) that gives us the name of the program file, a practice we will adhere to throughout the book. The use of two double forward slashes causes the compiler to ignore the succeeding text on a line. Had we wanted a comment that ran more than one line this could have been accomplished using either // at the start of each line or by bracketing the comment encompassing possibly multiple lines by /* and */.

The comment in Listing 2.1 is followed by the statement #include <iostream> which is a directive to the preprocessor to include information about the input/output classes in the C++ Standard Library. The actual code for the library has already been compiled and will automatically be linked with our program during the compilation process. There are many other tools available from this library beyond just those for input and output. They can be accessed similarly using include directives and the specific files that are included are called *header* files. Such files will be discussed in more detail in the next chapter. But, as one possible example, the use of #include <cmath> provides access to many of the standard mathematical functions such as the logarithm, trigonometric functions, etc.

The next statement in the code, using namespace std;, ensures that there will be no ambiguity in referring to functions and classes that have any of the specific names that have been given to those provided in the C++ Standard Library. Specifically, Listing 2.1 uses a cout object and the output manipulator endl that are declared in the iostream header file. The names of these operators are made globally available with the using namespace std; statement. Omission of this line from our code would have caused the compiler to generate an error message such as

```
storageSize.cpp: In function 'int main()':
storageSize.cpp:6: error: 'cout' was not declared in this scope
storageSize.cpp:7: error: 'endl' was not declared in this scope
```

Basically, a *namespace* is a collection of definitions of variables, functions and other key components associated with a library or program that have been gathered together for various possible reasons. Among other things namespaces provide a mechanism for reusing desirable names for functions and classes that may have already been employed in some other context. It is possible to bypass the **using** directive via use of the *scope resolution operator* :: as discussed in Section 3.8. For now we will travel the simpler path of stating up front that all references are to the Standard Library.

We now come to the start of the program signaled by the appearance of the word **main**. Every C++ program must have one and only one **main** function. It serves the purpose of directing the flow of activity within a program. While there may be many functions and objects whose actions are intricately interlaced throughout a program, whatever takes place must have been initiated by directions from **main**. The body of **main** (as well as that of any other function) is enclosed by the matching curly braces { and }.

In terms of content, **main** in Listing 2.1 is just a collection of explanatory text to be printed with values that are returned by the **sizeof** operator. An application of **sizeof** to a particular data type (supplied as its argument) returns the number of bytes that are used for its storage.

The symbols << and >> are the output and input insertion operators that work with the standard output and input stream objects **cin** and **cout**, respectively, that relay information to and from (typically) a shell. The insertion operators provide the facility to chain together input/output operations. The output *manipulator* **endl** that appears at the end of each output line produces a carriage return (or starts a new line) and flushes (or cleans out) the output buffer. You will see the effect of all this when we look at some output from the program. Note that each full line of code in **main** is ended with a semicolon.

The final line of the program is **return 0;**. In general the **return** statement has the effect that one might expect; it returns control to a calling function. In the case of how it is used here in **main**, it transfers control back to the operating system. An exit integer with a value of 0 is returned upon completion of the program. The fact that 0 is viewed as an integer is a consequence of the **int** designation that immediately precedes **main**. This is true more generally in that a function must return a value of the same type that appears in its definition with the exception of **main** functions and functions with return type **void**. The **return** statement could actually have been omitted from Listing 2.1 without causing compilation errors. The **void** return type will be discussed shortly.

The "storage size" source code was saved as storageSize.cpp. The cpp file extension is one of the permissible options for the GNU C++ compiler that is available in Unix environments. The others are .cc, .cxx and .C. To run the program, we first compile it using

```
$ g++ -Wall storageSize.cpp -o storageSize
```

The compilation command begins with g++ that invokes the GNU compiler. This has the effect of transforming the input file storageSize.cpp into machine language. In doing this, there were two compiler options that were employed: -Wall turns on all of the most commonly used compiler warnings while -o allows us to specify the name of the executable, as storageSize in this instance. Without the -o option the executable would be named a.out by default; the a.out name is an abbreviation of "assembler output".

To load and run the compiled program the name of the executable (i.e., storageSize) is entered on the command line prefaced by the ./ modifier that informs the shell where to look for the executable. In this particular instance, the result is

```
$ ./storageSize
The storage allocated for a char is 1 bytes
The storage allocated for an unsigned short integer is 2 bytes
The storage allocated for an integer is 4 bytes
The storage allocated for a long integer is 8 bytes
The storage allocated for an unsigned long integer is 8 bytes
The storage allocated for a float is 4 bytes
The storage allocated for a double is 8 bytes
The storage allocated for a long double is 16 bytes
```

The results obtained from the storageSize program will be machine-dependent. For the C++ implementation on the computer that was used to produce this output, a small integer with no attached sign as represented in **unsigned short** will be stored in two bytes or 16 bits of memory. Both the **long int** and **unsigned long int** storage types provide eight bytes of storage while the values for an **int** variable can take up no more than four bytes of memory.

One of the reasons for machine dependence is the flexibility that was written into the C++ Standard. Only minimum and maximum values are specified there for the different data types. For example, in the case of integers the requirement is that **long int** need only provide as much storage as **int**. The values that we just saw certainly satisfy this guideline since **long int** provides twice the storage capacity of **int**.

The **float** and **double** types are the irrational number analog of the **int** and **long int** storage types for integers. Another phrase that is used for **float** is *single precision* and **double** is short for *double precision*. This terminology becomes meaningful from the output of the storageSize program; a **double** is allocated 64 bits of memory or twice that which is allowed for a **float**. A **long double** is then stored in 128 bits of memory or twice that provided for a **double**.

2.3 Integers

Computer arithmetic is basically conducted with integers and floating-point values that are used to approximate the values of non-integers. Although floating-point storage has a more complex structure, it is not entirely dissimilar from what is employed for integers. In the end every number must be translated into a sequence of bits and the devil is in the details of how this is done. Integers are a good place to begin to think about such concepts and, accordingly, that is where this section begins.

To explore the way integers are stored in some specific cases of interest it will be helpful to have a program that returns the binary representation of a number. To accomplish this let us first deal with the **unsigned short int** or simply **unsigned short** data type. As the name suggests, these are small nonnegative integers. As seen in the previous section, they occupy 2 bytes of storage with values ranging from 0 to $2^{16} - 1 = 65535$. This makes them ideal for our current purpose where we want to extract the binary representation of numbers in a simple setting. A program that accomplishes this is provided by Listing 2.2.

Listing 2.2 *binaryRep.cpp*

```
//binaryRep.cpp
#include <iostream>

using namespace std;

void printBinary(unsigned short val);

int main(){
  unsigned short inVal;
  cout << "Enter a number between 0 and 65535: ";
  cin >> inVal;
  cout << "Your number in binary is ";
  printBinary(inVal);
  cout << endl;
  return 0;
}

void printBinary(unsigned short val){
```

```
   for(int i = 15; i >= 0; i--)
      if(val & (1 << i))
         cout << "1";
      else
         cout << "0";
}
```

Listing 2.2 employs some elementary bit-wise operations in a function called printBinary to pick off the internal binary representation of an integer. The first thing we see in the listing is a *prototype* or *declaration* of the basic form for printBinary that appears prior to the function main. The *return type* of the function (i.e., the keyword that immediately precedes the function's name) is designated as void which means it does not return a value to the calling program. In C/C++ it is necessary to tell the compiler about the essential details (e.g., return type and type for the arguments) in a function before it can be used. Because printBinary is to be used in the function main, we need to either define it prior to main or furnish a prototype or sketch of the function prior to calling it in main. The second option was chosen here. But, the program would have compiled and produced the same results if the entire definition of printBinary was moved from the bottom of the listing and placed before main.

The first line in main concerns the variable inVal that will hold the integer whose binary representation will be determined. The line consists of a *type declaration* which states that inVal is of storage type unsigned short. In C++ every variable's type must be explicitly stated when it is introduced into the program. Failure to do so will produce a compilation error. Indeed, omission of the unsigned short inVal; line from our program would have generated a message such as

```
binaryRep.cpp: In function    int  main()     :
binaryRep.cpp:10: error: 'inVal' was not declared in this scope
```

from the compiler.

The main function in Listing 2.2 also illustrates the use of cout's input partner cin to read a value for inVal from the shell command line. The printBinary function is then applied to inVal with the results being written to the shell before the program terminates.

We now arrive at the actual definition of the printBinary function. It takes as input an unsigned short integer that is two bytes in size. Such numbers can be viewed as having 16 possible slots corresponding to 2^0 all the way to 2^{15}. The program then steps through each of these slots using a for loop that begins with a look at slot 16 (i.e., the 2^{15} slot) and works backward to the first (or 2^0) slot.

The C++ version of a do or for loop consists of specifying three quantities. First, a starting value is given for the index variable whose value governs the iteration. In this case the index variable is i whose type is first designated as int before setting it to 15. Then, there is logical condition (i.e., i >= 0) that terminates the loop when it evaluates to false. Finally, direction is provided on how the loop should move through the values of the index variable. This is accomplished here with the i-- syntax which says that i should be decremented by one at each step through the loop. Section 3.4 provides more details on the use of loops in C++.

The printBinary function picks off the binary representation of the unsigned int variable supplied for its argument by first shifting the integer 1 ($= 2^0$) over i = 15 slots using the 1 << 15 operation. We have already seen the symbol << used in a very different context in conjunction with cout. The question this raises is how one operator can be used for two completely different purposes. The answer is that this is an example of polymorphism accomplished via a feature called operator overloading that will be developed in more detail in the next chapter. By using this facility it becomes possible to have operators that adjust automatically to the situation where they are applied.

Basically, the operation for `i = 15` at the beginning of the `for` loop in Listing 2.2 multiplies 1 (or 0000000000000001) by 2^{15} to obtain 1000000000000000. The `if` statement then relies on a conditional comparison of `val` to 1000000000000000 using the bit-wise AND operator `&`. Here `&` takes the binary representation of two numbers as input and returns a binary representation that has ones in all the places where *both* numbers have ones and zeros elsewhere. So, in terms of the way we have set this up, the argument in the `if` statement (i.e., `val & (1 << 15)`) will evaluate to zero (or `false` in terms of how it is viewed by the `if` conditional) unless `val` has a 2^{15} term in its binary representation. In the latter case `val & (1 << 15)` evaluates as `true` causing a 1 to be written to the shell with a 0 being written otherwise. This same process proceeds across descending powers of 2 as a result of the decrement operator `i--` that reduces the value of `i` by one after the comparison in the `if` statement is executed.

After compilation using

```
$ g++ -Wall binaryRep.cpp -o binaryRep
```

the program will produce results like

```
$ ./binaryRep
Enter a number between 0 and 65535: 2
Your number in binary is 0000000000000010
$ ./binaryRep
Enter a number between 0 and 65535: 65535
Your number in binary is 1111111111111111
```

The output suggests that the code is working as expected.

Up to this point, only nonnegative integers have been considered. To distinguish between negative and positive integers, one must retain information about the sign. For example, we could just let the last bit correspond to the sign. This is called the *signed magnitude* approach. To illustrate the idea, suppose for simplicity we are working with a one byte or eight-bit integer. Then, for example, the binary representation for 12 will be 00001100 and -12 would be represented by 10001100. By reserving the last bit for the sign we have left ourselves with only 7 "working" bits. Accordingly, the numbers that can be represented fall between $-(2^7 - 1) = -127$ and $2^7 - 1 = 127$.

Another way to store signed integers is the *two's complement* rule. Here the left-most bit is allowed to represent -2^{m-1} in an m bit storage allocation. For the case of $m = 8$ in our example, the last bit corresponds to $-2^7 = -128$. One can work backward from this to get other negative integers by allowing the other seven (more generally, $m - 1$) bits to provide coefficients for positive powers of two as before. So, with a slight abuse of notation, we can write

$$-12 = -128 + 64 + 32 + 16 + 4 = 11110100.$$

The two's complement approach with $m = 8$ allows us to represent all the negative integers between -128 and $-1 (= -128 + (2^7 - 1))$ and, hence, the integers that can be represented are now all those in the range -128 to 127. That is, one additional integer has been gained over the signed magnitude method. This result extends directly to the general case of m bits with the consequence that under the two's complement approach it is possible to represent all integers between

$$\overbrace{100\cdots0}^{m\text{ times}} = -2^{m-1} \tag{2.4}$$

and

$$\overbrace{0111\cdots1}^{m\,\text{times}} = 2^{m-1} - 1. \tag{2.5}$$

To adapt Listing 2.2 to handle signed integers it is only necessary to drop the `unsigned` designation from `inVal` and `val` in Listing 2.2. The allowable input values for a two's complement storage scheme will be $2^{15} = -32768$ to $2^{15} - 1 = 32767$. Using the corresponding modified version of Listing 2.2, the output below allows us to deduce that the computer on which this code was compiled and executed uses the two's complement rule to represent `int` variables.

```
$ ./binaryRep
Enter a number between -32768 and 32767: -32768
Your number in binary is 1000000000000000
$ ./binaryRep
Enter a number between -32768 and 32767: 32767
Your number in binary is 0111111111111111
```

2.4 Floating-point representation

Storage types with fixed positional representations work fine for integers. This basic idea could even be used, at least in principle, to represent rational numbers or fractions as they can be expressed as k/m with k and m both integers. However, the utility of the fixed point framework ends here and leaves unresolved the problem of storing and working with irrational numbers, which comprise the bulk of the real number system.

At this point it must be realized that a general real number cannot be stored in its entirety and, as a result, in most cases the stored value will represent only an approximation to the truth. Errors are created in computer arithmetic with real numbers due to both the rounding of the numbers for storage as well as further manipulations. These issues will be discussed in the next section. For the present it suffices to recognize that there is a limit to the precision that can be achieved from any computer representation that might be employed for irrational numbers. We will express the precision by the number of significant digits of agreement between the true value of a number and its floating-point representation. A good storage system is one that attempts to minimize losses in precision subject to the constraints that have been imposed on the allowed amount of storage.

Roughly speaking, significant digits are obtained by removing leading and trailing zeros from a number as would take place for conversion to scientific notation, for example. This is, in fact, an apt analogy for the developments in this and the next section. To effectively store irrational numbers on a computer the decimal point must be allowed to float. By this we mean that the decimal is always placed after the number's first significant digit. To recover the actual value of the number this decimal relocation is accompanied by multiplication by the base (e.g., 2 or 10) raised to an appropriate power. The result is an adaptive representation for numbers that allows for considerable storage flexibility relative to the integer case.

To be somewhat more specific, let us return to the base 10 scenario where the floating decimal concept is already familiar in the form of scientific E-notation. For example, under this format the values 21.237 and .021237 would be written as 2.1237E+1 and 2.1237E−2. The use of "E" in this context stands for the exponent of 10 and all that is really being stated is that $21.237 = 2.1237 \times 10^1$ and $.021237 = 2.1237 \times 10^{-2}$. This idea works quite generally in that $m + 1$ significant figures can be retained for a real number x by writing it as

$$x \doteq \pm a_0.a_1 a_2 \cdots a_m 10^p, \tag{2.6}$$

where $\doteq$ indicates approximate equality, $a_0, \ldots, a_m$ are all integers between 0 and 9 with $a_0 \neq 0$ and p is a signed integer.

The idea behind (2.6) is that any real number can be written as

$$x = \pm \sum_{j=0}^{\infty} a_j 10^{p-j} \tag{2.7}$$

and (2.6) comes from retaining only $m + 1$ terms in the series (2.7). Since we only want to include significant digits in the approximation to x, cases where $a_0 = 0$ can be excluded. Thus, subject to a rounding rule that determines a_m (e.g., augment a_m by one if $a_{m+1} \geq 5$), the approximation (2.6) is uniquely determined. In the case of 21.237, $m = 4$ and $p = 1$ while $m = 4$ and $p = -2$ for .021237.

There is nothing special about base 10 and the analog of (2.6) for base 2 looks like

$$x \doteq \pm b_0.b_1 b_2 \cdots b_m 2^p \tag{2.8}$$

with $b_0, \ldots, b_m$ all having values of either 0 or 1 and p, again, being a signed integer. As was the case for (2.6), (2.8) derives from writing

$$x = \pm \sum_{j=0}^{\infty} b_j 2^{p-j}$$

and then keeping the leading $m + 1$ terms from the series. It is only necessary to deal with cases where $b_0 = 1$ which means that a binary representation can always be written in the form

$$x \doteq \pm 1.b_1 b_2 \cdots b_m 2^p. \tag{2.9}$$

This has the practical consequence that the value of b_0 need not actually be stored. Numbers represented as in (2.9) are said to have been *normalized*. The remaining part of the approximation $.b_1 b_2 \cdots b_m$ is called the *mantissa* or *significand*.

Let us now see how an expression such as (2.9) might be translated into a specific storage scheme. We will begin with a simple, nonrealistic, illustration and then describe what can be expected on a typical machine.

Suppose now that a number x is to be stored in a floating-point representation that is allocated one byte of storage. There are essentially four things that must be accounted for: the sign of the number, the value of the exponent for 2 in (2.9) as well as its sign and, finally, the values for $b_1, \ldots, b_m$. To account for the sign for x, take the left-most bit to be a sign bit with value 0 for a positive number and 1 for a negative value. The next three bits can then be used to hold a two's complement representation of the exponent (that can range from -4 to 3). This leaves four bits (i.e., $m = 4$) to hold the significand. To apply this idea to, e.g., the number -12.5, first observe that

$$\begin{aligned} 12.5 &= 2^3 + 2^2 + 2^{-1} \\ &= 1.1001 \times 2^3. \end{aligned}$$

Hence, -12.5 would be represented as

$$\underbrace{1}_{\text{for the } -} \quad \underbrace{011}_{\text{for 3}} \quad \underbrace{1001}_{\text{for the significand}}$$

The leading 1 in the significand has been dropped because it is assumed *a priori* to be 1, and the remaining part, 1001, that is stored represents coefficients for $2^{-1}, 2^{-2}, 2^{-3}$ and 2^{-4}, respectively.

Now consider adding .25 to -12.5. We can represent .25 (exactly) in our storage plan with

$$.25 \; = \; 2^{-2}$$
$$= \; 1.0000 \times 2^{-2}.$$

But, to add it to -12.5 a common denominator is needed which, in this case, entails viewing .25 as

$$.25 = .00001 \times 2^3.$$

However, in this form it cannot be stored exactly in four bits and instead must be rounded to a significand of .0000 if we chop off the last digit or to .0001 if we round up. As a result, addition returns the sum of -12.5 and .25 as -12.5 under truncation and -12 if the answer is rounded up. This is an illustration of the types of problems that can occur when performing basic arithmetic operations using numbers that have very different magnitudes. In fact, in this case the addition of a number a to -12.5 leaves it unchanged for any $|a| \leq 2^{-3}$.

Finally, let us see what happens when we try to store $1/5$ using our simple system. This fraction cannot be stored exactly and, instead, the best that can be obtained is an approximate representation based on the fact that

$$\frac{1}{5} \; \doteq \; 2^{-3} + 2^{-4} + 2^{-7} + 2^{-8}$$
$$= \; .19921875.$$

The sign bit for .19921875 will have value 0 and the two's complement representation for its exponent for 2 (namely, -3) is 101. The significand cannot be stored in its entirety and rounding is required to make it fit into four bits of memory. The value is rounded up to 1010 and that is what is retained. The final result is that $1/5$ receives the floating-point representation

$$\underbrace{0}_{\text{for the -}} \qquad \underbrace{1001}_{\text{for the exponent}} \qquad \underbrace{1010}_{\text{for the significand}}$$

i.e., we approximate $1/5$ by $2^{-3} + 2^{-4} + 2^{-6} = .203125$.

There is a slight problem with our simple storage scheme. Since the lead bit of the significand was not stored the natural way to represent 0 has been lost. To account for this 0 can simply be represented by some value such as 01000000 which would have the effect of eliminating 2^{-4} as a possible value.

In the real world storage for floating-point numbers can be expected to abide by the IEEE 754 standard discussed, for example, in Stevenson (1981) and Goldberg (1991). In the case of, e.g., `float` types the 754 standard specifies four bytes of storage. The first of the 32 available bits is a sign bit with the next eight bits being allocated for storage of the exponent. The significand is then stored in the remaining 23 bits.

In contrast to the way the exponent was handled in our simple storage scheme, the 754 standard employs a *biased* exponent. With eight bits there are only 256 possible integers (e.g., 0 to 255 or -128 to 127) that can be stored whether they are signed or unsigned. Biasing then works with the unsigned integers 0 to 255 that can be stored in eight bits and transforms or biases the values by subtracting off 127 to produce exponents (in base 2) that range from $-127(= 0 - 127)$ to $128 = (255 - 127)$.

To illustrate the IEEE storage scheme, consider the number -193.625 as a four byte `float`. Now, note that

$$193.625 = 1 * (2)^7 + 1 * (2)^6 + 1 * (2)^0 + 1 * (2)^{-1} + 1 * (2)^{-3}$$

which means that the value in binary is

$$11000001.101 = 1.1000001101 * 2^7.$$

The first bit in the floating-point representation for -193.625 will be 1 to indicate that the value is negative. The next eight bits represent the exponent: i.e., 7. Instead of 7, the biasing approach leads to storage of

$$
\begin{aligned}
134 &= 7 + 127 \\
&= 2^7 + 2^2 + 2^1
\end{aligned}
$$

which is the binary sequence 10000110. The final result is the normalized representation

$$
\underbrace{1}_{\text{for the } -} \quad \underbrace{10000110}_{\text{for } p} \, \underbrace{10000011010000000000000}_{\text{for the significand}}
$$

As was true for our simple storage scheme, normalization entails the loss of the natural choice for 0. The problem is resolved by defining 0 as 2^{-127}: i.e., 0 is

$$
\underbrace{0}_{\text{sign}} \quad \underbrace{00000000}_{\text{exponent}} \, \underbrace{00000000000000000000000}_{\text{significand}}
$$

which translates to 2^{-127} after biasing of the exponent.

Unlike the integer case, extracting the binary representation of a floating-point number is a little tricky. This is due to the fact that the bit-wise operators used in the integer case cannot be used directly with floating-point numbers. A standard recommendation for an end run around this difficulty looks something like the code in the listing below.

```
#include <iostream>

using namespace std;

void printbinary(char val){
  for(int i = 7; i >= 0; i--)
    if(val & (1 << i))
      cout << "1";
    else
      cout << "0";
}

int main(){
  float f;
  cout << "enter a number: ";
  cin >> f;
  char* pf = reinterpret_cast<char*>(&f);
  cout << "your number in binary is ";
  for(int i = sizeof(float) - 1; i >= 0; i--)  printbinary(pf[i]);
  cout << endl;
  return 0;
}
```

This program illustrates two important aspects of C/C++ that will become essential for later chapters: namely, pointers and addresses. These quantities will be discussed in some detail in the next chapter. For now let us merely mention their general purpose and how they are used in this particular program.

The syntax `char* pf` in the program says that `pf` is a pointer to `char`. This means it is a variable whose value is the address in memory that is occupied by a `char` variable. The address that has been assigned to `pf` is essentially that of the floating-point variable `f` which is obtained through the syntax `&f`. The problem is that `&f` is the address of a `float` while `pf` is expecting to receive the address of a `char` for its value. To make the transition from a `float` to a `char` address, a *cast* is used. In general, casting is the process of changing a variable from one data type into another in terms of how it is viewed by the compiler. C++ performs implicit casting in standard cases such as transforming an integer into a floating-point value and similar types of conversions. More generally, an explicit cast of a variable x to a new data type newType is accomplished with syntax of the form

```
newType y = (newType)x;
```

provided the conversion is possible. For example, the ensuing code segment has the effect of creating a variable of type double from one of type int.

```
int x = 1;
double y = (double)x;
```

Our particular problem goes beyond the capability of a simple explicit cast such as this. We need to transform the memory address of a float into one that is treated as holding a char. For this purpose it is necessary to use the reinterpret_cast<char*> operator. This basically tells the compiler to forget the original type of data that was stored at &f and allow this location in memory to be viewed as memory for variables of type char. Our storageSize program revealed that variables of type char are allocated 1 byte of memory. Thus, after the cast the memory located at &f will be viewed by pf as representing the starting point for sizeof(float) (or, 4 in this instance) one-byte blocks of memory. The resulting pattern of on/off positions for the sequence of transistors found at these locations can be interpreted in a variety of ways. The goal is to figure out which bits are set to 1 and which are set to 0. This can be done by applying our bit-wise shift operator, that works equally well with char variables, on each one-byte block of memory. The new main function also illustrates the dereferencing of the pf pointer when the argument pf[i] is passed to the printBinary function. This is a vector-type indexing property of pointers wherein syntax such as pf[i] gives access to the contents of the ith block of memory rather than the value of the pointer itself.

An example of output from the floating-point representation program is

```
Enter a number: -193.625
Your number in binary is 11000011010000011010000000000000
```

The binary representation agrees with what was done before which implies that the IEEE standard for float is being employed on the machine that ran the program.

To conclude this section let us briefly discuss the mechanics of addition/subtraction and multiplication of floating-point numbers. First consider the multiplication of x_1 times x_2 with

$$x_i = 1.b_{i1} \cdots b_{im} 2^{p_i} \qquad (2.10)$$

for $i = 1, 2$. To compute the product $x_1 x_2$ exactly there are 2 steps: namely, 1) add the two exponents and 2) multiply the two significands $1.b_{11} \cdots b_{1m}$ and $1.b_{21} \cdots b_{2m}$. This latter step results in a $2m + 2$ binary decimal number of the form $1.c_1 \cdots c_{2m+1}$. The answer is then

$$x_1 x_2 = 1.c_1 \cdots c_{2m+1} 2^{p_1 + p_2}.$$

Of course, if only m decimal bits are available for storage, this answer must be rounded and even the evaluation of the coefficient c_{m+1} that would be needed to accomplish the rounding can be seen as problematic. The use of extended precision for doing the computation prior to rounding is one way to solve this problem. But, other approaches are used in practice as indicated in the next section.

Now consider adding x_1 and x_2 in (2.10). This process is more involved as a result of the common denominator that must be used to carry out the addition process as was illustrated with the addition of .25 to -12.5 under our eight-bit floating-point scheme. Assume that

$p_1 > p_2$ and then shift x_2 to have the form

$$x_2 = 0.0 \cdots 01 b_{21} \cdots b_{2m} 2^{p_1} \tag{2.11}$$

with $p_1 - p_2 - 1$ bits being 0 counting from the first bit on the right side of the decimal. The sum or difference can now be computed through binary addition or subtraction of the original x_1 and shifted x_2 significands with the result being rounded to have m bits to the right of the decimal. There are practical considerations that arise here in terms of how to store the shifted version of x_2 and problems arise if only m decimal bits are used to store the shifted value. Guard bits (e.g., Kaneko and Liu 1991) can be used to deal with such issues.

2.5 Errors

As seen in the last section, most numbers cannot be represented exactly in floating-point format. In this section we will explore the consequences of this for the precision of computations that are carried out on computers in binary arithmetic.

First let us address the losses that can arise from conversion of a number to its floating-point representation. It is possible to place bounds on the amount of error that can be incurred. The bounds are, of course, dependent on the type of "rounding" that is employed. The simplest approach is truncation or chopping.

Sign plays no role in this development. So, suppose that x is a positive real number that can be written as

$$x = \sum_{j=0}^{\infty} b_j 2^{p-j}$$

with $b_0 = 1$. Then, *chopping* replaces x by the approximation

$$\tilde{x} = 2^p + \sum_{j=1}^{m} b_j 2^{p-j}$$

for some integer m. The error that is incurred by this approach is

$$\begin{aligned}
x - \tilde{x} &= \sum_{j=m+1}^{\infty} b_j 2^{p-j} \\
&\leq 2^p \sum_{j=m+1}^{\infty} 2^{-j} \\
&= 2^p 2^{-(m+1)} \sum_{j=0}^{\infty} 2^{-j} \\
&= 2^{p-m}.
\end{aligned}$$

Instead of chopping an alternative approximation $\tilde{x}$ can be produced by rounding to the value that is closest to x in the sense of minimizing $|x - \tilde{x}|$. This cuts the bound in half to $2^{p-(m+1)}$ as we now demonstrate.

First observe that the *round-to-closest value* choice for $\tilde{x}$ is obtained by increasing b_m by one if $b_{m+1} = 1$ and using the chopped approximation otherwise. Consequently,

$$\tilde{x} = 2^p \left[\sum_{j=0}^{m} b_j 2^{-j} + \tilde{b}_{m+1} 2^{-(m+1)} \right]$$

with $\tilde{b}_{m+1}$ being either 2 or 0 depending on whether or not b_{m+1} is 1 or 0, respectively.

Thus, when $b_{m+1} = 1$ the error is

$$|x - \tilde{x}| = 2^p \left| 2^{-(m+1)} - \sum_{j=m+2}^{\infty} b_j 2^{-j} \right| \leq 2^{p-(m+1)}.$$

On the other hand, if $b_{m+1} = 0$, $x - \tilde{x} \leq 2^{p-(m+1)}$ from our analysis of rounding by chopping. To translate the previous bounds to relative absolute error, observe that, since $|x| \geq 2^p$,

$$\frac{|x - \tilde{x}|}{|x|} \leq 2^{-(m+1)} \tag{2.12}$$

for round-to-closest value and

$$\frac{|x - \tilde{x}|}{|x|} \leq 2^{-m}$$

for chopping. These values are sometimes referred to as the *machine epsilon*. An application of (2.12) to the IEEE floating-point format gives error bounds on the order of $2^{-24} \doteq .6(10^{-7})$ for single precision and $2^{-53} \doteq 10^{-16}$ for double precision arithmetic (Exercise 2.8).

Define the relative error from floating-point approximation to be

$$E = \frac{\tilde{x} - x}{x} \tag{2.13}$$

with $\tilde{x}$ the approximation to x obtained from either one of the rounding schemes. The relative error satisfies

$$\begin{aligned} \tilde{x} &= x + \tilde{x} - x \\ &= x \left(1 + \frac{\tilde{x} - x}{x} \right) \\ &= x(1 + E) \end{aligned} \tag{2.14}$$

which gives it a simple interpretation in terms of the way it measures the disparity between x and $\tilde{x}$. The value of $|E|$ pertains only to the significand and in that respect gives the number of significant digits of accuracy for the approximation. From (2.12), $|E| \leq 2^{-(m+1)}$ under rounding to the nearest value and $|E| \leq 2^{-m}$ in the case of chopping.

As an example, let us revisit the approximation for $1/5$ that was obtained using the simple eight-bit floating-point scheme of the previous section. In that case we had initially replaced $1/5$ by $.19921875$ where

$$.19921875 = 1.10011 \times 2^{-3}.$$

This led to the representation of $1/5$ as 01011010. The first sign bit is set to 0 and the two's complement representation of -3 (i.e., 101) occupies the next three bits. The final four bits, 1010, are used to approximate 10011. Chopping would replace this by 1001 while round-to-nearest value is what gave us 1010. The result is that chopping approximates .2 by .1953125 with an error of .0046875 $(< .0078125 = 2^{-7})$ while rounding to the nearest value approximates .2 by .203125 with an absolute error of .0031025 $(< .00390625 = 2^{-8})$. The relative absolute error is $.003125/.2 = .015625$ $(< .03125 = 2^{-5})$ for rounding to the nearest value and $.0046875/.2 = .0234375$ $(< .0625 = 2^{-4})$ for chopping.

We now wish to consider how the effect of rounding will propagate when performing the basic arithmetic operations of multiplication, division, addition and subtraction. Processors (at least those that comply with IEEE standards) use algorithms that perform arithmetic operations in ways that produce *exactly rounded* answers in a sense that will now be described. Details concerning these types of algorithms can be found in Koren (2002) and Lu (2004).

Let $\tilde{}$ denote the result of floating-point approximation so that, for example, $\tilde{x}$ is the floating-point approximation to a number x. Similarly, $\tilde{+}, \tilde{-}, \tilde{\times}$ and $\tilde{/}$ will be used to denote the floating-point implementation of the arithmetic operators $+, -, \times$ and $/$, respectively. Then, if $\tilde{x}$ and $\tilde{y}$ are floating-point representations of x and y, the operators $\tilde{+}, \tilde{-}, \tilde{\times}$ and $\tilde{/}$ produce exactly rounded results if

$$
\begin{aligned}
\tilde{x}\tilde{+}\tilde{y} &= \widetilde{\tilde{x}+\tilde{y}} \\
\tilde{x}\tilde{-}\tilde{y} &= \widetilde{\tilde{x}-\tilde{y}} \\
\tilde{x}\tilde{\times}\tilde{y} &= \widetilde{\tilde{x}\times\tilde{y}} \\
\tilde{x}\tilde{/}\tilde{y} &= \widetilde{\tilde{x}/\tilde{y}}.
\end{aligned}
$$

In words these relations mean that the operations on $\tilde{x}$ and $\tilde{y}$ produce the same answer as if the nonfloating-point operators $+, -, \times, /$ had been applied to the number $\tilde{x}$ and $\tilde{y}$ and the outcome was then rounded to obtain the final m-digit binary representation. If the operations are now assumed to produce exact rounding to the nearest value this leads to relations such as

$$
\tilde{x}\tilde{+}\tilde{y} = (\tilde{x}+\tilde{y})(1+E) \tag{2.15}
$$

and

$$
\tilde{x}\tilde{\times}\tilde{y} = (\tilde{x}\times\tilde{y})(1+E) \tag{2.16}
$$

for a generic relative error E satisfying $|E| \le 2^{-(m+1)}$.

Let us use (2.16) to bound the error incurred in computing the product $P_n = \prod_{i=1}^{n} x_i$ of real numbers $x_1, \ldots, x_n$. In this regard, there are two quantities to consider: the product $\tilde{P}_n = \prod_{i=1}^{n} \tilde{x}_i$ of floating-point approximations to the x_i and the approximation to $\tilde{P}_n$ (and P_n) provided by $\hat{P}_n = \tilde{x}_1 \tilde{\times} \cdots \tilde{\times} \tilde{x}_n$. An expression that relates $\hat{P}_n$ to $\tilde{P}_n$ can be obtained from the recursion that begins with $\hat{P}_1 = \tilde{x}_1$ and has the general step

$$
\hat{P}_j = \tilde{x}_j \tilde{\times} \hat{P}_{j-1}
$$

for $j = 2, \ldots, n$. From (2.16) one may conclude that

$$
\hat{P}_n = \tilde{P}_n \prod_{j=2}^{n} (1 + E_i)
$$

for $|E_i| \le 2^{-(m+1)}, i = 2, \ldots, n$. By now writing

$$
\prod_{j=2}^{n} (1 + E_i) = 1 + E
$$

bounding the relative error $|\hat{P}_n - \tilde{P}_n|/|\tilde{P}_n|$ becomes tantamount to placing a bound on $|E|$.
Following Sterbenz (1974, Section 3.5) we have

$$
-1 + \left(1 - 2^{-(m+1)}\right)^{n-1} \le E \le -1 + \left(1 + 2^{-(m+1)}\right)^{n-1}
$$

with

$$
\begin{aligned}
-1 + \left(1 - 2^{-(m+1)}\right)^{n-1} &= \sum_{j=1}^{n-1} \binom{n-1}{j} (-2)^{-j(m+1)} \\
&\ge -\sum_{j=1}^{n-1} \binom{n-1}{j} 2^{-j(m+1)} \\
&= -\left\{ -1 + \left(1 + 2^{-(m+1)}\right)^{n-1} \right\}.
\end{aligned}
$$

Thus,

$$|E| \leq -1 + \left(1 + 2^{-(m+1)}\right)^{n-1}$$

$$= -1 + \sum_{j=0}^{n-1} \binom{n-1}{j} 2^{-j(m+1)}$$

$$\leq -1 + \sum_{j=0}^{\infty} \frac{(n2^{-(m+1)})^j}{j!}$$

$$= -1 + \exp\{n2^{-(m+1)}\}.$$

So, if n is small compared to 2^{m+1}, the relative error in approximating $\tilde{P}_n$ by $\hat{P}_n$ will be similarly small and on the order of $n2^{-(m+1)}$.

A similar analysis can be applied using (2.14) to see that

$$\tilde{P}_n = P_n(1 + \bar{E})$$

with $|\bar{E}| \leq -1 + \exp\{n2^{-(m+1)}\}$. Putting both approximations together produces

$$\hat{P}_n = P_n(1 + \bar{E} + E + \bar{E}E)$$

giving a relative absolute error bound on the order of $-1 + \exp\{n2^{-(m+1)}\}$ for approximation of P_n. To the first order this bound would behave like $n2^{-(m+1)}$. Taking $n = 2^k$ suggests a loss of one (binary) decimal of accuracy for about every two multiplications in a worst-case scenario.

The errors incurred from floating-point division can be bounded similarly to the case of multiplication (Exercise 2.5). However, the same is not true for floating-point addition and subtraction. In fact, as noted by Wilkinson (1963, page 17), it is not possible to obtain a bound for the relative errors associated with such calculations as the floating-point sum/difference can be zero when the true sum is not.

An indication of how problems can arise from subtraction is provided by taking x and y to be floating-point numbers with m digit significands having $x = 1.0 \cdots 02^p$ and $y = b_0.b_1 \cdots b_m 2^{p-1}$ with $b_0 = \cdots = b_m = 1$. In this instance the difference between x and y is $2^{p-(m+1)}$. This difference is approximated by 2^{p-m} for chopping and 0 under rounding to the closest value. The absolute relative error for approximating $x - y$ is one in either case because $(2^{p-(m+1)} - 2^{p-m})/2^{p-(m+1)} = 1$ and $(0 - 2^{p-(m+1)})/2^{p-(m+1)} = -1$. This implies that the absolute error in subtracting x from y using floating-point calculations can be as large in magnitude as the target quantity $x - y$. Put another way, there may be no digits of accuracy in the output unless steps are taken to provide additional accuracy in the calculation.

Suppose that the floating-point addition operator produces exactly rounded results and again let $\tilde{x}_1, \ldots, \tilde{x}_n$ be floating-point approximations of numbers $x_1, \ldots, x_n$. To compute an approximation $\tilde{S}_n$ to $S_n = \sum_{j=1}^{n} \tilde{x}_j$ use the recursion that starts with $\tilde{S}_1 = \tilde{x}_1$ and has the general step

$$\tilde{S}_j = \tilde{x}_j \tilde{+} \tilde{S}_{j-1}$$

for $j = 2, \ldots, n$. At the jth step of the recursion

$$\tilde{S}_j = (\tilde{x}_j + \tilde{S}_{j-1})(1 + E_j)$$

with $|E_j| \leq 2^{-(m+1)}$ under rounding to the closest value. This leads to

$$
\begin{aligned}
\tilde{S}_n &= \tilde{x}_n(1 + E_n) + \tilde{S}_{n-1}(1 + E_n) \\
&= \tilde{x}_n(1 + E_n) + \tilde{x}_{n-1}(1 + E_{n-1})(1 + E_n) + \tilde{S}_{n-2}(1 + E_{n-1})(1 + E_n) \\
&= \sum_{j=1}^{n} \tilde{x}_j(1 + \bar{E}_j),
\end{aligned}
$$

where $\bar{E}_n = E_n$,

$$
1 + \bar{E}_j = \prod_{k=j}^{n} (1 + E_k)
$$

for $j = 2, \ldots, n-1$ and $\bar{E}_1 := \bar{E}_2$. (The notation $:=$ that appears here and elsewhere throughout the text indicates that the expression on the right hand of $:=$ replaces or overwrites the one on the left.) Hence, $S_n - \tilde{S}_n = \sum_{j=1}^{n} \tilde{x}_j \bar{E}_j$ and, as before

$$
|\bar{E}_j| \leq -1 + \exp\{(n - j)2^{-(m+1)}\}.
$$

But, unlike multiplication, the values of the $\tilde{x}_j$ are inextricably linked into the relative approximation errors with the consequence that the bound depends on the order of summation. This suggests that larger errors will accumulate as multiples of the earlier terms that are entered into the summation. From this perspective the best strategy would seem to require that addition should proceed with values being summed in inverse order of their magnitude to minimize the effect of size disparities on the summation process.

To see an example of the effect of order on addition, consider summing the series $1/j^2$ from $j = 1$ to n for some integer n. The code below represents an attempt to carry out this calculation numerically.

Listing 2.3 series.cpp

```
//series.cpp
#include <cstdlib>
#include <iostream>
#include <iomanip>

using namespace std;

int main(int argc, char* argv[]){
  float sumL = 0., sumU = 0.;
  int n = atoi(argv[1]);
  for(int j = 1; j <= n; j++){
    sumL = sumL + 1/(float)(j*j);
    sumU = sumU + 1/(float)((n + 1 - j)*(n + 1 - j));
  }
  cout << setprecision(8) << "Direct and reverse sums are "
       << sumL << " and " << sumU << endl;
  return 0;
}
```

The first new thing we see in this listing is the inclusion of the header files cstdlib and iomanip that, respectively, allow us to access some useful functions from C and output manipulators for formatting the printed output. In this instance iomanip furnishes the setprecision function that is used with argument 8 to have eight-decimal numbers written to standard output. This code also contains the first appearance of **argc** (for argument count) and **argv** (for argument vector) in the arguments to the **main** function. These are quantities that correspond to command line input. In particular, one can think of `char*` `argv[]` as being an array of character strings (actually, a pointer to memory that holds arrays of `char` variables) with each string being a non-white space component of the white-space delimited text that was entered on the shell command line. The first array element `argv[0]` contains the (./ prefaced) name of the executable while `argc` is the number of strings held in `argv`. Another new twist is the use of the function `atoi` that is made available with the cstdlib header. This function is used to transform a string of character values into an integer variable. The argument to `atoi` in this case is assumed to represent a sequence of digits that may be preceded by a sign. If the string cannot be interpreted as a number, `atoi` returns 0. This is a method of dealing with the fact that command line input necessarily comes in as character strings but will often be used to represent numeric (in this case integer) values. The function `atof` performs the same type of operation except that the transformation is to a floating-point representation.

To compute the sums in Listing 2.3 we need to calculate reciprocals of the expression `j*j` for a variable `j` of type `int` with `*` indicating multiplication. Unless precautions are taken to the contrary, such ratios will be evaluated using integer arithmetic with the consequence that a value of 0 will be returned for `1/(j*j)` for every `j` that exceeds 1. The way to circumvent this behavior is through the use of another cast. Here this takes the form `(float)(j*j)` which computes the integer product and converts the outcome to type `float`.

Our program sums the series in direct and reverse order to produce the two sums: `sumL` and `sumU`. The two values will necessarily agree if they can be evaluated without error. This is not the case for finite precision arithmetic as indicated by the results from running the program (compiled with the name series).

```
$ ./series 5000
Direct and reverse sums are 1.6447253 and 1.644734
```

As noted above, for type `float` the best one might expect for accuracy is seven decimal digits. But, the two ways of computing the sum have differences that appear in the fifth decimal place when there are $n = 5000$ terms in the sum. To determine which answer is closer to the actual sum, the computations can be redone in double precision: i.e., every `float` designation in series.cpp is replaced with `double`. This altered set of code (compiled as dseries) produced the result

```
$ ./dseries 5000
Direct and reverse sums are 1.6447341 and 1.6447341
```

As expected, computing the sum in reverse order where smaller values are added in first provides the most accuracy. The approximate sums can be compared to $\sum_{j=1}^{\infty} j^{-2} = \pi^2/6 \doteq 1.644934$.

2.6 Computing a sample variance

One of the earliest calculations carried out in an elementary statistics methods course is computing a sample mean and standard deviation from a set of data values $x_1, \ldots, x_n$. The sample standard deviation is defined to be

$$S = \sqrt{\mathrm{RSS}/(n-1)},$$

where RSS is the residual sum-of-squares

$$\mathrm{RSS} = \sum_{j=1}^{n} (x_j - \bar{x})^2$$

with $\bar{x} = n^{-1} \sum_{j=1}^{n} x_j$ the sample mean.

A direct application of the definitional formula would suggest a two-pass algorithm where $\bar{x}$ is obtained first and then used to calculate the RSS. A pseudo-code description of the recursive process is given below.

Algorithm 2.1 Two-pass algorithm for the standard deviation

$\bar{x} = 0$
for $j = 1$ to n **do**
$\qquad \bar{x} := \bar{x} + x_j$
end for
$\bar{x} := \bar{x}/n$
$\text{RSS} = 0$
for $j = 1$ to n **do**
$\qquad \text{RSS} := \text{RSS} + (x_j - \bar{x})^2$
end for
return $\sqrt{\frac{\text{RSS}}{n-1}}, \bar{x}$

As an alternative to the two-pass formula, it is standard pedagogical practice to describe the use of a "computational" formula for RSS that stems from the identity

$$\text{RSS} = \sum_{j=1}^{n} x_j^2 - n^{-1} \left(\sum_{j=1}^{n} x_j \right)^2.$$

The resulting computations then proceed along the lines of our next algorithm.

Algorithm 2.2 "Computational" algorithm for the standard deviation

---

$T = \text{RSS} = 0$
for $j = 1$ to n **do**
$\qquad T := T + x_j$
$\qquad \text{RSS} := \text{RSS} + x_j^2$
end for
$\text{RSS} := \text{RSS} - T^2/n$
return $\sqrt{\frac{\text{RSS}}{n-1}}, T/n$

This latter approach has the advantage of requiring only one pass through the data. The disadvantage is that $\sum_{i=1}^{n} x_i^2$ and $(\sum_{i=1}^{n} x_i)^2/n$ can agree across a number of significant digits. This produces inaccuracies when carrying out the subtraction at the end of the `for` loop for reasons that were discussed in the previous section.

Chan and Lewis (1979) define the condition number associated with S to be

$$\kappa = \sqrt{1 + n\bar{x}^2/\text{RSS}}.$$

If S is small relative to $|\bar{x}|$, κ is approximately $\bar{x}/S$. This latter quantity is recognized as the inverse of the sample coefficient of variation that provides a scale-free measure of data variation. In this setting small values for the coefficient of variation can be seen as indicative of instances where the "computational formula" may be problematic. In such cases the data will consist of values that are all (relatively) close to their mean value with the consequence that $\sum_{j=1}^{n} x_j^2$ and $n^{-1} \left(\sum_{j=1}^{n} x_j \right)^2$ may be sufficiently similar to create cancellation problems.

Suppose now that the original data $x_1, \ldots, x_n$ are replaced by $x_1(1+E_1), \ldots, x_n(1+E_n)$ as might occur on a computer through the use of floating-point approximations. Then, Chan and Lewis (1979) establish that the standard deviation $\tilde{S}$ computed (without round-off error) from the altered data will satisfy $\tilde{S} = S(1 + \delta)$ with

$$|\delta| \leq \kappa\gamma + O\left(n\left(\kappa\gamma\right)^2\right) \tag{2.17}$$

for

$$\gamma = \max_{1 \leq j \leq n} |E_j|.$$

For small γ the lead term in the bound on $|\delta|$ is linear in κ which indicates that an increase in the round-off error in the data will be met by a proportional increase in the (bound for the) relative absolute error for computing S with the condition number representing the proportionality factor. We can then view κ as a measure of the inherent sensitivity of S for a particular data set to the effect of floating-point approximation error.

The analysis that led to (2.17) was in the ideal setting where the computation of $\tilde{S}$ can be carried out without error. Of course, the ideal case does not hold in practice and additional errors will be introduced through whatever algorithm is used to compute an approximation to $\tilde{S}$. To measure this let $\tilde{S}_c$ be the approximation to S that is obtained by applying a computational algorithm to $x_1(1 + E_1), \ldots, x_n(1 + E_n)$. We then define the relative absolute error to be

$$\text{RAE} = |\tilde{S}_c - S|/S. \tag{2.18}$$

This quantity will be used to compare the accuracy of the two-pass and "computational" algorithms for computing a sample standard deviation that were discussed above. The condition number κ will play a central role in these comparisons.

Chan and Lewis (1979) give an approximate (ignoring higher-order terms) upper bound of the form

$$\text{RAE} \leq 2\kappa\gamma + \left(\frac{n}{2} + 1\right)\gamma$$

for the two-pass algorithm. A similar bound for the "computational formula" is

$$\text{RAE} \leq \gamma + \left(\frac{3}{2}n + 1\right)\kappa^2\gamma.$$

Figures 2.1–2.2 give plots of the (base 10 logarithm of the) bounds for the two-pass and "computational" formula, respectively. In doing this we took $n = 100$, $\gamma = 2^{-24}$ (as would be expected for single precision round-off errors), chose the true mean for the data to be one and then let the true standard deviation range from 10^{-5} to 1. For simplicity κ is approximated here by σ^{-1} and the horizontal axis is in terms of $-\log_{10}\sigma$ in the figures.

The suggestion from Figures 2.1–2.2 is that the "computational formula" will be less accurate for data sets with small coefficients of variation or large values of κ. This is a consequence of the fact that the "computational" formula's bound involves κ^2. In fact, the difference between the bounds for the "computational" and two-pass algorithms is

$$\frac{1}{2}n\gamma[\kappa^2 - 1] + \kappa\gamma[(n + 1)\kappa - 2].$$

This is always nonnegative because $\kappa \geq 1$. Of course, the fact that only upper bounds are involved means that comparisons of this nature are not conclusive.

A small empirical study was conducted to further investigate the relevance of the upper bound comparisons for use of the two computing algorithms. Data were generated from normal distributions with means of one and standard deviations σ for which $\log_{10}(\sigma) = (j - 10)/2, j = 0, \ldots, 10$. To accomplish this the Wichmann-Hill algorithm from Section 4.5 was used to generate uniform random deviates that were then transformed to normality using the Box-Muller transformation treated in Section 4.7. For each value of $\kappa = 1/\sigma$, 100 replicate samples of size $n = 100$ were generated and their standard deviations were

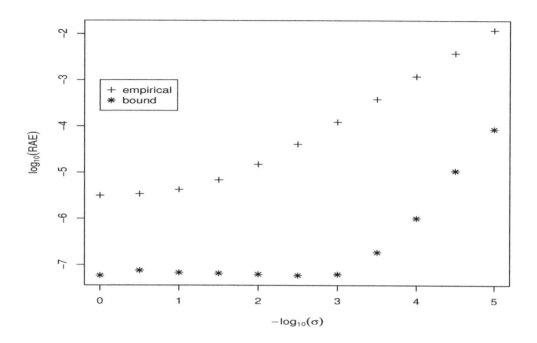

Figure 2.1 *Bounds and empirical results for the two-pass algorithm*

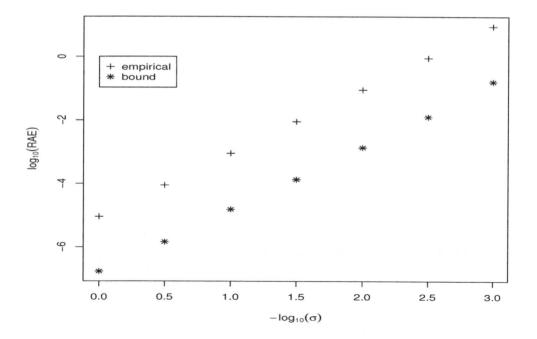

Figure 2.2 *Bounds and empirical results for the "computational" algorithm*

computed via the two-pass and "computational" algorithms. The calculations were carried out in single precision and then compared to a "true" value that was computed in double precision using the two-pass approach. This gives 100 values for RAE for each of the two methods for computing S which were then averaged to produce the results depicted in the plots. From this we can see that the bounds are quite conservative but are accurate in predicting the general form of the relationships between RAE and κ. In the case of the "computational" algorithm, only those RAEs for values of κ between 1 and 10^3 are reported; negative values for RSS began to arise for cases with $\kappa > 10^3$. The basic conclusions derived from the upper bounds and empirical work therefore seems to coincide with our original intuition in suggesting that the two-pass algorithm is more accurate and should be preferred for computations.

There are also updating algorithms that can be used to compute RSS. For example, an algorithm developed by West (1979) and others (see, e.g., the discussion in Chan, et al. 1983) takes the form

---

Algorithm 2.3 West algorithm for the standard deviation

$\bar{x} = x_1$
RSS $= 0$
for $j = 2$ to n **do**
    RSS $:= $ RSS $+ \frac{j-1}{j}(x_j - \bar{x})^2$
    $\bar{x} := \bar{x} + (x_j - \bar{x})/j$
end for
return $\sqrt{\frac{\text{RSS}}{n-1}}, \bar{x}$

The Chan/Lewis upper bound for this method is

$$RAE \leq \left(\frac{n}{2} + 2\right)\gamma + \left(\frac{\sqrt{2}}{3}n + 7\sqrt{n} + 1\right)\kappa\gamma.$$

Like the two-pass method, the error bound for West's algorithm involves only κ rather than κ^2. The analog of Figures 2.1–2.2 that applies to the West algorithm is shown in Figure 2.3. The empirical results shown in the figure were obtained using the same simulation methods that were employed to produce Figures 2.1–2.2. The two-pass method is clearly less sensitive to growth in κ although both it and the West algorithm appear to behave similarly when κ becomes large.

2.7 Storage in R

To conclude this chapter let us mention a few things about the data types and precision of computations in R. First, the "primitive" data types in R include character, double, integer and logical. The double and integer types are analogs of the types with the same names in C++ while logical corresponds to the C++ `bool` data type. The character designation in R indicates a variable that holds character strings rather than just a single character as in C++. The double type in R also goes by the equivalent, and more frequently used, name of *numeric*. The most basic data structure in R is an array comprised of one of the primitive data types that is referred to as an *atomic vector*.

To access the storage mode of a given object, one uses either the `mode` or `storage.mode` functions. In terms of storage of numeric, non-integer values, R purports no single precision data type and all real numbers are stored in a double precision (eight-byte) format. Machine specific details concerning storage, etc., are held in the R list variable `.Machine`. For

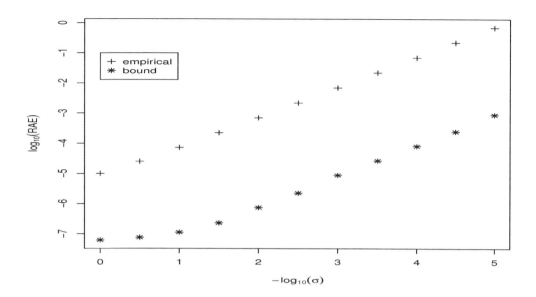

Figure 2.3 *Bounds and empirical results for the West algorithm*

example, for the machine that was used to create portions of the material in this chapter the .Machine information appears as

```
> noquote((format(.Machine)))
            double.eps          double.neg.eps             double.xmin
          2.220446e-16            1.110223e-16            2.225074e-308
          double.xmax             double.base            double.digits
         1.797693e+308                       2                       53
      double.rounding           double.guard        double.ulp.digits
                    5                       0                      -52
 double.neg.ulp.digits        double.exponent          double.min.exp
                  -53                      11                   -1022
       double.max.exp             integer.max             sizeof.long
                 1024              2147483647                       4
      sizeof.longlong      sizeof.longdouble          sizeof.pointer
                    8                      16                       4
```

Some of the names for the list components are familiar from our discussions in Section 2.2 while the meanings of others can be deciphered by looking at the R help page for .Machine. In particular, the value of 53 for the double.digits variable indicates that the significand for a double is composed of 53 bits (i.e., 52 bits plus an implied unit bit for a normalized significand) which corresponds to the IEEE floating-point standard.

There were a couple of useful features of the R language that were used to produce the printed representation of the information in .Machine. First, the format function arranges its input for "pretty printing". The noquote function then suppresses the use of quotes in the printed output.

2.8 Exercises

2.1. Show that the largest integer that can be stored in m bits is $2^m - 1$.

2.2. Write, compile and run C++ code that will write "Hello World!" to the shell. Then, alter your program to where it will greet the world using your first name. To accomplish this use argv as in Listing 2.3. If you enter your first name on the command line, it will be available in argv[1] so that you can string "Hello World from" and argv[1] together using the output insertion operator <<. How would you modify this to use your whole name in the greeting?

2.3. Let x be a number in $[0, 1)$ and let $\lfloor y \rfloor$ denote the greatest integer less than or equal to the real number y.

a) Show that the following algorithm will return the first m terms $b_1, \ldots, b_m$ in the binary representation for x. (Conte and de Boor 1972)

Algorithm 2.4 Binary representation algorithm

$c = x$
   for $j = 2$ to m **do**
   $b_j = \lfloor 2c \rfloor$
   $c := 2c - b_j$
   end for

---

b) The C++ function for $\lfloor \ \rfloor$ is floor. Use this to create a program that will take the value of a float variable x with a value in $[0, 1)$ as input and return the first five coefficients in its binary expansion (2.2). Note, to use floor you will need to include the math library with the statement #include <cmath>.

2.4. Another bit-wise operator is the exclusive OR or XOR operator ^. If k1 and k2 are two integers, the bits for k1^k2 are 1 whenever one or the other of the corresponding bits in k1 or k2 is 1 and are 0 otherwise.

a) Develop a binary addition algorithm based on XOR. [Hint: Incorporate a carry bit into the addition of two bits.]

b) Implement your algorithm from part a) in C++. Assume that you are dealing with numbers that are one byte in length and carry out your calculations using unsigned short integers.

2.5. Consider the division of two sets of floating-point numbers $\tilde{x}_1, \ldots, \tilde{x}_n$ and $\tilde{y}_1, \ldots, \tilde{y}_k$ with m-bit significands.

a) Assuming exact rounding, show that

$$(\tilde{x}_1 \cdots \tilde{x}_n)\tilde{/}(\tilde{y}_1 \cdots \tilde{y}_k) = [(\tilde{x}_1 \cdots \tilde{x}_n)/(\tilde{y}_1 \cdots \tilde{y}_k)] (1 + E)$$

with

$$\left(1 - 2^{-(m+1)}\right)^{(k+n-1)} \leq 1 + E \leq \left(1 + 2^{-(m+1)}\right)^{(k+n-1)}.$$

b) Use the bound for $1 + E$ obtained in part a) to show that the relative error from division should behave like a constant multiple of $(n+k)2^{-(m+1)}$ if $n+k$ is small relative to 2^{m+1}.

2.6. Let $\tilde{\ }$ denote the result of floating-point conversion. Prove or disprove that $(\tilde{x} \tilde{+} \tilde{y}) \tilde{+} \tilde{z} = \tilde{x} \tilde{+} (\tilde{y} \tilde{+} \tilde{z})$.

2.7. Let $y = 1 + x$ for a positive number x. If y is stored as the float value $\tilde{y}$ according to the IEEE standard, under what conditions will $\tilde{y} = 1$?

2.8. Under the IEEE standard the significand for a double is allocated 52 bits of storage.

a) Give bounds for the corresponding relative error associated with rounding to the nearest value and chopping for doubles.

b) How many decimal places of accuracy can be expected for a `double` under this storage scheme?

c) Can you conclude from this storage specification (why or why not) that summing the series $1/j^2$ in reverse order as in Section 2.5 except with double precision is accurate to seven decimals?

2.9. Show that upon completion Algorithm 2.3 will return the sample residual sum-of-squares.

2.10. Show that the algorithm below, due to Youngs and Cramer (1971), computes the sample residual sum-of-squares.

Algorithm 2.5 Youngs and Cramer algorithm

$T = x_1$
RSS $= 0$
for $j = 2$ to n **do**
   $T := T + x_j$
   RSS $:=$ RSS $+ \frac{(jx_j - T)^2}{j(j-1)}$
end for
return $\sqrt{\frac{\text{RSS}}{n-1}}, \frac{T}{n}$

2.11. Consider the quadratic polynomial $q(x) = x^2 - 111.11x + 1.2121$.

a) Find the roots of q using the quadratic formula but using only five decimal digits with chopping in all your calculations (addition, multiplication, square root, etc.).

b) For a quadratic polynomial $ax^2 + bx + c$ show that the roots x_1 and x_2 satisfy $x_1 x_2 = c/|a|$. Use this latter relationship to evaluate the root having smaller absolute value for q under the same conditions for digit retention and rounding and compare the result with your previous answer.

(Conte and de Boor 1972)

Chapter 3

A sketch of C++

3.1 Introduction

This chapter presents some of the essential ingredients that will be needed for code development in C++. We will begin with a few basics that will allow us to get started writing programs and then progress fairly quickly to incorporate tools that will allow for more effective access to the power of the language. The treatment is uneven at best. More detailed and exhaustive coverage can be obtained from Stroustrup (1997), Eckel (2000) and Prata (2005) as well as many other excellent language references that are available in print and on-line.

As seen in the previous chapter, schematically a C++ program will look like

```
include statements
declarations and definitions
int main(int argc, char* argv[]){
    body of program
}
```

Each include statement must be preceded by a # and the body of `main` has to be enclosed by curly braces. Every completed statement in the body of the program must end with a ; and the same is true for declarations and some definitions that describe variables and functions as will be explained in more detail shortly. The `argv` and `argc` arguments to `main` are optional but useful for specification of additional parameters from the command line as seen in Listing 2.3 of Section 2.5.

The body of the program will consist of variable definitions, arithmetic expressions, loops, conditional execution statements, etc. These will be discussed in the forthcoming sections. Perhaps the first thing to note is the fact that was mentioned in Section 2.3: every variable's type must be declared explicitly when it is introduced into the code. As an example, the code snippet

```
int x;
x = 0;
```

begins with a declaration that informs the compiler that a variable x will be used that is of type `int`. The next line of code is an assignment statement where x is given the value 0. This could all have been accomplished in one step using

```
int x = 0;
```

where the type of x is declared in the same statement where it is assigned a value. Both of these ways of introducing x into a program are syntactically correct. But, an attempt to compile

```
//typeError.cpp

int main(){
    x = 3;
    return 0;
}
```

will cause the compiler to lodge a complaint of the form

```
typeError.cpp: In function    int main()    :
typeError.cpp:4: error: 'x' was not declared in this scope
```

Note that this is the typical form of an error message returned by the Gnu C++ compiler. It tells us the name of the file where the error was found (i.e., typeError.cpp), the name of the function in the file which contains the error (i.e., main) and gives the line number in the program (specifically, line 4 in the file typeError.cpp) where the error was detected.

In C there is a distinction between a *declaration* and a *statement* with the former having the purpose of informing the compiler of storage requirements while the latter executes an action (or actions) on a variable (or variables), performs writing operations, etc. Under the older C standards, declarations and statements cannot be mixed. There is more flexibility in C++ in that declarations are also viewed as statements which allows them to be mixed with executable statements. For example, the code snippet

```
int main(){
   int x, y;
   x = 0; y = 0;
   int z = x + y;
return 0;
}
```

is legal in C++. However, the fourth line of code int z = x + y; is illegal in some versions of C.* The ability to mix declarations and code will feature prominently in C++ for loops as will be seen in Section 3.4.

As another example of some C++ code, the listing below is the obligatory "Hello World!" program that writes a greeting to standard output (typically the shell).

```
//hello.cpp
#include <iostream>
#include <string>

using namespace std;

int main(){
   string s;
   s = "Hi ya'll!";
   cout << s << endl;
   return 0;
}
```

The optional arguments (argv and argc) to main, that were not needed in this instance, have been omitted. Then, a variable (actually, more precisely, a data object in this case) of type string is used to hold the output phrase before it is directed to the shell by the output stream object cin introduced in Section 2.2. The string class will be discussed in more detail in Section 3.8.5. For now we merely note that an include statement has been used to make the class available and that the double quotes are needed to define a string object.

As in Chapter 2, to run the program from a Unix environment the Gnu C++ compiler is invoked from the shell command line with

```
$ g++ -Wall hello.cpp -o hello
```

---

* Actually this code will compile without error using, e.g., the Gnu C compiler gcc. Using gcc with the -pedantic option will produce the warning that the ISO (International Organization for Standardization) C90 standard "forbids mixed declarations and code".

An explanation of the compiler options that were used here was given in Section 2.2. In particular, the -o option that precedes hello in this case assigns the name hello to the executable version of hello.cpp that is created by the compiler. To execute the program from the current directory enter ./hello on the command line to obtain

```
$ ./hello
Hi ya'll!
```

3.2 Variables and scope

An example of a function was given in Listing 2.2. The body of this, as well as that of any other C++ function (such as **main**), must be contained inside curly braces. The use of braces extends more generally to the encapsulation of blocks of code that are nested inside functions. In this as well as other contexts they serve the purpose of defining the life and influence of variables inside a program.

In most cases the *scope* of a variable is the block of code encompassed by the innermost pair of curly braces that contain the variable's declaration. A variable exists only within its particular scope and ceases to exist (i.e., is no longer accessible to the rest of the program) when it goes out of scope. The listing below illustrates this point.

Listing 3.1 *scope.cpp*

```
//scope.cpp
#include<iostream>

using namespace std;

int main(){
  int x = 3;
  cout << "x = " << x << endl;
  {
    int x = 4;
    int y = 7;
    cout << "x = " << x << " y = " << y << endl;
  }
  cout << "x = " << x << " y = " << y << endl;
  return 0;
}
```

An attempt to compile Listing 3.1 produces the error message

```
scope.cpp: In function 'int main()':
scope.cpp:14: error: 'y' was not declared in this scope
```

which indicates the presence of an error on line 14

```
  cout << "x = " << x << " y = " << y << endl;
```

The problem is that y lives inside the second interior set of braces in the program **main** and goes out of scope once we pass back into the exterior set of braces and try to output its value. Thus, it no longer exists and, in fact, from the perspective of what takes place outside of the interior braces, the y variable never existed at all.

To fix Listing 3.1 replace the offending line of code with, e.g.,

```
  cout << "x = " << x << endl;
```

With this alteration the output from the program is

```
x = 3
x = 4 y = 7
x = 3
```

This reveals that a local version of x has been defined inside the interior braces that masks the variable of the same name that was defined in the outside code block. The interior x exists independently of the exterior one in that it can take on different values and the values it assumes have no effect on its exterior namesake.

An exceptional case occurs when a variable is defined outside of any curly braces thereby giving it global or program-wise scope. In that case it is available to all subsequent parts of the program and only goes out of scope when the program ends. Such global definitions can be useful for defining constants that can be shared by several different functions. This feature is used, for example, in Section 4.6 to provide a global definition of π.

3.3 Arithmetic and logical operators

Much of the code we will write for the body of **main** (and, more generally, for the bodies of functions or methods) will involve the use of simple arithmetic expressions being applied to primitive variable types such as those that were discussed in Chapter 2. Some of the tools for constructing such expressions are

+   addition

−   subtraction

*   multiplication

/   division

%   integer modulus (remainder after integer division)

and

=   assignment (e.g., x = y)

+=   addition assignment (x += y means x = x + y)

−=   subtraction assignment (x −= y means x = x − y)

*=   multiplication assignment (x *= y means x = x*y)

/=   division assignment (x /= y means x = x/y)

The assignment shortcuts +=, −=, *= and /= prove useful for streamlining source code.

Similar to the assignment shortcuts, simplified syntax for incrementing or decrementing the value of a variable (e.g., i in this case) by one have been provided in the form of

++i   increment i and then use it

−−i   decrement i and then use it

i++   use i and then increment it

i−−   use i and then decrement it

These will be used repeatedly in the creation of **for** loops as in Listing 2.3.

There are also the basic comparison operators that play a prominent role in carrying out conditional execution. In this regard, C++ includes

==   logical equal

!=   logical not equal

>, <, >= and <= for, respectively, logical greater than, less than, greater than or equal and less than or equal

These operators appear in expressions that return values `false` (or 0) or `true` (or 1). An expression such as (x >= 3) evaluates as either `true` or `false` according to whether or not the variable x is greater than or equal to 3 (Exercise 3.1).

Finally, there are the logical operators that correspond to intersection, union and negation: namely,

&& logical AND

|| logical OR

! logical NOT

The expression A && B evaluates to `true` if both A and B are `true` while A || B evaluates to `true` if either A or B are `true`. The ! operator negates the argument to its right: for example, !(x $\geq$ 3) evaluates as `true` if x is less than 3.

3.4 Control structures

The power of computing comes from the ability to adaptively carry out relatively simple tasks a large number of times. The adaptive feature is provided by conditional if-then-else structures and the repetition is achieved using loops. The structure of the simplest block of code containing a C++ conditional is

```
if(testExpression){
   statements and expressions
}
```

Here `testExpression` may be any syntactically valid expression. C++ will cast the result to `bool` and return either a `false` or a `true`. When `testExpression` in the argument to `if` evaluates to `true`, the first line or block (i.e., a collection of statements contained in curly braces) of code after `if` is evaluated. Otherwise the control passes to the next line of code after this block.

Somewhat more involved constructs allow for branching as in

```
if(testExpression){
   statements and expressions
}
else{
   statements and expressions
}
```

where the `else` block is executed when `testExpression` evaluates to `false`. Still more generally, multiple branches of the form

```
if(testExpression1){
   statements and expressions
}
elseif(testExpression2){
   statements and expressions
}
        .
        .
        .
else{
   statements and expressions
}
```

can be utilized.

Looping ability in C++ is provided by `for` and `while` loops. The former looks like

```
for(initializingExpression; testExpression; increment){
   statements and expressions
}
```

The loop begins with the initializing statement `initializingExpression` that will generally give the starting value for an index variable. After the first ; a condition represented by `testExpression` is evaluated as either `true` or `false`. Execution of the related block of code (i.e., the statements between the two curly braces) or body of the loop continues until `testExpression` evaluates to `false`. Each time the body of the loop is executed, a control expression (usually the value of an index variable) is updated or incremented and `testExpression` is evaluated again.

A `while` loop appears like

```
while(testExpression){
   statements and expressions
}
```

Unlike the `for` loop, the while loop will continue to force execution of its associated block of code as long as `testExpression` evaluates to `true`. Thus, the code

```
while(1){
   cout << "I'm in a loop!" << endl;
}
```

will print "I'm in a loop!" to standard output indefinitely until execution is forcefully terminated (see Section 4 of Appendix A).

Actually, the `for` and `while` loops are equivalent in the sense that

```
for(; testExpression;){
   statements and expressions
}
```

will emulate a `while` loop. Similarly,

```
initializingExpression;
while(testExpression){
   statements and expressions
   increment;
}
```

emulates a `for` loop using a `while` construct. The version of a `for` loop used here illustrates that the arguments for a `for` loop are optional although the semicolons are mandatory. In particular,

```
for(;;)
   cout << "I'm in a loop!" << endl;
```

will produce the same effect as the previous `while` loop that printed the "I'm in a loop!" message and must also be manually terminated by the user.

The program listed below adds integers from a given starting point until the sum passes a specified bound. The summation is accomplished using both `for` and `while` loops.

Listing 3.2 *forWhileEx.cpp*

```
//forWhileEx.cpp
#include <iostream>
#include <cstdlib>

using namespace std;
```

```
int main(int argc, char* argv[]){
  int start = atoi(argv[1]);
  int bound = atoi(argv[2]);
  //compute the sum with a for loop
  int sum = 0;
  for(int i = start; sum <= bound; i++)
    sum += i;
  cout << "The sum from the for loop is " << sum << endl;
  //compute the sum with a while loop
  sum = 0;
  int i = start;
  while(sum <= bound){
    sum += i;
    i++;
  }
  cout << "The sum from the while loop is " << sum << endl;
  return 0;
}
```

Values for the bound and the starting point for the summation are passed into the program as the second and third character strings that are entered on the command line after the `./` prefaced name of the executable. They are converted from character to integer values using the `atoi` function that was used previously in Listing 2.3. The sum is then initialized to 0 and incremented iteratively in each of the two loop structures. Results from using the program might appear something like

```
$ g++ -Wall forWhileEx.cpp -o forWhileEx
$ ./forWhileEx 10 100
The sum from the for loop is 108
The sum from the while loop is 108
```

There are two other conditional statements in C++ that can provide alternatives to if/else blocks: `switch` and the *conditional operator* `?:`. The `switch` statement allows selection among different sections of a set of code depending on the value of an integral expression. As such, it can provide a convenient alternative to if/else blocks when dealing with logical conditions that produce multiple branches. The syntax for the conditional operator is

```
    testExpression ? expression1 : expression2
```

Here `testExpression` is a logical construct. When it evaluates to `true`, `expression1` is executed and `expression2` is enacted otherwise. Detailed treatments of `switch` and the conditional operator can be found, for example, in Chapter 6 of Prata (2005).

It is sometimes useful to terminate a loop when some unusual condition arises. A `break` statement will terminates the execution of the most interior enclosing loop or conditional statement in which it appears. For example,

```
//breakEx.cpp
#include <iostream>

using namespace std;

int main(){
  for(int i = 0; i < 3; i++)
    for(int j= 0; j < 10; j++){
      if(j == 2) break;
      cout << j << " ";
    }
```

```
    cout << endl;
    return 0;
}
```

produces the output

```
$ g++ -Wall breakEx.cpp -o breakEx
$ ./breakEx
0 1 0 1 0 1
```

3.5 Using arrays and pointers

To this point we have primarily dealt with primitive data types. These are the building blocks for compound data types that include arrays and classes. Arrays will be treated in this section while classes are the topic of Section 3.7.

An important part of most programming languages is the use of arrays for data storage and manipulation. A one-dimensional array is a vector that, in C++, may contain objects of any one data type, either primitive or user defined.

Arrays are declared in C++ code by using square brackets []. There is nothing special about the one-dimensional case. The declaration

```
    int x[nRows][nCols];
```

will reserve the memory space for a two-dimensional integer array or matrix with nRows rows and nCols columns. The (i, j)th array entry is accessed using x[i][j]. Note that arrays in C/C++ use 0-offset indexing, which means that the first element of a one-dimensional array x is x[0] rather than x[1]. Similarly, the first component of a two-dimensional array x is x[0][0].

The ensuing lines of code illustrate various ways to initialize and assign values to elements of an array.

```
    double a1[3] = {1, 2, 3};
    int a2[7] = {1, 2};
    float a3[2][3] = {{1, 3, 5}, {2, 4, 6}};
```

The first statement initializes a three-element array a1 of doubles with a1[0] = 1, a1[1] = 2 and a1[2] = 3. The second statement initializes only two of the seven elements in the integer array a2. The noninitialized entries are set to 0 by default; i.e., a2[2] = $\cdots$ = a2[6] = 0. Finally, a3 is a 2×3 array with first row a3[0][0] = 1, a3[0][1] = 3, a3[0][2] = 5 and second row a3[1][0] = 2, a3[1][1] = 4, a3[1][2] = 6.

Another way to work with arrays is through the use of *pointers*. The value of an object of type pointer is a memory location. Pointers have a myriad of applications beyond just their connection to arrays that make them an essential ingredient of any C++ programmer's repertoire.

Whenever we execute a program it is loaded into the computer's memory. One may think of this memory as being sequentially arranged, one-byte blocks that are numbered starting from zero and going up to an integer that represents the total amount of available memory, although this description may not be strictly true in reality. Now, each part of the program occupies one or more of these memory blocks. The address of a program component can be accessed in C/C++ using the *address operator* &. An illustration of this is the following simple program.

```
//addresses.cpp

#include <iostream>

using namespace std;

int main(){
  double v[4] = {1., 2., 3., 4.};
  cout << "Address of v[0] = " << &v[0] << endl;
  cout << "Address of v[3] = " << &v[3] << endl;
  return 0;
}
```

which produces the (machine-dependent) output

```
$ g++ -Wall addresses.cpp -o addresses
$ ./addresses
Address of v[0] = 0xbffff9d0
Address of v[3] = 0xbffff9e8
```

The addresses returned from the program are in hexadecimal format with the letters a, b, c, d, e and f corresponding to the numbers 10–15, respectively. The output shows that, as will always be true for arrays in C++, the storage for v has been allocated in contiguous blocks of eight bytes of memory. To see this, first recall that a **double** is allocated eight bytes of storage. Then, note that the memory allocated to the first component of v at 0xbffff9d0 is 24 (i.e., 3×8) bytes (counting in hexadecimal) from where the memory for its last component begins at 0xbffff9e8.

Pointers are variables that allow us to work with addresses. A pointer takes a memory location address as its value and provides the facility to access (i.e., read from and write to) that location. A pointer **p** to an object of type dataType is declared via the syntax

```
dataType* p;
```

So, for example,

```
int* p;
```

declares a pointer p that can be used to hold the address of a variable of type **int**. To carry this a bit further, if we write

```
int* p;
int i = 1;
p = &i;
```

the result is that p "points" to the address of i in memory. To access the number that is stored in this location, the pointer is dereferenced using the syntax *p. This is illustrated by the code example below.

```
//simplePointer.cpp
#include <iostream>

using namespace std;

int main(){
  int* p = 0;
  int i = 1;
  p = &i;
  cout << "Address of i = " << &i << endl;
  cout << "Address of p = " << &p << endl;
```

```
cout << "Value of p = " << p << endl;
cout << "Value stored in p = " << *p << endl;
cout << "Value stored in p = " << p[0] << endl;
return 0;
}
```

Apart from the specific addresses returned as output, the results from the compilation and execution of this program will look like

```
$ g++ -Wall simplePointer.cpp -o simplePointer
$ ./simplePointer
Address of i = 0x7fff5fbff74c
Address of p = 0x7fff5fbff740
Value of p = 0x7fff5fbff74c
Value stored in p = 1
Value stored in p = 1
```

The last line of output from the program was produced using an alternative way to dereference the pointer p: namely, p[0] rather than *p. This syntax is indicative of the connection between pointers and arrays that will be discussed shortly.

Note that the pointer p in simplePointer.cpp was initialized with 0 which has the effect of producing a so called *null pointer*: i.e., a pointer that points to no location in memory and, as a result, cannot be dereferenced. An alternative, equivalent, approach is to initialize the pointer with the "value" NULL that in C++ is an alias for 0. We will use the 0 option throughout this text and view explicit initialization of pointers with 0 as a good programming practice. It makes it clear to ourselves and others that the pointer does not point to a valid memory location and cannot be dereferenced until that has been changed.

Null pointers are actually quite useful. Although they cannot be dereferenced, they can be compared using == and != to other pointers. No other valid pointer (i.e., one that points to a valid location in memory) will compare as equal to the null pointer and this can be used in conditional statements to determine if a memory allocation problem has occurred, if a function has executed successfully, etc. Null pointers play an essential role in many of the basic operations that are performed with the abstract data structures of Chapter 9.

Pointers are the key to dynamic memory allocation for vectors and arrays and for passing variables to functions in a manner that will allow their values to be altered. To address the first point note that in C++ there are three memory storage types: automatic storage (or "the stack"), static storage and the *free-store* or the *heap*. The free-store is a large block of memory used for allocation of memory at run-time. It is what should be used if there is any uncertainty about precisely how much memory will be needed when you run your program. This, of course, is a common occurrence in many statistical applications where code should be able to adapt to data sets of general sizes that may not be determined until run time. Even in the context of our random number generation programs in the next chapter, the size of the vector of random numbers they produce needs to be determined at run-time and could be arbitrarily large depending on the needs of the user.

The operator new is used to allocate memory on the free-store. The syntax looks like

```
dataType* p = new dataType[size];
```

In this instance the operator new returns a pointer to dataType whose value is the beginning location of size contiguous blocks of memory with each block being large enough to hold a dataType object. The code

```
double* p = 0;
p = new double[10];
```

will give a value for the pointer p that points to the first of 10 blocks of eight byte segments of memory. To write to or read the values in each block of memory the pointer is dereferenced using the operator []; i.e., values stored in our 10 blocks of memory are accessed by p[0],..., p[9]. This form of dereferencing looks essentially the same as what was done above for a one-dimensional array and, for all practical purposes, it is. Given the very close relationship between pointers and arrays, it is not surprising that the subscripting conventions in C++ generalize to the pointer context.

Free-store memory should be released whenever it is no longer needed. This is accomplished with the **delete** operator. For instance, if a pointer **p** is obtained from

```
int* p = new int;
```

this memory would be released by simply using

```
delete p;
```

If a block of memory has been allocated for an array as in the previous example, using

```
p = new double[10];
```

this memory can be released with

```
delete[] p;
```

Failure to release memory used by an object that was created using the **new** operator can lead to "memory leaks" that can potentially exhaust all the available free-store memory.

In terms of how they are used in standard arithmetic (and even in their behavior when passed to functions) there is little difference between pointers and arrays. However, the two types are not the same and the reason this is true can be appreciated by understanding the C++ **const** designation.

The use of a **const** modifier to preface the type of a variable has the effect that its value cannot be altered. For example, by using the inverse cosine function, **acos**, available through the cmath header, we can define

```
const double pi = 2*acos(0.);
```

This gives us a symbolic equivalent (namely **pi**) for the (floating-point approximation of the) value of π that can now be conveniently used anywhere in the code block that contains the statement. Prefacing **double** with **const** guarantees that **pi** will hold the same value whenever it is used. Symbolic constants that are defined in preprocessor directives are a standard feature from C. The **const** designation serves a similar purpose but with more functionality in the sense that the result acts just like a variable; it can be passed as the argument to a function and will cease to exist when it goes out of scope. Moreover, any data type can be made **const**. When doing so, a value must be assigned in the declaration of the variable.

In the case of pointers there are two ways **const** can be used. A pointer to a **const** variable means that the value of the variable cannot be changed. On the other hand, the pointer itself may be declared **const** and this will imply that the memory location used in its initialization cannot be changed. The syntax for creating a pointer to a **const** variable (i.e., one whose value cannot be changed) is

```
const dataType* p;
```

The way to read this is that the data type is **const dataType** with the effect that * creates a pointer to a variable of type **dataType** whose value cannot be changed as a result of the **const** designation. In contrast, a **const** pointer to a variable of type **dataType** is created with

```
dataType* const p;
```

which says that p is a const pointer to dataType; that is, the value of the pointer itself cannot be changed. The idea is illustrated in the elementary program that appears below.

Listing 3.3 *constPointer.cpp*

```
//constPointer.cpp

int main(){
  int x = 1, y = 1;
  const int* p1 = &x;
  int* const p2 = &x;
  p1 = &y;
  p2 = &y;//error
  p1[0] += 1;//error!
  p2[0] += 1;
  return 0;
}
```

The attempt to make the const pointer p2 point to the address of y in line 8 of the listing produces a compile time error because once p2 has been initialized it cannot be reassigned to point to another memory location. Similarly, the value stored in the memory location that is pointed to by p1 has been declared as that of a const integer variable, and therefore the attempt to change that value in memory on line 9 leads to a compilation error. But, it is perfectly legal to have p1 point to a new address or change the value in memory accessible through p2 as can be seen by commenting out lines 8 and 9.

An array is effectively just a const pointer. That is, once an array is dimensioned in a program it cannot be redimensioned nor can the memory it occupies be utilized for any other purpose until it goes out of scope. With this in mind we can think of a statement that dimensions an n-element array like

```
double z1[n];
```

as being similar to

```
double* z2 = new double[n];
```

in that both require allocation of the memory that is necessary to store n doubles. However, in the first case the memory for z1 is allocated automatically on the stack and released automatically whenever z1 goes out of scope. Because we have no control over this memory, z1 cannot be redimensioned. In the second case the memory for z2 is obtained dynamically on the free-store. It must then be freed explicitly via

```
delete[] z2;
```

But, once the memory is freed, a new block of memory, possibly of different size (although still for doubles), can be pointed to by the z2 pointer.

So far we have only dealt with how to represent vectors (one-dimensional arrays) using pointers. This is not sufficient and it is also necessary to have an analogous way of working with two-dimensional arrays or matrices. This can be accomplished through the use of a vector or one-dimensional array of pointers. The memory for the rows of the matrix is then allocated to each of the pointers in the array. Of course, the way to get a vector of pointers is to use a pointer as well. That is, to create an array with nRows rows and nCols columns, we first create a pointer that points to a block of memory containing nRows pointers, one corresponding to each row of the array. Then, each of these memory locations contains the address of the block of memory that is allocated to the corresponding row of the matrix.

This pointer-to-pointer storage scheme is illustrated in the diagram below.

$$p \longrightarrow \&p[0] \qquad\qquad p[0] \longrightarrow \&p[0][0]$$

$$p[1] \longrightarrow \&p[1][0]$$

$$\vdots \qquad\qquad\qquad \vdots$$

$$p[\mathtt{nRows} - 1] \longrightarrow \&p[\mathtt{nRows} - 1][0]$$

The diagram shows a pointer p that stores the address of the first of nRows contiguous memory locations that will each hold a pointer to the data type that will be stored in their respective rows. The value written to each of the row pointers is the address of the first of the nCols blocks of memory allocated to hold the actual data values that will be placed in the corresponding row of the array. It should be noted that the memory required to hold an address will depend on the computer's architecture with four and eight bytes being typical values. The important point is that, whatever the size, it will be independent of the size of the block of memory to which it points.

There are two steps that are needed to obtain a pointer-to-pointer memory allocation to hold an nRows × nCols "matrix" of components of type dataType. First, obtain the memory needed to store an array of pointers to dataType. The elements of this array are therefore of type dataType* so that the necessary memory may be requested from the free-store with something like

```
dataType** p = new dataType*[nRows]
```

The outside * in the dataType declaration indicates that p is a pointer while the inside * is really part of the data type name of dataType*.

At this point memory has been allocated only for the pointers that will hold the locations of the memory that will be allocated for the rows of our array. The second step is to obtain the memory for the rows via a loop such as

```
for(int i = 0; i < nRows; i++)
   p[i] = new dataType[nCols];
```

Since p[i] is a pointer to dataType this works perfectly well and values can now be assigned to the memory locations through dereferencing. Specifically, the syntax p[i][j] implicitly carries out two steps. First p[i] dereferences p to obtain the pointer to the ith row of the "matrix". Then, p[i][j] dereferences the pointer p[i] to obtain access to the memory for the (i, j)th array element.

A specific example that uses the pointer-to-pointer approach to store a matrix is provided in the next listing.

Listing 3.4 *matrixPointer.cpp*

```
//matrixPointer.cpp
#include <iostream>

using namespace std;

int main(){
```

```
int nRows = 2, nCols = 3;
double** pM = 0;
pM = new double*[nRows];
for(int i = 0; i < nRows; i++)
  pM[i] = new double[nCols];
cout << "The value of pM is " << pM << endl;

for(int i = 0; i < nRows; i++)
  cout << "The address for pM[" << i << "] is "<<
    &pM[i] << endl;

for(int i = 0; i < nRows; i++){
  cout << "The value of pM[" << i << "] is " << pM[i] << endl;
  for(int j = 0; j < nCols; j++){
    cout << "  The address of pM[" << i << "]["
         << j << "] is "<< &pM[i][j] << endl;
  }
  cout << endl;
}

for(int i = 0; i < nRows; i++){
  for(int j = 0; j < nCols; j++){
    pM[i][j] = i*j;
    cout << "pM[" << i << "][" << j << "]=" << pM[i][j] << " ";
  }
  cout << endl;
}

//memory clean up
for(int i = 0; i < nRows; i++)
  delete[] pM[i];
delete[] pM;
return 0;
}
```

This code will produce output that resembles

```
The value of pM is 0x100150
The address for pM[0] is 0x100150
The address for pM[1] is 0x100154
The value of pM[0] is 0x100160
  The address of pM[0][0] is 0x100160
  The address of pM[0][1] is 0x100168
  The address of pM[0][2] is 0x100170

The value of pM[1] is 0x100180
  The address of pM[1][0] is 0x100180
  The address of pM[1][1] is 0x100188
  The address of pM[1][2] is 0x100190

pM[0][0]=0 pM[0][1]=0 pM[0][2]=0
pM[1][0]=0 pM[1][1]=1 pM[1][2]=2
```

Listing 3.4 creates a 2×3 matrix of double precision variables. It does this by first obtaining a pointer that can hold (i.e., take on the value of) the memory location of a pointer to double. The first for loop carries out the allocation of memory to each of the two rows of the matrix. The resulting output from our program then confirms what was indicated in the diagram above; that is, pM points to (i.e., holds the address of) the pointer pM[0] which, in turn, points to pM[0][0], etc. Specifically, the second `for` loop in Listing 3.4 demonstrates that memory is laid out in contiguous blocks of four bytes to hold the values of pM[0] and pM[1] (because 0x100150 and 0x100154 are four bytes apart). The third, nested `for` loop reveals that the values of pM[0] and pM[1] are the same as the addresses of pM[0][0] and pM[1][0] and that contiguous blocks of 8 bytes of memory have been allocated to hold pM[0][0], pM[0][1], pM[0][2] and pM[1][0], pM[1][1], pM[1][2] (since, e.g., 0x100160 and 0x100168 are eight bytes apart).

In general, there is no reason (or guarantee) that memory locations will be contiguous from row to row. In fact, there appears to be a gap between the memory blocks for the first and second row in our array due to the presence of eight bytes of memory between the end of the block for pM[0][2] and the one that begins at pM[1][0]. Actually, this is a feature of C/C++ where an additional "one-past-the-end" block of memory is always provided for a pointer. This additional block of memory can be pointed to but cannot be dereferenced.

Listing 3.4 demonstrates that data stored in the pointer-to-pointer format can be manipulated in essentially the same manner as for an ordinary array. To assign values to or print the value in the (i, j)th memory location for the pointer pM we use the pM[i][j] syntax.

The final `for` loop in the listing demonstrates how to release the memory for pointer-to-pointer array storage. In the present case this is technically unnecessary as all the memory will be released automatically when the program terminates. But, in general, pointers may exist inside of other functions called from `main` as discussed in the next section.

The correct route to releasing memory from a pointer-to-pointer storage format proceeds along the lines of the one used in Listing 3.4. The memory for each row must be released before that of the pointer to the row pointers. A look at the example should be enough to see that this is the only route that will really work. Releasing the memory for pM first will only free up two storage locations (i.e., the ones allocated to pM) and would produce a memory leak where we would no longer have access to the six storage locations allocated to the rows of the array; that particular memory would remain allocated and unusable until the program terminated.

3.6 Functions

The C/C+ language would be very restrictive if it only allowed for source code that could be put in the body of `main`. However, we already know from Chapter 2 that this is not the case and it is possible to include other functions in our programs. To accomplish this a prototype and/or definition of the function must be provided prior to the code for `main`. This can be accomplished by placing the prototype in the same file as `main` or through the use of an `#include` directive. We will restrict attention to the first option for the moment.

The general structure of a function definition has the form

```
returnType functionName(arguments){
   body of function
   return statement
   }
```

Here the `returnType` can be any of the primitive data types, a pointer, etc. The quantity returned in the `return` statement must be of the same type as `returnType`.

The code listing below is for a simple program that calls a function `square` in the body of `main` to square the value of some number that is provided as input through the shell interface using the input stream object `cin` introduced in Section 2.3.

Listing 3.5 *square.cpp*

```
//square.cpp
#include <iostream>

using namespace std;

double square(double x);

int main(){
  double x;
  cout << "Enter a number ";
  cin >> x;
  cout << square(x) << endl;
  return 0;
}

double square(double x){
  return x*x;
}
```

Notice the prototype that was provided for **square** prior to its use in **main**. If this had not been done and the definition of **square** was also not given prior to **main** a compiler error would have been generated that looked something like

```
square.cpp: In function   int main()    :
square.cpp:10: error: 'square' was not declared in this scope
```

Compilation and execution of **square.cpp** produces, e.g.,

```
$ g++ -Wall square.cpp -o square
$ ./square
Enter a number 4
16
$ ./square
Enter a number 1.278
1.63328
```

In C/C++ arguments to functions are passed by value. This means that a copy of each argument is created for use in the function. The function then works with the copies with the consequence that the actual values of the arguments cannot be altered. The copying process incurs both a memory and time overhead that should be managed carefully.

Among other things, the pass-by-value paradigm restricts what functions can accomplish as they can usually only return one object. An end run around this restriction can be obtained by using pointers or references. If a pointer is passed by value to a function the value of the pointer itself cannot be changed: i.e., the function cannot change the address the pointer points to in memory. But, it does have access to the memory location pointed to by the pointer. In particular, it can change the values that are stored in the memory locations associated with a pointer. In addition, passing a memory address means that creating the copy needed for use in the function will incur very little memory or time overhead. Pointers are generally stored in four or eight bytes of memory regardless of what type of variable they point to. In contrast, passing composite data types such as the class objects discussed in Section 3.6 can be unnecessarily wasteful of memory and time resources.

The listings below and their corresponding output illustrate the implications of the previous discussion. First, the code

```cpp
//passByVal.cpp
#include <iostream>

using namespace std;

void f(int i){
  i += 1;
}

int main(int argc, char* argv[]){
  int i = 0;
  f(i);
  cout << "The value of i is " << i <<endl;
  return 0;
}
```

leads to

```
$ g++ -Wall passByVal.cpp -o passByVal
$ ./passByVal
The value of i is 0
```

The result demonstrates that the value of the variable i is not changed by passing it to the function f. In contrast, the program

```cpp
//passByPoint.cpp
#include <iostream>

using namespace std;

void f(int* pi){
  pi[0] += 1;
}

int main(int argc, char* argv[]){
  int i = 0;
  int* pi = &i;
  f(pi);
  cout << "The value of i is " << i <<endl;
  return 0;
}
```

gives us

```
$ g++ -Wall passByPoint.cpp -o passByPoint
$ ./passByPoint
The value of i is 1
```

thereby showing that the value of i has been successfully altered.

The same effect that was realized by passing a pointer to a function can be obtained by passing an object by *reference*. A reference variable corresponding to a variable of type dataType is obtained using the syntax dataType&; for example,

```cpp
int i = 0;
int& j = i;
```

produces a reference to int variable j that refers to the same location in memory as the one occupied by i.

References provide objects that behave in most ways as if they were just alternative names or aliases for the objects they reference. A reference also has the property of a pointer in that if an object is passed under its reference "name" to a function the function can actually alter the object being referenced. A reference can be viewed as being similar to a pointer that can only point to one location in memory: the one that is used when it was initialized. Then, dereferencing is accomplished by simply using the name of the referenced variable.

The code below illustrates some of the features of references.

```cpp
//refEx.cpp
#include<iostream>

using namespace std;

void f(int& j){
  j++;
  cout << &j << endl;
}

int main(){
  int i = 2;
  cout << &i << endl;
  f(i);
  cout << i << endl;

  int& j = i;
  int k = j;
  cout << i << " " << j << " " << k << endl;
  cout << &i << " " << &j << " " << &k << endl;

  k++;
  j = k;
  cout << i << " " << j << " " << k << endl;
  cout << &i << " " << &j << " " << &k << endl;
  return 0;
}
```

The output from execution of this program might look something like

```
0x7fff5fbffa1c
0x7fff5fbffa1c
3
3 3 3
0x7fff5fbffa1c  0x7fff5fbffa1c  0x7fff5fbffa18
4 4 4
0x7fff5fbffa1c  0x7fff5fbffa1c  0x7fff5fbffa18
```

The two addresses and the lone integer 3 that appear in the first three lines of the output demonstrate that both the variable i from main and the reference to i provided by j in the function f share the same address in memory and that altering j in f will also alter i. We next create a variable j in main that is a reference to int and initialize it to i. A new integer variable k is then assigned the value of j. But, k is seen to be located in a different memory location. This is a consequence of how the assignment operator = behaves when used with primitive data types: i.e., a new memory location is created for the variable on the left-hand side of = and the value of the right-hand side variable is written into this location. Thus, the effect of assigning the reference variable j to k is the same as would be obtained if a pointer pi was created that pointed to i (because j is a reference to i) and then the assignment k = *pi was executed. So, in that sense j is like a pointer where simply using its name carries out dereferencing. The final two lines of output result from switching k and j to be on the right-hand and left-hand sides of = with the result that now the value of i is changed but memory locations are not altered.

Another useful property of C++ is that it allows for function overloading. That is, it allows for two or more functions to have the same name provided only that their argument types differ. The code below illustrates this with a modification of the square program.

```
//squareOverload.cpp
#include <iostream>

using namespace std;

double square(double x);
int square(int x);

int main(){
  double x;
  cout << "Enter a number ";
  cin >> x;
  cout << square(x) << endl;
  int y;
  cout << "Enter an integer ";
  cin >> y;
  cout << square(y) << endl;
  return 0;
}

double square(double x){
  cout << "From double version" << endl;
  return x*x;
}

int square(int x){
  cout << "From int version" << endl;
  return x*x;
}
```

The output from this program below verifies that the appropriate version of square is being called to deal with the cases of double and int argument types.

```
$ g++ -Wall squareOverloaded.cpp -o squareOverloaded
$ ./squareOverloaded
Enter a number 12.9
From double version
166.41
Enter an integer 12
From int version
144
```

Note that, when overloading a function, it is only differences in the argument types that matter. A difference in the return type alone will be reported as a compile-time error. In particular, the use of a function such as

```
int square(double x){
  return (int)(x*x);
}
```

in our square program, although perhaps reasonable conceptually, would produce an error with the compiler telling us that this new function "ambiguates" the first `square` function that had a `double` argument and return type.

3.7 Classes, objects and methods

The real power of C++ relative to C stems from its flexibility in defining new, user-created, data types. To a certain extent this facility existed in C through the ability to create a data structure or struct. Basically, a struct was an aggregate of various data types that could include a mix of arrays, character strings, integers, other structs, etc. The components of a struct were called members

C++ generalizes the struct concept by allowing data structures to include functions or methods as they are usually termed in this context. This generalization is called a class and, as noted in Chapter 1, specific instances of a class are referred to as objects.

To define a class in C++ we use a syntactic framework of the general form

```
class className{
private:
    declarations of members, method prototypes/definitions
public:
    declarations of members, method prototypes/definitions
};
```

Note that a `};` must appear at the end of a class definition. The keywords `public` and `private` are designations that determine how other classes and functions may access class members. Basically, `private` members and methods are for internal use inside the class and cannot (apart from some special exceptions) be accessed or used by other classes or functions. In particular, a class method can access and change the private members of its own class but not those of another class. Methods and class members are taken to be `private` unless specifically designated otherwise by placing them after the `public:` line of the class declaration. As a result, explicit use of the `private:` statement is not necessary.

The easiest way to clarify and solidify the class concept is through an example. The code listing below extends our square.cpp program to compute general integer powers of a given double precision number. In contrast to what we did before, the implementation now uses an OOP approach through the creation of the class `Power`.

Listing 3.6 *power.cpp*

```
//power.cpp
#include <iostream>

using namespace std;

class Power{
  int K;//integer power

public:
  Power(int k);
  double xToTheK(double x);
};

Power::Power(int k){
  if(k >= 1)
    K = k;
  else
    cout << "The power must be an integer >= 1!" << endl;
```

```
}                                                            19
                                                             20
double Power::xToTheK(double x){                             21
  double y = 1.;                                             22
  for(int i = 1; i <= K; i++)                                23
    y = x*y;                                                 24
  return y;                                                  25
}                                                            26
                                                             27
int main(){                                                  28
  int k;                                                     29
  double x;                                                  30
  cout << "Enter an integer power ";                         31
  cin >> k;                                                  32
  Power* pPow = new Power(k);                                33
  cout << "Enter a number to raise to this power ";          34
  cin >> x;                                                  35
  cout << pPow->xToTheK(x) << endl;                          36
  cout << "Enter another number to raise to this power ";    37
  cin >> x;                                                  38
  cout << pPow->xToTheK(x) << endl;                          39
  delete pPow;                                               40
  return 0;                                                  41
}                                                            42
```

Listing 3.6 begins with a prototype description of the class similar in nature to what we did for the function square in Listing 3.5. This is also sometimes called the class *interface* as it provides sufficient information for a programmer to use the class and its methods without having to see the actual details of the class implementation. Detailed definitions of the member functions are then provided outside of this prototype by using the class name along with the scope resolution operator ::. Thus, for example, line 21 of the listing states that xToTheK is a function or method that belongs to class Power and returns a double precision value while taking a double precision argument. In this simple setting, the definitions of member functions for the class could have actually been included in the class declaration or prototype. But, in general the present approach leads more directly towards readable object-oriented code.

The class Power has only one data member: the integer K that represents the power we want to use for computations. This value is determined explicitly through the function Power. A function whose name is the same as the name of the class is called a *constructor*. Because C++ allows for function name overloading, there can be more than one constructor provided that their arguments differ. The purpose of a constructor is to produce a class object: i.e., a specific instance of the class. As seen from line 33 of Listing 3.6, the constructor can be called using the new operator to produce a pointer to an object of the class.

If no constructor is included in a class the compiler will automatically supply one; this *default constructor* is a function with an empty body and no arguments. For our Power class it would have the (implicit) form

```
Power(){};
```

that would return an "empty" Power object. This behavior would not be satisfactory or useful in the applications we have in mind for this class. By supplying our own (explicit) constructor, we ensure that the C++ default constructor will not be implicitly added to the class. That is, given the explicit constructor, a declaration for a class object without a constructor argument will result in a compile-time error. In particular, statements such as

```
Power Pow;
```

or

```
Power pPow[5];
```

will produce such errors due to the absence of any mechanism for creating Power class objects without a specified value for the integer power K. If a default constructor is required when an explicit constructor has been provided, it must also be defined explicitly. For our Power class, a reasonable default constructor might look like

```
Power(){K = 1;}
```

With this added to class Power, a statement such as Power Pow; will produce a Power object whose K member is set equal to 1.

The value of a double raised to the power K is computed iteratively using the member function xToTheK. In general, to use a class method function methodName with a pointer pointerName to an object of a class, we employ the syntax pointerName->methodName. This is illustrated on lines 36 and 39 of Listing 3.6 for the xtoTheK method. Line 39 demonstrates that the Power object can be used for any given choice of K as many times as one desires without having to respecify that particular power. At the end of the listing we go through the formality of explicitly releasing the memory that was allocated to the pointer even though it would have been automatically released when pPow went out of scope after the } was encountered on the last line of the program. An example of output from the program is shown below.

```
$ g++ -Wall power.cpp -o power
$ ./power
Enter an integer power 3
Enter a number to raise to this power 1.6
4.096
Enter another number to raise to this power 2.7
19.683
```

There is an alternative to using the new operator for obtaining a member of a class. Specifically, one can invoke the class constructor directly by writing the object's name succeeded by a specification of appropriate values for the arguments in the constructor. The resulting object will be allocated memory on the main stack instead of the free-store. For the Power class this can be accomplished by replacing lines 33–40 of the Listing 3.6 with

```
Power powObj(k);
cout << "Enter a number to raise to this power ";
cin >> x;
cout << powObj.xToTheK(x) << endl;
cout << "Enter another number to raise to this power ";
cin >> x;
cout << powObj.xToTheK(x) << endl;
```

Here a Power class object powObj (rather than a pointer to a class object) calls xToTheK using the *member access operator* represented by a dot or period. In general an object objectName invokes a method methodName (or member called memberName) via the syntax objectName.methodName (or objectName.memberName).

The development so far has been successful in creating and using a class. However, the combination of class prototypes, method definitions and main all in the same file is needlessly cluttered, unwieldy and discourages code reuse. It is generally better to separate the collection of code that declares and defines a class from other code that actually uses an object from the class. This can be readily accomplished through the use of header files and include directives. Header files were discussed in Chapter 2 as a means of making available various aspects of the C++ standard library. We will expand on that idea in the sense of creating our own header files to facilitate access by other programs to code that we have created for various specific purposes.

Basically, a header file for a class should consist of the class prototype or declaration. So, for the Power class, the header file would be constructed using essentially lines 6–12 of the listing for power.cpp. There is a slight caveat here concerning include guards and we need to deal with this first before giving the exact listing for the header.

Header files provide the connective medium between classes and other programs that use them. For this reason, they must be included in any program that intends to use a member of a class. As a result, when multiple programs are being combined it is possible that a particular header file could be included more than once into the final compilation process. This would be viewed by the compiler as constituting a redefinition of the class and tagged as an error. *Include guards* provide a way to avoid this type of problem. The procedure is to use preprocessor directives to conditionally define a variable the first time the header is encountered. Once the variable is defined, it is then used to force the preprocessor to ignore the header file if it is encountered through an `include` directive from any other program.

A listing of the header file for class Power complete with an include guard is given below. The first time the header is encountered by the preprocessor, the variable POWER_H is defined and the code in the file is processed. Any other requests to include the header will have `#ifndef POWER_H` evaluate to `false` and the remainder of the file will be ignored.

```
//powerClass.h
#ifndef POWER_H
#define POWER_H

class Power{
  int K;//integer power

public:
  Power(){K = 1;}//default constructor
  Power(int k);
  double xToTheK(double x);
};
#endif
```

In addition to the include guard, a default constructor has been placed *inline* by giving its definition (rather than just declaration) in the header file. Any function defined within a class declaration is automatically inline with the effect that it will be expanded in place by the compiler (i.e., the body of the function will be used directly) in lieu of a function call. For small functions this can improve the speed of a program.

The definitions for the member functions for class Power are now placed in a separate file called powerClass.cpp which looks like

```
//powerClass.cpp
#include <iostream>
#include "powerClass.h"

using namespace std;

Power::Power(int k){
  if(k >= 1)
    K = k;
  else
    cout << "The power must be an integer >= 1!" << endl;
```

```
}

double Power::xToTheK(double x){
  double y = 1.;
  for(int i = 1; i <= K; i++)
    y = x*y;
  return y;
}
```

Note that an include directive has been used to bring in the Power class prototype contained in powerClass.h. The use of quotation marks instead of < > tells the preprocessor to begin looking for the header file in the current directory.

Finally, a program that actually uses class Power is

```
//powerClassDriver.cpp
#include <iostream>
#include "powerClass.h"

using namespace std;

int main(){
  int k;
  double x;
  cout << "Enter an integer power ";
  cin >> k;
  Power powObj(k);
  cout << "Enter a number to raise to this power ";
  cin >> x;
  cout << powObj.xToTheK(x) << endl;
  return 0;
}
```

Observe that the header file for class Power has been included in this driver program and that a Power class object is now being used to call the xToTheK method.

In looking back at the header file and associated cpp file for class Power, one will see that the namespace directive was employed only in the cpp file. This was done on purpose and it is, in general, not advisable to place namespace directives in header files. The reason for this is that the preprocessor essentially inserts the content of a header file into any source code file (which could even be another header file) that imports it using an #include statement. This means that a namespace designation in a header file can propagate unintentionally if the header file is being used several times in a multiple source code file project. The result can be name clashes that are hard to diagnose. Therefore, it is best to use explicit namespace qualifications via the scope resolution operator in header files and restrict the use of namespace directives to the cpp files for a class. We will demonstrate this approach in the next section.

The only thing that remains is dealing with how to create an executable program using our three separate files for the code. This can be accomplished directly via

```
$ g++ -Wall powerClass.cpp powerClassDriver.cpp -o powerClass
$ ./powerClass
Enter an integer power 3
Enter a number to raise to this power 1.2
1.728
```

which is a direct extension to multiple source files of our previous approach to compilation. Alternatively, one can use the -c option of the GNU compiler to create intermediate *object files* from the two cpp files. These files have a .o extension and can be linked to produce an executable. For our specific example this takes the form

```
$ g++ -c -Wall powerClass.cpp
$ g++ -c -Wall powerClassDriver.cpp
$ g++ -Wall powerClass.o powerClassDriver.o -o powerClass
$ ./powerClass
Enter an integer power 3
Enter a number to raise to this power 1.2
1.728
```

The first two invocations of g++ with the -c option produce the object files powerClass.o and powerClassDriver.o. These are then linked by the final call to g++ that produces the executable named powerClass. Further discussion of the use and options for g++ can be found in Gough (2005).

Linking object files to compile multiple source files can become rather tiresome if there are many source files and/or the process has to be repeated frequently. A program that allows us to efficiently automate the compilation of each set of code in a project is the make utility. Among other features, this program checks to see if any files in the project have changed since the last compilation and then recompiles only the ones that have been altered. To use make it is necessary to have a makefile in the directory that contains the code to be compiled. An illustration of such a file, named powerClass.mk in this case, that was written to compile our **Power** class example is shown below.

Listing 3.7 powerClass.mk

```
powerClass : powerClassDriver.o powerClass.o                          1
        g++ -Wall powerClassDriver.o powerClass.o -o powerClass        2
                                                                       3
powerClass.o : powerClass.cpp powerClass.h                             4
        g++ -c -Wall powerClass.cpp                                    5
                                                                       6
powerClassDriver.o : powerClassDriver.cpp powerClass.h                 7
        g++ -c -Wall powerClassDriver.cpp                              8
```

This is a very simple makefile. But, it illustrates the basic idea. The first item on line 1 of the makefile is the name for a *target*. After the colon, the required *dependencies* are listed and then the second line gives the gcc command line that should be used. The ensuing blocks of the makefile recursively describe how to produce each of the dependencies. Thus, lines 4–5 of Listing 3.7 give instructions for generating the object file powerClass.o, etc. The new feature is that there is now additional information provided about the dependence of each of our cpp files on other files which is what make uses to determine whether recompilation is necessary. An important requisite for a makefile is that *the lines of code after each set of dependencies must be preceded by a tab* and not just white space. To be specific, in our example every line of code that contains g++ starts with a tab.

To actually use our makefile apply the make utility via

```
$ make -f powerClass.mk
g++ -c -Wall powerClassDriver.cpp
g++ -c -Wall powerClass.cpp
g++ -Wall powerClassDriver.o powerClass.o -o powerClass
```

By default, make looks in the current directory for a file with the specific name makefile. The -f option that was employed here is a way to specify the file you want make to access. If we repeat the make command we receive the message

```
$ make -f powerClass.mk
make: 'powerClass' is up to date.
```

The updating feature is realized by, e.g., making a change to powerClassDriver.cpp. This will produce

```
$ make -f powerClass.mk
g++ -c -Wall powerClassDriver.cpp
g++ -Wall powerClassDriver.o powerClass.o -o powerClass
```

Make only recompiles the one object file that needed to be changed as the result of the editing. One may check that if, instead, powerClass.h had been changed, then all the object files would have been recreated because both powerClassDriver.cpp and powerClass.cpp have been designated as depending on powerClass.h in powerClass.mk. More detail on the make utility can be obtained from Stallman, et al. (2004).

3.8 Miscellaneous topics

In this section a few somewhat unrelated concepts will be examined that will arise in subsequent developments. Let us begin with the discussion of a slightly simplified version of the class construct that derives from the C++ implementation of the C struct data type.

3.8.1 Structs

As mentioned at the beginning of Section 3.7, in C it is possible to construct data types called structs that are similar to classes in the sense that they have data members. Unlike classes the C version of a struct has no associated methods (or member functions) and the data members are public. For compatibility reasons, in C++ a struct is simply a class where all data members and methods are public (rather than private) by default.

The code below illustrates the construction and use of a struct.

```
//structEx.cpp
#include <iostream>

using namespace std;

struct X{
  int K;
};

int main(){
  X x;
  x.K = 1;
  cout << x.K << endl;
  return 0;
}
```

If we run this code, it simply writes the integer 1 to the shell. What this demonstrates is that the member K of the X object x can be accessed directly from a calling program (or `main` in this case). One should check that this will no longer be the case if `struct X` is replaced by `class X` in this listing.

3.8.2 The this pointer

Every class contains a "hidden" or implicit member in the form of a pointer named `this`. When an object of a class is created the value of `this` is the address of that object in memory. The object is therefore "aware" of its own memory location and when necessary or useful it can directly access that information. This turns out to be particularly useful in the context of creating assignment type operators (e.g., =, +=, etc.) that work correctly when objects are using dynamically allocated memory. We will explore this particular issue in the next section. For now let us concentrate on understanding the properties of the this pointer in a simple setting.

Listing 3.8 carries out some experiments with the this pointer.

Listing 3.8 thisEx.cpp

```cpp
//thisEx.cpp
#include <iostream>

using namespace std;

class X{
public:
  int K;
  X(){};
  X(int k){K = k;}
  X* getThis(){return this;}
  X& getThisRef(){return *this;}
};

int main(){
  X w(1);
  cout << "w address = " << &w << endl;
  cout << "w value = " << w.K << endl;

  X* x = w.getThis();
  cout << "x value = " << x << endl;
  x->K += 1;
  cout << "w.K = " << w.K << endl;

  X& y =  w.getThisRef();
  cout << "y address = " << &y << endl;
  y.K += 1;
  cout << "w.K = " << w.K << endl;

  X z = *w.getThis();
  cout << "z address = " << &z << endl;
  z.K =+ 1;
  cout << "w.K = " << w.K << endl;
  return 0;
}
```

A simple class X with one public member K and two methods is created. One class method (getThis) returns the value of the this pointer while the other (getThisRef) returns a reference to an object constructed from class X. The way references are returned makes the syntax for the latter method a bit subtle. The fact that this is a pointer to the object's address means that the dereferenced pointer (as in Section 3.5) *this is the object itself. So, to return a reference to the object, we dereference this by returning *this and the function return type (i.e., X&) dictates that the address of the reference is that of *this: namely, the value of this.

The main function in Listing 3.8 instantiates an X object named w and then creates several

other X objects using w. The first object x is a pointer to data type X which we create using
the this pointer for w. The second object y is a reference variable for X objects that is
initialized with a reference to *this or, equivalently, a reference to w. The final object z is
of type X but is created by dereferencing the this pointer for w. The addresses are printed
out for all four objects and x, y and z are each used to change the value of their K data
member. After each change the value of K for w is reported. Upon execution this program
produces the output

```
w address = 0xbffffa24
w.K = 1
x value = 0xbffffa24
w.K = 2
y address = 0xbffffa24
w.K = 3
z address = 0xbffffa20
w.K = 3
```

As expected, x and y serve as aliases for w. This is not true for z which lies in a different
memory location and exists independently of w. Thus, augmenting its value for K does not
affect the value of w.K. What has happened in the creation of z is that the use of the =
operator has caused the creation of a new X object. This new object is obtained under the
default behavior of element-by-element copying. Specifically, a new memory location is set
aside for z and then the value of the K member for the X object *w.getThis() is copied
into this location.

Note that this is a const pointer in the sense of Section 3.5 as illustrated in Listing 3.3.
It cannot be reassigned to point to another object.

3.8.3 const correctness

We briefly discussed the C++ keyword const in Section 3.5. The use of this modifier
imposes an additional level of type checking that furnishes compile-time protection against
inadvertent changes being made to objects in one's code. The phrase *const correctness* means
that const has been used anyplace where it is appropriate. It is a practice that has value
not only in terms of error protection but also by the way it forces a programmer to examine
code to determine when the use of const is indicated. In general, using const appropriately
incurs no loss in functionality while providing safeguards and code clarity. Our discussion of
the const keyword in this section will focus on four cases where it will arise in later sections
and chapters: const pointers, const references, const class members and const member
functions.

Due to the efficiency of passing pointers or addresses over (copies of) objects, the use
of addresses or pointers for function arguments is quite common and preferable in many
situations. However, the use of addresses and pointers as function arguments carries with
it the danger that the function may alter the object being pointed to in cases when this
is not desirable. To avoid such events, a const designation can be applied for a pointer or
reference. Syntax of the form

```
const dataType* p = &x;
```

creates a pointer p to a variable x where the value of x cannot be changed. Similarly,

```
const dataType& y = x;
```

accomplishes the same thing with a reference; that is, y is just an alias for x except that
the value that y refers to in memory cannot be changed and, hence, y cannot be used to
change the value of x.

As noted in Section 3.6, references can be used in program statements in exactly the same way as if they had been variable names. This can result in inadvertent errors. A useful "rule" that avoids such issues and makes code easier to read is to pass objects by const reference as the default method of relaying arguments to functions and use pointers as arguments for cases where the object being passed will be altered by the function.

The nonoperative code below illustrates the fact that using the const modifier does, in fact, prevent us from writing on a location in memory that is passed by const reference.

```
//constRef.cpp
#include <iostream>;

using namespace std;

struct X{
  int K;
  void printK(){
    cout << K << endl;
  }
};

void g(const X& x){
  x.K = 1;
}

int main(){
  X x;
  g(x);
  x.printK();
  return 0;
}
```

If one attempts to compile this code the result will be an error message such as

```
constRef.cpp: In function 'void g(const X&)':
constRef.cpp:14: error:assignment of data-member 'X::K' in read-only structure
```

Removing the const modifier in the argument for the function g will solve the "problem".

The next section begins the development of a class Matrix that uses the pointer-to-pointer storage format discussed in Section 3.5. A class constructor will take a pointer to double* as input and it is imperative that the constructor not be allowed to alter the data that would generally be accessible through such a pointer. The way to accomplish this is with syntax of the form

```
    const dataType* const* p
```

This expression is read using two basic rules: namely,

• const applies to whatever is on its immediate left or,

• if there is nothing to the left, const applies to whatever is on its immediate right.

Thus, the syntax says that p is a pointer to a const pointer to a const dataType object. It has the desired effect of ensuring that neither the values of the p[i][j] nor the addresses pointed to by the p[i] can be altered. The subsequent listing demonstrates our point.

```
//constPToP.cpp
#include <iostream>

using namespace std;
```

```
void f(const int* const* y){
  int* z = new int[1];
  *z = 2;
  //*y = z;
  cout << **y << endl;
  //**y += 1;
}

int main(){
  int k = 0;
  int* x = &k;
  int** y = &x;
  f(y);
  return 0;
}
```

The first commented line in constPToP.cpp attempts to change the memory location being pointed to by the pointer-to-pointer variable y from the address of x defined in main to a location created with new inside f. The second commented statement tries to change the value of k defined in main that is being pointed to by x and therefore available as the value of the dereferenced pointer *y (i.e., as **y) in the function f. Uncommenting either statement will produce error messages stating that a value cannot be assigned to a read-only location. These errors are resolved if the argument for f is changed to type int**.

The const modifier arises in classes in three forms: const class members, static const class members and const member functions. The use of const for a class member means that once a class object has been created, the value of that member cannot be changed. Initialization of const members requires a special type of syntax involving a constructor *initializer list.*

An initializer list is a comma separated list of argument and value pairs that occurs after the constructor's argument list and a colon but before the opening curly brace that prefaces the constructor's body. For a constructor for a class className with two arguments that provide values for the member variables memberName1 and memberName2 the syntax might look something like

```
className(argument1, argument2) : memberName1(argument1),
  memberName2(argument2){
    body of method
}
```

In general the values for only a subset (including none) of the member variables can be set using an initializer list. On the other hand, the values for all member variables can be initialized in this manner in which event the body of the constructor may be empty. Even if that is the case, the curly braces must still be present.

The value of any const member variable must be set in an initializer list as in the example below

```
//constEx1.cpp
#include <iostream>

using namespace std;

struct X{
  const int I;
  int J;

  X(int i, int j) : I(i) {
```

```
    J = j;
  }
  int getI(){return I;}
  int getJ(){return J;}
  //void setI(int i){I = i;}
};

int  main(){
  X x(1, 2);
  cout << x.getI() << " " << x.getJ() << endl;
  x.J = 3;
  //x.I = 4;
  return 0;
}
```

The program will print the integers 1 and 2 to standard output. Note that removal of the comments from before the **setI** function in the **X** struct definition will produce a compile time error. The value of a **const** member variable can be set only once within a constructor initializer list. Because we are working with a struct, member variables are **public** by default and their values can be changed. Thus, it is legal to change the value of the **J** variable from inside main. The same is not true for **I** and removing the comment from the next to last line in **main** will also produce a compilation error,

An equally serviceable constructor for the **X** class is

```
  X(int i, int j) : I(i), J(j) {}
```

that sets the values of both member variables in the initializer list. But, an attempt to use

```
  X(int i, int j){
    I = i;
    J = j;
  }
```

will produce a compiler error message stating that the **I** class member is "read only".

Once its value has been set, a **const** member variable cannot be changed; but, it can vary from object to object depending on the value that is furnished in an initializer list. To produce a constant that is the same for every class member, **static** needs to be placed in front of the **const** keyword. The **static** keyword in this case means that there is only one instance of the variable that will have the same value for every object created from the class. The idea is illustrated in the next listing.

```
//constEx2.cpp
#include <iostream>

using namespace std;

class X{
  const int I;
  static const int J = 2;

public:

  X(int i) : I(i) {}
  int getI(){return I;}
  int getJ(){return J;}
};
```

```
int  main(){
  X x1(1);
  cout << x1.getI() << " " << x1.getJ() << endl;
  X x2(2);
  cout << x2.getI() << " " << x2.getJ() << endl;
  return 0;
}
```

Here the X class has both static const and const class members. In main instances of the class are created for different values of the const data members and then the values of both variables are written to standard output producing

```
1 2
2 2
```

This demonstrates that the value of the static const class member is the same across instances of the class.

We discussed const pointers and references earlier in this section. In general, any object can be declared as const with the meaning that none of its data members can be modified. To ensure this is the case, ordinary member functions are forbidden from working with const objects. This leads to a member function dichotomy wherein there are ordinary member functions as well as others that are guaranteed not to modify an object's data members. The latter functions are given a special const designation that takes the form

```
    returnType functionName(arguments) const {}
```

The const keyword is actually a part of the function's signature. Consequently, it must appear in both the prototype and definition for the function and one can have overloaded functions with the same name and arguments that differ only in terms of the const modifier. Both the compiler and linker will make sure that only const member functions are called by const objects.

The listing below illustrates the difference between const and non-const member functions.

```
//constEx3.cpp
#include <iostream>

using namespace std;

struct X{
  int I;

  X(int i) : I(i) {}
  int getI(){return I;}
  int getIConst() const {return I;}
};

int  main(){
  X x1(1);
  const X x2(2);
  cout << x1.getI() << endl;
  cout << x1.getIConst() << endl;
  cout << x2.getIConst() << endl;
  //cout << x2.getI() << endl;
  return 0;
}
```

The X struct has two different accessor functions for its I member variables: a version with the const designation (i.e., getIConst) and one without (getI). This program will compile and run in its current form thereby illustrating that a const member function can be called by either const or non-const objects. However, if the next to the last line in main is uncommented, the compiler will respond with the admonishment

```
passing const X as this argument of int X::getI() discards qualifiers
```

This refers to the fact that the x2 object is const with the consequence that its this pointer becomes a const pointer to a const object. In particular, this entails that it cannot be used with a non-const member function.

A somewhat more meaningful example that has implications for developments in Section 3.9 is provided by the next listing.

Listing 3.9 *constEx4.cpp*

```cpp
//constEx4.cpp
#include <iostream>

using namespace std;

struct X{
  int I, J;

public:

  X(int i, int j) : I(i), J(j) {}

  X operator*(int a);
  X operator*(int a) const;
};

X X::operator*(int a){
  X x(a*I, a*J);
  return x;
}

X X::operator*(int a) const {
  X x(a*I, a*J);
  cout << "const version of *" << endl;
  return x;
}

X operator*(int a, const X& x){
  return x*a;
}

int  main(){
  X x(1, 2);
  cout << (x*2).I << " " << (x*2).J << endl;
  cout << (2*x).I << " " << (2*x).J << endl;
  return 0;
}
```

In this case, the same basic struct as in the previous examples is being used that has two int data members I and J. The difference is that a form of multiplication has been defined for the struct's objects through a process called *operator overloading* that will now be described.

We have already seen that C++ allows function overloading. The same is true for the basic arithmetic operators such as +, -, /, *, the shortcut operators +=, -+, /=, *= and many others. To create an overloaded operator use the syntax

```
returnType operator operatorName(arguments)
```

As a case in point, two overloaded versions of the multiplication operator have been defined for the X struct in Listing 3.9: one that is designated as const and one that is not. Both operators perform a type of scalar multiplication wherein their int argument a multiplies an X object's two int data members.

Both of the multiplication operators in the X struct work for expressions of the form x*a with x an X object and a an integer. It is expressions like a*x that are problematic; this would only make sense if integers were themselves a class with an overloaded multiplication operator for dealing with X objects. Such is not the case and, as a result, it is necessary to instead define * directly as a binary function with int and X object arguments. Once inside the function, the order of the int and X objects can be reversed and the multiplication methods from the X class can be used. However, the two-argument version of * is not an X class method and, as a result, its X argument should be passed in as a const reference to make certain that it cannot be modified by the multiplication operation. This means that only the const version of the * operator in the X class can be used to carry out the computation.

The code in Listing 3.9 was compiled and executed to produce

```
2 4
const version of *
const version of *
2 4
```

The output shows that, as expected, the non-const version of * was used for right-hand multiplication of the X object while the const version was used to carry out left multiplication. Actually, the presence of two multiplication operators is redundant here in that the const version can be used exclusively without loss of functionality for the class.

3.8.4 Forward references

As programs grow, more complex relationships can develop between different classes. It is commonly the case that objects of one class will produce and operate on objects of another. A simple example of this is provided by the next listing.

Listing 3.10 circular.cpp

```
//circular.cpp
struct Y;//forward declaration
struct X{
  Y* A;
  X(){}
  X(Y* a){A = a;}
  Y& funcX(Y& y){return y;}
};

struct Y{
  X B;
  Y(){}
  Y(X b){B = b;}
  X funcY(X x){return x;}
};
```

```
int main(){
  X x; Y y;
  return 0;
}
```

The initial line of code, `struct Y;`, that appears in Listing 3.10 would seem to be of questionable value and not particularly informative. This type of statement is known as a *forward declaration* and actually serves an important purpose. It tells the compiler that a struct named Y exists that will be defined at a later point. In its present form this listing compiles and executes without a problem. Commenting out the forward declaration will produce a compile time error that contains a statement such as

```
circular.cpp:5: error: ISO C++ forbids declaration of 'Y' with no type
```

In order for the X struct to use Y objects the compiler simply must know that the Y struct exists and the forward reference solves this problem. Of course, one solution might be to just define the Y struct first, rather than simply declaring it. But, the Y struct in this case uses X objects. So, this merely moves us around the "circle" and a forward declaration of the X struct would now be required before this alternative arrangement could be made operable.

There is more going on in Listing 3.10 than just the use of a forward reference. Notice that Y objects arise in the X struct in the form of pointers or references. If this is changed the compiler will issue messages stating that the Y struct is "incomplete". This is certainly the case seeing that no details of Y have been supplied when the compiler encounters the definition of the X struct. If an actual Y object is to be a member of the X struct or a method argument or return type, the compiler needs to determine how much memory to allocate for the object. The reason that pointers and references work is that the amount of memory they require is known *a priori*: namely, the four or eight bytes that is needed to hold an address. For similar reasons, if the Y struct definition were moved ahead of that for the X struct the resulting code would fail to compile even with the inclusion of a forward reference to the X struct. This happens because Y objects require X objects by value.

While the positioning of the X and Y structs is somewhat inconsequential here, the problems that arise in a more general context can be realized through restructuring the Listing 3.10 code into separate files with separate headers. For the X struct we use

```
//x.h
#ifndef X_H
#define X_H

struct Y;//forward declaration

struct X{
  Y* A;
  X(){}
  X(Y* a){A = a;}
  Y& funcX(Y& y){return y;};
};
#endif
```

while the Y struct has header file

```
//y.h
#ifndef Y_H
#define Y_H
#include "x.h"

struct Y{
```

```
    X B;
    Y(){}
    Y(X b){B = b;}
    X funcY(X x){return x;};
};
#endif
```

Both of these headers are then used by

```
//xy.cpp
#include "y.h"
#include "x.h"

int main(){
    X x; Y y;
    return 0;
}
```

Since the X struct uses only pointers and addresses to Y objects, it suffices to use a forward reference for the Y struct in its header file. But, the Y struct needs X objects by value making it necessary to include the header file x.h in the header file for the Y struct. One may check that this code compiles without difficulty and, more to the point, the arrangement of the include directives in the xy.cpp file has no effect on this. However, if the include directive is removed from y.h or even replaced with a forward reference to the X struct, the success of compilation for xy.cpp will, in fact, depend on the order of inclusion of the X and Y header files (Exercise 3.17).

The previous example leads to an important conclusion: if class Y uses objects of class X by value, the Y class header must have an include directive to bring in the header file from the X class. Otherwise, if the Y class involves only references or pointers to X objects, a forward reference will do the job.

3.8.5 Strings

Scientific computing generally involves numerical variables. Nonetheless, there are instances where statistical data contains text. Even with strictly numerical data, input and other processing may require manipulation of character representations of numbers; this was seen, for example, in our use of atoi (e.g., Listing 3.2) to translate command line input into numerical values and we will deal with similar problems in Chapter 10. Thus, it is worthwhile to have at our disposal some of the (many) C++ tools that are available for dealing with character strings.

A *string* is an ordered collection of character variables that occupy a contiguous region of memory; that is, a string is basically an array whose elements are all character variables. There are essentially three ways to deal with a string in C++: directly as an array of characters, as a C-style string and using the C++ string class. The array and C-style string approaches are essentially the same except that C-style strings have a null character \0 as their last element. The C++ string class provides a simpler way of dealing with strings than the array perspective and, accordingly, we will progress along that route.

The string class contains methods that are useful for string input and manipulation of string objects. It is accessed by inclusion of the string header file. Listing 3.11 contains several examples of the string class features.

Listing 3.11 *stringEx1.cpp*

```
//stringEx1.cpp                                      1
#include <iostream>                                  2
#include <string>                                    3
                                                     4
using namespace std;                                 5
                                                     6
int main(){                                          7
  string s1;                                         8
  s1 = "Hi ";                                        9
  string s2("y'all");                                10
  s1 += s2;                                          11
  cout << s1 + "!" << endl;                          12
                                                     13
  cin >> s1;                                         14
  //cin.get();                                       15
                                                     16
  getline(cin, s2);                                  17
  cout << s1 + " " + s2 + "!" << endl;               18
  return 0;                                          19
}                                                    20
```

Lines 8 and 10 of Listing 3.11 use two different **string** class constructors to create **string** objects; the first object **s1** is created using the default **string** class constructor while the second object **s2** is initialized using the string literal **"y'all"**: i.e., words surrounded by quotation marks. The "empty" **string** object **s1** is "filled" using an overloaded version of the = operator on line 9. On line 11 of the listing the two **string** objects are concatenated via an overloaded **+=**. Another string is then appended to **s1** (using an overloaded + operator) with the result being written to standard output via an overloaded version of the output insertion operator. The exclamation point that is appended at output is enclosed in quotation marks. This raises the question of how to create a **string** object that contains a quotation mark. The answer is to use \". For example, **string s = "\"";** produces a **string** object with a quotation mark as its "value". To obtain a string object that contains \ you must use \\.

Things become more interesting on line 14 of Listing 3.11. Here the intent is to first read the value of the string **s1** in from standard input and then read a string of input (e.g., several words) using the **getline** function that becomes available through inclusion of the **iostream** header. The result from our first attempt at this is

```
$ g++ -Wall stringEx1.cpp -o stringEx1
$ ./stringEx1
Hi y'all!
Hi
Hi !
```

The first message was expected. But, upon entering "Hi" and a carriage return from the keyboard, the program simply prints out **Hi** ! and seemingly terminates without executing the last group of commands. What has happened here is that **>>** reads until it reaches a space, newline or tab and stops while leaving the character (i.e., space, newline or tab) in the input stream. This space, newline or tab will then be read by the next input operation. That is what has transpired in this case; **getline** reads the newline "character" that was left by **cin**, discards it and returns **s2** as a blank **string**. There are several ways to fix this. One is to use the **get** method associated with the **istream** object **cin**. As used here in the commented statement on line 15 of the listing, **get** will simply extract and discard the next character in the input stream. When the comments are removed the output becomes

```
$ ./stringEx1
Hi y'all!
Hi
there ya'll
Hi there ya'll!
```

The use of `getline` for reading file input is further examined in Exercise 3.38.

The `string` class has a multitude of methods that can be used for comparing strings, swapping strings, replacing elements of strings, etc. Detailed descriptions of these functions can be found in, e.g., Prata (2005, Appendix F) and Stroustrup (1997, Chapter 20). The ensuing code highlighting a few of the methods will suffice for our purposes.

<hr>

Listing 3.12 *stringEx2.cpp*

```cpp
//stringEx2.cpp
#include <iostream>
#include <string>

using namespace std;

int main(){
  string s;
  getline(cin, s);
  cout << "The length of s is " << s.size() << endl;
  cout << "The fifth element of s is " << s[4] << endl;
  cout << "The letter a first occurs at " << s.find("a") << endl;
  cout << "The last a occurs at " << s.find_last_of("a") << endl;
  cout << "The letter z first occurs at " << s.find("z") << endl;

  const char* str;
  str = s.c_str();
  for(int i = 0; i < s.size(); i++)
    cout << str[i];
  cout << endl;

  if(str[s.size()] == '\0')
    cout << "str is a C-style string" << endl;
  return 0;
}
```

<hr>

Output from this program looks like

```
$ g++ -Wall stringEx2.cpp -o stringEx2
$ ./stringEx2
Hi there all ya'll!
The length of s is 19
The fifth element of s is h
The letter a first occurs at 9
The last a occurs at 14
The letter z first occurs at 18446744073709551615
Hi there all ya'll!
str is a C-style string
```

As the program begins to execute we enter the string `Hi there all y'all!` on the command line and this is read into the `string` variable s. The length of s is assessed by the `string` class member function `size`. This value is written out followed by the fifth component of s and the locations of the first and last occurrence of the letter "a" in the string. The fifth entry for s is accessed with the syntax s[4] to illustrate that string variables possess an array structure. The first and last occurrences of a were located using the string class member functions find and find_last_of. There are several relatives of these two functions that include find_first_of, find_first_not_of and find_last_not_of that perform similar functions as indicated by their names. The find function was also used to attempt to locate the letter z in the input string. As z was not in this particular string, the value that was returned was the constant string::npos which corresponds to the maximum possible number of characters in a string. For the computer that ran this program this value is $2^{64} - 1 = 18446744073709551615$.

The last section of Listing 3.12 deals with translating a string object into a C-style string. This is useful, for example, when dealing with file input or output where the file names that are employed to open input/output connections must be in the C-style string format. The string class method c_str converts a string object to a C-style string. It returns a const pointer to char that points to the new memory for data type char that is sufficient to hold the contents of the calling string object plus the \0 character. This transformation is illustrated using the string s which shows that the result is a C-style string by first printing out all but the last member of the character array corresponding to the char pointer and then checking that the last stored value is the termination character \0.

3.8.6 Namespaces

Let us briefly return to the consideration of namespaces. Although the using namespace std; directive is a convenient way to obtain access to all the names in the C++ standard library, it is generally a tremendous overkill in that only a few (sometimes only one as in cin) names from the library will be used in a program. In such instances it is best to access the requisite names with the scope resolution operator. For example,

```
std::cout << x << std::endl;
```

will do the job for printing out the value of a variable x without requiring the entire standard namespace. An alternative approach can be obtained through a specific using directive as illustrated in the next listing.

```
//namespaceEx.cpp
#include <iostream>;

using std::cout; using std::endl;

int main(){
  cout << "You don't need the entire standard namespace to use cout "
       << "or end a line." << endl;
  return 0;
}
```

The program's claim is verified when it is compiled and executed.

3.8.7 Handling errors

The C++ language provides the facility for managing run-time errors through a sophisticated process called *exception handling*. The exit function and assert macro that are available through inclusion of the cstdlib and cassert header files provide useful, albeit less refined, tools that can also be used to deal with run-time errors. We will discuss both approaches briefly in this section.

The code below uses `exit` to terminate the program whenever the user inputs an integer of 5 or larger.

```cpp
//exitEx.cpp
#include <iostream>
#include <cstdlib>

using std::cout; using std::endl; using std::cin;

int main(){
  int n;
  cout << "Enter an integer < 5" << endl;
  cin >> n;
  if(n >= 5){
    cout << "Please follow instruction! " << endl;
    exit(1);
  }
  cout << "n is " << n << endl;
  return 0;
}
```

The value of the argument that is supplied to `exit` is returned to the shell. A zero return value for a program indicates successful execution while other values (such as 1) signify an error condition.

An example of output from the program is

```
$ g++ -Wall exitEx.cpp -o exitEx
$ ./exitEx
Enter an integer < 5
3
n is 3
$ ./exitEx
Enter an integer < 5
7
Please follow instruction!
```

Notice that `exit` is called and the input integer value is not printed when the user supplies invalid input.

The C macro `assert` can be used in a similar manner as `exit`. The argument for `assert` is an expression to be "tested". If the expression is `true` program execution is allowed to continue. If it is `false`, `assert` writes the name and line number of the file where the error occurred, the function that contained the error and the expression that failed the "test" to standard output before calling `abort` to terminate the program. The `abort` and `exit` functions both provide ways to exit a program that is executing. The difference is that `exit` performs cleanup operations that include closing streams and temporary files, calling destructors, etc., before termination while `abort` does not.

The listing below uses `assert` rather than `exit` to enforce rule compliance on integer selection.

```cpp
//assertEx.cpp
#include <iostream>
#include <cassert>

using std::cout; using std::endl; using std::cin;

int main(){
  int n;
```

```
  cout << "Enter an integer < 5" << endl;
  cin >> n;
  assert(n <= 5);
  cout << "n is " << n << endl;
  return 0;
}
```

Some output from the program is

```
$ g++ -Wall assertEx.cpp -o assertEx
$ ./assertEx
Enter an integer < 5
3
n is 3
$ ./assertEx
Enter an integer < 5
7
Assertion failed: (n <= 5), function main, file assertEx.cpp, line 11.
Abort trap
```

In contrast to our use of `exit`, information is given here on the location of the error in our code.

It is possible to obtain the same information provided by `assert` and still use `exit`. This can be accomplished with the predefined macros __FILE__, __FUNCTION__ and __LINE__ that become strings during processing containing the current file and function names and an integer that gives the current line number. Note, all three macro names involve prepended and appended double underscores. Using this approach the `main` function for our example program that used `exit` becomes

```
int main(){
  int n;
  cout << "Enter an integer < 5" << endl;
  cin >> n;
  if(n >= 5){
    cout << "Please follow instruction! " << endl;
    cout << "The error occurred on line " << (__LINE__ - 2)
         << " of file " << __FILE__ << " in function "
         << __FUNCTION__ << endl;
    exit(1);
  }
  cout << "n is " << n << endl;
  return 0;
}
```

As __LINE__ is an integer, we can subtract two to move the output value back to the line where the error will actually occur. An example of output from the program is

```
Enter an integer < 5
7
Please follow instruction!
The error occurred on line 11 of file exitEx.cpp in function main
```

In C++ the phase *exception* is used to describe an unexpected or exceptional event that occurs in a program at run-time. The C++ exception mechanism provides a means to deal with such events when they occur. The basic components of this system are `try` and `catch` blocks. These are segments of code that are contained within curly braces. The `try` block encapsulates the program segment where the error may occur. If an error is detected, the `try` block *throws* an exception that is *caught* by a `catch` block whose corresponding code is then executed. A reformulation of the program we used to illustrate the `exit` and `assert` functions demonstrates the basic idea.

```
//tryCatchEx1.cpp
#include <iostream>
#include <string>

using std::cout; using std::endl; using std::cin;

void intCheck(const int i){
  if(i >= 5)
    throw std::string("bad intCheck() integer >= 5 is not allowed");
  cout << "Your integer is " << i << endl;
  cout << "Enter a new integer < 5  (q to quit)" << endl;
}

int main(){
  int n;
  cout << "Enter an integer < 5 (q to quit)" << endl;
  while(cin >> n){
    try{
      intCheck(n);
    }
    catch(std::string s){
      cout << s << endl;
      cout << "Enter a new integer < 5  (q to quit)" << endl;
      continue;
    }
  }
  cout << "All done" << endl;
  return 0;
}
```

The first step is to read in what is presumably an integer. If, for example, a character is provided, `cin` will register an error that will force `cin >> n` to evaluate as `false` in the `while` loop. Thus, entering, e.g., q, will either bypass or terminate the `while` loop. As long as integers are entered on the command line, the while loop will continue to execute and the function `intCheck` is in charge of printing out the value of the input integer. Since the call to `intCheck` is placed inside a `try` block, an input integer value of 5 or more will cause the function to throw an exception; in this case the exception is a variable of type `string` containing the relevant error message that is preceded by the `throw` key word. The value of this string variable is caught by the `catch` block that, as required, has been placed after the `try` block. When an exception is caught, the action of `catch` is to print the error message and query for input of another integer value. The `continue` statement causes the program to end that segment of the `while` loop and evaluate its defining conditional expression again; i.e., it will execute `cin >> n` again in an attempt to acquire input from the shell. Output from running the program is shown below.

```
$ g++ -Wall tryCatchEx1.cpp -o tryCatchEx1
$ ./tryCatchEx1
Enter an integer < 5 (q to quit)
3
Your integer is 3
Enter a new integer < 5  (q to quit)
7
bad intCheck() integer >= 5 is not allowed
```

```
Enter a new integer < 5   (q to quit)
2
Your integer is 2
Enter a new integer < 5   (q to quit)
q
All done
```

The `try/catch` mechanism as seen in the previous example certainly takes error handling to a higher technical level. But, from a practical perspective there is little that distinguishes our tryCatchEx1 program from what was accomplished previously with `exit` or `assert`. This is mostly a function of the simplicity of the example and, perhaps, makes the point that when simple errors are expected a simple error handling method will suffice. In more complicated situations, the `try/catch` approach has many features that may make it worth considering. For example, a function can throw objects as exception. This provides a convenient way to package diagnostic information that can be processed later in the `catch` block. In this latter respect, there are many built-in exception classes that are used by C++ functions and these can also be adapted to deal with some of the standard errors that are produced by user-defined functions. In general, an exception class can be expected to have a member function `what` that reports the form of the error. When creating new objects from one of these classes, the argument supplied to the class constructor will determine the output from `what`.

As a case in point, consider the `out_of_range` exception class that is made available by including the stdexcept header file. Objects of this type are thrown by some member functions of the `string` class and the `vector` and `deque` container classes of Chapter 10 when an attempt is made to access elements whose indices are not valid. The program below illustrates this with the `string` class `insert` method while providing yet another spin on using our `intCheck` function.

```cpp
//tryCatchEx2.cpp
#include <iostream>
#include <string>
#include <stdexcept>

using std::cout; using std::endl; using std::string;

void intCheck(const int i) throw(std::out_of_range){
  if(i >= 5)
    throw std::out_of_range("intCheck");
  cout << "Your integer is " << i << endl;
}

int main(){
  string s;
  s = "Hi ya'll!";
  int n;
  std::cin >> n;
  try{
    cout << s.insert(n, "there ") << endl;
    intCheck(n);
  }

  catch(std::out_of_range e){
    cout << "Out of range " << e.what() << endl;
  }
  cout << "All done" << endl;
```

```
    return 0;
}
```

The **insert** method will insert the **string** object supplied as its second argument into the calling **string** object at the position specified by its first integer argument provided the position is valid. Otherwise, it throws an exception object from the out_of_range class. The **intCheck** function has also been reworked to throw an out_of_range exception. As a slight variation we have indicated this through an *exception specification* that uses the **throw** keyword in the function's definition to notify both the compiler and user of the type of exception that it can produce. If the user supplied integer is 5 or more, **intCheck** throws an out_of_range object that is created using the class constructor with the function's name as its argument. Finally, the **catch** block retrieves any out_of_range exception that may have been thrown and prints the information from its **what** method to standard output.

Some experimentation with the tryCatchEx2 program produced the output

```
$ g++ -Wall tryCatchEx2.cpp -o tryCatchEx2
$ ./tryCatchEx2
3
Hi there ya'll!
Your integer is 3
All done
$ ./tryCatchEx2
6
Hi ya'there ll!
Out of range intCheck
All done
$ ./tryCatchEx2
11
Out of range basic_string::insert
All done
```

Our first attempt to use the tryCatchEx2 program with the integer 3 produces the desired (at least by us) results. Choosing the integer as 6 generates a jumbled greeting with **insert**; but, in spite of this grammatical faux pas, the method has worked correctly. Instead, an exception has been thrown by **intCheck** because 6 exceeds 5 as indicated by the output from the **catch** block. Our last attempt with the integer 11 as input produces an exception from **insert** since the string "Hi ya'll!" has only nine "characters" including the space. Of course, 11 will also generate an exception if it is used in **intCheck**. The output indicates this has not occurred. Instead, only the first function to throw an exception is evaluated and, after the exception is caught, only the code in the **catch** block is executed.

In Section 3.5 we requested memory on the free-store using the **new** operator. In general, there is no guarantee that it will be possible to fill a memory request and a program will typically crash if the memory is not available. In days gone by, the behavior of **new** was to return a null pointer which could easily be checked to see if an error had occurred. This is no longer true and, instead, when memory allocation problems are encountered the **new** function throws an exception of type **bad_alloc**. Problems of this nature can therefore be managed adaptively with **try/catch** blocks. In the absence of other alternatives, the usual response to a memory allocation failure would likely be to simply stop execution with an error message for debugging purposes. In such instances there is a simpler option that uses an overloaded version of **new** available through the new header file that takes an additional argument **nothrow**. The code snippet below illustrates how the **nothrow** version of **new** can be used to deal with a memory allocation failure.

```
p = new(nothrow) double[n];
if(!p){
  cout << "allocation failure" << endl;
  exit(1);
}
```

The new operator is used to acquire the memory to store n doubles. If the memory is not available, the nothrow version of new returns a null pointer that will evaluate as false when used in a conditional statement. So, if p is null, the effect of the if block will be to invoke exit to terminate the program.

3.8.8 Timing a program

To conclude this section we will touch on a topic that often arises when evaluating numerical methods: computation time. The idea is that one has written a program and wants to determine how long it takes to execute. The tools for accomplishing this are accessed through inclusion of the ctime header file. This makes available the function clock that returns an object of type clock_t whose values represent elapsed processor time relative to some arbitrary starting point that excludes any time spent waiting for input, output or other processes that may be running. The constant CLOCKS_PER_SEC tells us the number of clock ticks that occur per second and can be used to translate an interval of clock time into seconds. To avoid truncation from integer arithmetic the values of clock_t types should be cast to double (see Section 2.4) when using them to evaluate run-time for a program.

The listing below uses the clock function to calculate the time required to carry out computations in a while loop.

```
//clockEx.cpp
#include <iostream>
#include <ctime>

using std::cout; using std::endl;

int main(){

  double x, sec;
  clock_t b = clock();
  while(x < 1000)
    x += .00001;
  sec = (double)(clock() - b)/(double)CLOCKS_PER_SEC;
  cout << "The while loop took " << sec << " seconds to compute x = "
      << x << endl;
  return 0;
}
```

Upon execution the output this code produced for the particular machine where it was employed was

```
The while loop took 0.346182 seconds to compute x = 1000
```

Thus, about a third of a second was needed to compute a sum of 1000 when it is accumulated in increments of size 10^{-5}.

3.9 Matrix and vector classes

In this section the discussion of classes will be expanded through the creation of a class framework for conducting matrix operations. In many respects the idea is to create some-

thing similar to the C++ **string** class from the previous section that added functionality
to character arrays. Here the base storage scheme is the pointer-to-pointer construct of Section 3.4 and the aim is to create a corresponding class structure that has the properties of
a matrix such as (accessible) row and column dimensions while incorporating methods for
matrix addition, subtraction, etc.

At the outset let us first consider how a matrix might be created. There are numerous
types of matrices that can be expected to arise in practice such as diagonal matrices with
specified constants on the diagonal (which allows for creation of an identity matrix) or full
matrices created from data stored in a pointer-to-pointer memory allocation. The matrix
elements could then be composed of integers, floating-point numbers, etc. For simplicity,
the current focus will be directed toward only a couple of possible scenarios with the understanding that other options can be handled in a similar manner.

First, the obvious choices for members of a class **Matrix** are the row and column size of the
matrix and the pointer-to-pointer variable that is needed for dynamic memory allocation to
store the elements of a two-dimensional array. With this as a starting point, two (overloaded)
constructors will be provided: a constructor for a diagonal matrix and a constructor for
data stored in the pointer-to-pointer format. A simple class that implements these ideas is
described by the header file below.

Listing 3.13 simpleMatrix.h

```
//simpleMatrix.h
#ifndef SIMPLEMATRIX_H
#define SIMPLEMATRIX_H

class Matrix{
  //class members
  int nRows, nCols;
  double** pA;
  void pointerCheck() const;
  void pointerCheck(int i) const;

 public:
  //constructors and destructor
  Matrix(int nrows = 0, int ncols = 0, double a = 0);
  Matrix(int nrows, int ncols, const double* const* pa);
  ~Matrix();

  void printMatrix() const;
  int getnRows() const {return nRows;}
  int getnCols() const {return nCols;}
};
#endif
```

Listing 3.13 describes a class with three data members: the number of rows for the matrix
nRows, the number of columns for the matrix nCols and a double precision pointer-to-pointer for the memory locations that will hold the matrix entries. The class comes with
two constructors: a constructor for a diagonal matrix and a constructor for data that is
stored in a pointer-to-pointer scheme. Note that the **const double* const*** syntax from
Section 3.8.3 was used in the pointer-to-pointer constructor to guard against the possibility
of modifying the input data.

In general, default arguments can be supplied for any member (or other) function with the
caveat that once a default argument is given all the subsequent arguments must be given
default values as well. The default values should be specified in the method declaration

in the header file and not in the method definition. We have done this for the diagonal matrix constructor in class `Matrix` with all three arguments being given default values of 0. This has the consequence that the diagonal matrix constructor will serve as the default constructor because it can respond to a nonspecific request for a `Matrix` object such as

```
Matrix A;
```

As will be seen below, in this case the action of the constructor will be to simply return a `Matrix` object A with `A.nRows = A.nCols = 0` and `A.pA` the null pointer.

The first new feature in the `Matrix` class header is the specification of two private, overloaded functions named `pointerCheck`. These are simple utility functions that will check that the memory allocations for the class element `pA` and its associated matrix row pointers are successful. They are intended for internal use within the class and would be of no direct interest for users. It therefore seems most fitting to designate them as private. They will (and should) not modify class data members and, accordingly, have been designated as `const` member functions.

The next new function is the destructor `~Matrix()` and we need to discuss this concept a bit before proceeding further. All classes have destructors that are called every time a class object goes out of scope. Thus, for example, our `Power` class had a default destructor implicitly supplied by the compiler even though one was never explicitly defined. The compiler's destructor simply releases the memory for each of the class members (i.e., for the integer power K in the `Power` class) whenever a `Power` object goes out of scope. While this works fine for `Power` objects, for objects from classes with dynamically allocated memory acquired with the `new` operator this behavior is problematic. As a case in point, for an object of class `Matrix` a compiler supplied destructor would release the memory that has been allocated to the pointer `pA`. At the instance when this happens `pA` points to memory locations that hold the values of pointers which, in turn, hold the addresses for blocks of memory that have been allocated to hold the row elements of a two-dimensional array. Deleting `pA` releases none of this latter memory and has the side effect of making it impossible to access any of the memory that was previously accessible by dereferencing `pA`. To avoid this type of memory leak it is necessary to provide an explicit destructor for the `Matrix` class that releases memory correctly as described in Section 3.5. This will be done in Listing 3.14 that provides the method definitions for the `Matrix` class. The general rule concerning destructors is that an explicit destructor should be supplied for any class that uses dynamic memory allocation.

The remainder of the `Matrix` header file gives the *accessor* functions `getnRows`, `getnCols` and the prototype for a print utility method `printMatrix`. The purpose of accessor functions is to provide noninvasive access to private members. The definitions for `getnRows` and `getnCols` have been given inside the header file which, as noted in Section 3.7, has the effect of making them inline functions. Functions of this nature tend to be only a line or two of code which makes them suitable for inline treatment and we will generally use this option for the accessor functions that are encountered throughout the text. The accessors and the print utility should not alter a `Matrix` object and have accordingly been designated as `const` member functions.

The next step is to create the definitions for the functions that are not defined in the header. These are given in Listing 3.14 below.

Listing 3.14 *simpleMatrix.cpp*

```
//simpleMatrix.cpp
#include <iostream>
#include <cstdlib>
#include <new>
#include "simpleMatrix.h"
```

```cpp
using std::cout; using std::endl;

Matrix:: Matrix(int nrows, int ncols, double a){
  nRows = nrows;
  nCols = ncols;
  //set pA to null pointer in default case
  if(nRows == 0 || nCols == 0){
    pA = 0;
    return;
  }
  pA = new(std::nothrow) double*[nRows];
  pointerCheck();

  for(int i = 0; i < nRows; i++){
    pA[i] = new(std::nothrow) double[nCols];
    pointerCheck(i);
    for(int j = 0; j < nCols; j++){
      if(i == j)
        pA[i][j] = a;
      else
        pA[i][j] = 0.;
    }
  }
}

Matrix:: Matrix(int nrows, int ncols, const double* const* pa){
  nRows = nrows;
  nCols = ncols;
  pA = new(std::nothrow) double*[nRows];
  pointerCheck();

  for(int i = 0; i < nRows; i++){
    pA[i] = new(std::nothrow) double[nCols];
    pointerCheck(i);
  }

  for(int i = 0; i < nRows; i++)
    for(int j = 0; j < nCols; j++)
      pA[i][j] = pa[i][j];
}

Matrix::~Matrix(){
  if(pA != 0){
    for(int i = 0; i < nRows; i++)
      delete[] pA[i];
    delete[] pA;
  }
}
void Matrix::printMatrix() const {
  for(int i = 0; i < nRows; i++){
    for(int j = 0; j < nCols; j++)
      cout << " " << pA[i][j] << "   ";
    cout << endl;
  }
}
```

```
void Matrix::pointerCheck() const {
  if(pA == 0){
    cout << "Memory allocation for pA failed" << endl;
    exit(1);
  }
}

void Matrix::pointerCheck(int i) const {
  if(pA[i] == 0){
    cout << "Memory allocation for pA[" << i << "] failed" << endl;
    exit(1);
  }
}
```

The listing begins with the definitions for the two constructors. The first one creates a diagonal matrix corresponding to a specified number of rows and columns and a specified constant to go on the diagonal. Since this serves as the default constructor, a check is made to see if the default values are in effect: i.e., if nRows and nCols are zero. In that case there is nothing else to do. So, pA is assigned the null pointer and control is passed back to the calling function via a **return** statement. The second constructor takes a pre-existing two-dimensional array stored in the pointer-to-pointer layout and transfers its content into a **Matrix** class object. The next method is the destructor that begins the destruction process with a check to see if memory has been allocated by determining whether or not pA is the null pointer; one of the two constructors must be called when creating a **matrix** object and the only way memory will not be allocated is if the default constructor is used giving pA = 0. In that event there is no memory to release and the destructor's job is finished. On the other hand, if memory has been allocated it is released correctly in the manner that was discussed in Section 3.4.

A utility function printMatrix has been included to output the data corresponding to the pA pointer for a **Matrix** object. Note that **const** appears in the definition as well as in the class header. Had this not been true the compiler would have reported an error as the version of printMatrix in the header file that used **const** would be viewed as a different function than the one for which the definition is given.

The final two functions in Listing 3.14 stem from discussions in Section 3.8.7 concerning the **nothrow** version of the **new** operator. They provide checks for memory allocation failures and will terminate the program with a message indicating the source of the difficulty should problems arise. As both functions perform the same basic purpose it is natural to use function overloading to allow them to share the same name.

A simple driver program that uses the matrix class is

```
//simpleMatDriver.cpp
#include "simpleMatrix.h"
#include <iostream>

using std::cout; using std::endl;
int main(){
  Matrix m1(2, 2, 1);
  cout << "Here is an identity matrix " << endl;
  m1.printMatrix();
  double** p = new double*[3];
  for(int i = 0; i < 3; i++){
    p[i] = new double[2];
    for(int j = 0; j < 2; j++)
      p[i][j] = (double)(i + 1)*(j + 1);
```

```
   }
   Matrix* m2 = new Matrix(3, 2, p);
   cout << "Here is a matrix initialized by a pointer-to-pointer"
        << endl;
   m2->printMatrix();
   return 0;
}
```

This program creates an identity matrix and another matrix with nonzero off-diagonal elements using the pointer-to-pointer storage format. Both a `Matrix` object and a pointer to a `Matrix` object are used to represent the matrices thereby illustrating the syntax that is used in the two cases.

An associated makefile for our `Matrix` class code might look like

```
simpleMat : simpleMatDriver.o simpleMatrix.o
        g++ -Wall simpleMatDriver.o simpleMatrix.o -o simpleMat

simpleMatrix.o : simpleMatrix.cpp simpleMatrix.h
        g++ -c -Wall simpleMatrix.cpp

simpleMatDriver.o : simpleMatDriver.cpp simpleMatrix.h
        g++ -c -Wall simpleMatDriver.cpp
```

The output obtained from execution of our code is then seen to be

```
Here is an identity matrix
   1    0
   0    1
Here is a matrix initialized by a pointer-to-pointer
   1    2
   2    4
   3    6
```

Our `Matrix` class has limited utility in that it basically can only be used to print out an object from the class. The next step is to expand on this skeleton structure to create more functionality including the provision of methods for matrix addition, multiplication, etc. It is possible to accomplish this through operator overloading as will eventually be demonstrated. But, it is first necessary to deal with the concept of a *copy constructor* for our class.

We will eventually want to use matrix objects as arguments to functions or in assignments (via =) after carrying out some type of numeric calculation. In particular, this means that there will be occasions where a copy of a matrix object is created. This is accomplished using the *default copy constructor* that is implicitly provided by C++ for every class unless an explicit copy constructor is defined. The implicit copy constructor makes an element-by-element copy of the members of the class and this is not always desirable.

An illustration of how the default copy constructor will behave for our class `Matrix` can be obtained by attempting to run

```
//copyEx.cpp
#include "simpleMatrix.h"
#include <iostream>

using std::cout; using std::endl;
void g(Matrix A){};

int main(){
  Matrix m(2, 2, 1.);
```

```
  g(m);
  cout << "The destructor will now be called." << endl;
  return 0;
}
```

A similar outcome will be obtained from naive use of the assignment or = operator in

```
//equalEx.cpp
#include "simpleMatrix.h"
#include <iostream>

using std::cout; using std::endl;

int main(){
  Matrix m1(1., 2, 2);
  Matrix m2;
  m2 = m1;
  cout << "The destructor will now be called." << endl;
  return 0;
}
```

Specifically, either of these two programs produces run-time errors that return output including a statement such as

```
The destructor will now be called
equalEx(4147) malloc: *** error for object 0x100100090:
                      pointer being freed was not allocated
*** set a breakpoint in malloc_error_break to debug
Abort trap
```

This error message is about a memory problem related to the C function `malloc` for dynamic memory allocation that lies under the hood of the C++ `new` operator. It suggests the problem may arise from an attempt to delete the memory for a pointer (using the C function `free` that is called by `delete`) that has not actually been allocated memory on the free-store. That is precisely what has occurred.

To understand what has transpired we need to think about when a destructor for a class is called. This occurs when an object for that class goes out of scope: i.e., when the end of a block of code that contains the object is reached. For the first nonoperative driver program copyEx.cpp a copy of the `Matrix` object m is created to pass by value to the function g. This copy then goes out of scope when control passes back to `main` and the copy is destroyed by the destructor function `~Matrix` that deletes the memory allocated to the pointers associated with the `pA` member for the copy. But, since the copy was made element-for-element, this means that the memory pointed to by m.pA (i.e., the pA pointer associated with the `Matrix` object m in `main`) is what is actually deleted. So, at this point no memory is allocated to m.pA. When m goes out of scope at the end of `main` the class destructor is called again and attempts to deleted the memory for the pointer m.pA that no longer points to the allocated memory.

A similar problem arises for the second nonoperative driver program equalEx.cpp. In this latter instance the default behavior of = is to make an element-by-element copy m2 of the `Matrix` object m1 when m2 is equated to m1. At the end of `main` the destructor for m2 is called thereby also deleting the memory allocated to m1.pA. When an attempt is made to delete this memory again, this time for the object m1, an error occurs.

The above discussion makes it clear that when dynamic memory allocation is involved two things must be defined explicitly: namely, a copy constructor and an overloaded version of the assignment operator. Let us first deal with the issue of creating a copy constructor.

Copy constructors are a special type of constructor that have a `const` reference to a class object as their argument. Specifically, they employ the syntax

```
className(const className& objectName)
```

In particular, for class `Matrix` a copy constructor can be implemented using

```
Matrix::Matrix(const Matrix& A){
  nRows = A.nRows;
  nCols = A.nCols;
  pA = new(std::nothrow) double*[nRows];
  pointerCheck();

  for(int i = 0; i < nRows; i++){
    pA[i] = new(std::nothrow) double[nCols];
    pointerCheck(i);
  }

  for(int i = 0; i < nRows; i++)
    for(int j = 0; j < nCols; j++)
      pA[i][j] = A.pA[i][j];
}
```

If this is included in the definitions for the `Matrix` class methods along with insertion of

```
  Matrix(const Matrix& A);
```

in the associated header file, the output produced by copyEx.cpp is just

The destructor will now be called.

Apparently, the copying problem has been resolved.

To handle the problem with using assignment in class `Matrix` the solution is to define an overloaded version of the = operator along the lines of developments in Section 3.8.3. An overloaded = operator for our matrix class would have the form

```
  Matrix& operator=(const Matrix& A);
```

Here, and more generally, assignment operators need to return a reference to allow for composite expressions that use assignment in other function calls.

The definition for the assignment operator might look something like

```
Matrix& Matrix::operator=(const Matrix& A){

  if(this == &A) //avoid self assignment
    return *this;
  //if dimensions match we can just overwrite; otherwise....
  if(nRows != A.nRows||nCols != A.nCols){

    if(pA != 0)//check first before releasing memory
      this->~Matrix();

    //define/redefine object's members
    nRows = A.nRows;
    nCols = A.nCols;
    pA = new(std::nothrow) double*[nRows];
    pointerCheck();

    for(int i = 0; i < nRows; i++){
      pA[i] = new(std::nothrow) double[nCols];
      pointerCheck(i);
    }
  }
```

```
  for(int i = 0; i < nRows; i++)
    for(int j = 0; j < nCols; j++)
      pA[i][j] = A.pA[i][j];

  return *this;
}
```

This code is similar to what was used for our copy constructor. The main difference is that it now allows for the possibility that we are working with an existing object. In that case = must be taken to mean that the `Matrix` object on the left-hand side of = is to be overwritten with the contents of the `Matrix` object `A` in the operator's argument that corresponds to the quantity on the right-hand side of the = operator. If memory has already been allocated for that object's `pA` member there are two possibilities: either the dimensions of the two matrices are the same or they are not. If they are the same, we can simply overwrite the existing memory. If the dimensions differ, the existing memory must be released with the destructor and then reallocated so that the contents of `A` can be stored. A final possibility that can arise in this instance is that the left-hand side object was created with the default constructor in which case there is no memory to release. This condition is checked first before proceeding with memory release operations. One may check that inclusion of our overloaded = operator in the code for the `Matrix` class resolves our prior difficulties and assignment will no longer produce a run-time error.

The definition of the overloaded = operator also represents our first real use of the hidden pointer `this` that was discussed in Section 3.8.2. The = operator has return-type `Matrix&` which means it must return the address of a matrix object: namely, the address of the calling matrix object that resides on the left-hand side of the = operator. This latter object needs to actually "know" its own address for us to be able to return the reference and it keeps that information in the value of `this`.

Operators overloading can be used to produce operators for all the standard matrix operations for members of class `Matrix`. With that in mind, a somewhat more complete development of a class `Matrix` that incorporates addition, multiplication and other new features has the header file in Listing 3.15.

Listing 3.15 *simpleMatrix.h* (expanded version)

```
//simpleMatrix.h
#ifndef SIMPLEMATRIX_H
#define SIMPLEMATRIX_H

class Matrix{
  //class members
  int nRows, nCols;
  double** pA;
  void pointerCheck() const;
  void pointerCheck(int i) const;

 public:

  //constructors and destructor
  Matrix(int nrows = 0, int ncols = 0, double a = 0.);
  Matrix(int nrows, int ncols, const double* const* pa);
  Matrix(const Matrix& A);
  ~Matrix();

  Matrix& operator=(const Matrix& B);
```

```
    Matrix operator+(const Matrix& B) const;
    Matrix& operator+=(const Matrix& B);
    Matrix operator*(const Matrix& B) const;
    Matrix operator*(double b) const;
    const double* operator[](int i) const {return pA[i];}

    Matrix trans() const;
    void printMatrix() const;
    int getnRows() const {return nRows;}
    int getnCols() const {return nCols;}
};

Matrix operator*(double, const Matrix& A);
#endif
```

The class now contains a method **trans** for computing the transpose of a matrix as well as overloaded operators for addition, multiplication and element access. In particular, the subscripting operator [] is defined inline. Although it may not be entirely obvious, this gives access to the matrix elements. To see this, suppose that A is a **Matrix** object. Then, from the operator's definition, A[i] = pA[i] for a given index i. But, that means that A[i][j] = pA[i][j] which verifies our claim. The const modifier before the **double*** return type in Listing 3.15 is to ensure that the elements of the matrix cannot be modified; that is, with **const double*** as the return type a statement such as A[i][j] = b for indices i and j and some constant b will produce an error message.

Implementations of the two methods for matrix addition whose prototypes appear in the class **Matrix** header are provided in the listing below.

```
Matrix Matrix::operator+(const Matrix& B) const {
  if(nRows != B.nRows || nCols != B.nCols){
    cout << "Bad row and/or column dimensions in +!" << endl;
    exit(1);
  }
  Matrix temp(nRows, nCols);
  for(int i = 0; i < nRows; i++)
    for(int j = 0; j < nCols; j++)
      temp.pA[i][j] = this->pA[i][j] + B.pA[i][j];

  return temp;
}


Matrix& Matrix::operator+=(const Matrix& B){
  if(nRows != B.nRows || nCols != B.nCols){
    cout << "Bad row and/or column dimensions in +=!" << endl;
    exit(1);
  }
  for(int i = 0; i < nRows; i++)
    for(int j = 0; j < nCols; j++)
      pA[i][j] += B.pA[i][j];

  return *this;
}
```

Both addition methods first check that the row and column dimensions of the two matrices agree so that addition is possible. The + operator uses the diagonal **Matrix** constructor to create a matrix **temp** of all zeros that is of the right size to hold the matrix sum. Note that only the row and column dimensions are specified in the constructor call. The reason this works is that when the compiler evaluates the expression `Matrix temp(nRows, nCols)` it looks for any constructor that is completely determined through specification of its first two integer arguments. The diagonal `Matrix` constructor has a default value for its third argument so that specification of the number of rows and columns is enough to provide an argument match with the diagonal constructor for the class. The goal of + is to create a new matrix object that is the sum of its B argument and the calling `Matrix` object: i.e., A + B translates to `A.operator+(B)`. We do not want B to be altered by this operation which is why it is passed in as a `const` reference. The same protection is needed for the calling `Matrix` object and this is provided by making + a `const` method. Similar considerations apply for * and the `tran` method.

To understand the return types for the overloaded addition operators one must consider how these functions are to work in practice. With + the goal is to create a brand new `Matrix` object (hence, the `Matrix` return type) that is the sum of the matrix (corresponding to the hidden pointer `this`) that will appear on the left-hand side of the + sign and the `Matrix` object B that appears as the argument to the function. In contrast to + the += operator must overwrite the contents of the matrix on the left-hand side of += with the sum of the calling object and the object B that is listed as the argument of the function. Accordingly, no new memory needs to be allocated and the contents of the `Matrix` object (corresponding to the pointer `this`) is overwritten using the scalar version of += on an element-by-element basis. In that the `Matrix` being overwritten (i.e., the one on the left-hand side of +=) is presumed to already exist, a reference is returned to its location in memory rather than a new `Matrix` object. The explicit use of `this->pA[i][j]` in the + operator accumulation loop is unnecessary and just `pA[i][j]` would have worked equally well. This particular choice of syntax merely helps to distinguish between the three pointers that are active in the loop.

There are three overloaded versions of the * operator that have been specified for multiplication: one for ordinary matrix multiplication and then two for scalar multiplication. The following code gives the definitions for these operators.

```
Matrix Matrix::operator*(const Matrix& B) const {
  if(nCols != B.nRows){
    cout << "Bad row and column dimensions in *!" << endl;
    exit(1);
  }
  Matrix C(nRows, B.nCols);//matrix of all zeros
  for(int i = 0; i < nRows; i++)
    for(int j = 0; j < B.nCols; j++)
      for(int k = 0; k < B.nRows; k++)
        C.pA[i][j] += pA[i][k]*B.pA[k][j];

  return C;
}
Matrix Matrix::operator*(double b) const {
  Matrix C(nRows, nCols);//matrix of all 0s
  for(int i = 0; i < nRows; i++)
    for(int j = 0; j < nCols; j++)
      C.pA[i][j] = b*pA[i][j];

  return C;
}

Matrix operator*(double b, const Matrix& A){
  return A*b;
}
```

The matrix multiplication version of * is just the two-loop accumulation process that would be expected here. The presence of two methods for scalar multiplication is for the same reason as in Listing 3.9 in Section 3.8.3. The first scalar multiplication method corresponds to multiplication of a `Matrix` object A on the right by a scalar b. In this case A*b translates to A.operator*(b); that is, the method operator* belongs to the `Matrix` class and the `Matrix` object A is using this method with argument b. This works as expected. Problems arise when an attempt is made to use syntax such as b*A. If scalars had a class structure this would look like b.operator*(A) and the operator could be defined as an overloaded version of multiplication for that class. But, there is no scalar class and even if there were one it would not make sense for something so fundamental to the `Matrix` class to be specified in isolation from the other `Matrix` methods. As in Listing 3.9, a work around is obtained by defining a binary overloaded * operator that is not a class method. Once inside the function the outcome can simply be defined as A*b. Since scalar multiplication should not alter the original matrix, the `Matrix` argument is passed to * as a const reference. Consequently, the right-hand version of * needs to be a const member function to allow it to work with the one for multiplication on the left.

Finally, the definition of the trans method is

```
Matrix Matrix::trans() const {
  Matrix B(nCols, nRows, 0.);
  for(int i = 0; i < nCols; i++)
    for(int j = 0; j < nRows; j++)
      B.pA[i][j] = pA[j][i];
  return B;
}
```

A new `Matrix` of all zero elements is created first using the diagonal matrix constructor. Then, the contents of the calling object are transferred to the new one while reversing the roles of rows and columns.

A program that uses some of the new features of the `Matrix` class is given below.

```
//matDriver.cpp
#include "simpleMatrix.h"
#include <iostream>

using std::cout; using std::endl;

void g(const Matrix& A){
  cout << "The function g was called" << endl;
};

int main(){
  double** pMat = new double*[2];
  for(int i = 0; i < 2; i++){
    pMat[i] = new double[2];
    for(int j = 0; j < 2; j++)
      pMat[i][j] = (double)((i + 1)*(j + 1) + j);
  }

  Matrix m1(2, 2, pMat);
  cout << "A matrix initialized from a pointer-to-pointer" << endl;
  m1.printMatrix();

  for(int i = 0; i < 2; i++)
    delete[] pMat[i];
  delete[] pMat;
```

```
    g(m1);
    Matrix m2;
    m2 = m1;
    cout << "A copy of the matrix obtained with =" << endl;
    m2.printMatrix();
    cout << "The sum of the last 2 matrices is" << endl;
    (m1 + m2).printMatrix();
    cout << "The product of the first matrix and its transpose is"
         << endl;
    (m1*m1.trans()).printMatrix();
    cout << "The first matrix left multiplied by 2 is" << endl;
    (2*m1).printMatrix();
    cout << "The (1, 1) element of the first matrix is " <<
       m1[1][1] << endl;
    return 0;
}
```

The output produced by this program is

```
A matrix initialized from a pointer-to-pointer
  1    3
  2    5
The function g was called
A copy of the matrix obtained with =
  1    3
  2    5
The sum of the last 2 matrices is
  2    6
  4    10
The product of the first matrix and its transpose is
  10    17
  17    29
The first matrix left multiplied by 2 is
  2    6
  4    10
The (1, 1) element of the first matrix is 5
```

Class Matrix represents our first attempt at creating a framework for manipulating two-dimensional arrays. The developments in this section are just the beginning and this topic will arise again in Chapter 7. Our final version of the class is summarized in Appendix D.

A two-dimensional array with one column is a vector which means that the Matrix class also encompasses the vector case in a certain sense. The code that evolves from this approach (e.g., the use of A[i][0] for working with a two-dimensional representation of a vector A) can be cumbersome. There are also certain entities like the scalar inner product of two vectors that are vector-specific and do not translate particularly well to a two-dimensional environment. Such obstacles can be overcome by creating methods that are specialized to the case of nCols being 1. The fact that such specialization is needed suggests that vector objects should have their own identity rather than be encapsulated in the Matrix class. Accordingly, in the remainder of this section we will provide a sketch of how a class Vector can be developed. The task of providing the class with more functionality is the subject of Exercise 3.22.

Listing 3.16 represents the header file for our first attempt at creating a class Vector.

Listing 3.16 *simpleVector.h*

```
//simpleVector.h
#ifndef VECTOR_H
#define VECTOR_H

class Vector{
  double* pA;
  int nRows;
  void pointerCheck() const;

 public:

  Vector(int nrows = 0, double b = 0.);
  Vector(int nrows, const double* pa);
  Vector(const Vector& v);
  ~Vector();

  Vector& operator=(const Vector& v);
  double operator[](int i) const {return pA[i];}
  void printVec() const;
  int getnRows() const {return nRows;}

  //friends
  friend class Matrix;
};
#endif
```

Many of the prototypes in Listing 3.16 are familiar from our previous experience with class Matrix. The class has two elements: a pointer to double, pA, that will hold the location of the memory that contains the values of the vector's elements and an integer variable nRows that will contain the number of rows in the stored vector. A **private** utility function pointerCheck performs the same task as in the Matrix class in checking for memory allocation problems. Both default and pointer initialization type constructors are provided as well as a copy constructor, a destructor and an overloaded = operator. The last line of code represents another useful feature of C++: namely, the ability to give other classes **friend** status that allows them access to private class members and functions. This concept and its motivation will be examined in more detail subsequently.

Listing 3.17 gives an illustration of definitions that could be used for the member functions in the Vector class.

Listing 3.17 *simpleVector.cpp*

```
//simpleVector.cpp
#include <iostream>
#include <cstdlib>
#include <new>
#include "simpleVector.h"

using std::cout; using std::endl;

Vector::Vector(int nrows, double b){
  nRows = nrows;
  if(nRows == 0){
    pA = 0;
    return;
```

```
    }

  pA = new(std::nothrow) double[nRows];
  pointerCheck();

  for(int i = 0; i < nRows; i++)
    pA[i] = b;
}

Vector::Vector(int nrows, const double* pa){
  nRows = nrows;
  pA = new(std::nothrow) double[nRows];
  pointerCheck();

  for(int i = 0; i < nRows; i++)
    pA[i] = pa[i];
}

Vector::Vector(const Vector& v){
  nRows = v.nRows;
  pA = new(std::nothrow) double[nRows];
  pointerCheck();

  for(int i = 0; i < nRows; i++)
    pA[i] = v.pA[i];
}

Vector::~Vector(){
  if(pA != 0)
    delete[] pA;
}

Vector& Vector::operator=(const Vector& v){

  if(this == &v)//avoid self assignment
    return *this;

  //define/redefine object's members
  if(nRows != v.nRows){

    if(pA != 0)
      delete[] pA;
    nRows = v.nRows;
    pA = new(std::nothrow) double[nRows];
    pointerCheck();
  }

  for(int i = 0; i < nRows; i++)
    pA[i] = v.pA[i];

  return *this;
}

void Vector::pointerCheck() const {
  if(pA == 0){
    cout << "Memory allocation for pA failed" << endl;
```

```
      exit(1);
   }
}

void Vector::printVec() const {
   for(int i = 0; i < nRows; i++)
      cout << " " << pA[i] << " " << endl;
}
```

There is nothing conceptually new in Listing 3.17 beyond what was used in class Matrix. The principal difference is the simplifications that are realized through reduction of the array dimension from two to one. For example, the destructor now requires a single statement, rather than a loop, to release the dynamic memory allocated for a Vector object.

Let us now discuss the C++ friend relationship that can be created between two classes. To motivate the idea consider the method below that carries out matrix-vector multiplication.

```
Vector Matrix::operator*(const Vector& v) const {
   double temp;
   Vector c(nRows);//vector of all 0s
   for(int i = 0; i < nRows; i++){
      temp = 0;
      for(int j = 0; j < nCols; j++)
         temp += pA[i][j]*v[j];

      c.pA[i]=temp;
   }

   return c;
}
```

Without the friend designation for class Matrix in Listing 3.16, insertion of this function in class Matrix will result in a compilation error. The problem occurs in accumulating the sums for the elements of the product in the Vector object c. As pA is a private member of the Vector class, the compiler forbids access to c.pA by objects from another class. The friend designation has the effect of allowing Matrix objects to have access to the private members (and member functions should such exist) of Vector objects which is what is needed to make the matrix-vector multiplication code operable.

The overloaded * operator for matrix-vector multiplication in class Matrix returns a Vector object by value. Thus, as discussed in Section 3.8, the compiler needs to know specifics about the Vector class that require placing an include statement in the Matrix class header to bring in the header file simpleVector.h. The statement

```
   Vector operator*(const Vector& v) const;
```

must also be added to the Matrix class declaration.

The driver program below gives a test for our new overloaded * operator.

```
///matVecDriver.cpp
#include <iostream>
#include "simpleVector.h"
#include "simpleMatrix.h"

using std::cout; using std::endl;

int main(){
```

```
    double** pMat = new double*[2];
    double* pVec = new double[3];
    for(int i = 0; i < 2; i++){
      pMat[i] = new double[3];
      for(int j = 0; j < 3; j++){
        pMat[i][j] = (double)((i + 1)*(j + 1) + j);
        pVec[j] = (double)(j + 1);
      }
    }
    Matrix m(2, 3, pMat);
    cout << "The matrix m" << endl;
    m.printMatrix();
    Vector v(3, pVec);
    cout << "The vector v" << endl;
    v.printVec();
    cout << "The product of m and v is" << endl;
    (m*v).printVec();
    return 0;
}
```

The output from the program is

```
The matrix m
1    3    5
2    5    8
The vector v
1
2
3
The product of m and v is
22
36
```

3.10 Input, output and templates

Up to now all the input and output to our programs has been directed through the shell interface. This is unlikely to be sufficient in any general sense and certainly not for the purposes of statistical analysis where input and output to files is essential. Thus, in this section we will address the issue of reading from and writing to files. The solutions that we obtain will be satisfactory for many applications but are nonetheless constrained to handling file structures that have very specific formats. A more flexible treatment is postponed until Chapter 10.

The problems that will be considered are the input and output of arrays of data that contain values of one of the C/C++ primitive data types. Of the two, the output problem is usually the more straightforward in that one can expect to have created the array to be written to a file during the course of computation and therefore have knowledge of its structure.

The C++ input and output functionality involves *stream* classes that are abstractions of input and output devices. We have used the iostream class to send and receive information from the shell. Including the iostream header file in our programs automatically provided access to the two stream objects cout and cin that were used for this purpose. File input and output works similarly. The header file fstream gives the declaration of the ifstream class for file input and the ofstream class for output. Objects from these two classes then

perform the functions for files that were carried out by cin and cout for standard input
and output.

To create an ofstream object with name outFile corresponding to an existing file whose
name is stored in the C-style string variable fileName, use either

```
std::ofstream outFile;
outFile.open(fileName);
```

or

```
std::ofstream outFile(fileName);
```

Note that the "name" of the file must include a path specification if it is not in the same
directory as the program. The scope resolution operator has also been used to specify that
ofstream refers to the definition in the std namespace.

To actually write to the file the output insertion operator << is used in much the same
manner as for writing to standard output. The only syntactic difference is that cout must
be replaced with the name of the ofstream object or outFile in this particular case. Upon
conclusion of writing to a file it may be closed, thereby freeing it for other uses (e.g., reading
or appending), with

```
outFile.close();
```

The basic ideas behind C++ file output are illustrated in the following program.

Listing 3.18 *simpleOut.cpp*

```
//simpleOut.cpp
#include <cstdlib>
#include <fstream>

int main(int argc, char** argv){

  std::ofstream outFile;
  outFile.open(argv[1]);
  for(int i = 0; i < atoi(argv[2]); i++){
    for(int j = 0; j < atoi(argv[3]); j++)
      outFile << (i + 1)*(j + 1) << " ";
    outFile << std::endl;
  }

  outFile.close();

  outFile.open(argv[1], std::ios_base::app);
  outFile << "End of file" << std::endl;
  outFile.close();
  return 0;
}
```

Listing 3.18 opens an output file stream to a file name that is input from the command line
in argv[1]. If a file with the argv[1] name does not exist it will be created and if it does
exist its contents will be overwritten. An array is written to argv[1] with row and column
dimensions determined by the command line arguments argv[2] and argv[3], respectively.
The connection is closed after which a second connection is made to argv[1] using outFile.
In doing this a second argument std::ios_base::app has been passed to the open method

that will result in output being appended (in lieu of overwriting) if a file already exists.[†]
The simpleOut program produces results such as

```
$ g++ -Wall simpleOut.cpp -o simpleOut
$ ./simpleOut out.txt 3 4
$ cat out.txt
1 2 3 4
2 4 6 8
3 6 9 12
End of file
```

To effectively handle output problems of practical interest an output class should be constructed. An object from the class could then be used to write an array to a specified file. The resulting methods that might be created for writing or object construction would depend on the data type for the array. This would require writing different methods to handle integer and floating-point data, for example. Thus, it would be convenient if we could instead create a class that uses a generic data type that would allow arrays constructed from, e.g., any of the primitive data types to be written to files with only a single method. This can be accomplished using a class template and that is the solution that will be developed here.

The syntax for creating a template class looks like

```
template<class T> class className{
   member declarations and method prototypes
};
```

Here T represents a "wild card" of sorts that must be explicitly specified by the calling program. In the general context of template classes T could be any data type including one that was user created. For statistical purposes T will most often be one of the primitive data types such as char, int, float, double, etc., and there is no harm in viewing it from that perspective throughout this section.

The declarations for members and methods as well as specification of public and private access for template classes are exactly the same as for the non-template case. To specify method definitions outside the class declaration use

```
template<class T> returnType className<T>::methodName(arguments){
   body of method
}
```

There is a caveat here; the method definitions *must* reside in the same file as the class declaration.

Previously class declarations have been physically separated from the method definitions in header and cpp files. The cpp files were then compiled into object files that were linked at the end to produce an executable. The problem with this approach for templates is the generic T that is used as a place holder for the actual data type that will be substituted for T to produce a specific object. The compiler has no way to allocate storage for T when an object file is created without an actual data type being specified. This can be done only when a specific choice is made for T at which point the compiler essentially goes through both the source (i.e., cpp file) code and declarations for the class (that reside in the header) and substitutes in the "value" for T. However, if the cpp file was compiled separately the compiler does not have access to the source code and, instead, it assumes that definitions for all the methods with the necessary "values" for T will exist elsewhere as in other precompiled object files. The task of putting the code together then falls to the linker which looks through the object files to find method definitions using the specified value (or values) of T with no

---

[†] The value of this second argument std::ios_base::app is a static const member of the C++ ios_base class that provides constants and functions used by all the other stream classes.

success and states that there are "unresolved references". There is a keyword **export** that can be used to resolve this difficulty on compilers that support its use. Since g++ does not support **export** we will not discuss it here. Details on its use are given, for example, in Prata (2005).

Listing 3.19 below provides a simple example that illustrates some of the basic ideas behind construction of a template class.

Listing 3.19 simpleTemp.cpp

```cpp
//simpleTemp.cpp
#include <iostream>
using std::cout; using std::endl; using std::cin;

template<class T> class simpleTemp{
  T x;
public:
  simpleTemp(){
    cout << "Enter a value for x " << endl;
    cin >> x;
  }
  T getX();
};

template <class T> T simpleTemp<T>::getX()
{
    return x;
}

int main (){
  simpleTemp<int> y;
  cout << sizeof(y.getX()) << " " << y.getX() << endl;
  simpleTemp<double> z;
  cout << sizeof(z.getX()) << " " << z.getX() << endl;
  return 0;
}
```

The template class **simpleTemp** in Listing 3.19 contains one member x of the generic class T whose value is read in from the command line. The **main** program then creates two **simpleTemp** objects, y and z, with T chosen to be **int** and **double**, respectively, and prints out the values of their x members along with the number of bytes of storage that was allocated to them by the compiler. An example of the use of **simpleTemp** is

```
$ g++ -Wall simpleTemp.cpp -o simpleTemp
$ ./simpleTemp
Enter a value for x
11
4 11
Enter a value for x
11.2
8 11.2
```

Seeing that an **int** is allocated four bytes of memory and a **double** has eight, the output demonstrates that the class has successfully adapted to the two different data types that were used in creation of the objects y and z.

The basic ideas behind the **simpleTemp** class are readily generalized to construction of something useful. In particular, a template class that can be used for file output is provided by Listing 3.20.

Listing 3.20 fileOut.h

```
//fileOut.cpp
#include <iostream>
#include <fstream>
#include <cstdlib>

template<class T = double> class fileOut{
  char* fName;
  bool App;

public:
  fileOut(char* fileName, bool app = 0);
  ~fileOut(){};

  void write(int nrows, int ncols, const T* const* pData) const;
};

//method definitions
template <class T> fileOut<T>::fileOut(char* fileName, bool app){
  fName = fileName;
  App = app;
}

template<class T> void fileOut<T>::write(int nrows,
       int ncols, const T* const* pData) const {

  if(App == 0){
    std::ofstream outFile;
    outFile.exceptions(std::ofstream::failbit |
        std::ofstream::badbit);
    try{
      outFile.open(fName);
    }
    catch(std::ofstream::failure e){
      std::cout << "Error opening output file: " << e.what()
              << std::endl;
    }

    for(int i = 0; i < nrows; i++){
      for(int j = 0; j < ncols; j++)
        outFile << " " << pData[i][j] << " ";
      outFile << std::endl;
    }
    outFile.close();
  }

  else{
    std::ofstream outFile(fName, std::ios::app);
    if(!outFile.is_open()){
      std::cout << "Error opening output file!" << std::endl;
      exit(1);
    }
```

```
    for(int i = 0; i < nrows; i++){
      for(int j = 0; j < ncols; j++)
        outFile << " " << pData[i][j] << " ";
      outFile << std::endl;
    }
    outFile.close();
  }
}
```

The class has two members: the output file name `fileName` and a Boolean variable `App` that indicates whether or not the output is to be appended to `fileName`. A default value of `false` is used for `App` in the class constructor. The operative class method is `write` which takes the row and column dimensions of a pointer held array as well as the array pointer as arguments. Notice the use of the syntax `const T* const*` from Section 3.8.3 that will ensure that the method does not modify the data corresponding to the `pData` pointer. The `write` method opens an output file in either create/overwrite or append mode depending on the value of `App`. In either case an explicit check is made to make sure that the target file is open. For the first case with `App = 0`, the check is made using the `try/catch` mechanism from Section 3.8.7. The instance of `App = 1` is handled with the `ofstream` member function `is_open`. This function returns the value of a Boolean variable that evaluates as `false` if problems are encountered in making a connection to the specified output file. The use of two different ways to check for file access error is for illustration purposes. In general, only one of the two methods would likely be used.

It is possible to give default values to template parameters. We have used this option in Listing 3.20 and made `double` the default data type. This has the consequence that a specific value for the template parameter need not be specified when instantiating a `fileOut` object.

Now let us explore some of the details behind the use of exceptions in Listing 3.20. There are flags corresponding to input and output stream objects that provide information about the stream's state. These are one-bit variables with the names `eofbit`, `failbit` and `badbit`. The value of `eofbit` is set to 1 if the end of a file is reached and is 0 otherwise. Unit values for `failbit` indicate problems when reading or writing operations encounter an unexpected value while `badbit` is set to 1 in the case of corruption of an input or output stream. It might be hoped that an exception would be thrown if any of the three bits evaluates to 1. This is not true by default; but, it can be made so by using the `exceptions ofstream` class method with one or more of the bits as arguments. The bits are combined using the bit-wise OR operator | that evaluates to 1 if any of the bits are 1. The first write operation in Listing 3.20 illustrates the idea. An `ofstream` object is created that is then used to set up the conditions for throwing an exception. The argument `std::ofstream::failbit | std::ofstream::badbit` that is given to the `exceptions` method will evaluate to 1 if either `failbit` or `badbit` is 1. In that case, an object of the exception class `failure` that corresponds to stream objects will be thrown. The actual call to the `open` method is made inside a `try` block. If a `failure` exception is thrown, it is retrieved by the `catch` block and the information from its `what` method is output.

The next listing uses the template output class to write both integers and doubles to specified files.

```
//outDriver.cpp
#include "fileOut.h"

int main(int argc, char** argv){

  int** pI = new int*[2];
```

```
    double** pD = new double*[2];
    for(int i = 0; i < 2; i++){
      pI[i] = new int[3];
      pD[i] = new double[3];
      for(int j = 0; j < 3; j++){
        pI[i][j] = (i + 1)*(j + 1);
        pD[i][j] = (double)(i + 1)/(double)(j + 1);
      }
    }

    fileOut<int> outInt(argv[1], 0);
    outInt.write(2, 3, pI);

    //Now append to argv[2]
    fileOut<> outDouble(argv[2], 1);
    outDouble.write(2, 3, pD);
    return 0;
}
```

To use a `fileOut` object with integer data, the template parameter must be given explicitly as `int`. However, the second write operation uses data of type `double` in which case the default parameter value will suffice. The angle brackets `<>` are still required in this instance; but, an explicit value for T is unnecessary. Output from this program looks like

```
$ g++ -Wall outDriver.cpp -o outDriver
$ ./outDriver out.txt out.txt
$ cat out.txt
  1   2   3
  2   4   6
  1   0.5   0.333333
  2   1   0.666667
$ chmod u-w out.txt
$ ./outDriver out.txt out.txt
Error opening output file: basic_ios::clear
terminate called after throwing an instance of 'std::ios_base::failure'
  what():  basic_ios::clear
Abort trap
$ ./outDriver out1.txt out.txt
Error opening output file!
$ cat out1.txt
  1   2   3
  2   4   6
```

Note that only the outDriver.cpp file was compiled here because fileOut.h contains the entire set of source code for the class rather than just declarations/prototypes. In essence, the compiler works with one large file that begins with the class declaration and the method definitions before concluding with the **main** function. The first use of the outDriver executable produces the expected results. Once the file write permission is changed for out.txt, both write and append operations to that file produce error messages.

To conclude this section, let us briefly discuss creating a file input parallel of our `fileOut` class. In many ways the same concepts apply in that we want to create a template class to allow for general forms of input data. In fact, if the dimensions of the input array can be presumed to be known, then an input class can be obtained through a simple modification of the `fileOut` formulation. Specifically, one need only replace `ofstream` with `ifstream` to obtain

```
    ifstream inFile;
    inFile.open(fileName);
```

or

```
ifstream inFile(fileName);
```

and then use the input (rather than output) insertion operator >> with inFile analogously
to the use of the istream object cin. The details are left as an exercise (Exercise 3.33).

3.11 Function templates

Class templates provide a way of producing generic functions in the sense that class methods
can be made applicable to any data type (either primitive or user defined) that is used in
the creation of the class object. However, once a template class object has been created
with a specific data type, its methods become data type specific. For example, objects from
the fileOut class will all have an associated method write. But, a fileOut object created
for working with data type int will balk if an attempt is made to use it to write out data
of type double.

Function templates are a way to obtain more flexibility than what is available from class
template methods. They are standalone functions that are not tied to class objects and
can be applied to different data types within the same program. The syntax for creating a
function template looks like

```
template<class T>
returnType functionName(T t, additional arguments)
{
    body of function
}
```

Here both the function's return type returnType as well as the additional arguments can
be any data type as well as an object of the generic class T. Unlike template classes, default
values cannot currently be given for the template parameters in a template function.

The succeeding program performs the same basic operations as the write method for our
fileOut template class in the previous section using a template function.

```
//funTemp.cpp
#include <iostream>
#include <fstream>

template <class T>
void write(const T* pT, int nrows, int ncols, char* fName, bool app){
  if(app){
    std::ofstream fileOut(fName, std::ios::app);
    for(int i = 0; i < nrows; i++){
      for(int j = 0; j < ncols; j++)
        fileOut << pT[i][j] << " ";
      fileOut << std::endl;
    }
    fileOut.close();
  }
  else{
    std::ofstream fileOut(fName);
    for(int i = 0; i < nrows; i++){
      for(int j = 0 ; j < ncols; j++)
        fileOut << pT[i][j] << " ";
      fileOut << std::endl;
    }
    fileOut.close();
  }
}
```

```
int main(int argc, char** argv){
  int** pI = new int*[2];
  for(int i = 0; i < 2; i++){
    pI[i] = new int[2];
    for(int j = 0; j < 2; j++)
      pI[i][j] = (i + 1)*(j + 1);
  }
  write(pI, 2, 2, argv[1], 0);

  double** pD = new double*[2];
  for(int i = 0; i < 2; i++){
    pD[i] = new double[3];
    for(int j = 0; j < 3; j++)
      pD[i][j] = (double)((i + 1)*(j + 1))/6.;
  }
  write(pD, 2, 2, argv[1], 1);

  for(int i = 0; i < 2; i++){
    delete[] pI[i]; delete[] pD[i];
  }
  delete[] pI; delete[] pD;
  return 0;
}
```

The syntax in `write` is a bit subtle. The function is implicitly working on two-dimensional arrays. The first argument is of type T* which suggests only a one-dimensional array is being passed into the function. The trick is that T is itself a pointer to some other generic class T_0 which gives us a pointer to a pointer to T_0 objects; abusing notation a bit, the relation can be expressed as T* = T_0**. The template function `write` contains additional arguments that specify the number of rows (i.e., `nrows`) and columns (i.e., `ncols`) of the array, the name for the file that is to be used for writing and a logical variable that determines whether or not to append the output to an existing file. The `main` function in the listing takes an output file name from a command line argument and then writes a two-dimensional array of integers to the file using `write` specialized to data of type `int`. The output file is then reopened in append mode and `write` is used to output data of type `double`. Some results obtained by using the program are

```
$ g++ -Wall funTemp.cpp -o funTemp
$ ./funTemp outFile.txt
$ cat outFile.txt
1 2
2 4
0.166667 0.333333
0.333333 0.666667
```

3.12 Exercises

3.1. Evaluate and print the values of the logical expressions (x < 1), (x > 1), (x >= 1), (x <= 1), (x == 1), (x != 1) when x is an integer variable with value 1. Display the output as zeros and ones as well as in `true/false` format using the `boolalpha` format flag; for example

```
cout << boolalpha << (x == 1) << endl;
```

will print out `true` if x has the value 1.

3.2. Use a **for** loop to produce the same results as those obtained from

```
while (1){
    cout << "I'm in a loop!" << endl;
}
```

Warning: You should probably read Section 4 of Appendix A before attempting to run your program.

3.3. Explain the output from the forWhileEx.cpp program. Specifically, as the bound for the sum was 100, why was the final value of 108 produced for the sum in Section 3.4? How should the program be altered to make certain that the sum will never exceed the bound?

3.4. Write a program that takes in values of doubles from keyboard input. It should allow for termination of input at any point after which it reports the average of the input data and the largest in magnitude residual about the mean: i.e., the largest absolute difference between an input value and the mean of the data.

3.5. A prime number is one that can only be divided (without remainder) by 1 and itself. Write a program that takes as input the number of prime numbers to be calculated and then computes and displays the values for the primes. In doing this it is useful to note that

a) 1 is not a prime,

b) 2 is the only even prime and

c) the C++ modulus function % returns the remainder from integer division. If i and j are two integers, i%j == 0 if and only if i == k*j for some integer k.

3.6. Print out the integers between 1 and a user-supplied upper bound in argv[1] in increments of size **step** with the value of **step** being obtained through a request for command line input from your program.

3.7. In C++ it is possible for a function to call itself. The act of doing so is called *recursion*. Code of this nature might look something like

```
returnType functionName(arguments1){
    statements
    if(booleanExpression)
    functionName(arguments2)
    statements
}
```

Use this approach to provide an alternative version of the xToTheK method in the Power class of Section 3.7.

3.8. Use the same approach as that of Exercise 3.7 to create a function that computes $n! = n(n-1)!$ for any integer $n \geq 1$ with $0! := 1$. Would your factorial function work effectively for computing the binomial coefficient directly from its definition

$$\binom{n}{k} = \frac{n!}{k!(n-k)!}?$$

Why or why not? If not, then how should a binomial coefficient be computed?

3.9. Define the Fibonacci numbers $F_0 = F_1 = 1$ and

$$F_j = F_{j-1} + F_{j-2}, j = 2, \ldots.$$

a) Create a C++ function that uses recursion (in the sense of Exercise 3.7) to evaluate a portion of the Fibonacci sequence with the number of terms to use in the evaluation being determined from user input.

b) Create an analog of your function from part a) that computes Fibonacci numbers directly via a **for** loop.

c) Which of the functions from parts a) and b) is faster? Use the C++ `clock` function from Section 3.8.8 to carry out the comparison.

d) Explain the results from part c). What does this tell you about recursion?

e) Rewrite the recursive Fibonacci function so that whenever it evaluates F_i for some i it stores its value in an array that persists between calls to the function. Make sure that on each call the function firsts checks if the relevant value has already been evaluated and, if so, returns it without any additional work. How does the efficiency compare to naive recursion? To the iterative version? (This technique is called memoization.)

3.10. Consider a game where there are three white balls and one red ball in an urn. Balls are drawn at random from the urn without replacement. If at any point the red ball is drawn, all the balls are replaced in the urn and the drawings continue. Let X be the number of times a red ball is drawn over the course of n draws.

a) Write a C++ program that uses recursion (in the sense of Exercise 3.7) to evaluate the probability distribution of X.

b) How is the speed of computation affected by n? [Hint: Experiment with small values of n.]

c) From your answer to part b) can you project how long it would take to evaluate the probability distribution for $n = 100$? $n = 1000$? $n = 10000$?

d) How does this problem differ from computation of the Fibonacci numbers in Exercise 3.9? Do you expect recursion to be computationally efficient in this setting? Why or why not?

3.11. Assume that $x \geq 1$ and let $\lfloor y \rfloor$ denote the greatest integer less than or equal to the real number y.

a) Develop an algorithm that employs `while` loops to compute the binary expansion (2.2) of $\lfloor x \rfloor$.

b) The C++ function for $\lfloor \ \rfloor$ is `floor`. Use this to create a program that will take the value of a `float` variable x with value of one or larger as input and return the coefficients in the binary expansion of `floor(x)`.

c) Combine the results of part b) of this problem with those from Exercise 2.3 to create code that will compute the binary representation for a floating-point number.

3.12. Consider the code segment

```
int* p;
*p = 1;
```

Write a program that attempts to execute these two lines of code and explain the problem you encounter. Then, rework your program to make it functional.

3.13. Write code that will dynamically allocate and delete memory to hold a three-dimensional array that stores the values of the probability density

$$f(x, y, z) \propto \begin{cases} z(x - y)^2 & x, y = 1, \ldots, 10, \ z = 1, \ldots, \max(x, y), \\ 0, & \text{otherwise.} \end{cases}$$

a) Compute the proportionality constant that makes f a density.

b) Compute the covariances between all three pairs of variables.

c) Compute the marginal densities for each of the three variables.

d) Compute the conditional density of (x, y) when $z = 3$ and find the corresponding (conditional) covariance of the x and y variables. How does this result compare to the unconditional case?

3.14. Create a function that can be used to switch the values of three double precision input variables x, y and z in a circular fashion: i.e., x → y, y → z and z → x.

3.15. Construct a function that takes a pointer for a one-dimensional array as its argument and reverses the order of the array's elements in place.

3.16. Syntax of the form

```
returnType functionName(dataType a[], arguments)
```

can be used to pass an array a to a function. Write a function that passes an array of type double to a function that computes the average of the array's elements. Can the function change the elements of the array? If so, how can you ensure that this will not happen?

3.17. Remove the include directive from the y.h header file in Section 3.8.4. Then, verify that the order of the include directives in the xy.cpp file will determine whether or not the program will compile.

3.18. Write a program that will test the nothrow version of new from Section 3.8.7.

3.19. A very useful operator for overloading is the output insertion operator. A prototype for << is

```
ostream& operator<<(ostream& out, const dataType& object)
```

The first argument is a reference to an ostream object and the second is a reference to the object for output.

The simple program below illustrates how the output insertion operator can be overloaded to provide a print utility for a class.

```
#include <iostream>

using std::ostream; using std::cout; using std::endl;

class X{
  int I;

public:
  X(int i) : I(i) {}
  friend ostream& operator<<(ostream& out, const X& x);
};

ostream& operator<<(ostream& out, const X& x) {
  out << x.I;
  return out;
}
int main(){
  X x(7);
  cout << x << endl;
  return 0;
}
```

First, << is made a friend function of the X class. This serves the same purpose as making a class a friend: namely, it allows the function to have access to the private members of the class. Our definition for << produces a binary operator that is not owned by (i.e., not a member of) either the X or ostream class. Thus, in particular, it will be defined outside the X class. By giving it friend status we have made it possible to access the I member of the X class for printing purposes.

Replace the printMatrix utility in class Matrix with an overloaded output insertion operator.

3.20. Create methods for class `Matrix` that provide analogs of the R `cbind` and `rbind` functions discussed in Appendix B.

3.21. Create functions for class `Matrix` that will

a) insert a specified `Vector` object into a specified column of a `Matrix` object and

b) extract a specified column from a `Matrix` object and return it as an object of class `Vector`.

3.22. Expand the class `Vector` from Section 3.9 to include

a) an overloaded + operator for vector addition,

b) an overloaded += operator for vector addition,

c) an overloaded * operator that computes a dot (or inner) product,

d) overloaded * operators that compute the left- and right-hand products of a vector and a scalar,

e) overloaded *= operators that compute the left- and right-hand products of a vector and a scalar,

f) an overloaded ^ operator that raises every component of a vector to a specified integer power,

g) an overloaded << operator that provides a print utility function for the class. [Hint: See Exercise 3.19.]

Demonstrate through an example that all the operators/methods you have created for the class work as expected.

3.23. Develop a complex number class that has overloaded operators for addition, subtraction, multiplication, division and conjugation (using ~) as well as methods that return the real and complex parts of a complex number and compute its modulus. [Note: A file `complex.h` is likely to already exist. So, it is safest to avoid this name.]

3.24. Create a class with three overloaded methods for computing the sum of a one-dimensional array of numbers stored in a pointer format. The three methods should compute sums for pointers to `int`, `float` and `double`.

3.25. Repeat Exercise 3.24 except use a single sum method in a template class.

3.26. Repeat Exercise 3.25 except use a template function.

3.27. Write code for a struct X that has two integer member elements a and b. One method for the struct should have the prototype

```
bool compare(const& X)
```

The `compare` method should compare two X objects and return the object that has a smaller value for the a member. How could you accomplish the same goal using operator overloading?

3.28. What happens if you remove the & from the statement that defines y in Listing 3.8? Explain why this occurs.

3.29. Construct a program that will take two integers as input (in arbitrary order) and sum all the integers (inclusive) between the two. The program should terminate if it requires more than .01 second to compute the sum.

3.30. Create a template class version of class `Matrix` that can, for example, deal with matrices composed of integer, float or double precision elements. Note that the template parameter must also be used when the return type is a `Matrix` object or a reference or pointer to a `Matrix` object. Thus, for example, the return type for the overloaded = operator will be `Matrix<T>&`.

3.31. Similar to Exercise 3.30, create a template class version of your class Vector from Exercise 3.22. As in Exercise 3.30, the template parameter must be used when the return type is a Vector object or a reference or pointer to a Vector object. For example, the return type for the overloaded + operator is Vector<T>.

3.32. Lists provide useful constructs for holding and managing data. As a minimal level of functionality a list should

a) hold objects of a particular type and expand or contract as objects are inserted or removed,

b) maintain a count of the number of objects in the list and

c) allow for access to any element in the list.

Create a list class that can hold objects of a common (but arbitrary) type with the maximum size of the list being set in the class constructor. List elements are identified by the value of a string variable that holds their names. Overload [] to provide access to the list elements by their names.

3.33. Create a file input class fileIn that can read in an array whose elements are of any of the primitive data types and the dimension of the array is either read in from the command line or provided in the first line of the input file. Demonstrate via an example that your class performs as expected.

3.34. Create a data summary class that will produce summary statistics for an nRows × nCols array of doubles that are stored in the form of a pointer to double* with variables as columns. In particular, provide a class method that uses the West algorithm from Section 2.6 to compute both the means and standard deviations for the variables.

3.35. Create a template function that will read and write arrays of arbitrary specified sizes consisting of any of the primitive data types from and to specified files.

3.36. Create a template function add that performs addition for objects of any class that supports the + operator. Demonstrate that it works with objects of type Matrix as well as those from your class Vector developed in Exercise 3.22.

3.37. Create a template function that will swap two objects of a given class (i.e., the objects' member elements are switched from one to the other) on the basis of the relational operators < and ==. Apply your function to a struct X with two int data members a and b where ordering is in terms of only the a member.

3.38. In Section 3.8 getline was used with standard input cin to read lines of input from the shell. This function also works with ifstream objects. For example, if fileIn is an ifstream object and s is a string, getline(fileIn, s) will read a line of input from the file connected to fileIn until it reaches either i) the end-of-file or ii) the newline character. In the latter instance, the newline character is discarded effectively moving the next read operation to the next line of the file. Write a program that will read in a file containing 23 numbers: four on the first five rows of the file and three on the last row. Return the results to the shell in a manner that will produce output consisting of three rows with eight numbers on the first row, 11 on the second and four on the third.

3.39. Create a class that has methods for computing the standard statistical tests associated with location parameters such as

a) the one-sample t-statistic,

b) the Wilcoxon signed-rank statistic,

c) the paired t-statistic,

d) the two-sample t-statistic with a pooled variance estimator and

e) the two-sample Mann-Whitney U statistic.

3.40. A simple pairwise swapping algorithm for sorting a one-dimensional array is provided by *bubblesort*. The primary **for** loop for this algorithm is given in Algorithm 3.1.

Algorithm 3.1 Bubblesort algorithm: Sort the array $a = (a_1, \ldots, a_n)$

---

for $i = 1$ to $n - 1$ **do**
   for $j = 1$ to $n - i$ **do**
      if $a_j > a_{j+1}$ **then**
         $temp = a_{j+1}$
         $a_{j+1} = a_j$
         $a_j = temp$
      end if
   end for
end for

---

a) Implement the bubblesort algorithm in C++.

b) Create a sort class that contains a method to sort (in-place) the rows of an nRows × nCols array corresponding to a user-selected column: i.e., rows of the array are to be swapped in a way that the values for the chosen column will be in ascending order. The array can be assumed to contain all **doubles** that are stored in the form of a pointer to **double***.

3.41. Suppose that values $x_1, \ldots, x_n$ have been realized from a random variable X. Estimators of the percentiles for X can be obtained from the order statistics: i.e., the data arranged in numerically ascending order. If $x_{(1)} \leq \cdots \leq x_{(n)}$ denote the order statistics, the uth quantile (or $100u$th percentile) is defined to be

$$\tilde{Q}(u) = \begin{cases} x_{(nu)}, & \text{if } nu \text{ is an integer,} \\ x_{(\lfloor nu \rfloor + 1)}, & \text{otherwise.} \end{cases}$$

Of particular importance are the quartiles and median that are used to produce three-number summaries that arise in a variety of exploratory data analysis applications. The median is just $\tilde{Q}(.5)$ while the quartiles are $\tilde{Q}(.25)$ and $\tilde{Q}(.75)$.

Assume that data values are contained in an nRows × nCols array of doubles that are stored in the form of a pointer to **double***. Use your work from Exercise 3.40 to create a method for your summary class in Exercise 3.34 that will return a 3 × nCols array pointer that provides access to the three-number summary values for the nCols variables in the data.

Chapter 4

Generation of pseudo-random numbers

4.1 Introduction

The *random sample* concept represents a cornerstone in the theory and application of statistics. The usual definition for this term would be

Definition 4.1. *A collection of random variables $X_1, \ldots, X_n$ is a random sample if they are all independent and have the same probability distribution.*

The idea seems simple enough. But, the physical creation of collections of numbers that behave like realized values from random samples turns out to be a problem that is anything but simple. The solution that will be studied in this chapter is the pseudo-random number generator (PRNG). This approach has a long history with overviews provided, for example, in Devroye (1986), L'Ecuyer (1990), Monahan (2001, Chapter 11) and Gentle (2003). The artificial data sets obtained from PRNGs have a myriad of uses that span the scientific disciplines. In particular, they represent the primary tool that statisticians use for testing new methodology. Thus, it is important to have some understanding of i) how PRNGs work, ii) the properties a "good" PRNG should possess, iii) the types of PRNGs that exist and iv) where high-quality PRNG implementations can be found for use in practice. These topics are among those that will be addressed in the present chapter.

At the outset, we should explain the reason for the "pseudo" part of the PRNG terminology. PRNGs typically rely on starting values or *seeds* to initialize the recursion that produces the desired random sample. If the seeds are produced by some random process,[*] then the numbers produced by the generator will also be random. But, given any specific values for the seeds, the outcomes produced by a generator are deterministic or exactly reproducible; that is, each time the random number generator is started using the same seeds, exactly the same sequence of values will be returned. This is a good property in the sense that it allows the user to repeat portions of experiments that may have been lost due to disk errors, accidentally erased, invalidated by coding errors, etc. It also allows the results of an entire experiment to be reproduced to obtain alternate or enhanced output and summary statistics. On the other hand, deterministic sequences of numbers can hardly be considered "random". For this reason, the modifying phrase "pseudo" is usually prepended to "random" when discussing algorithmic methods for generating random numbers. While this tells us what a PRNG is not, it leaves the meaning of the PRNG terminology in a state of limbo. To be precise we will therefore use the definition for a PRNG employed, e.g., by Robert and Casella (2004).

Definition 4.2. *A PRNG is an algorithm that, starting from an initial seed (or seeds), produces a sequence of numbers that behaves as if it were a random sample from a particular probability distribution when analyzed using statistical goodness-of-fit tests.*

Although there are a variety of probability distributions that are of practical interest to statisticians, the problem of generating random numbers from a distribution can always be reduced to the canonical problem of simulating from a uniform distribution over the interval

---

[*] Creative solutions include the use of "machine noise" (i.e., electronic or thermal noise that is produced by electron movement in some electrical conductor) to produce seeds and human intervention (e.g., typing a character from the keyboard or moving the mouse to initiate the selection process for a generator that has already been started with nonrandom seeds).

$[0, 1]$. To see why this is so, suppose that a random variable X has cumulative distribution function (cdf) F defined by

$$F(x) = \text{Prob}(X \leq x).$$

The *quantile function* for the distribution is then defined as

$$F^{-1}(u) = \inf\{x : F(x) \geq u\}. \tag{4.1}$$

The function F^{-1} is well defined due to the fact that F is nondecreasing and right-continuous. The utility of the quantile function for simulation purposes is a consequence of Theorem 4.1.

Theorem 4.1. *If U is a random variable that has a uniform distribution over the interval $[0, 1]$, $F^{-1}(U)$ is a random variable with cdf F.*

Theorem 4.1 has the consequence that a random sample from a random variable X with cdf F can be obtained through Algorithm 4.1.

Algorithm 4.1 Inversion method

Generate a random sample $U_i, i = 1, \ldots, n$, from a uniform distribution on $[0, 1]$
return $X_i = F^{-1}(U_i), i = 1, \ldots, n$

The validity of Theorem 4.1 stems from the fact (Exercise 4.1) that for any $u \in (0, 1)$ and any x for which $F(x) > 0$

$$F^{-1}(u) \leq x \text{ if and only if } u \leq F(x). \tag{4.2}$$

Thus,

$$\begin{aligned} \text{Prob}(F^{-1}(U) \leq x) &= \text{Prob}(U \leq F(x)) \\ &= F(x) \end{aligned}$$

because the distribution function for a uniform random variable is the identity function.

The transformation of a uniform random variable U to $F^{-1}(U)$ in Theorem 4.1 is known as the *probability integral transform*. The way it has been employed here to obtain a random sample from a random variable with cdf F is referred to as the *inversion* method. We will return to this approach later in the chapter. But, it is worth mentioning now that this is not always the best path to producing a pseudo-random from a computational perspective. In general there will be no closed form for F^{-1} and a somewhat more subtle approach such as the accept-reject method discussed in Section 4.6 can lead to more easily implemented generation methods.

Perhaps the most important message to be gleaned from Theorem 4.1 and its accompanying discussion is the paramount importance of producing high quality pseudo-random numbers from a uniform distribution. As a result, a large part of this chapter will be spent on that specific problem.

4.2 Congruential methods

The random numbers that were used in early computers came from random number tables that had been stored in memory. To streamline the generation process John von Neumann proposed a *middle-square method* whereby numbers could be produced directly by the computer. To generate d-digit integers via a middle-square recursion one begins with a d-digit integer seed or starting value. The seed is squared and the middle d digits of the square are retained. The process is repeated until the desired number of "random" numbers has been obtained.

As an example, suppose that $d = 4$ and the initial 4-digit seed is $x_0 = 1234$. Then, $x_0^2 = 1522756 = 01522756$ giving $x_1 = 5227$ as the next integer in the sequence. This leads to $x_1^2 = 27321529$ that produces $x_2 = 3215$, etc.

The middle-square algorithm does not generate numbers that are "random" in any sense. But, it does produce numbers with enough of the characteristics of random numbers to have made it useful for von Neumann's particular application.

Streams of numbers created with the middle-square technique would not meet today's standards for random number generation. It was the idea of using a deterministic algorithm to produce numbers that possess many of the qualities of random numbers that was the breakthrough that set the course for modern generation methods. The next step down this path was made by Lehmer (1949, 1954) who pioneered the development of the *linear congruential* random number generators that represent the topic of this section.

A linear congruential generator requires a single integer starting seed that will be denoted by x_0. A sequence of pseudo-random integers is then generated via the recursion

$$x_i = (ax_{i-1} + c) \bmod m, \ i = 1, \ldots, \tag{4.3}$$

where a, c and m are all integers. The integer m is called the *modulus* of the generator while a and c are referred to as the *multiplier* and *increment*, respectively.

For two integers a and b the expression $a \bmod b$ refers to the integer remainder after integer division of a by b. For example,

- $5 \bmod 2 = 1$; i.e., $5 = 2 \times 2 + 1$

- $6 \bmod 3 = 0$

- $5 \bmod 6 = 5$

In general, $(rm + k) \bmod m = k$ for r an integer and $k = 0, 1, \ldots, m - 1$. So, for any integer r, $r \bmod m \leq m - 1$.

Generators with $c \neq 0$ are sometimes called *mixed-congruential* generators while those with $c = 0$ are referred to as *multiplicative-congruential* generators. To produce pseudo-random uniform [0, 1] numbers (also referred to as *random deviates*) from generators of either variety, we simply make a transformation such as $u_i = x_i/m$ that produces values in the unit interval.

Construction of a linear congruential generator seems relatively straightforward in that only the values for a, c and m need to be chosen. Unfortunately, not every choice will produce sequences of numbers that are satisfactory. An example of what can happen by making a poor choice for a, c and m is given in Gentle (2003). He considers a multiplicative congruential generator with $m = 31, a = 3, c = 0$ and uses $x_0 = 9$ for the starting value. This produces the sequence

$$27, 19, 26, 16, 17, 20, 29, 25, 13, 8, 24, 10, 30, 28, 22, 4, 12, 5, 15, 14, 11, 2, 6, 18, 23, 7, 21, 1, 3, 9$$

at which point the series repeats. The first observation is that there are not many unique numbers that can be produced by the generator. Second, a plot of consecutive pairs of these integers as shown in Figure 4.1 reveals a strong pattern that is not obvious from an examination of just the integers alone.

Actually, the pattern in Figure 4.1 stems from a characteristic that is shared by all congruential generators: successive k-tuples of values produced by such generators possess a lattice structure. Specifically, Marsaglia (1968) showed that all such k-tuples will fall on at most $(k!m)^{1/k}$ hyperplanes in $\mathbb{R}^k$. For our example with $k = 2$ and $m = 31$, the bound is $\sqrt{62}$ which rounds down to 7. It takes only three hyperplanes (or lines, in this case) to capture all the points in Figure 4.1 and that is part of the problem with the generator. Another generator that required more than three lines would likely produce pairs that were less dependent and appeared more random than those in Figure 4.1. For example, using a

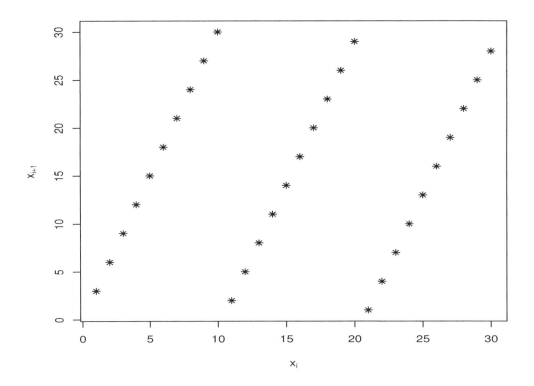

Figure 4.1 *Pairs of "random" digits*

multiplier $a = 12$ gives successive pairs of values that fall on six lines and thereby provides some improvement over $a = 3$ (Exercise 4.4).

Guidelines for making "good" choices for the parameters of linear congruential generators are given in Chapter 3 of Knuth (1998a). The first consideration should probably be the choice of the modulus as that provides an upper bound for the number of unique values that can be produced by a generator. Thus, m should be large with $m = 2^p$ and $p = 31, 32, 63, 64$ being obvious choices for 32 and 64 bit architectures. It turns out that better results are obtained by using $2^{31} - 1$ and $2^{63} - 1$ (Exercise 4.7). Even if m is large, there is no assurance that the *period* of a generator, or the length of the pseudo-random number sequences it produces before they repeat, will also be large. Conditions under which the maximum possible period can be attained stem from results in number theory concerning prime and relatively prime numbers.

A *prime number* is a positive integer or natural number for which there are only two natural number divisors that produce another natural number as the quotient: i.e., for which the division has zero remainder. If there are only two divisors, then these numbers must be 1 and the number itself. Examples of prime numbers are 2, 3, 5 and 7. The integer 1 is not considered to be a prime.

Two natural numbers are *relatively prime* if they have only 1 as a common divisor. For example, 7 and 9 are relatively prime with only 7 being a prime. The numbers 9 and 10 are relatively prime and neither of these is a prime which demonstrates that primality and relatively primality are distinct concepts.

A result from Knuth (1998a) summarizes what can be expected for a mixed-congruential generator.

Theorem 4.2. *If $c \neq 0$, the period of the generator (4.3) is equal to m if and only if*

a) c is relatively prime to m,

b) $a - 1$ is a multiple of every prime number that divides m and

c) $a - 1$ is a multiple of 4 if m is a multiple of 4.

As an example, take $m = 2^p, c = 3$ and $a = 5$ (Exercise 4.8).

Results for multiplicative generators where $c = 0$ are somewhat more involved. A detailed presentation of this case is provided in Section 3.2.1.2 of Knuth (1998a). When $m = 2^p$ for some $p \geq 4$, it turns out that the maximum possible period for the generator is 2^{p-2} and this period will be attained when $a \bmod 8 = 3$ or 5 and the starting value x_0 is odd. The so-called SUPER-DUPER generator of Marsaglia (1972) that has $p = 32$, $a = 69069$ and x_0 odd falls into this category since $69069 \bmod 8 = 5$.

A popular choice in practice has been $m = 2^p - 1$. An application of Theorems B and C of Knuth (1998a, page 20) will often suffice to deal with this and other similar cases. In combination these two results produce

Theorem 4.3. *If $c = 0$ and $m > 2$ is a prime number in (4.3),*

a) the maximum possible period is $m - 1$ and

b) the maximum period is achieved if $a \bmod m \neq 0$ and $a^{(m-1)/q} \bmod m \neq 1$ for every prime divisor q of $m - 1$.

Consequently, generators with maximum period can be constructed for the case of $m = 2^p - 1$ if m is a prime and a is chosen appropriately. The key consideration is primality of m which turns out to be equivalent to asking whether m is a Mersenne prime (e.g., Andrews 1971). At the time of this writing there were only 44 known Mersenne prime numbers that include $m = 2^{31} - 1$ and $m = 2^{61} - 1$ which are of particular interest in the context of congruential generators.

Of course, even if a generator can achieve a long period, this is no assurance that the resulting numbers will be satisfactory from a statistical perspective. For example, taking $a = c = 1$ in (4.3) gives a generator that attains the maximum period m (Exercises 4.6). Yet, the resulting numbers cycle sequentially through the integers between 0 and $m - 1$. Thus, although long periods are desirable, they are not sufficient to guarantee that high quality pseudo-random number streams will be forthcoming from the generator. This begs the question of how one can evaluate the quality of a PRNG. As suggested by Definition 4.2, the answer is through statistical tests of goodness-of-fit and independence. For example, standard goodness-of-fit tests should be applied to the output of a generator that has been transformed to the interval $[0, 1]$. These include the Kolmogorov-Smirnov, Cramér-von Mises and (binned) chi-square tests to assess if the marginal distribution of the generated numbers departs from uniformity. Beyond just uniformity of the marginals, an appearance of independence is necessary and tests, e.g., for serial correlation should be employed, as well. Fishman and Moore (1982, 1986) and Knuth (1998a) provide detailed overviews of different statistical tests that are commonly used for evaluation of random number generators.

Packages that are available for testing PRNGs include the Diehard collection of tests that was developed by George Marsaglia circa 1995. Discussions of some of the tests that are used in the Diehard package and their extensions are provided in Marsaglia and Zaman (1993) and Marsaglia and Tsang (2002). Another group of tests developed at the National Institute of Standards and Technology are described in Rukhin (2001). L'Ecuyer and Simard (2007) provide a suite of tests in their TestU01 package.

4.3 Lehmer type generators in C++

There are many linear congruential random number generators that have been studied in great detail in the literature. In this section our attention will be focused on a particular variety of multiplicative generator deriving from a proposal in Lehmer (1949) that was later implemented in Payne, et al. (1969). The idea behind such generators is to choose m as a large Mersenne prime and then take a to satisfy the conditions of Theorem 4.3 and so ensure that the maximum period is attained.

Let us now focus on the particular case of $m = 2^{31} - 1$. One finds that

$$m - 1 = 2^{31} - 2 = 2 \times 3^2 \times 7 \times 11 \times 31 \times 151 \times 331.$$

All the numbers in this product are prime numbers and the *fundamental theorem of arithmetic* (cf. Andrews 1971) entails that this factorization is unique: i.e., there are no other prime numbers that divide m. The second condition of Theorem 4.3 can now be checked to see if any particular a is a multiplier that will produce a full period. This approach allows us to verify that the generator specified by

$$x_i = 630360016 x_{i-1} \bmod (2^{31} - 1) \tag{4.4}$$

will have the full $2^{31} - 2$ period. The choice of $a = 630360016$ was suggested by Payne, et al. (1969). It appears as generator II in the Fishman and Moore (1982) comparison study and, for that reason, will hereafter be referred to as FM2.

An implementation of the FM2 generator is provided in the code listing below.

Listing 4.1 basicFM2.cpp

```
//basicFM2.cpp
#include <iostream>

using std::cout; using std::endl;

unsigned long power(int i, int k){
  unsigned long iout = (unsigned long)i;
  if(k > 1)
    iout = iout*power(iout, k - 1);
  return iout;
}

int main(int argc, char* argv[]){
  unsigned long a = 630360016;//multiplier
  unsigned long seed = 123;//seed
  unsigned long m = power(2, 31) - 1;//modulus
  cout << "Your random number is ";
  seed = (a*seed)%m;
  double u = ((double)seed)/((double)m);
  cout << u << endl;
  return 0;
}
```

This code can be compiled and executed using

```
$ g++ -Wall basicFM2.cpp -o basicFM2
$ ./basicFM2
Your random number is 0.104714
$ ./basicFM2
Your random number is 0.104714
```

Listing 4.1 contains a simple utility function that computes the value of an integer raised to a specified integer power. This function gives an illustration of using the C++ recursion capability that allows a function to call itself (Exercises 3.7–3.10). Notice that the function power has return type unsigned long and, accordingly, must return an unsigned long integer via a return statement at (or before) the end of the function's *scope*: i.e., it must return a value somewhere in the range of statements that appear between the two curly braces containing the body of the function. The unsigned long int data type will give us eight bytes of storage which is what is needed to retain (exactly) all the integer values that are used in the generator. One may check that the four bytes available with the int type is not enough and the resulting code would produce unusable output.

As far as PRNG implementations go, this one leaves a lot to be desired. First, it only returns one number and, second, as seen from the results of running the program, it gives us the same "random" number every time. These two shortcomings are easily fixed by allowing for user input. This can be done either by using command line arguments that will be collected in argv and/or by cin from the iostream library. A possible implementation that allows for user input of both the seed and a value for a sample size might appear as in Listing 4.1 with the alternative main program section given below.

```cpp
int main(int argc, char* argv[]){
  unsigned long a = 630360016;
  unsigned long seed = 0;
  int n = 0;
  cout << "Choose a value for the sample size:" << endl;
  std::cin >> n;
  cout << "Choose a value for the seed:" << endl;
  std::cin >> seed;
  unsigned long m = power(2, 31) - 1;

  cout << "Your random numbers are " << endl;
  for(int i = 0; i < n; i++){
    seed = (a*seed)%m;
    double u = ((double)seed)/((double)m);
    std::cout << " " << u;
  }

  cout << endl;
  return 0;
}
```

Some results produced with this modified program are

```
Choose a value for the sample size:
5
Choose a value for the seed:
123
Your random numbers are
 0.104714 0.723072 0.156521 0.32855 0.635906
```

Our modification of Listing 4.1 certainly represents an improvement over what the program could do before. But, it still falls short of what might be needed from something whose purpose is to generate a collection of random numbers. First, although the actual numbers can be accessed or saved by redirecting the program's output to a file rather than the shell, the numbers are not retained in the program and, as a result, cannot be used in further calculations. In the grand scheme of things, a program like this would typically be used as a utility to generate random uniform deviates for use in other settings. So, it really needs to be of a reusable nature that allows it to be easily called up whenever it is needed.

The issue of data retention can be fixed fairly easily using either arrays or pointers and both of these options will be discussed below. The OOP features of the C++ language are used in the next section as the means for achieving a simple form of code reuse.

To allow for the storage of a vector of n random numbers, the loop of Listing 4.1 can be replaced by

```
double u[n];
std::cout << "Your random numbers are" << std::endl;
for(int i = 0; i < n; i++){
   seed = (a*seed)%m;
   u[i] = ((double)seed)/((double)m);
   std::cout << "u[" << i <<"]= " << u[i] << " ";
}
```

An equivalent implementation with pointers would be to instead use

```
double* pU = new double[n];
std::cout << "Your random numbers are:" << std::endl;
for(int i = 0; i < n; i++){
   seed = (a*seed)%m;
   pU[i] = ((double)seed)/((double)m);
   std::cout << "u[" << i <<"]= " << pU[i] << " ";
}
```

In this case an additional line of code

```
delete[] pU;
```

could also be included for memory clean-up purposes. The output from the pointer variant of the FM2 generator code looks like

```
Choose a value for the sample size:
5
Choose a value for the seed:
123
Your random numbers are:
u[0]= 0.104714 u[1]= 0.723072 u[2]= 0.156521 u[3]= 0.32855 u[4]= 0.635906
```

To get some idea of what our PRNG is doing, 500 pseudo-random uniform values were generated using the array formulation of our program without the text in the output and with the seed 123. The output was redirected to a file that was then imported into R using the scan function as described in Section 5.2. A histogram of the data that was created with the R function hist (e.g., Appendix B) is shown in Figure 4.2. By using the freq = FALSE option the result is a probability histogram in the sense that the height of each bar in the histogram is chosen so that the area of the bar is equal to the relative frequency of the number of values that fall in its bin.

One possible assessment of whether this data derives from a uniform distribution can be based on the Kolmogorov-Smirnov statistic (e.g., Conover 1998). The R function ks.test computes and returns the value of this statistic. An application of this function produces

```
> ks.test(uDat, punif)

        One-sample Kolmogorov-Smirnov test

data:  uDat
D = 0.0404, p-value = 0.3884
alternative hypothesis: two-sided
```

The first argument uDat that is supplied to the ks.test function in this instance is the array of uniform random deviates that was imported into R. The second argument is the

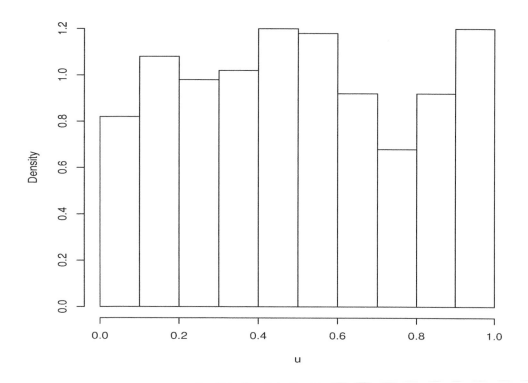

Figure 4.2 *Histogram for simulated data*

uniform distribution function `punif` that is to be used for the null model. More generally, one can use the cdf `pnorm`, etc., to obtain goodness-of-fit statistics for other distributions that are available in R. The value of the Kolmogorov-Smirnov statistic is returned as D in the output. As expected, the reported p-value does not suggest a statistically significant departure from uniformity.

4.4 An FM2 class

Let us conclude our discussion of the FM2 random number generator by implementing it in the class structure of a true C++ program. The code below accomplishes this and we will spend the rest of this section working our way through the details of this listing.

Listing 4.2 *FM2Class.cpp*

```
//FM2Class.cpp                                                              1
#include <iostream>                                                         2
#include <cstdlib>                                                          3
#include <new>                                                              4
                                                                           5
using std::cout; using std::cin; using std::endl;                          6
                                                                           7
class FM2{                                                                  8
    unsigned long a;//multiplier                                           9
    int expFor2;//exponent for 2 in modulus                               10
```

```
  unsigned long power(int i, int k) const;

public:

  FM2();//default constructor
  ~FM2(){};//destructor
  void rangen(unsigned long seed, int n, double* pu) const;
};

unsigned long FM2::power(int i, int k) const {
  unsigned long iout = (unsigned long)i;
  if(k > 1)
    iout = iout*power(iout, k - 1);
  return iout;
}

FM2::FM2(){
  a = 630360016;
  expFor2 = 31;
}

void FM2::rangen(unsigned long seed, int n, double* pU) const {
  unsigned long m = power(2, expFor2) - 1;
  for(int i = 0; i < n; i++){
    seed = (a*seed)%m;
    pU[i] = ((double)seed)/((double)m);
  }
}

int main(int argc, char* argv[]){
  int n = 0;
  unsigned long seed = 0;
  cout << "Choose a value for the sample size:" << endl;
  cin >> n;
  cout << "Choose a value for the seed:" << endl;
  cin >> seed;
  FM2* pFM2 = new(std::nothrow) FM2();
  if(pFM2 == 0){
    cout<< "Memory allocation for pFM2 failed"<<endl;
    exit(1);
  }
  double* pU = new(std::nothrow) double[n];
  if(pU == 0){
    cout<< "Memory allocation for pu failed"<<endl;
    exit(1);
  }
  pFM2->rangen(seed, n, pU);
  cout << "Your random numbers are" << endl;
  for(int i = 0; i < n; i++)
    cout << "u[" << i << "]= "<< pU[i] << " ";
  cout << endl;
  delete pFM2;
  delete[] pU;
  return 0;
}
```

Recall from Chapter 3 that the C/C++ compiler requires us to provide a sketch or prototype information about functions or classes before they can be used. Accordingly, statements 8–9 of Listing 4.2 tell the compiler that our class FM2 has two private data members: namely, a and the exponent for two that is to be used in the modulus named expFor2. The two members are declared to have types unsigned long and int, respectively. Statement 10 then gives the prototype for the now familiar recursive power function. This function is only used internally in the method rangen for the class and is therefore designated as private.

The next group of statements (i.e., 13–14) give two special public functions: namely the class constructor FM2 and the class destructor ~FM2 that are used to create and destroy objects from the class. Finally, a prototype is provided for the actual random number generation function that has been called rangen. The arguments for this function will be the seed, the sample size n and a pointer to an "array" of n double precision storage locations. The class prototype ends with the delimiter };

The functions or methods that appear in the class (with the exception of the class destructor) are defined in detail next. The compiler is informed that these functions belong to the class FM2 through use of the scope resolution operator ::. The (default) class constructor (i.e., the function FM2) sets the value for the constant a and the exponent for two in the modulus in a way that produces the FM2 generator as defined in (4.4). The FM2 class could, of course, be structured to provide much more freedom in the creation of a generator by making different choices for a and expFor2. This could be incorporated into the class by including an additional constructor that would take their values as arguments (Exercise 4.11). Of course, if the FM2 generator is all we wish to use the value of $m = 2^{31} - 1$ should be defined as a global constant to avoid having to (repeatedly) compute it.

The purpose of the class destructor is, among other things, to release any memory that has been dynamically allocated in constructing an instance of the class or an object. There is no such commodity that arises in this implementation which means that ~FM2() can be just the trivial function with an empty body between the curly braces. The effect will be that the memory will be deleted on an element-wise basis whenever an FM2 object goes out of scope. This is actually the default behavior for the destructor that is automatically used when none has been specified. Consequently, our explicit specification of the destructor on line 14 of Listing 4.2 was unnecessary in this instance.

To actually use an object of type FM2 a pointer to a class FM2 object is created on line 45 of main. In this instance, the new operator calls the FM2 class constructor and through this allocates memory that is sufficient to hold all the aspects of the resulting FM2 object. Once this pointer is available, the rangen method can be accessed via the -> operator. On lines 46–49 and 51–54 checks are carried out to see that the requested memory allocation was successful. If allocation fails, the pointer returned from the nothrow version of new will be null which evaluates to the Boolean value false. In that case, an error message will be output and the C++ system function exit (available via the cstdlib header) from Section 3.8.7 will be used to immediately terminate the execution of the program.

One possible variation on the design of this code involves allocation of the memory that holds the actual random numbers produced by the program. The tactic taken here has been to allocate the memory in main thereby making the process more obvious. One could just as easily have made the vector of random numbers a (private) member of class FM2. In that case the destructor would actually need to perform the memory release process and it would be necessary to add another (const) class method that would provide access to the values of the random numbers.

While Listing 4.2 effectively illustrates the class concept, it still comes up short in terms of our idealized "plug-in" code construct. After all, we do not want to have to copy class prototypes, definitions of methods, etc., into every program where they might be used. The result would be both unwieldy and redundant. Fortunately, this problem can be easily solved by proceeding as in Chapter 3 to break the code into subsegments that are connected to each other as needed via include directives for header files. To accomplish this with the FM2Class.cpp file, a header file FM2.h is first created that appears as

```
//FM2.h
#ifndef FM2_H
#define FM2_H

class FM2{
  unsigned long a;//multiplier
  int expFor2;//exponent for 2 in modulus
  unsigned long power(int i, int k) const;

public:

  FM2();//default constructor
  ~FM2(){};//destructor
  void rangen(unsigned long seed, int n, double* pu) const;
};
#endif
```

As was true for the power class in Chapter 3, the FM2.h file essentially consists of just the class prototype that appeared at the beginning of FM2Class.cpp. The only real difference is the presence of the include guard preprocessor directives discussed in Section 3.7.

The header file for a class contains all the information that is needed to employ objects of the class. It represents a simple, public interface for the class and one does not need to know any of the details about the class methods, etc., in order to be able to use an object of the class. In fact, the code in the body of the method definitions can even be changed without affecting the class interface. This is an important feature in that the effects of altering the internal aspects of a method (in response to, e.g., new computational methodology) will not propagate into other programs that might use the class. In addition, casual users of a class need not and, in general, should not alter method definitions. This latter fact is a particularly compelling reason why, in general, the actual details of the class should be implemented in a separate source file.

As a case in point, the method definitions for the FM2 class will now be placed in a file FM2.cpp that has the contents

```
//FM2.cpp
#include "FM2.h"

unsigned long FM2::power(int i, int k) const {
  unsigned long iout = (unsigned long)i;
  if(k > 1)
    iout = iout*power(iout, k - 1);
  return iout;
}

FM2::FM2(){a = 630360016; expFor2 = 31;}

void FM2::rangen(unsigned long seed, int n, double* pU) const {
  unsigned long m = power(2, expFor2) - 1;
  for(int i = 0; i < n; i++){
    seed = (a*seed)%m;
    pU[i] = ((double)seed)/((double)m);
  }
}
```

The only "new" feature here is the `#include` statement that brings in the class prototype in `FM2.h`. Notice that quotes appear here instead of `< >` thereby telling the preprocessor to look for the header file in the current directory. Finally, the code that actually uses the `FM2` class is just the `main` function in Listing 4.2 apart from having the initial statements

```
#include <iostream>
#include "FM2.h"
#include <new>
using std::cout; using std::cin; using std::endl;
```

that, e.g., bring in the necessary header file. The corresponding code was collected in the file driverFM2.cpp.

At this point the original set of code in Listing 4.2 has been split into three distinct parts. By doing this the FM2 class has been furnished with an independent identity that allows it to be employed in other programs. All the pieces are cobbled back together in a way that produces an executable program through the use of the makefile shown below.

```
driverFM2 : FM2.o driverFM2.o
        g++ -Wall FM2.o driverFM2.o -o driverFM2

FM2.o : FM2.cpp FM2.h
        g++ -Wall -c FM2.cpp

driverFM2.o : driverFM2.cpp FM2.h
        g++ -Wall -c driverFM2.cpp
```

This produces the executable `driverFM2` that can then be used, as before, to obtain output from the program.

4.5 Other generation methods

The periods of linear congruential generators are too small to be effective in many modern applications. This section describes other PRNGs with much longer periods that are better suited for general use.

Before proceeding further it is perhaps worthwhile to clarify what is meant by the phrase "longer period". To put this in the proper context we can begin by thinking of linear congruential generators. This type of generator produces a sequence of integers (that are then transformed into values in the unit interval) such that every number in the sequence is unique. If, for example, the modulus of a multiplicative generator is $2^{31} - 2$, having full period means that from an initial seed the generator can march through $2^{31} - 2$ numbers without ever repeating a previously generated value. In contrast, the generators discussed in this section can have periods on the order of 2^p where p is essentially arbitrary and, in particular, independent of a computer's largest storage format or *word length*. But, if the maximum word length is w bits (e.g., $w = 64$ in a 64-bit machine) the computer can only produce numbers that can be represented with w bits (e.g., the largest integer is $2^{64} - 1$ in a 64-bit storage environment). To achieve a period larger than $2^w - 1$ it is therefore necessary that individual integer values appear multiple times in the pseudo-random number stream. The result will then be a sequence of integers in which the position (rather than value) of each number relative to others in the sequence is unique until the entire sequence begins again. The length of the sequence up until it begins to repeat is now what will be called the period of the generator.

Tausworthe (1965) proposed a class of random number generators that are generally referred to as *linear feedback shift register* generators. Although these particular generators are not held in high regard, their extensions are widely used and provide some of what are arguably the best generators that are currently available.

Linear feedback shift registers work on a bit-wise level to produce numbers in binary representation. Suppose that r-bit integers are to be generated. Then, for some integer p, we start with a p-bit seed of binary form $b_1 \cdots b_p$ with the b_i all being 0 or 1. Subsequent bit values are produced via the recursion

$$b_i = b_{i-p} \oplus b_{i-p+q} \qquad (4.5)$$

for some integer $q < p$ with $\oplus$ indicating the bit-wise exclusive OR operator XOR that returns a 1 only if one of the bits is 1 and the other is 0.

An illustration of the use of (4.5) is provided by an example with $r = 3$ from Lewis and Payne (1973). Take $p = 5$, $q = 2$ and choose the initial seed bits as $b_1 = b_2 = b_3 = b_4 = b_5 = 1$. This produces $b_6 = b_1 \oplus b_3 = 1 \oplus 1 = 0, b_7 = b_2 \oplus b_4 = 1 \oplus 1 = 0, b_8 = b_3 \oplus b_5 = 1 \oplus 1 = 0, b_9 = b_4 \oplus b_6 = 1 \oplus 0 = 1, b_{10} = b_5 \oplus b_7 = 1 \oplus 0 = 1, \ldots$. The entire sequence is 1111100011011101010000100101100 after which the bit pattern repeats.

The bits produced from (4.5) are grouped appropriately to provide r-bit integers for some $2 \leq r \leq p$. The multi-step approach of Tausworthe (1965) carries out the grouping using contiguous, nonoverlapping blocks of size r. In terms of our example this produces

$$x_1 = 111 \text{ in binary or } x_1 = 7, \quad x_2 = 110 \text{ in binary or } x_2 = 6,$$
$$x_3 = 001 \text{ in binary or } x_3 = 1, \quad x_4 = 101 \text{ in binary or } x_4 = 5,$$
$$x_5 = 110 \text{ in binary or } x_5 = 6, \quad x_6 = 101 \text{ in binary or } x_6 = 5,$$
$$x_7 = 000 \text{ in binary or } x_7 = 0, \quad x_8 = 010 \text{ in binary or } x_8 = 2,$$
$$x_9 = 010 \text{ in binary or } x_9 = 2, \quad x_{10} = 110 \text{ in binary or } x_{10} = 6.$$

At this point we are seemingly at an end even though one bit (the 31st) remains. However, there are more numbers that can be mined from our 31 bits by combining the last bit with two bits from the start of the bit sequence and cycling through the sequence again in three-bit blocks. The result is

$$x_{11} = 011 \text{ in binary or } x_{11} = 3, \quad x_{12} = 111 \text{ in binary or } x_{12} = 7,$$
$$x_{13} = 000 \text{ in binary or } x_{13} = 0, \quad x_{14} = 110 \text{ in binary or } x_{14} = 6,$$
$$x_{15} = 111 \text{ in binary or } x_{15} = 7, \quad x_{16} = 010 \text{ in binary or } x_{16} = 2,$$
$$x_{17} = 100 \text{ in binary or } x_{17} = 4, \quad x_{18} = 001 \text{ in binary or } x_{18} = 1,$$
$$x_{19} = 001 \text{ in binary or } x_{19} = 1, \quad x_{20} = 011 \text{ in binary or } x_{20} = 3.$$

This process can be repeated once more before the cycle starts again which produces the 31 three-bit integer sequence 7615650226370672411317433520454. These values were obtained by allowing (4.5) to run through its entire period $r = 3$ times while picking off groups of 3 digits. This idea works for general p, q, r as seen in Exercise 4.12.

The maximum period for the shift register generator (4.5) is attained if the polynomial $t^p + t^q + 1$ is a primitive polynomial modulo 2 (e.g., Niederreiter and Lidl 1986). Extensive tables of such trinomials can be found in Zierler and Brillhart (1968, 1969) and Zierler (1969). (Note that the values of q, or k in their notation, in Table 1 of Zierler and Brillhart 1968 that make the corresponding trinomial primitive are indicated by italics rather than by underlines as was stated in the paper.) In the case where bits are allocated via the multi-step approach the maximum period of the generator is $(2^p - 1)/\gcd(r, 2^p - 1)$ with $\gcd(r, 2^p - 1)$ the greatest common divisor of r and $2^p - 1$ (e.g., Niederrieter 1992). For our example the primitive, fifth-degree polynomial $t^5 + t^2 + 1$ from Table 1 of Zierler and Brillhart (1968) was used which gives the maximum period $(2^5 - 1)/\gcd(3, 2^5 - 1) = 31$.

The ordinary shift register methods that have been described thus far work with an initial set of p bits that have 2^p possible values and then with additional groups of r bits that can be arranged in 2^r ways. Thus, there are 2^{pr} possibilities for bit representations that are actually available and, in that sense, the feedback shift register approach does a poor job of mining its available resources.

Suppose that we now think of an r-bit integer as being represented by a vector v of length r having all 0 or 1 entries. Vectors of this type will be called *binary vectors*. The generalized feedback shift register generator of Lewis and Payne (1973) then produces integers via the recurrence

$$v_i = v_{i-p} \oplus v_{i-p+q}, \tag{4.6}$$

where the v_j are binary vectors of length r and $\oplus$ now denotes component-wise application of the XOR operator. The maximum period for this type of generator is $2^p - 1$ and this is attained if $t^p + t^q + 1$ is a primitive polynomial modulo 2.

To illustrate (4.6) we will again take $p = 5, q = 2, r = 3$ and choose

$$v_1 = \begin{bmatrix} 1 \\ 1 \\ 1 \end{bmatrix}, v_2 = \begin{bmatrix} 0 \\ 1 \\ 1 \end{bmatrix}, v_3 = \begin{bmatrix} 1 \\ 0 \\ 0 \end{bmatrix}, v_4 = \begin{bmatrix} 1 \\ 0 \\ 1 \end{bmatrix}, v_5 = \begin{bmatrix} 0 \\ 1 \\ 1 \end{bmatrix}. \tag{4.7}$$

The recursion $v_i = v_{i-5} \oplus v_{i-3}$ then produces

$$v_6 = v_1 \oplus v_3 = \begin{bmatrix} 0 \\ 1 \\ 1 \end{bmatrix}, v_7 = v_2 \oplus v_4 = \begin{bmatrix} 1 \\ 1 \\ 0 \end{bmatrix}, \ldots.$$

The entire sequence of $2^5 - 1 = 31$ integers obtained from this particular seed of five binary vectors is 7615663735102253271743054464120.

Listing 4.3 provides one possible implementation of the generator in (4.6).

Listing 4.3 *gfsr.cpp*

```cpp
//gfsr.cpp
#include <iostream>

void intToBin(unsigned long val, bool* pb, int r){
  for(int i = 0; i < r; i++){
    if(val & (1 << i))
      pb[r - i - 1] = 1;
    else
      pb[r - i - 1] = 0;
  }
}

unsigned long power(int i, int k){
  if(k == 0)
    return 1;
  else{
    int iout = i;
    if(k > 1)
      iout = iout*power(iout, k - 1);
    return iout;
  }
}

unsigned long binToInt(bool* pb, int r){
  unsigned long val = 0;
  for(int i = 0; i < r; i++)
    val += pb[i]*power(2, r - i - 1);
```

```cpp
    return val;
}

void binMult(bool** pb, int p, int q, int r){
  //use the extra row of pb here
  for(int j = 0; j < r; j++){
    pb[p][j] = pb[0][j]^pb[q][j];
  }
  //Update pb
  for(int i = 0; i < p; i++)
    for(int j = 0; j < r; j++)
      pb[i][j] = pb[i + 1][j];
}

int main(){
  int r;
  std::cout << "Input the value for r:" << std::endl;
  std::cin >> r;
  int p;
  std::cout << "Input the value for p:" << std::endl;
  std::cin >> p;
  int q;
  std::cout << "Input the value for q:" << std::endl;
  std::cin >> q;
  unsigned long* x0 = new unsigned long[p];
  std::cout << "Input a " << p << " integer seed:" << std::endl;
  for(int i = 0; i < p; i++)
    std::cin >> x0[i];

  unsigned long modulus = power(2, p) - 1;

  //matrix of Booleans to hold the bits
  bool** pb = new bool*[p + 1];
  //rows will hold the integer bit-wise representations
  //extra row added for temporary storage
  for(int i = 0; i < (p + 1); i++)
    pb[i] = new bool[r];

  //initialize the bit pattern using x0
  for(int i = 0; i < p; i++)
    intToBin(x0[i], pb[i], r);

  //now generate the "data"
  for(int i = 0; i < (modulus - p); i++){
    binMult(pb, p, q, r);
    std::cout << " " << binToInt(pb[p - 1], r) << " ";
  }
  std::cout << std::endl;

  return 0;
}
```

Beginning in the main segment of Listing 4.3 the user is queried for values of the parameters p, q and the number of bits r that are to be used in constructing the generated integers. The p integers that will provide the seed for the generator are also required as user input. In particular, the choice of p = 5, q = 2, r = 3 and x0[0] = 7, x0[1] = 6, x0[2] = 1, x0[3] = 5, x0[4] = 6 as the seed integers is equivalent to the vector seeds (4.7) and reproduces the previous random number sequence.

The operative part of the generalized feedback shift register code works with a $(p+1) \times r$ array of Boolean variables that is manipulated in the form of the pointer-to-pointer pb. The first p rows of the array hold the binary vectors that are needed to carry out the recursion (4.6) while the last row is used for temporary storage of the output of the recursive formula on each step of the algorithm. The function binMult implements (4.6) using the C++ bit-wise exclusive OR operator ˆ and stores the result in the $(p+1)$st row of pb. The first row of pb is then deleted and the remaining contents of pb are shifted back one row to prepare for the next iteration.

Listing 4.3 also contains functions intToBin and binToInt that convert integers to binary and back again, respectively. The intToBin function uses the bit-wise shift operator along with the bit-wise AND operator to pick off the binary representation of the integers that are provided for the seed in a similar fashion to what was done in Listing 2.2. The array pb is initialized using intToBin while the binToInt function transforms the output from the generator into integers directly from binary representations with the help of power for computing powers of 2.

A variant of the generalized linear feedback shift register is the *twisted feedback shift register generator* (e.g., Matsumoto and Kurita 1992). In this case binary vectors of length r are produced using

$$v_i = (Av_{i-p}) \oplus v_{i-p+q} \tag{4.8}$$

with A an $r \times r$ matrix having all 0 and 1 entries. This approach can produce generators with periods as long as $2^{pr} - 1$ provided that $t^p + t^q + 1$ is primitive modulo 2 and the characteristic polynomial of A is irreducible. In this respect the generators realize the potential of shift register methods. Matsumoto and Kurita (1992) suggest using

$$A^T = \begin{bmatrix} 0 & 1 & 0 & \cdots & 0 \\ 0 & 0 & 1 & \cdots & 0 \\ \vdots & \vdots & \vdots & \ddots & 0 \\ 0 & 0 & 0 & \cdots & 1 \\ b_1 & b_2 & b_3 & \cdots & b_r \end{bmatrix}$$

with $b = (b_1, \ldots, b_r)^T$ a vector of zeros and ones and provide specific choices for b that produce the maximum period. The *Mersenne twister* developed by Matsumoto and Takuji (1998) is a popular variation of the twisted feedback shift register generator that has period $2^{19937} - 1$. The original *Mersenne twister* proposal has been modified in various ways (e.g., Saito and Matsumoto 2006) to improve its speed and ease of initialization.

There are many other varieties of random number generators that extend congruential methods. One natural extension is to use a recursion of the form

$$x_i = (a_1 x_{i-1} + a_2 x_{i-2} + \cdots + a_k x_{i-k}) \bmod m, \tag{4.9}$$

where k is some integer, $a_1, \ldots, a_k$ are integers between 0 and $m - 1$ and the recursion is initialized with k starting values. Generators of this type are often termed *multiple recursive*. When m is a prime, the maximum period length for a multiple recursive generator is $m^k - 1$ which can be attained if and only if the *characteristic polynomial* $P(z) = z^k - a_1 z^{k-1} - \cdots - a_k$ is primitive modulo m (see, e.g., Section 3.2.2 of Knuth 1998a).

A relatively simple way to produce "good" random number generators with longer periods is to use linear combinations of the output from several multiple recursive generators. A general set-up would have G such generators with the gth one producing the values

$$x_{gi} = (a_{g1} x_{g(i-1)} + a_{g2} x_{g(i-2)} + \cdots + a_{gk} x_{g(i-k)}) \bmod m_g \tag{4.10}$$

for $g = 1, \ldots, G$. Integer coefficients $c_1, \ldots, c_G$ are then chosen to produce numbers on the interval $[0, 1]$ via

$$u_j = \left(c_1 \frac{x_{1j}}{m_1} + c_2 \frac{x_{2j}}{m_2} + \cdots + c_G \frac{x_{Gj}}{m_G} \right) \bmod 1$$

with $x \bmod 1$ being the fractional part of x. Under conditions detailed in L'Ecuyer and Tezuka (1991) and L'Ecuyer (1996) the period length for the generator can be as large as the product of the periods for the individual generators. Tables containing good parameter choices for generators of this nature are given in L'Ecuyer (1999).

An example of a combined generator is the one proposed by Wichmann and Hill (1982) that will be referred to as the WH generator in the sequel. The algorithm uses the linear congruential (and, hence, multiple recursive with $k = 1$) generators

$$\begin{aligned} x_i &= 171x_{i-1} \bmod 30269 \\ y_i &= 172y_{i-1} \bmod 30307 \\ z_i &= 170z_{i-1} \bmod 30323 \end{aligned}$$

with the resulting uniform random deviate being obtained from

$$u_i = \left(\frac{x_i}{30269} + \frac{y_i}{30307} + \frac{z_i}{30323} \right) \bmod 1.$$

The period of this generator exceeds 2.78×10^{13}.

Our implementation of a C++ WH class for the WH generator has the header file

---
Listing 4.4 *WH.h*

```
//WH.h
#ifndef WH_H
#define WH_H

class WH{
  unsigned long a1, a2, a3;
  unsigned long m1, m2, m3;

 public:

  WH();
  ~WH(){};
  void rangen(unsigned long seed1, unsigned long seed2,
        unsigned long seed3, int n, double* pu) const;
};
#endif
```

with the method definitions given in

Listing 4.5 *WH.cpp*

```
// WH.cpp
#include "WH.h"

WH::WH(){
  a1 = 171; m1 = 30269; a2 = 172; m2 = 30307; a3 = 170; m3 = 30323;
}
void WH::rangen(unsigned long seed1, unsigned long seed2,
        unsigned long seed3, int n, double* pu) const {
  double temp = 0;
```

```
for(int i = 0; i < n; i++){
   seed1 = (a1*seed1)%m1;
   seed2 = (a2*seed2)%m2;
   seed3 = (a3*seed3)%m3;
   temp = (((double)seed1)/((double)m1)
           + ((double)seed2)/((double)m2)
           + ((double)seed3)/((double)m3));
   if(temp > 2.) temp -= 2.;
   if(temp > 1.) temp -= 1.;
   pu[i] = temp;
}
}
```

Here a1, a2, a3, m1, m2 and m3 represent the multipliers and moduli for the three generators. Although they have been specified explicitly in the constructor definition to produce the WH generator, other choices could be obtained as input to another constructor to provide a more general implementation. The operative class method takes as arguments the seed, sample size and a pointer to `double`. The memory locations corresponding to the pointer are then filled with random deviates produced by the generator.

The WH generator provides an appreciable increase in period length beyond that of the Fishman-Moore generator of Section 4.4. Even so, its period is still too short for many applications that require very long streams such as random number generation in parallel computing environments. One source for generators that can be used for such purposes is the Rngstreams package of L'Ecuyer, et al. (2001, 2002). The source code for the package in C, C++, C# and Java can be downloaded from

$$\text{http://www.iro.umontreal.ca/~lecuyer/myftp/streams00}$$

In particular, the C++ implementation of RngStreams furnishes an OOP framework where objects are created that have an associated method `RandU01` to produce uniform random deviates.

The "backbone" generator for RngStreams is called Mrg32k3a. It is a combination of the two multiple recursive generators that produce the states

$$x_{1,n} = (1403580 \times x_{1,n-2} - 810728 \times x_{1,n-3}) \mod (2^{32} - 209), \qquad (4.11)$$

$$x_{2,n} = (527612 \times x_{2,n-1} - 1370589 \times x_{2,n-3}) \mod (2^{32} - 22853) \qquad (4.12)$$

at the nth step of the recursion given initial seeds $\tilde{x}_{i,0} = (x_{i,0}, x_{i,-1}, x_{i,-2})^T, i = 1, 2$. The two states are combined to produce the uniform random deviate u_n via the rule

$$z_n = (x_{1,n} - x_{2,n}) \mod (2^{32} - 209),$$
$$u_n = \begin{cases} z_n/(2^{32} - 208), & \text{if } z_n > 0, \\ (2^{32} - 209)/(2^{32} - 208), & \text{if } z_n = 0. \end{cases}$$

In (4.10) this corresponds to $G = 2, k = 3, m_1 = 2^{32} - 209, a_{11} = 0, a_{12} = 1403580, a_{13} = -810728, m_2 = 2^{32} - 22853, a_{21} = 527612, a_{22} = 0$ and $a_{23} = -1370589$. The period for this generator is approximately 2^{191} provided that $x_{1,0}, x_{1,-1}, x_{1,-2}$ are not all 0 and are all less than $(2^{32} - 209)$ and $x_{2,0}, x_{2,-1}, x_{2,-2}$ are all less than $(2^{32} - 22853)$ and not all 0.

The initial state of the RngStreams package is set using the function `SetPackageSeed` that has prototype

```
static bool SetPackageSeed (const unsigned long seed[6])
```

with `seed` containing the six initial integer seeds for the generator that must be supplied by the user. The `static` modifier has the consequence that the function can be called without using an object from the `RngStream` class. The actual random uniforms are then obtained from the `RandU01` method using an object from the `RngStream` class. Listing 4.6 below illustrates the process.

```cpp
//rngStreamEx.cpp
#include <iostream>
#include "RngStream.h"

using std::cout; using std::endl;

int main(){
  unsigned long seed[6] = {1, 2, 3, 4, 5, 6};
  RngStream::SetPackageSeed (seed);
  RngStream rngObj;
  cout << "Your random number is ";
  cout << rngObj.RandU01() << endl;
  return 0;
}
```

The key step is inclusion of the header file RngStream.h at the start of the program. The initial generator seeds $x_{1,-2} = 1, x_{1,-1} = 2, x_{1,0} = 3, x_{2,-2} = 4, x_{2,-1} = 5$ and $x_{2,0} = 6$ are set with the `SetPackageSeed` function whose *static* designation allows it to be used directly without a class object. An `RngStream` object is instantiated and used to generate a random number that is printed to standard output. The program is compiled and executed with

```
$ g++ -Wall RngStream.cpp rngStreamEx.cpp -o rngStreamEx
$ ./rngStreamEx
Your random number is 0.0010095
```

4.6 Nonuniform generation

As noted in the introduction, the generation of uniform random numbers is the key to producing pseudo-random samples from nonuniform distributions. The most direct approach is to use the probability integral transform of Theorem 4.1 as implemented in Algorithm 4.1. A case where this can be done in closed form is the exponential distribution with probability density function

$$f(x) = \begin{cases} \lambda \exp\{-\lambda x\}, & x, \lambda > 0, \\ 0, & \text{otherwise.} \end{cases}$$

The distribution function corresponding to f is $F(x) = 1 - \exp\{-\lambda x\}$ for $x > 0$ for which the inverse or quantile function is $F^{-1}(u) = -\ln(1 - u)/\lambda$ with $u \in (0,1)$. The driver program below uses this function to produce a pseudo-random sample from the exponential distribution using the WH generator as in Listings 4.4–4.5.

```cpp
//driverExp.cpp
#include <iostream>
#include <cstdlib>
#include <cmath>
#include <new>
#include "WH.h"

using std::cout; using std::cin; using std::endl;
```

```
double Finv(double u, double lambda){
  return (-log(1 - u)/lambda);
}

int main(int argc, char* argv[]){
  double lambda; int n;

  cout << "Input the value for lambda:" << endl;
  cin >> lambda;

  unsigned long seed1, seed2, seed3;
  cout << "Choose a value for the sample size:" << endl;
  cin >> n;
  cout << "Choose a value for seed1:" << endl;
  cin >> seed1;
  cout << "Choose a value for seed2:" << endl;
  cin >> seed2;
  cout << "Choose a value for seed3:" << endl;
  cin >> seed3;

  WH* pWH = new(std::nothrow) WH();
  if(pWH == 0){
    cout << "Memory allocation for WH failed" <<endl;
    exit(1);
  }
  double* pU = new(std::nothrow) double[n];
  if(pU == 0){
    cout <<"Memory allocation for double pointer failed" <<endl;
    exit(1);
  }
  pWH->rangen(seed1, seed2, seed3, n, pU);
  cout << "Your random numbers are " << endl;
  for(int i = 0; i< n; i++)
    cout << Finv(pU[i], lambda) << " ";
  cout << endl;
  delete pWH;
  delete[] pU;
  return 0;
}
```

The only new aspect in Listing 4.7 is the specification of the F^{-1} function at the beginning of the program and its use in the for loop to transform the uniform random numbers obtained from the WH object pointer pWH. It is also necessary to include the cmath header in order to use the natural logarithm function log in our Finv function. Upon compilation into the executable driverExp, the program produces output such as

```
$ ./driverExp
Input the value for lambda:
1
Choose a value for the sample size:
5
Choose a value for seed1:
123
Choose a value for seed2:
456
Choose a value for seed3:
```

```
789
Your random numbers are
1.22472 2.50259 0.159843 1.07643 0.44995
```

Similar to what was done for the FM2 generator, `driverExp` was used with seeds 123, 456, 789 to generate a random sample of 500 values from the exponential distribution with unit mean that was imported into R as a `numeric` vector object called `eDat`. The following information was then obtained about the data.

```
> summary(eDat)
     Min.  1st Qu.    Median      Mean  3rd Qu.      Max.
 0.001526 0.258000 0.657300 0.980200 1.300000 8.784000
> sd(eDat)
[1] 1.066776
> IQR(eDat)
[1] 1.042139
> pnorm(sqrt(length(eDat))*(mean(eDat) - 1))
[1] 0.3286831
> ks.test(eDat, pexp)

        One-sample Kolmogorov-Smirnov test

data:  eDat
D = 0.0402, p-value = 0.3954
alternative hypothesis: two-sided
```

The R function `summary` was invoked in this R session to calculate some of the basic sample statistics for the imported data. This information is augmented with the sample standard deviation (using the function `sd`) and the sample inter-quartile range (with the function `IQR`). Seeing as the data was produced using an input value of 1 for λ, both the mean and the variance of the target parent population are one while the median and inter-quartile range are, approximately, .693 and 1.1 (Exercise 4.18). As a result, the sample statistics are more or less in agreement with what might be expected. Although the outcome is obvious here, a normalized (through multiplication by the square root of the sample size) difference between the sample mean and 1 was computed that should behave approximately like a number drawn from the standard normal distribution if the values come from an exponential distribution with $\lambda = 1$. The p-value obtained using the R cumulative distribution function `pnorm` for the standard normal distribution is not significant and the Kolmogorov-Smirnov goodness-of-fit test also detects no significant departure from the exponential model. The histogram of the data shown in Figure 4.3 exhibits the J-shape that would be expected from exponentially distributed data.

If F has a simple closed form, but there is no closed form for F^{-1}, Algorithm 4.1 can still be used with the quantile function being evaluated by numerical methods (e.g., Exercise 8.26). When there is no closed form for F, as is true for the normal distribution, the cdf itself can be approximated numerically using, e.g., numeric integration and the same basic approach can be applied to the approximation. There is an alternative, less involved, strategy that we will now describe.

As before, the problem is that of simulating values from a continuous random variable with density f. However, now it is assumed that there is another available density g with the property that for all real numbers x

$$f(x) \leq cg(x) \tag{4.13}$$

for some constant $c \geq 1$. The g density is sometimes referred to as the *instrumental* density with f being the *target* density. The basic *accept-reject* algorithm then takes the form

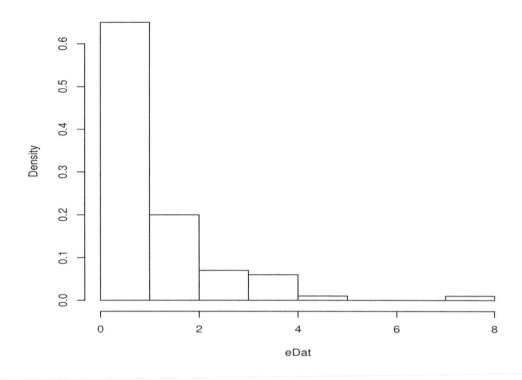

Figure 4.3 *Histogram for simulated data*

Algorithm 4.2 Accept-reject algorithm

$r = u = a$ (for some arbitrary $a > 1$)
while $ru > 1$ **do**
    Generate y from the distribution corresponding to g
    Generate u from a uniform distribution on $[0, 1]$
    $r = cg(y)/f(y)$
end while
return y

The mechanics involved here are quite simple. We first generate pairs of random numbers. Then, in each step of the while loop, a value u from a uniform distribution is paired with a value y sampled from the distribution corresponding to the instrumental density g. The process of generating pairs continues until a pair u, y for which $u \leq f(y)/cg(y)$ is found at which point y is returned as the value that was simulated from the distribution with density f. Assuming that y behaves as a value sampled from the density f, this approach can potentially solve the problem of f not having a closed form cdf and/or quantile function. Specifically, if a density g can be found that satisfies (4.13) for which the corresponding quantile function G^{-1} has a closed form, it is easy to produce samples from the g distribution and the entire process can be driven by a uniform random number generator.

The justification for the accept-reject approach is provided by Theorem 4.4.

Theorem 4.4. *Let Y have density g satisfying (4.13) and let U conditional on $Y = y$ have a uniform distribution on $[0, cg(y)]$. Then, $X = Y|U < f(Y)$ has density f.*

Proof. Proofs of Theorem 4.4 can be found in Devroye (1986) and Robert and Casella (2004), for example. The idea is that if A is any open interval of the real line

$$
\begin{aligned}
\mathrm{Prob}(X \in A) \;&=\; \mathrm{Prob}(Y \in A|U \le f(Y)) \\[2mm]
&=\; \frac{\int_A \int_0^{f(y)} \frac{1}{cg(y)} g(y)\,du\,dy}{\int_{-\infty}^{\infty} \int_0^{f(y)} \frac{1}{cg(y)} g(y)\,du\,dy} \\[2mm]
&=\; \int_A f(y)\,dy
\end{aligned}
$$

which proves the theorem.   □

   Note that Theorem 4.4 says to first generate a value y from the density g and then generate a uniform random number on $[0, cg(y)]$. But, to obtain a uniform random number from $[0, cg(y)]$ it suffices to generate from the interval $[0, 1]$ and then multiply by $cg(y)$ since the quantile function for a uniform random variable on $[0, b]$ is $Q(u) = bu$ for $u \in [0, 1]$ and $b > 0$. Thus, the plan is i) generate y from a distribution with density g, ii) generate u from a uniform distribution on $[0, 1]$, iii) transform to $cg(y)u$ and then iv) take y as the desired value if $cg(y)u \le f(y)$ or, equivalently, $ru \le 1$ as in our original formulation.

   A measure of performance of an accept-reject algorithm is provided by the number of rejections that are required to actually generate a random number from the target distribution. Let U and Y be as in Theorem 4.4. Then, on any given trial the probability of acceptance is

$$
\mathrm{Prob}\,(U \le f(Y)) = \int_{-\infty}^{\infty} \int_0^{f(y)} \frac{1}{cg(y)} g(y)\,du\,dy = 1/c
$$

and the number of trials before the first acceptance has a geometric distribution with success probability c^{-1}. The value of c reflects the extent to which g is able to approximate f. Consequently, the accept-reject approach will benefit from a careful choice of g that allows us to take c close to one.

   An illustration of the accept-reject method can be obtained by using the Cauchy density to generate standard normal random variables. The Cauchy density with scale parameter $\sigma > 0$ has the form

$$
g(x) = \frac{1}{\sigma \pi} \frac{1}{1 + x^2/\sigma^2}, \quad -\infty < x < \infty, \; \sigma > 0, \tag{4.14}
$$

for which the quantile function is

$$
G^{-1}(u) = \sigma \tan\left(\pi(u - .5)\right).
$$

 To use g for simulating from the standard normal distribution it is necessary to find a value $c \ge 1$ for which $f(y)/g(y) \le c$ with

$$
f(x) = \frac{1}{\sqrt{2\pi}} \exp\{-x^2/2\}.
$$

Proceeding as in Section 3.2 of Devroye (1986) we will find the c that will provide the tightest bound. That is, c will be chosen as the smallest number so that $f(x) \le cg(x)$. That choice will be the one that, on the average, will lead to the fewest rejections and thereby to faster sample generation.

   The ratio $f(x)/g(x)$ will achieve a maximum when $(1 + x^2/\sigma^2)\exp\{-x^2/2\}$ does. It is equivalent to maximize the natural logarithm of this ratio which is tantamount to finding the solutions of $2x = (\sigma^2 + x^2)x$: namely, $x = 0$ and $x = \pm\sqrt{2 - \sigma^2}$. The latter two roots are only applicable when $\sigma \le \sqrt{2}$ in which case the maxima are at $\pm\sqrt{2 - \sigma^2}$ corresponding to a maximum value of $c = \sqrt{2\pi} \exp\{\sigma^2/2 - 1\}/\sigma$ for f/g. For $\sigma > \sqrt{2}$ the maximum for f/g is $c = \sigma\sqrt{\pi/2}$. If c is now viewed as a function of σ it is minimized at $\sigma = 1$ with minimum value $\sqrt{2\pi/e} \doteq 1.520347$.

The code listing below is the header file for a `WHNCauchy` class that uses the accept-reject method to generate pseudo-random standard normals using the Cauchy distribution with the optimal bounding constant and with uniform random deviates produced by the WH generator.

Listing 4.8 *WHNCauchy.h*

```
// WHNCauchy.h
#ifndef WHNCAUCHY_H
#define WHNCAUCHY_H

class WHNCauchy{
  unsigned long a1, a2, a3;
  unsigned long m1, m2, m3;
  double g(double x) const;
  double Ginv(double x) const;
  double f(double x) const;

 public:

  WHNCauchy();
  ~WHNCauchy(){};
  void rangen(unsigned long seed1, unsigned long seed2,
        unsigned long seed3, int n, double* pNorm) const;
};
#endif
```

The functions g and `Ginv` are the Cauchy density and quantile function while f is the normal density. As there is no obvious need for access to these functions by a typical user, they have been made `private`.

The method definitions for the `WHNCauchy` class are provided in the next listing.

Listing 4.9 *WHNCauchy.cpp*

```
//WHNCauchy.cpp
#include <cmath>
#include <iostream>
#include "WHNCauchy.h"

using std::cout; using std::endl;

double pi = 2.*acos(0.);
double sqrtwopi = sqrt(2.*pi);
double c = sqrt(2*pi/exp(1.));

WHNCauchy::WHNCauchy(){
  a1 = 171; m1 = 30269; a2 = 172; m2 = 30307; a3 = 170; m3 = 30323;
}


void WHNCauchy::rangen(unsigned long seed1, unsigned long seed2,
        unsigned long seed3, int n, double* pNorm) const {
  double temp, y;
  double u[2];
```

```
  int nFill = 0, iAccept, nReject = 0;

  while(nFill < n){
    iAccept = 0;
    while(iAccept == 0){
      //generate two uniforms
      for(int i = 0; i < 2; i++){
        seed1 = (a1*seed1)%m1; seed2 = (a2*seed2)%m2;
        seed3 = (a3*seed3)%m3;
        temp = (((double)seed1)/((double)m1)
                 + ((double)seed2)/((double)m2)
                 + ((double)seed3)/((double)m3));
        if(temp > 2.) temp -= 2.;
        if(temp > 1.) temp -= 1.;
        u[i] = temp;
      }
      y = Ginv(u[1]);
      if(c*g(y)*u[0] <= f(y)){//accept the result
        pNorm[nFill] = y;
        nFill += 1;//augment the number of samples
        iAccept = 1;//end the interior while loop
      }
      else nReject += 1;
    }
  }
  cout << "The Number of rejections was " << nReject << endl;
}

double WHNCauchy::g(double x) const {
  return 1/((1 + x*x)*pi);
}

double WHNCauchy::Ginv(double u) const {
  return tan(pi*(u - .5));
}

double WHNCauchy::f(double x) const {
  return exp(-.5*x*x)/sqrtwopi;
}
```

There are a few features that merit discussion in this listing. First, at the beginning of the listing three recurring constants are defined: $\pi, \sqrt{2\pi}$ and the optimal constant for use with the Cauchy instrumental density. This endows them with global scope as discussed in Section 3.2. That is, by defining pi, sqrtwopi and c outside the scope of any function in the program they become available to all functions that may need them thereby removing the need to compute them more than once. The algorithm itself works through a while loop that terminates when nFill, the number of samples that have been accepted, coincides with the desired sample size. A variable that counts the number of rejected samples has also been included to assess the efficiency of using the Cauchy distribution for generation of samples from the normal distribution.

It turns out that most of the distributions that arise in statistics can be tied to the chi-square and/or normal distributions. For example, a random variable from the F-distribution can be obtained as the ratio of two independent chi-square random variables divided by their degrees-of-freedom while a random variable from the t-distribution can be created as the ratio of a standard normal random variable and the square root of an independent chi-square random variable divided by its degrees-of-freedom.

To get samples from the chi-square distribution with an integer degrees-of-freedom ν we could use the fact that such a random variable has the same distribution as the sum of ν squared standard normal random variables. So, one could generate independent standard normal random variables $Z_1, \ldots, Z_\nu$ and then $X = \sum_{j=1}^{\nu} Z_j^2$ could be taken as the desired value. This is not the most computationally efficient approach and a superior alternative is provided by Kinderman and Monahan (1977) (see also Section 11.2C of Monahan 2001) and Cheng and Feast (1979). The idea is a consequence of our next theorem.

Theorem 4.5. *Let f be a density on the real line and suppose that the random variables U_1, U_2 have a joint distribution that is uniform over*

$$A = \{(u_1, u_2) : 0 \le u_1 \le \sqrt{f(u_2/u_1)}\}. \tag{4.15}$$

Then, U_2/U_1 is a random variable with density f.

Proof. Let us first evaluate the area of the region A. To do so, make the change of variable from u_1 and u_2 to $x_1 = u_2/u_1$ and $x_2 = u_1$. The Jacobian for this transformation is x_2 so that

$$\int_{-\infty}^{\infty} \int_0^{\sqrt{f(x_1)}} x_2 dx_2 dx_1 = \frac{1}{2} \int_{-\infty}^{\infty} f(x_1) dx_1 = \frac{1}{2}.$$

The joint density for $X_1 = U_2/U_1$ and $X_2 = U_1$ is therefore

$$h(x_1, x_2) = \begin{cases} 2x_2, & 0 \le x_2 \le \sqrt{f(x_1)}, \\ 0, & \text{otherwise.} \end{cases}$$

The marginal density for X_1 is now seen to be f which establishes the theorem. □

To use Theorem 4.5 in practice we need to be able to bound the region A where U_1 and U_2 reside. Then, uniforms can be simulated over the region laid out by the bounds. Now, at worst $U_1 \le \max_x \sqrt{f(x)}$ and, since $U_2 = (U_2/U_1)U_1$, it is also true that $\min_x x\sqrt{f(x)} \le U_2 \le \max_x x\sqrt{f(x)}$. Thus, random variables $\tilde{U}_1$ and $\tilde{U}_2$ can be generated with $\tilde{U}_1$ and $\tilde{U}_2$ uniformly distributed on $\left[0, \max_x \sqrt{f(x)}\right]$ and $\left[\min_x x\sqrt{f(x)}, \max_x x\sqrt{f(x)}\right]$, respectively. If $\tilde{U}_1^2 \le f(\tilde{U}_2/\tilde{U}_1)$, $X = \tilde{U}_2/\tilde{U}_1$ is accepted as a sample from the distribution corresponding to f and otherwise the process is repeated.

As in Cheng and Feast (1979), we will apply this idea to the case of a gamma random variable with the density

$$f(x) = \begin{cases} \frac{1}{\Gamma(\alpha)} x^{\alpha-1} \exp\{-x\}, & 0 \le x, \\ 0, & \text{otherwise.} \end{cases}$$

This will allow us to simulate chi-square random variables as well because a random variable X with a gamma distribution can be transformed to a chi-square random variable Y with $\nu = 2\alpha$ degrees-of-freedom by taking $Y = 2X$.

If $\alpha > 1$, which we shall assume to be the case, $x^{\alpha-1} \exp\{-x\}$ and $x^{\alpha+1} \exp\{-x\}$ are maximized at $x = \alpha - 1$ and $x = \alpha + 1$, respectively. Thus, in our previous formulation $\max_x \sqrt{f(x)} \le [(\alpha - 1)/e]^{(\alpha-1)/2}/\sqrt{\Gamma(\alpha)}$, $\min_x x\sqrt{f(x)} = 0$ and $\max_x x\sqrt{f(x)} \le [(\alpha + 1)/e]^{(\alpha+1)/2}/\sqrt{\Gamma(\alpha)}$. Gamma random deviates can therefore be generated using Algorithm 4.3 (Exercise 4.23).

Algorithm 4.3 Gamma generation algorithm

$b_1 = [(\alpha - 1)/e]^{(\alpha-1)/2}, b_2 = [(\alpha + 1)/e]^{(\alpha+1)/2}$
$c = b_2/\{b_1(\alpha - 1)\}$
Generate U_1, U_2 from a uniform distribution on $[0, 1]$
while $\frac{2}{\alpha-1} \ln(U_1) - \ln(cU_2/U_1) + cU_2/U_1 - 1 > 0$ **do**
   Generate U_1, U_2 from a uniform distribution on $[0, 1]$
end while
return $(\alpha - 1)cU_2/U_1$

The implementation of this algorithm is the subject of Exercise 4.24.

4.7 Generating random normals

Given the importance of the normal distribution to statistics it is no surprise that a number of methods have been developed for the generation of normal random deviates. Notable examples include those proposed in Box and Muller (1958), Ahrens and Dieter (1973) and Kinderman and Ramage (1976). We will focus attention here on the Box-Muller approach.

The basic result as stated in Box and Muller (1958) is that if U_1 and U_2 are random variables with a uniform distribution of $[0, 1]$ then

$$Z_1 = \sqrt{-2\ln(U_1)}\cos(2\pi U_2) \qquad (4.16)$$

$$Z_2 = \sqrt{-2\ln(U_1)}\sin(2\pi U_2) \qquad (4.17)$$

are independent standard normal random variables (Exercise 4.34). This has the implication that pairs of random normals can be generated from pairs of random uniforms through the simple transformations in (4.16)–(4.17). A polar coordinates perspective leads to yet another formulation that employs the accept-reject method while avoiding the use of trigonometric functions.

Theorem 4.6. *Let (U_1, U_2) have a uniform distribution over the unit circle: i.e., the joint (U_1, U_2) density is*

$$f(u_1, u_2) = \begin{cases} \frac{1}{4\pi}, & -1 \leq u_1, u_2 \leq 1, u_1^2 + u_2^2 \leq 1, \\ 0, & \text{otherwise.} \end{cases}$$

Then, if $R = U_1^2 + U_2^2$ and $V = U_1/\sqrt{R}$,

$$Z_1 = \sqrt{-2\ln(R)}V, \qquad (4.18)$$

$$Z_2 = \sqrt{-2\ln(R)}\sqrt{1 - V^2} \qquad (4.19)$$

are independent standard normal random variables.

Proof. The proof is relatively straightforward, but somewhat tedious. The result will be established in two steps. First, it will be shown that if the bivariate random variable (U_1, U_2) is uniformly distributed over the unit circle, R and V are independent with R having a uniform distribution on $[0, 1]$.

To accomplish our first objective, let $r = u_1^2 + u_2^2$ and take $v = u_1/\sqrt{r}$. This gives $u_1 = \sqrt{r}v$ and $v = \sqrt{r(1 - v^2)}$ from which the Jacobian is found to be $(1 - v^2)^{-1/2}/2$. The joint (R, V) density can now be written as

$$f(r, v) = \begin{cases} \frac{1}{4\pi}\frac{1}{1-v^2}, & 0 \leq r \leq 1, -1 \leq v \leq 1, \\ 0, & \text{otherwise,} \end{cases}$$

which proves that R is uniformly distributed on $[0, 1]$ and independent of V.

Now take $z_1 = \sqrt{-2\ln(r)}v$ and $z_2 = \sqrt{-2\ln(r)}\sqrt{1-v^2}$. The inverse of this transformation is $r = \exp\{-(z_1^2 + z_2^2)/2\}$ and $v = (1 + (z_1/z_2)^2)^{-1/2}$ which produces

$$\frac{2z_2}{\sqrt{z_1^2 + z_2^2}}\exp\{-(z_1^2 + z_2^2)/2\}$$

as the Jacobian. Thus, the joint Z_1, Z_2 density is

$$f(z_1, z_2) = \frac{1}{2\pi}\exp\{-(z_1^2 + z_2^2)/2\}, \quad -\infty < z_1, z_2 < \infty$$

and the theorem is proved. □

Theorem 4.6 suggests a simple strategy for generating numbers from the normal distribution. Generate two random uniforms U_1 and U_2 from the interval $[-1, 1]$. If $U_1^2 + U_2^2 \le 1$, accept the pair and obtain two pseudo-random normal deviates using (4.18)–(4.19). Otherwise, generate another pair and repeat the accept-reject evaluation. This approach discards $100\left(\frac{4}{\pi} - 1\right) = 27.32395$ percent of the uniform random number pairs (Exercise 4.35).

The code listings below correspond to an implementation of the Box-Muller method that employs the WH generator for the generation of uniform random deviates. The header file for the resulting WHBM class is

```
#ifndef WHN_H
#define WHN_H

class WHBM{
  unsigned long a1, a2, a3;
  unsigned long m1, m2, m3;

  public:

  WHBM();
  ~WHBM(){};
  void rangen(unsigned long seed1, unsigned long seed2,
         unsigned long seed3, int n, double* pu) const;
};
#endif
```

The values of a1, m1, a2, m2, a3, m3 are the same as those in Listing 4.9. They are defined explicitly in the class constructor whose definition is given, along with that of the rangen method in the listing below.

```
//WHBM.cpp
#include "WHBM.h"
#include <cmath>

WHBM::WHBM(){
  a1 = 171; m1 = 30269; a2 = 172; m2 = 30307; a3 = 170; m3 = 30323;
}

void WHBM::rangen(unsigned long seed1, unsigned long seed2,
         unsigned long seed3, int n, double* pNorm) const {
  double temp = 0, spare = 0, mult = 0, u[2] = {0, 0}, Rsqr = 0;

  int nFill = 0, iAccept = 0;

  while(nFill < n){
    if(iAccept == 1){//if we have one left, use it now
      pNorm[nFill] = spare;
      iAccept = 0;//next time we need to generate another pair
```

```
    nFill += 1;
}
else{
  while(iAccept == 0){
    //generate two uniforms
    for(int i = 0; i < 2; i++){
      seed1=(a1*seed1)%m1;
      seed2=(a2*seed2)%m2;
      seed3=(a3*seed3)%m3;
      temp = (((double)seed1)/((double)m1)
               + ((double)seed2)/((double)m2)
               + ((double)seed3)/((double)m3));
      if(temp > 2.) temp -= 2.;
      if(temp > 1.) temp -= 1.;
      u[i] = 2*temp - 1; //translate into [-1, 1]
    }
    Rsqr = u[0]*u[0] + u[1]*u[1];
    if(Rsqr < 1 && Rsqr != 0){//accept the result
      mult = sqrt(-2.*log(Rsqr)/Rsqr);
      pNorm[nFill] = u[0]*mult;
      spare = u[1]*mult;//save this one for next time
      nFill += 1;
      iAccept = 1;//now we have one left over
} } } }
```

The code in **rangen** is relatively straightforward apart from some bookkeeping that is needed to deal with the production of pairs of random numbers rather than singletons. An indicator variable or flag **iAccept** is set to 1 or 0 depending on whether or not there is still one of the pair of normals available to avoid using the uniform PRNG more than necessary. The decision variable in the loop **nFill** has to be augmented carefully to make certain that it is incremented both when a new pair of numbers is generated as well as when a remaining value from a previous pair is added to the sample.

The **WHBM** class was used with seeds 123, 456 and 789 to generate 500 standard normal random deviates. These values were imported into R as a **numeric** vector named **nDat** that was used to produce the histogram shown in Figure 4.4. The histogram exhibits some of the bell shape that would be expected from normal data.

Kernel density estimators provide alternatives to histograms that give smooth estimators of the parent density for a set of data. If $x_1, \ldots, x_n$ are observed sample values, such estimators have the form

$$f_h(x) = \frac{1}{nh} \sum_{i=1}^{n} K\left(\frac{x - x_i}{h}\right),$$

where the kernel K is a density function and $h > 0$ is the bandwidth that controls the amount of averaging or smoothing that is performed on the data. For kernel estimators to perform effectively the bandwidth must be chosen with care and there are a number of bandwidth selection methods that can be used for this purpose.

Figure 4.4 also shows a kernel density estimator that was fit to the data using the R function **density** by the command

```
> den<-density(nDat, kernel = "gaussian")
```

This estimator employs a "rule-of-thumb" bandwidth **nrd0** for the normal distribution discussed, for example, in Silverman (1986). There are a number of other bandwidth choices that can be perused by examining the help file for **bw.nrd**.

Summary statistics were also computed for the simulated "normal" data with the following results.

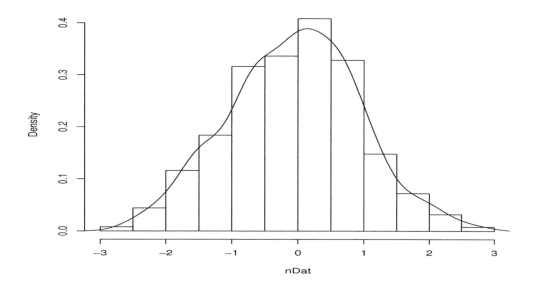

Figure 4.4 *Histogram for simulated data*

```
> summary(nDat)
     Min.   1st Qu.    Median      Mean   3rd Qu.      Max.
-2.60400  -0.71500  -0.01388  -0.05790   0.63210   2.82700
> sd(nDat)
[1] 0.9918201
> IQR(nDat)
[1] 1.347105
> pnorm(sqrt(length(nDat))*mean(nDat)/sd(nDat))
[1] 0.09588755
> ks.test(nDat, pnorm)

        One-sample Kolmogorov-Smirnov test

data:  nDat
D = 0.0346, p-value = 0.5865
alternative hypothesis: two-sided

> shapiro.test(nDat)

        Shapiro-Wilk normality test

data:  nDat
W = 0.9967, p-value = 0.4145
```

Since the data is supposed to have come from a normal distribution, the Shapiro-Wilk tests statistic was computed as well as the Kolmogorov-Smirnov statistic. On the basis of these two measures the hypothesis of normality cannot be rejected at any reasonable significance level.

4.8 Generating random numbers in R

As described in Appendix B, the R language provides the facility for obtaining random deviates from most of the standard distributions used in statistics. Random numbers for a distribution MyDist are produced with syntax such as

```
rMyDist(n, ...)
```

with n being the desired sample size and ... indicating other arguments that might be needed for a particular distribution of interest.

The type of random number generator that is used in the rMyDist functions can be determined and set using the functions RNGkind. Entering RNGkind() will return two values: the generator that is used to generate uniforms and the generator that is used to generate random normals. The default uniform generator is the Mersenne twister. There are five other options, including the WH generator, that are listed in the R help page for RNGkind. The default method for generating normal random deviates is direct inversion. There are four other options one of which is the Box-Muller method. A user-supplied method can also be employed for generating either random uniforms or normals.

To choose one of the other options for generating uniforms or normals, simply specify a value for one or both of the two arguments for RNGkind. These are kind and normal.kind and determine the uniform and normal generator, respectively. The ensuing code segment illustrates querying for the current generator choices and then changing them to the Wichmann-Hill and Box-Muller methods.

```
> RNGkind()
[1] "Mersenne-Twister" "Inversion"
> RNGkind(kind = "Wichmann-Hill", normal.kind = "Box-Muller")
> RNGkind()
[1] "Wichmann-Hill" "Box-Muller"
```

Although the Mrg32k3a generator from the RngStreams package discussed in Section 4.5 is not one of the options that can be chosen with RNGkind, it can still be accessed through the rstream package. After loading the package with

```
> library("rstream")
```

the random number generator for the current R session can be set as Mrg32k3a via the command

```
rstream.RNG(stream = new(''rstream.mrg32k3a'', seed = SEED))
```

with SEED a vector of six integers. A portion of an R session that illustrates the idea is shown below.

```
> library(rstream)
> rstream.RNG(new("rstream.mrg32k3a", seed = c(1, 2, 3, 4, 5, 6)))
> runif(1)
[1] 0.001009498
```

The seed argument was chosen to be the same as the one that was used to initialize the Mrg32k3a generator from RngStreams in Section 4.5. The output from runif indicates that the Mrg32k3a generator with this seed is now being used to generate uniforms in R.

In order to exactly recreate data that has been previously produced by a random number generator we need to know the seed or seeds that were used in the generator. The function set.seed provides a simple mechanism for ensuring this in R. The argument for set.seed is an integer variable seed that is then used to set whatever other seeds are required for the chosen generation method. In the example shown below seed is first set to 123. Then two sets of five random normals are generated. The second set of five is different from the first as would be expected. Then `seed` is reset to 123 and the `rnorm(5)` command produces the values that were originally obtained.

```
> set.seed(123)
> rnorm(5)
[1] -0.4542997  0.1593962  0.8977403  1.1607478 -0.4748856
> rnorm(5)
[1] -1.5676364  0.6444441  1.0574895  0.1692509  0.6725876
> set.seed(123)
> rnorm(5)
[1] -0.4542997  0.1593962  0.8977403  1.1607478 -0.4748856
```

Note that this exercise was carried out using the WH generator for the uniforms that were required for the Box-Muller method. This generator requires three seeds. The `set.seed` function uses its single argument to create the three seeds that are needed for this particular generator and performs similar tasks for the other generator options that require multiple and/or complicated choices for their seeds.

4.9 Using the R Standalone Math Library

The R language is largely implemented in C. One consequence of this is that it is possible to create a code library that contains the C implementations of the standard functions that are used in R. These include the quantile functions, densities and distribution functions for the standard probability distributions as well as the associated random number generators. To obtain this library the following steps are required:[†]

1. Download the R source code from `http://cran.r-project.org` and save it in a directory, e.g., Rsrc.

2. From inside Rsrc enter ./configure on the shell command line.

3. If there are no errors from the previous step, enter make on the shell command line.

4. Move to the subdirectory src/nmath/standalone and again enter the make command.

If things have gone smoothly you will now have the library libRmath.a in the standalone directory. The corresponding header file Rmath.h is located in the subdirectory src/include of the directory where the R source code was downloaded. The library should be placed in the /usr/local/lib directory of the unix directory tree and the header file should be placed in the /usr/local/include directory. Subsequent compilation commands will assume that to be the case.

To evaluate the contents of the Rmath library one can peruse Rmath.h. The function prototypes that are provided there generally have exactly the same arguments in the same order as their analogs that are available from within an R session. For example, the prototype for the normal density function in Rmath.h is

```
double dnorm(double, double, double, int)
```

which lines up directly with the R prototype

```
function(x, mean = 0, sd = 1, log = FALSE)
```

that one sees by entering `dnorm` at the R command prompt. So, in this case the three `double` arguments for the `dnorm` function in the library are the evaluation point (i.e., `x`), the mean and the standard deviation, in that order. By looking at the R help file for `dnorm` it can be seen that the last `int` argument in the `dnorm` function corresponds to a logical variable for

---

[†] In some instances, (e.g., for Mac users) it may be necessary to first install the gfortran compiler.

which a value of 1 will result in the natural logarithm of the density being returned rather than the density value.

The next code listing uses the Rmath library to evaluate the distribution functions for the standard normal distribution and a chi-square distribution with one degree-of-freedom.

Listing 4.10 RmathEx1.cpp

```cpp
//RmathEx1.cpp
#define MATHLIB_STANDALONE
#include <iostream>
#include "Rmath.h"

using std::cout; using std::cin; using std::endl;

int main()
{
  double x1, x2;
  cout << "Enter an argument for the normal cdf:" << endl;
  cin >> x1;
  cout << "Enter an argument for the chi-square cdf:" << endl;
  cin >> x2;

  cout << "Prob(Z <= " << x1 << ") = " <<
    pnorm(x1, 0, 1, 1, 0)  << endl;
  cout << "Prob(Chi^2 <= " << x2 << ")= " <<
    pchisq(x2, 1, 1, 0) << endl;
  return 0;
}
```

The primary ingredient in Listing 4.10 is the use of #define MATHLIB_STANDALONE at the top of the file. This needs to be placed before the #include statement for Rmath.h because it determines how definitions will be set in the header file.

There is a question of how to determine the arguments for the pnorm and pchisq functions for evaluating the two distribution functions. To accomplish this we look at the two function prototypes in Rmath.h and find them to be

```cpp
double pnorm(double, double, double, int, int);
```

and

```cpp
double pchisq(double, double, int, int);
```

By comparing these prototypes to the descriptions on the help pages for the R functions pnorm and pchisq, it can be discerned that

a) the double arguments for pnorm are the evaluation point, mean and standard deviation. The two integer arguments are interpreted as logical variables. The first one is 1 or 0 depending on whether a lower or upper tail evaluation is desired and the second is 1 or 0 depending on whether or not the return value should be the natural log of the probability.

b) the double arguments for pchisq are the evaluation point and degrees-of-freedom. The two integer arguments are, again, logical variables with a value of 1 for the first one meaning that lower tail probabilities will be returned and a choice of 0 for the second producing probabilities as the return values.

Compilation of RmathEx1.cpp requires linking with the library libRmath.a as well as the C math library libm.a. This is accomplished as with

```
$ g++ RmathEx1.cpp -lRmath -lm -o RmathEx1
$ ./RmathEx1
Enter a argument for the normal cdf:
1.96
Enter a argument for the chi-square cdf:
3.8416
Prob(Z <= 1.96) = 0.975002
Prob(Chi^2 <= 3.8416)= 0.950004
```

The -lm and -lRmath terms in the compilation command instruct the compiler to look for the libraries libm.a and libRmath.a in the usual locations along the library search path.

As another example, let us consider use of random number generators from the R Standalone Math Library. First note that the uniform random number generator that underlies the R math library is a multiply-with-carry method due to Marsaglia (e.g., Marsaglia 2003) that has period in excess of 2^{60}.[‡] This generator requires two seeds that are set with the function set_seed through two **unsigned integer** arguments that are passed by value. The ending values for the two seeds can be retrieved by the function get_seed with arguments of type **unsigned integer*** that allows the function to write back into memory maintained by the calling program.

Listing 4.11 illustrates the use of the R math library for generating normal random deviates. Here a sample size and the two seeds are requested as input at which point the seeds for the uniform generator are initialized with set_seed. The sample is then generated and the ending seeds are retrieved and printed to standard output. The seeds are reset to their starting values and the process is repeated to illustrate that identical values will be produced by the generator.

Listing 4.11 RmathEx2.cpp

```
//RmathEx2.cpp
#define MATHLIB_STANDALONE
#include <iostream>
#include "Rmath.h"

using std::cin; using std::cout; using std::endl;

int main()
{
  unsigned int seed1, seed2, n, df;
  cout << "Enter a value for the sample size:" << endl;
  cin >> n;
  cout << "Enter a value for the first seed:" << endl;
  cin >> seed1;
  cout << "Enter a value for the second seed:" << endl;
  cin >> seed2;
  cout << "Enter a value for degrees-of-freedom:" << endl;
  cin >> df;
  set_seed(seed1,seed2);
  cout << "Here is a sample of standard normals" << endl;
  for(int i = 0; i < n; i++)
    cout << rnorm(0, 1) << " ";
  cout << endl;
  cout << "Here is a sample of chi-squares with " << df
```

[‡] A multiply-with-carry generator produces a sequence of integers via the recursion $x_n = (ax_{n-k} + r_{n-1}) \bmod m$, for specified integers a, k and m with $r_n = \lfloor (ax_{n-k} + r_{n-1})/m \rfloor$.

```
            << " degrees-of-freedom" << endl;
  for(int i = 0; i < n; i++)
    cout << rchisq(df) << " ";
  cout << endl;

  unsigned int seed3, seed4;
  get_seed(&seed3, &seed4);
  cout << "The first seed is now " << seed3 << endl;
  cout << "The second seed is now " << seed4 << endl;

  set_seed(seed1,seed2);
  cout << "Here is your sample of standard normals again " << endl;
  for(int i = 0; i < n; i++)
    cout << rnorm(0, 1) << " ";
  cout << endl;
  return 0;
}
```

The output produced by the program is

```
$ g++ RmathEx2.cpp -lRmath -lm -o RmathEx2
$ ./RmathEx2
Enter a value for the sample size:
5
Enter a value for the first seed:
123
Enter a value for the second seed:
456
Enter a value for degrees-of-freedom:
5
Here is a sample of standard normals
-0.293497 -0.334377 -0.411847 -0.346151 -0.952098
Here is a sample of chi-squares with 5 degrees-of-freedom
7.28352 9.95327 2.49237 2.96577 12.4712
The first seed is now 2386089332
The second seed is now 344712168
Here is your sample of standard normals again
-0.293497 -0.334377 -0.411847 -0.346151 -0.952098
```

4.10 Exercises

4.1. Verify identity (4.2).

4.2. Give the form of the quantile function F^{-1} in (4.1) for

a) the Weibull distribution with density

$$f(x) = \begin{cases} \frac{\alpha}{\beta^\alpha} x^{\alpha-1} \exp\left\{-\left(\frac{x}{\beta}\right)^\alpha\right\}, & x > 0, \ \alpha, \beta > 0, \\ 0, & \text{otherwise.} \end{cases}$$

b) the Pareto distribution with density

$$f(x) = \begin{cases} \alpha \frac{\beta^\alpha}{x^{\alpha+1}}, & x > \beta, \ \alpha, \beta > 0, \\ 0, & \text{otherwise.} \end{cases}$$

4.3. Give the form of the quantile function F^{-1} in (4.1) for the binomial distribution with probability mass function

$$f(x) = \begin{cases} \binom{k}{x} p^x (1-p)^{k-x}, & x = 0, 1, \ldots, k, \ 0 < p < 1, k \geq 1, \\ 0, & \text{otherwise.} \end{cases}$$

4.4. Consider the multiplicative congruential generator with $a = 12, m = 31$ and $x_0 = 9$. Does the resulting sequence of numbers represent an improvement over the case of $a = 3$? If so, why is it better?

4.5. Rework the FM2 generator code from Section 4.3 so that it will adaptively produce structured output for any given value of n.

4.6. Discuss the behavior of mixed congruential generators that have $a = 1, a = 0$ and $a = m$.

4.7. Let a, c and m be parameters for a mixed congruential generator having $x_n = (ax_{n-1} + c) \bmod m$ and let d divide m (without remainder).

a) If $y_{n-1} = x_{n-1} \bmod d$ show that $y_n = (ay_{n-1} + c) \bmod d$.

b) Use the previous result to show that when $m = 2^p$ the first r bits of x_n form a congruential sequence that has period of at most 2^r for $r > 1$ and the first bit is either constant or strictly alternating.

c) Show that the bit pattern that holds for $m = 2^p$ does not occur if $m = 2^p \pm 1$.

d) Conclude from the last two results (why?) that $m = 2^p \pm 1$ may be preferable to using $m = 2^p$.

(Knuth 1998a, Section 3.2.1.1)

4.8. Verify that the choices $m = 2^p, c = 3$ and $a = 5$ satisfy the conditions of Theorem 4.2. Can you suggest a general result based on this example?

4.9. Let a, c and m be the parameters of a mixed congruential generator with $x_n = (ax_{n-1} + c) \bmod m$. Show that

$$x_{n+k} = (a^k x_n + (a^k - 1)c/(a - 1)) \bmod m.$$

Conclude from this (why?) that the subsequence consisting of every kth term from $\{x_n\}$ represents a linear congruential generator with multiplier $a^k \bmod m$ and increment $(a^k - 1)c/(a - 1) \bmod m$.

(Knuth 1998a, Section 3.2.1)

4.10. Create an alternative version of the FM2 class where the generated numbers are held in a pointer to double that is a private member of the class. Provide an overloaded [] operator to give access to the random deviates.

4.11. Create an alternative version of class FM2 that allows for general specifications of m and a with the FM2 generator from (4.4) as the default. Other possible choices for a that might be considered when $m = 2^{31} - 1$ are $a = 742938285, 950706376, 1226874159$ and 1343714438 (Fishman and Moore 1986). Generate data using some of these choices and assess the goodness-of-fit of the uniform distribution using methods available in R.

4.12. Develop C++ code that implements a feedback shift generator for specified values of p, q and r. Do not store the bit sequence obtained from (4.5). Instead, use a p-bit bool pointer to retain the active bits that are needed for the recursion and adaptively select contiguous, nonoverlapping blocks of r bits from this bool working array. Make sure that your generator cycles through the bit sequence produced from (4.5) in a way that will allow it to realize a full period.

4.13. Apply your feedback register code from Exercise 4.12 to the case where $p = 9, q = 4$ and $r = 8$. Does the resulting generator have full period?

4.14. Verify the results that were obtained for the generalized feedback shift register generator when $p = 5, q = 2, r = 3$ and the seed is from (4.7).

4.15. Consider the use of Algorithm 4.4 for evaluation of a real number c raised to an integer power p.

Algorithm 4.4 Exponentiation algorithm

$a = 1, b = c$
while $p \neq 0$ **do**
   $r = p \bmod 2$
   $p = \lfloor p/2 \rfloor$
   if $r \neq 0$ **then**
      $a = b * a$
   else
      $b := b * b$
   end if
end while
return a

---

a) Show that this algorithm returns c^p. [Hint: Work with the binary representation for p.]

b) How many multiplications does this algorithm perform to compute c^p? [Hint: As in part a) work with the binary representation for p.] Compare this to the "naive" algorithm implemented, for example, in Listing 3.6 and the recursion based method of Exercise 3.7.

c) Implement Algorithm 4.4 in C++ and compare its performance to that of the C++ code created for Exercise 3.7.

Knuth (1998a)

4.16. Modify the RngStream.h and RngStream.cpp files to where the six seeds for the Mrg32k3a generator are obtained by supplying a single integer seed to the FM2 generator.

4.17. Refer to Exercise 3.10. Write a C++ program that will simulate from the distribution of the number of red balls selected in $n = 100$ draws and use this to approximate the distribution of the random variable to within four digits of accuracy with 95% confidence.

4.18. Derive the median and inter-quartile range for an exponential random variable with mean $1/\lambda$.

4.19. Use your results from Problem 4.2 to create a C++ class that will produce pseudo-random deviates for the Weibull and Pareto distributions.

4.20. Let X be the number of trials required to obtain a success in a series of independent Bernoulli trials. Create a C++ function that will generate a random sample from the X distribution.

4.21. Use your results from Exercise 4.3 to create a C++ class that will produce pseudo-random deviates for the binomial distribution. Provide an alternate method for this class that produces binomial random deviates using sums of Bernoulli random variables. Compare the performance of these two approaches in terms of computation time for various choices for the number of trials.

4.22. Let $S = 1, \ldots, n$ and consider the problem of selecting a random subset (e.g., for control or treatment groups) from S. Develop an algorithm that will select a k-element subset using only a single array for storage: i.e., the subset must be created in place using the integer array that holds the contents of S. [Hint: Successively swap elements in the original S array while appropriately rescaling the simulation subinterval as the subset is selected.]

4.23. Verify that Algorithm 4.3 will produce random variables that fall in the region (4.15).

4.24. Develop a C++ implementation of Algorithm 4.3 for generating gamma random deviates.

4.25. Let X have a gamma distribution with parameter $\alpha > 1$. Develop an accept-reject algorithm that will produce pseudo-random deviates for the X distribution using the exponential distribution for an instrumental density. What is the best choice for the mean of the instrumental distribution from the perspective of minimizing the expected number of iterations that is needed to produce a sample of size n?

4.26. Provide a C++ implementation of the algorithm in Exercise 4.25. How does this approach fare from a computation time perspective to the ratio based approach in Exercise 4.24?

4.27. Algorithm 4.5 is a standard for generating random deviates from a Poisson distribution with parameter λ.

Algorithm 4.5 Poisson random number generation

$sum = 0$
$X = 1$
while $sum < 1$ **do**
   Generate Y from an exponential distribution with parameter $1/\lambda$
   $sum = sum + Y$
   $X = X + 1$
end while
return $X - 1$

Show that this algorithm works in the sense that the returned value of X will behave as a pseudo-random Poisson deviate with parameter λ.

4.28. Show that the algorithm in Exercise 4.27 is equivalent to generating uniform random deviates $U_1, \ldots$ until a value of n is found where $\Pi_{i=1}^{n} U_i < e^{-\lambda}$.

4.29. Create a PRNG for the normal distribution using an accept-reject approach based on the double exponential distribution that has density

$$f(x) = (.5/\sigma) \exp\{-|x|/\sigma\}, \quad -\infty < x < \infty, \sigma > 0.$$

Derive an optimal choice for σ and compare this alternative to the one based on the Cauchy density that was developed in the text. Which produces the most rejections?
Devroye (1986)

4.30. Carry out experiments like those in Section 2.5 to evaluate the sensitivity of the Youngs/Cramer algorithm of Exercise 2.10 to round-off error. In particular create a version of Figure 2.3 showing the empirical bounds that result from your experiment.

4.31. Show how to use random samples from gamma and normal distributions to construct random samples from

a) a chi-square distribution with ν degrees-of-freedom,

b) a t-distribution with ν degrees-of-freedom,

c) an F-distribution with ν_1 numerator degrees-of-freedom and ν_2 denominator degrees-of-freedom and

d) a beta distribution.

4.32. Create a C++ "wrapper" class that implements random number generation for standard probability distributions using the R Standalone Math Library.

4.33. Create a portmanteau random number generation class that includes methods for generation of random deviates from all the distributions treated in this chapter.

4.34. Show that the random variables defined in (4.16)–(4.17) are independent and have standard normal distributions.

4.35. Show that the Box-Muller accept-reject method from Section 4.7 discards $100 \left(\frac{4}{\pi} - 1\right)$ percent of the uniform random number pairs that it generates.

4.36. Consider the problem of computing the integral

$$W = \int_B f(u_1, \ldots, u_d) du_1 \cdots du_d$$

of a square integrable function of d variables over a subset B of the unit d-cube $[0, 1]^d$. Let $U_1, \ldots, U_n$ be a random sample of size n from the uniform distribution over $[0, 1]^d$. Then, a Monte Carlo estimator of W is

$$\widehat{W} = n^{-1} \sum_{i=1}^{n} f(U_i) I_B(U_i)$$

with I_B the indicator function for the set B: i.e., $I_B(u)$ is 1 or 0 depending on whether or not $u \in B$.

a) Show that $E[\widehat{W}] = W$ and that $\text{Var}(\widehat{W}) = ||f||_B^2 / n$ with

$$||f||_B^2 = \int_B (f(u_1, \ldots, u_d) - W)^2 du_1 \cdots du_d.$$

b) A stratified sampling approach to estimation of W partitions B into k contiguous, nonoverlapping subsets $B_1, \ldots, B_k$ with $\cup_{j=1}^k B_j = B$. Random samples $U_{j1}, \ldots, U_{jn_j}, j = 1, \ldots, k$ are then taken from uniform distributions over each set in the partition that produce the integral estimator

$$\widehat{W}_S = \sum_{j=1}^{k} n_j^{-1} \sum_{i=1}^{n_j} f(U_{ji}) I_{B_j}(U_{ji}).$$

Show that $E[\widehat{W}_S] = W$ and that $\text{Var}(\widehat{W}_S) \leq \text{Var}(\widehat{W})$.

4.37. Theorem 4.7 is shown, e.g., in Ross (2006).

Theorem 4.7. *If $h(u_1, \ldots, u_n)$ is a monotone function of each of its arguments and $U_1, \ldots, U_n$ are independent random variables that are uniformly distributed on the interval [0, 1],*

$$Cov\left(h\left(U_1, \ldots, U_n\right), h\left(1 - U_1, \ldots, 1 - U_n\right)\right) \leq 0.$$

Now consider estimation of the integral

$$W = \int_0^1 f(u) du$$

for some monotone function f on $[0, 1]$ via

$$\widehat{W} = n^{-1} \sum_{i=1}^{n} f(U_i)$$

versus the antithetic variable based estimator

$$\widetilde{W} = n^{-1} \sum_{i=1}^{n} \frac{f(U_i) + f(1 - U_i)}{2}.$$

Show that $E\widehat{W} = E\widetilde{W} = W$ while $\text{Var}\left(\widehat{W}\right) \geq \text{Var}\left(\widetilde{W}\right)$.

4.38. Use the methods from Problems 4.36 and 4.37 to obtain simulation approximations of the integral $\int_0^1 x^2 dx$. For the stratified sample take a uniform partition of $[0, 1]$ into k subintervals. Repeat this experiment many times (e.g., 1000) using various values of k (e.g., 5, 10, 15, etc.) and keep track of the results from each of the methods to where you can report basic summary statistics (e.g., mean, median, standard deviation, inter-quartile range) for all three approaches at the end of your study.

Chapter 5

Programming in R

5.1 Introduction

C++ provides a powerful programming language with sufficient flexibility and built-in features to solve essentially all statistical computing problems. Nonetheless, it is not always the best or most convenient choice for addressing many of the computational issues that arise in statistics. In particular, the common C++ implementations are compiler-based which means that source code must first be parsed by a compiler into a machine interpretable form before it can be executed.

An alternative to the compiler system is provided by an immediate interpreter which consists of software that reads the program in the given language while parsing and then executing it along the way. These types of interpreted implementations have been translated into interactive programming with direct execution of commands either through the command line or by import of code from files.

R is an example of an interpreted language. The immediate response that can be realized through the R interactive setting is particularly helpful in the context of data analysis where exploration (both numerical and graphical) must generally be used in conjunction with inferential tools to obtain meaningful outcomes.

R is a functional language with object-oriented features that include a class structure. It comes equipped with native, "precompiled" classes some of which are designed for statistical analysis of data. It is also quite simple to import C++ code into R which allows one to realize the best from both worlds and, at the very least, does not preclude the use of one language over another.

Although both R and C++ have class structures, there are major differences between the OOP model of R relative to that of C++. The differences (and similarities) will be explored in more depth in the next chapter. The goal of the present chapter is to provide some of the essential language details and, in particular, to develop the tools that are needed for creating new R functions. The assumption here is that the reader is familiar with R's basic features. Some of the elementary aspects of the R language that are employed in this text are discussed in Appendix B. Since one of the primary uses for R is analysis of data, we begin in the next section by discussing some of R's input/output utilities.

5.2 File input and output

A detailed discussion of the various input/output options that are available in R is provided by the *R Data Import/Export* manual available from the CRAN website. This section provides a nonexhaustive overview of some of the most commonly used input/output tools and those that will be useful throughout the remainder of the text. More details on any of the R functions that appear in this section or elsewhere can be found by using the `help` or `?` commands with the function's name as the argument.

R has analogs of the basic Unix commands for managing and exploring a system's directory tree. These include `file.create`, `file.remove`, `file.append` and `file.copy` that, respectively, create a new file, remove an existing file, append one file (the second function argument) to another file (the first function argument) and create a copy of a file. These functions take file paths as arguments and otherwise assume that the named file or files reside in the current working directory. In this regard, the R functions getwd and setwd that determine and reset, respectively, the location of the current working directory are described in Appendix B.

The operations performed by file.create, file.remove and file.copy can also be obtained with the Unix commands cat, rm and cp discussed in Appendix A. The primary difference is that the former functions can be called directly from inside an R session while the latter commands require use of the system function from Appendix B to pass a command to a shell.

Let us begin by discussing the import of data into R. For the moment, assume that the data of interest resides in a text file fileName in the directory pathToFile and that the information in the file is formatted as a rectangular array with columns representing variables and the rows corresponding to the observations. Then, the simplest option is to proceed via the command

```
> A <- read.table(file = "pathToFile/fileName")
```

This reads in fileName as a data frame and then the assignment operator <- places the result in the data frame object A. The data is presumed to be white-space delimited. If this is not the case, another delimiter can be specified using the sep argument for read.table. For example,

```
> A <- read.table(file = "pathToFile/fileName", sep = "\t")
```

would produce the same outcome if fileName were tab rather than white-space delimited. Once A has been imported into R, it may be edited in a spreadsheet-type environment via the command edit(A).

To illustrate a few of the features of read.table consider the group1.txt file shown below.

```
$ cat group1.txt
25  26  30  38
25  34  29  25
31  24  28  39
27  20  25  29
31  22  28  34
35  29  20  40
```

This rectangular array, or portions of it, can be imported directly via read.table. To read in the entire file we can use

```
> A <- read.table("group1.txt")
> A
   V1 V2 V3 V4
1  25 26 30 38
2  25 34 29 25
3  31 24 28 39
4  27 20 25 29
5  31 22 28 34
6  35 29 20 40
```

By setting the nrows argument of read.table to 3 only the first three rows are imported as in

```
> A <- read.table("group1.txt", nrows = 3)
> A
   V1 V2 V3 V4
1  25 26 30 38
2  25 34 29 25
3  31 24 28 39
```

Only the name of the file was supplied to `read.table`. This works because the target file group1.txt is in the current, as viewed by R, working directory. If that had not been the case, it would have been necessary to provide the path to the file.

Notice from the example that the default action is to use the letter V along with the column numbers as names for the variables and integers for the row names. There are various ways this behavior can be customized. For example, to assign specific names to the rows and columns of the group1.txt data, one might use

```
> read.table("group1.txt",col.names = paste("Sample", 1:4,
+ sep = ""), row.names= paste("row", 1:6, sep = ""))
     Sample1 Sample2 Sample3 Sample4
row1      25      26      30      38
row2      25      34      29      25
row3      31      24      28      39
row4      27      20      25      29
row5      31      22      28      34
row6      35      29      20      40
```

There are two aspects of this operation that merit some additional comment. First, and foremost, this example illustrates how names can be given to the rows and columns by supplying values for the `col.names` and `row.names` arguments for `read.table`. Both arguments can be assigned an array of character values of the same length as the number of columns or rows. The other interesting feature is how the specific character vectors were constructed in this case as will now be explained.

The `paste` function provides a very convenient tool for creating character variables from a mix of numeric and character values. It performs a character conversion and concatenation operation wherein all the elements supplied as arguments are converted to character mode and then concatenated along with white-space separation into a single character object. Thus, for example,

```
> paste("Sample", 1)
[1] "Sample 1"
```

is a way to combine the string "Sample" and the integer 1 to create the string "Sample 1". The problem with this particular choice for, e.g., a column name for the group1.txt data is the presence of an embedded white space. In R syntactically valid names can use only letters, digits, dots or underline characters and must start with a letter or a dot that is not followed by a digit. Names which meet this requirement in our case can be obtained by simply avoiding blank spaces. This is readily accomplished by setting the `sep` argument for `paste` as "". For example, we could construct

```
>   v <- c(paste("Sample", 1, sep = ""), paste("Sample", 2, sep = ""),
+ paste("Sample", 3, sep = ""), paste("Sample", 4, sep = ""))
> v
[1] "Sample1" "Sample2" "Sample3" "Sample4"
```

and then supply v as the value for `col.names` in `read.table` to produce the desired results. Our particular approach uses an additional shortcut that exploits the way `paste` handles vectors; i.e., vector arguments are concatenated term-by-term to return a character vector. Consequently,

```
> paste("Sample", 1:4, sep = "")
[1] "Sample1" "Sample2" "Sample3" "Sample4"
```

creates the desired character vector in one command by concatenating "Sample" with the vector

```
> 1:4
[1] 1 2 3 4
```

produced by the sequence operator : from Appendix B.

Instead of assigning names to the rows and columns of an array on input, they may be specified in the text file that contains the data. For our running example, this might take the form of the group2.txt file listed below.

```
$ cat group2.txt
"Sample1" "Sample2" "Sample3" "Sample4"
"Row1" 25 26 30 38
"Row2" 25 34 29 25
"Row3" 31 24 28 39
"Row4" 27 20 25 29
#missing data in next 2 rows

"Row5" 31 n 28 34
"Row6" NA q 20 NA
```

Here the variable/column names are specified in a header and each row is given its own name in the file. If, as is the case here, the header contains one fewer entry than the first row, read.table automatically proceeds as if the first column contains row names.

There are three other new features that appear in the group2.txt file. First, there is a comment inserted as designated by the R comment symbol #. A blank line appears after the comment. We will see that read.table will ignore both comments and empty lines.

Of somewhat more substance is the presence of missing values in group2.txt. Missing data are a common occurrence that arise from errors, nonresponse, etc. A language such as R that provides data analysis tools must have effective means of dealing with missing observations. In this regard, many R functions have an na.action or na.rm argument that can be used to determine how missing values will be treated.

The default string that read.table uses for missing values is NA. Other values may also be specified through the argument na.string that can be given a vector of character values. To read in the group2.txt file one could therefore use

```
> A <- read.table("group2.txt", na.string = c("n", "q", "NA"))
> A
     Sample1 Sample2 Sample3 Sample4
Row1      25      26      30      38
Row2      25      34      29      25
Row3      31      24      28      39
Row4      27      20      25      29
Row5      31      NA      28      34
Row6      NA      NA      20      NA
```

Notice that the comment and blank line were ignored and that the symbols for all missing values were translated to the R logical constant NA. Operations can now be performed on the A data frame as usual provided the missing values are handled appropriately. For example, a rote application of the mean function to our imported data produces

```
> mean(A)
 Sample1  Sample2  Sample3  Sample4
      NA       NA 26.66667       NA
```

By instead setting the na.rm argument for mean to TRUE the missing values will be omitted from the calculations resulting in the output

```
> mean(A, na.rm = TRUE)
 Sample1   Sample2   Sample3   Sample4
27.80000  26.00000  26.66667  33.00000
```

The workhorse function that underlies `read.table` is `scan`. This function provides great flexibility in reading input at the expense of generally having to manually reshape the data that it imports. One handy use of `scan` is for reading command line input. To accomplish this `scan` is used with no arguments at which point data may be entered directly until the process is terminated by input of a blank line. As an illustration, a segment from an R session is shown where `scan` was used to place the integers from 1 to 10 in a numeric object v.

```
> v <- scan()
1: 1
2: 2 3
4: 4 5 6 7
8: 8 9 10
11:
Read 10 items
> v
 [1]  1  2  3  4  5  6  7  8  9 10
```

More generally, an abbreviated prototype for scan is

```
scan(file, what, n)
```

The `file` argument is the name of the file to be read with a path specification if it does not reside in the current working directory. The `n` argument determines the number of values to be read from the file; the entire file will be read when this argument is not specified. The `what` argument indicates the type of data that will be read and defaults to `double(0)`. The seemingly innocuous syntax for this default conveys more than it may seem and provides a key point of access to the power and flexibility of `scan`.

As noted in Section 2.7, the primitive data types in R include `character`, `integer` and `logical` as well as `numeric` that is also equivalent to `double`. These data types are also classes in R with constructors that are invoked through use of the class name. In particular, we see from

```
> integer(2)
[1] 0 0
> logical(1)
[1] FALSE
> numeric(3)
[1] 0 0 0
> character(2)
[1] "" ""
```

that calling one of these constructors with an integer n as its argument will produce a vector of length n consisting of the default values for that data type. The special cases `character(0)`, `integer(0)`, `logical(0)` and `numeric(0)` (or `double(0)`) produce length zero or "empty" objects of type `character`, `integer`, `logical` and `numeric`, respectively. Thus, we can understand the specification of `what = double(0)` in scan as being a means of indicating that the data to be read will be of type `numeric` and, similarly, that using the other options `character(0)`, `integer(0)` or `logical(0)` will serve the same basic purpose.

The following illustrates how `scan` can be used to read the data from our group1.txt file.

```
> A <- scan("group1.txt", what = integer(0))
Read 24 items
> storage.mode(A)
[1] "integer"
> A
 [1] 25 26 30 38 25 34 29 25 31 24 28 39 27 20 25 29 31 22 28 34 35 29 20 40
```

By setting `what` equal to `integer(0)` the data has been imported as type `integer` rather than `double`. Unfortunately, the rectangular structure in the file has been lost while it was maintained by the `read.table` function. The row/column form can be restored directly using the subscripting ability of the R language: e.g., with

```
> A <- cbind(A[seq(1, 20, by = 4)], A[seq(2, 20, by = 4)],
+ A[seq(3, 20, by = 4)], A[seq(4, 20, by = 4)])
```

that also uses the `seq` and `cbind` functions discussed in Appendix B.

A more direct route to preserving the structure in an input file is available through the `what` argument for `scan`. In general, `what` can be given a list of types that correspond to the different columns in the input file. For our example, this might take the form

```
> A <- scan("group1.txt", what = rep(list(integer(0)),4))
Read 6 records
> A
[[1]]
[1] 25 25 31 27 31 35

[[2]]
[1] 26 34 24 20 22 29

[[3]]
[1] 30 29 28 25 28 20

[[4]]
[1] 38 25 39 29 34 40
```

In specifying the value for `what` we used the R replication function `rep`. The prototype for this function looks like

```
  rep(x, times)
```

with x an R object and `times` the number of times the object is to be replicated. The `rep` function returns an object of the same type as its x argument containing `times` copies of x. For our case, using `what = rep(list(integer(0)),4)` is equivalent to

```
> scan("group1.txt", what = list(integer(0), integer(0), integer(0),
+ integer(0)))
```

Also, note that the object A returned by `scan` in this instance is a `list` which entails that its component integer vectors must be accessed using the `[[]]` operator. The matrix form for the data could be recovered with, e.g.,

```
> A <- cbind(A[[1]], A[[2]], A[[3]], A[[4]])
```

While `scan` reads one data value at a time from a file, the `readLines` function will read the file in one line at a time. An application of `readLines` to the group1.txt file produced

```
> A <- readLines("group1.txt", n = 2)
> A
[1] "25 26 30 38" "25 34 29 25"
```

Here the parameter n that determines the (maximum) number of lines to read has been set to 2 so that only the first two lines are read from the file. The readLines function is designed to read text input and, as a result, returns each line from the file as a quoted character string. In this particular case the result is a two-element character array. There are R functions for dealing with strings that can be used to convert the string input into numeric values as we will now demonstrate.

We have already considered the paste function that can be used to concatenate strings and numbers as a string. It is the opposite of this operation that is needed now; the input has been read as a string containing integers separated by white spaces and we need to extract the integer substrings. This can be accomplished with the strsplit function as demonstrated by

```
> strsplit(A, split = " ")
[[1]]
[1] "25" "26" "30" "38"

[[2]]
[1] "25" "34" "29" "25"
```

The desired result has been obtained by specifying the split argument for strsplit as a white space. The output from strsplit has been returned in a list structure. To convert the lists of character variables to integer values requires an application of the unlist function and coercion/casting of the result to integer type; e.g.,

```
> as.integer(unlist(strsplit(A[1], split = " ")))
[1] 25 26 30 38
```

Then,

```
> rbind(as.integer(unlist(strsplit(A[1], split = " "))),
+   as.integer(unlist(strsplit(A[2], split = " "))))
     [,1] [,2] [,3] [,4]
[1,]   25   26   30   38
[2,]   25   34   29   25
```

reproduces the two rows that were read from the group1.txt file. As an aside, it is worth mentioning that R also has functions nchar and substr that count the number of characters (including white spaces) and remove or replace a substring in a string.

The previous example marks our first exposure to the ubiquitous (in R) as function that performs the R analog of the C++ casting operations. The R term for this process is *coercion*. In general, an application of as.dataType to an object will cause R to disregard its current data type and instead view the object as being of data type dataType provided such a view is possible. The as.integer function we just employed is one example that can be used to have character strings, numeric (i.e., double precision) values, etc., viewed as being integer values. There are many choices for dataType with as.matrix (that coerces or casts an object into a matrix object) and as.Date (that we will use to have character strings treated as objects from the R Date class) representing examples that will be encountered in this chapter. An application of the function methods with argument as produces a list of over 100 associated functions.

It is also possible to import files with more "exotic" origins into R. As explained in Appendix B there are many add-on packages that have been written for R. Among these is the foreign package that contains functions for both reading and writing files that correspond to the statistics software packages Minitab, S-Plus, SAS, SPSS, Stata and Systat. A significant omission from this list is Excel which, although not a statistics package per se, is nonetheless a standard for storage and manipulation of data. Direct access of Excel files is apparently somewhat problematic from R. Rather than attempting to do so directly, the advice is to instead export the data from Excel in an R readable text format. For example, an Excel spreadsheet can be saved in a .csv (comma separated values) format and then read in via `read.table` with `sep = ","`. A specialized version of `read.table` named `read.csv2` is available that works specifically on .csv files.

Data that are created in R can be retained from session to session by saving and then reloading a workspace image as described in Appendix B. The end result will still be that the data in the workspace are loaded into memory having the consequence that special techniques are needed for R to work with very large or massive data sets. One option is to work directly with relational databases created by some database management system. The R packages RMySQL, ROracle, RPostgreSQL, RSQLite, DBI and RODBC can be used for this purpose. The filehash package discussed in Section 9.2.4.3 provides another option.

There is an output version of `read.table` called `write.table` that can be used to output a data frame or matrix to a text file. For example,

```
> write.table(A, "pathToFile/fileName")
```

writes a data frame or matrix A to a text file fileName in the directory pathToFile complete with row and column headings. The command

```
> write.table(A, "pathToFile/fileName",col.names = FALSE,
+ row.names = FALSE)
```

will write A to fileName sans the R row and column designations. An excerpt from an R session gives an illustration of both of these options.

```
> A <- read.table("group2.txt", na.string = c("n", "q", "NA"))
> write.table(A, "temp.txt")
> system("cat temp.txt")
"Sample1" "Sample2" "Sample3" "Sample4"
"Row1" 25 26 30 38
"Row2" 25 34 29 25
"Row3" 31 24 28 39
"Row4" 27 20 25 29
"Row5" 31 NA 28 34
"Row6" NA NA 20 NA
> write.table(A, "temp.txt", col.names = FALSE, row.names = FALSE)
> system("cat temp.txt")
25 26 30 38
25 34 29 25
31 24 28 39
27 20 25 29
31 NA 28 34
NA NA 20 NA
```

The output analog of `scan` is `cat`. This function outputs its arguments in a general sense that includes writing to files. An example of this feature is provided by

```
> A <- read.table("group1.txt")
> cat(as.matrix(A), "\n", file = "temp.txt")
> system("cat temp.txt")
25 25 31 27 31 35 26 34 24 20 22 29 30 29 28 25 28 20 38 25 39 29 34 40
```

The group1.txt file has been imported as a data frame using `read.table`. Then, `cat` is used to write it to the text file temp.txt. The fact that `cat` does not work with lists means it cannot be used directly with a data frame. To account for this, the data frame object A was converted to a `matrix` object using the `as.matrix` version of the `as` function. The Unix version of `cat` then reveals that the R `cat` function has ignored the array structure of the data when writing it to a file. The use of the linefeed character "\n" in the R `cat` function signals the end of a line so that the Unix `cat` will produce a carriage return when it is applied to temp.txt.

The function `write` is a `cat` "relative" for (primarily) writing `matrix` objects to files while retaining their array structure. In terms of our group1.txt example, using `write` produces results such as

```
> A <- read.table("group1.txt")
> write(t(as.matrix(A)), file = "temp.txt", ncolumns = 4)
> system("cat temp.txt")
25 26 30 38
25 34 29 25
31 24 28 39
27 20 25 29
31 22 28 34
35 29 20 40
```

The fact that `write` is based on `cat` makes it necessary to transpose the matrix `as.matrix(A)` (using the `t` function) in order to have the contents of the output file look like its R representation. Both `write.table` and `write` have a logical `append` argument that defaults to FALSE. To write to an existing file and preserve its existing contents, the `append` argument can be set to TRUE.

An issue that has so far been overlooked is the precision that R provides for output of data. The `write.table` function writes real numbers to the maximum possible precision. In contrast, `write` uses the precision of the current R session as can be seen from

```
> write.table(cos(1:3), file = "temp.txt", col.names = FALSE,
+ row.names = FALSE)
> system("cat temp.txt")
0.54030230586814
-0.416146836547142
-0.989992496600445
> getOption("digits")
[1] 7
> write(cos(1:3), file = "temp.txt")
> system("cat temp.txt")
0.5403023 -0.4161468 -0.9899925
> options(digits = 14)
> write(cos(1:3), file = "temp.txt")
> system("cat temp.txt")
0.54030230586814  -0.41614683654714  -0.98999249660045
```

Here the values of the cosine function at 1, 2 and 3 have been written to the file temp.txt using both `write.table` and `write`. Using `write` gives only seven significant digits of accuracy. This results from the `digits` option for the current environment being set at 7 as determined by the `getOptions` function. To alter the precision for `write`, one can change the `digits` options for the current environment as was done here to produce 14 digits of precision in the output. To see all the current option settings enter `options()`.

The use of `cat` is not restricted to writing data to files. It provides one means of writing output from, e.g., functions to (R) standard output as illustrated by

```
> cat("a", "b", 1:3, "\n")
a b 1 2 3
```

The linefeed character is also necessary here to produce a prompt on a new line for subsequent input. Note that the quotes have been dropped from the characters that appear in the output. In contrast, the R `print` function could have been used to give

```
> print(c("a", "b", 1:3))
[1] "a" "b" "1" "2" "3"
```

Quoted output is the default option in this latter instance. To suppress this behavior one can set the `quote` argument to `print` as `FALSE`.

The `save.image` function was discussed in Appendix B as a means to save an R workspace for later use. Actually, this is a special case of the more general R function `save` that writes a representation of an R object to a file where it can later be accessed using the `load` function. The following group of commands demonstrates how this works with a simple array of integers.

```
> A <- 1:5
> B <- 6:10
> save(A, B, file = "AB.RData")
> rm(A, B)
> temp <- load("AB.RData")
> temp
[1] "A" "B"
> A
[1] 1 2 3 4 5
> B
[1] 6 7 8 9 10
```

After creating the arrays A and B, an R decipherable representation for the two vectors is stored in the file AB.RData and the original arrays are destroyed using the remove function `rm`. The information about A and B is then successfully recovered from the file using the `load` function. Note that the return value from `load` is an array whose elements are character strings giving the name of the objects that were loaded.

In some instances it is useful to output code or functions that have been written inside of R, e.g., for editing. A command such as

```
> dump("functionName", "pathToFile/fileName")
```

can be used to accomplish this. (Note that the quotes around `functionName` are necessary here.) This will write a text representation for the function `functionName` into the file fileName that will be created (if it does not already exist) in the directory pathToFile. To recover or read such a file the command is `source`. So, in this instance

```
> source("pathTofile/fileName")
```

brings the contents of fileName into the current R workspace. This feature is designed to work with R code and the input file's content will also be parsed to check for syntax errors.

The output from an R session or program can be collected in a text file with the `sink` function. A condensed prototype for this function is

```
sink(file, append, split)
```

The `file` argument is the name of the file that will hold the output while `append` and `split` are logical variables that both default to `FALSE`. A specification of `append` as `TRUE` means that the output will be appended to any that already exists in the output file while a choice of `TRUE` for split means that output will be sent to standard output as well as the designated file. The use of `sink` is illustrated in the next example.

```
> sink("temp.txt")
> acos(0)
> sink()
> sink("temp.txt", append = TRUE, split = TRUE)
> cos(acos(0))
```

```
[1] 6.123234e-17
> sink()
> system("cat temp.txt")
[1] 1.570796
[1] 6.123234e-17
```

Initially, the output is written only to the target file temp.txt after which the connection to the file is closed by invoking `sink` with no arguments. Output is then appended to the temp.txt file while being echoed to the command line.

It is also possible to open a *connection* to a file that allows for reading, writing or appending of input and output. This approach is treated in depth in Chapter 10 of Chambers (1998) and Section 4.3 of Gentleman (2009). Using a connection is more efficient, e.g., when data is being repeatedly read from or written to a file. The `read.table` and `write.table` functions actually open connections to the files provided as their arguments that are closed upon conclusion of their assigned reading or writing task. More generally, a connection can be given, rather than a file name, for the `file` argument for these functions as well as for `scan` and `cat`.

The `file` command opens a connection that remains open for reading, writing or both until it is explicitly closed. In its simplest form an application of `file` might appear like

```
connectionName <- file("pathTofile/fileName", open = "r")
```

This would open a connection to fileName in the directory pathToFile for reading as a result of the r specification of the `open` argument. To write or append to fileName, `open` can be specified as w or a, respectively. A connection is closed with

```
close(connectionName)
```

where `connectionName` is the name that has been assigned to the connection being closed.

The `file` function was used along with the simple text file group1.txt to produce

```
> A <- read.table("group1.txt")
> write.table(A, file = "temp.txt", col.names = FALSE,
+ row.names = FALSE)
> fIn <- file("temp.txt", open = "r")
> read.table(file = fIn, nrows = 1)
  V1 V2 V3 V4
1 25 26 30 38
> scan(file = fIn, n = 3)
Read 3 items
[1] 25 34 29
> close(fIn)
> fOut <- file("temp.txt", open = "a")
> write.table(A[1,], file = fOut, col.names = FALSE,
+ row.names = FALSE)
> write.table(A[2,], file = fOut, col.names = FALSE,
+ row.names = FALSE)
> close(fOut)
> system("cat temp.txt")
25 26 30 38
25 34 29 25
31 24 28 39
27 20 25 29
31 22 28 34
35 29 20 40
25 26 30 38
25 34 29 25
```

First the data in group1.txt is written to the file temp.txt and then a connection is opened for reading (i.e., `open = "r"`) from this new file. A single line is read from the file using `read.table` followed by an application of `scan` to read three more elements. The output from `scan` reveals that new read operations begin at the point in the file where the previous read operation concluded. After closing the input connection `fIn` with the `close` function, `file` is used again to open an output connection for appending (i.e., `open = "a"`) to the temp.txt file. The first two lines of the file are then appended to its end and the connection is closed. The output illustrates that the append operations were successful and that, like reading, writing operations are performed sequentially starting at the end of the most recent insertion into a file.

5.3 Classes, methods and namespaces

R comes equipped with a large number of precompiled classes and associated methods. The command `class(objectName)` can be used to determine the class of an object `objectName`. To obtain a list of all classes that are available in the current R workspace use the command `getClasses()`.

Every R object has two *attributes*: namely *length* and *mode*. The meaning of the length characteristic is self-explanatory for most classes. Values for mode indicate storage types that include `numeric`, `logical`, `function`, `expression` and `list`. To determine the length or mode of an object `A` use `length(A)` and `mode(A)`.

As in C++, objects in R are created using R supplied constructors. The language does have a `new` command with, e.g.,

```
> A <- new("matrix")
> A
<0 x 0 matrix>
```

having the effect of creating a new matrix object `A` of the default 0×0 size. The use of `new` will be explored in the next chapter. For now the focus will be on using the class constructors that are already available for pre-existing classes.

The class constructors can be expected to have the same name as the class although that is not strictly necessary. The arguments that are needed for a constructor, or any other function, can be determined by using the `args` function with the name of the function of interest as its argument: e.g.,

```
> args(matrix)
function (data = NA, nrow = 1, ncol = 1, byrow = FALSE, dimnames = NULL)
NULL
```

More generally, in most cases the entire function listing is shown if only its name is entered on the command line. Noteworthy exceptions include arithmetic operators. The `get` function can also provide useful information as in

```
> get("%%")
function (e1, e2)  .Primitive("%%")
```

The output indicates that `%%` is a binary function (in view of the two arguments `e1` and `e2`). The presence of the `.Primitive` function indicates that `%%` is implemented within the internal C code on which R is based. This and the related `.Internal` function will be seen on occasion in R function listings. Both functions are the province of R developers and, as such, fall outside the realm of this text.

The previous output from `args` tells us that the `matrix` class constructor has five arguments: `data`, `nrow`, `ncol`, `byrow` and `dimnames` whose default values are `NA`, 1, 1, `FALSE` and `NULL`. We have already discussed the `NA` constant. The reserved words `TRUE` and `FALSE` are the R Boolean constants. As in C++, `TRUE` and `FALSE` have numerical values of 1 and

0. The default choice of NULL for the dimnames argument corresponds to another reserved word in R that represents a null object. A "value" of NULL is often used to initialize an object that will be assigned a meaningful value elsewhere in a program.

The argument list for the matrix constructor tells us the information that can be specified to create a nondefault matrix object. The purpose of the first three arguments is relatively clear with data representing a collection of one or more real numbers that are to provide the entries in the matrix, nrow being the number of rows and ncol the number of columns. Thus,

```
> A <- matrix(data = 0, nrow = 3, ncol = 3)
```

produces a 3×3 matrix of all zero entries. The same effect is obtained from

```
> A <- matrix(0, 3, 3)
```

which shows that explicit use of the argument names is not necessary provided they are entered in the same order they appear in the function definition.

The byrow argument for the matrix class constructor is used to specify whether elements of the data argument are to be used to fill in the matrix object by rows (when byrow = TRUE) or by columns (when byrow = FALSE).* Finally, dimnames is an argument giving names for the rows and columns of the matrix that will be discussed in more detail below.

As an example, the command line entry

```
> A <- matrix(scan("myFile"), ncol = 3, byrow = TRUE)
```

would read a three-column data set with an arbitrary number of rows from the file myFile in the current directory and place it in the matrix object A. Note that the scan function treated the data in myFile as a one-dimensional array of length 3×nrow. The array was then reshaped via the matrix class constructor.

Some of the other standard R classes with their associated constructors are listed below.

- Class vector: vector(mode, length), with mode an R storage type such as character, list, logical and numeric.

- Class data frame: data.frame(object). Here object must have components that are numeric vectors, matrices, lists or other data frames.

- Class expression: expression("expression")

- Class factor: factor(x, levels)

- Class formula: formula(object). The object in this case is generally an R expression with a call to the ~ operator that is used to separate the left- and right-hand sides in model formulae.

- Class lm (i.e., linear models): lm(formula, data)

The constructors generally have additional arguments beyond those that have been listed here that can be found, e.g., on the corresponding R help pages.

The list class provides another important R data structure where objects are obtained by combining elements from various classes. A function in R is allowed to return only one object as output.† As a result, lists turn out to be a key aspect of the language that provide a vehicle for bundling together different objects that might be produced in an analysis into a single object (namely, a list) that can be returned at the end of computation.

---

* The default choice of FALSE for byrows stems from the fact that matrices in R are stored in *column major order*. In contrast, C/C++ uses *row major order*. To see the difference, let $a_1, a_2, \ldots, a_{mn}$ be a sequence of numbers that comprise the elements of an $m \times n$ array. With column major order the (i, j)th element of the matrix will be $a_{(j-1)m+i}$ while for row major order it will be $a_{(i-1)n+j}$.

† This statement is not entirely true in that the R return function can return multiple arguments (separated by commas) that it automatically packages in a list structure. But this practice is discouraged with a warning message.

The code segment below creates two lists and gives names to their elements.

```
> a <- list("a", "b")
> names(a) <- c("a1", "a2")
> b <- list(b1 = "c", b2 = "d", b3 = "e")
> names(b)
[1] "b1" "b2" "b3"
> class(a[1])
[1] "list"
> class(a[[1]])
[1] "character"
> class(a$a1)
[1] "character"
```

The names for the list objects have been assigned in two ways: using the `names` function and directly in the `list` class constructor. As seen from this example, list elements may be accessed via the dereferencing operator `[[]]` (which is always possible) or by their names using `$` if names have been assigned. Dereferencing in this manner, returns the actual object that was stored in the list. Note that using the `[]` operator with a list, instead of `[[]]`, will return a sublist that corresponds to the specified indices. In this particular case, `a[1]` is a one-element list while `a[[1]]` is the `character` object that provides the sole component of the `a[1]` list.

Let us now return to the discussion of the `dimnames` argument for the `matrix` class. This is a list that consists of two character strings giving the names for the rows and columns of the matrix. The list can be supplied as an argument on creation of a matrix or specified after the fact via the `dimnames` function. This is illustrated by the next example where a 2×3 matrix is created and `dimnames` are assigned using both approaches.

```
> dimA <- list(c("R1", "R2"), c("C1", "C2", "C3"))
> A <- matrix(1:6, 2, 3, dimnames = dimA)
> dimnames(A)
[[1]]
[1] "R1" "R2"

[[2]]
[1] "C1" "C2" "C3"

> dimnames(A) <- list(c("C1", "C2"), c("R1", "R2", "R3"))
> A <- t(A)
> A
   C1 C2
R1  1  2
R2  3  4
R3  5  6
> A["R2", "C1"]
[1] 3
```

Notice in this example that using the `dimnames` function without an assignment returns the current value of `dimnames` for the object supplied as its argument. The values for `dimnames` are seen to appear in the display of the matrix `A` that was created in the example and were used to access the elements of the matrix by supplying them as arguments to the subscripting operator `[]`. It is also seen that `dimnames` really does correspond to the names of dimensions through the way they adjust automatically to the replacement of the matrix `A` with its transpose `t(A)`.

As will be seen in the next chapter, the class concept in R closely parallels that of C++ in terms of the way it deals with data members. The case of "member" functions is another matter. In a certain sense, there is no direct analog of the C++ member functions or methods for an R class. That is, functions are not segregated as belonging to only a single class. Instead, a function is itself an object with an independent existence that is free of such restrictive ties. This allows for the creation of something called *generic functions* where the R language interpreter determines which function is appropriate by examining the types of the arguments in the function call. There are two systems that produce generic functions in R: S3 and S4. Both of these will be discussed in Chapter 6.

For similar reasons to those for C++, R has a namespace system that is used by some packages. More generally, any R package can include variables with names that can mask those from other packages that are being used in a current R session. The way to resolve such name clashes is with the R scope/namespace resolution operator : : that provides access to exported variables from a package via syntax of the form

```
packageName::name
```

with `name` the name for a variable or function and `packageName` the name of the package where `name` resides. The term *exported variables* refers to variables/functions that a package creator has made available for package users.

5.4 Writing R functions

We now have enough background to begin to write our own R functions. An R function definition will have the basic form

```
functionName <-function(arguments){
     statements and expressions
}
```

A value can be explicitly returned from any point within a function (thereby terminating computation) by using a `return` statement similar to what transpires in C++. Otherwise, the return value is whatever is produced by the last line of the function definition. As an illustration, the function below takes a matrix argument `A` and returns a list containing the product of `A` with its transpose `t(A)` and the number of rows of the matrix.

```
myFunc <- function(A){
  list(A%*%t(A),length(A[,1]))
}
```

To apply this function to a particular matrix B we would use `myFunc(B)`.

The body of a function will generally consist of expressions using the basic R operators and class objects in conjunction with various execution control structures created by `for` and `while` loops and `if` and `if/else` statements. The `for` loop in R looks like

```
for(varName in sequence){
     statements and expressions
}
```

with `sequence` representing a vector of values that `varName` can assume. In contrast, a `while` loop has the form

```
while(testExpression){
     statements and expressions
}
```

with `testExpression` an expression that evaluates to `TRUE` or `FALSE`.

Additional control over loop iteration can be obtained by using `break` and `next` statements. When `break` is encountered the entire loop will terminate while `next` moves the loop to the next iteration step. The `repeat` construct provides a simple loop structure that involves no logical conditions. It must be terminated with a `break` statement.

Conditional execution statements have the usual form in R: e.g., either

```
if(testExpression){
        statements and expressions
}
```

or

```
if(testExpression){
        statements and expressions
   } else {
        statements and expressions
   }
```

There is also an R analog of the C/C++ `switch` statement that was mentioned in Section 3.4. It is discussed in Section 2.6 of Venables and Ripley (2000) and the corresponding R help page. The `ifelse` function in R is a vectorized relative of the C/C++ conditional operator (cf. Section 3.4). It has the syntax

```
   ifelse(testExpression, yes, no)
```

with `yes` and `no` vectors providing values that will be returned when the elements of the vector `testExpression` of Boolean variables evaluate to `TRUE` or `FALSE`, respectively. The values in `yes` and `no` will be recycled if either vector is too short.

As in C++ the curly braces in a loop or conditional expression are not needed if the corresponding code is a single line: e.g.,

```
x <- 3
for(i in 1:3)
   if(x < 5) x <- x + 1
```

will run without error. Of somewhat more importance is the syntax involving `else` in an `if/else` block. The `else` term must be on the same line as the closing brace for the preceding `if` phrase in an interactive session. For example, consider

```
> x <- 5
> if(x < 5){
+ x <- x + 3
+ } else
+ x <- sqrt(x)
> x
[1] 2.236068
```

that will replace the value of a variable `x` with the square root of `x + 3` or the square root of `x` depending on whether or not its value is less than 5. If we forget to place the `else` term on the same line as the last curly brace the outcome is

```
> x <- 5
> if(x < 5){
+ x <- x + 3
+ }
> x
[1] 5
```

Entering a new line after the last brace produces a syntactically complete expression and the interpreter has no way of knowing that there is more code yet to come. So, it simply executes the expression that it has to that point and returns the command prompt indicating that it is done.

Listing 5.1 illustrates some of the ideas that were discussed above.

Listing 5.1 *sumFunc.r*

```
#sumFunc.r
sumFunc <- function(x){
  n <- length(x)
  x <- as.integer(x)
  sum <- 0
  for(i in 1:n){
    if(!x[i]%%3)
      sum <- sum + x[i]
    else sum <- sum + 3*x[i]
    if(sum > 300){
      print(paste("Your sum of", sum, "exceeds 300"))
      break
    }
  }
  sum
}
```

The function in Listing 5.1 takes a vector as input and then performs some meaningless calculations that produce a return value that has the property of, for example, being an integer multiple of 3. The first step in the process is to determine the length of the input vector and convert all its elements to integers. Then, the program steps through all the elements of the transformed vector adding them to a running sum if they are 0 modulo 3 (i.e., if they are divisible by 3 without remainder) and adding in three times their value otherwise. If the sum exceeds 300 execution is terminated via a **break** statement. Some output from the sumFunc is

```
> sumFunc((10 + pi):20)
[1] 270
> sumFunc((30 + pi):50)
[1] "Your sum of 387 exceeds 300"
[1] 387
```

A somewhat more complicated (and meaningful) example is in Listing 5.2. This function carries out unprotected pairwise comparisons between group means using pooled standard deviation estimators. It takes as input a matrix (or data frame) whose columns correspond to samples from different populations. The program then employs the R **mean** and **var** functions discussed in Appendix B to compute the sample means and variances that correspond to the columns of the input array. Setting the logical parameter **na.rm** equal to **TRUE** causes the **mean** and **var** functions to ignore (or *remove*) missing values. The function is.na is applied to each column of the input matrix to count the number of missing observations in each sample. This returns vectors of Boolean variables having the same length as the input arguments. The entries in the vectors evaluate to **TRUE** for every **NA** that is encountered and are **FALSE** otherwise. The **sum** function treats the Boolean variables as numeric with value 0 for **FALSE** and 1 for **TRUE** with the result that summing such a vector produces the number of **TRUE** values. In this particular case the number of **TRUE**s is the number of missing values.

Listing 5.2 *meanComp.r*

```
#meanComp.r
meanComp <- function(A){
  B <- vector(length = ncol(A)*(ncol(A) - 1)/2)
  count <- 1
  for(i in 1:(ncol(A) - 1)){
    for(j in (i + 1):ncol(A)){
```

```
    n1 <- nrow(A) - sum(is.na(A[, i]))
    n2 <- nrow(A) - sum(is.na(A[, j]))
    m1 <- mean(A[, i],na.rm = TRUE)
    s1 <- var(A[, i],na.rm = TRUE)
    m2 <- mean(A[, j], na.rm = TRUE)
    s2 <- var(A[, j], na.rm = TRUE)
    n <- n1 + n2 - 2
    sdFactor <- sqrt(n1^{-1} + n2^{-1})
    sPooled <- sqrt(((n1 - 1)*s1 + (n2 - 1)*s2)/n)
    mDiff <- m1 - m2
    B[count] <- mDiff/(sPooled*sdFactor)
    count <- count+1
  }
 }
 B
}
```

An application of `meanComp` to the data from the group2.txt file discussed in Section 5.2 produces output that suggests there are significant differences between the sample mean for the fourth sample's parent "population" and those of the other three.

```
> A <- as.matrix(read.table("group2.txt",
+ na.string = c("n", "q", "NA")))
> meanComp(A)
[1]   0.5982930   0.5502351  -1.7391402  -0.2231713  -1.7602455  -2.1686961
```

The format used in printing the output from `meanComp` is the default for the R installation that was used for this example. This will suffice for our purposes here and elsewhere. More control over formatting can be obtained with the R functions `format`, `formatC`, `prettyNum` and `sprintf`.

The previous example could have been coded more directly using R's `apply` function from Appendix B and the next section. There are, of course, a number of pre-existing methods for carrying out mean comparisons that are either native to R or accessible by downloading the package multcomp. For example, the `TukeyHSD` function (available in the R base package that comes with the initial R installation) performs mean comparisons using the "Honest Significant Difference" method. It works with, e.g., an object returned from the R function `aov` that carries out an analysis of variance.

The R language has an exception or *condition* mechanism that bears some similarity to the C++ version discussed in Section 3.8.7. The R help page on this subject provides one source of information while the text by Gentleman (2009) gives an accessible introduction to the topic.

The simplest entry point to the R exception management facility is through the `tryCatch` function. Exceptions/conditions in R come in the form of `message`, `warning` and `error` objects. These are instances of classes that provide information about a condition that occurred in the execution of a program. In its simplest form, the `tryCatch` function is comprised of an expression to be evaluated and one or more handler functions that define the action or actions to be taken when a particular condition occurs. If there are no conditions generated by evaluating the expression, the results of the evaluation are returned. Otherwise, a search is made to find a handler that matches the condition.

One place where errors might occur is on data input. These could come in the form of errors in the data (Exercise 5.5) or perhaps an incorrect file name. For example, the excerpt below from an R session shows the results of an unsuccessful attempt to read the group1.txt file from Section 5.1.

```
> scan("grop1.txt", quiet = TRUE)
Error in file(file, "r") : cannot open the connection
In addition: Warning message:
In file(file, "r") :
  cannot open file 'grop1.txt': No such file or directory
```

Both a **warning** and an **error** object are generated by our spelling error. The `tryCatch` function can be used to recover from this mistake by "catching" either of the two conditions and instituting an appropriate fix. A function that does this is given below.

```
readFile <- function(fileName){
  out <- tryCatch(suppressWarnings(scan(fileName)),
       error = function(e){
    print(e)
    print("Please enter a file name followed by a return")
    newName <- scan( , what = character(0), quiet = TRUE)
    readFile(newName)})
  out
}
```

Here the use of `scan` is encapsulated inside `tryCatch`. The `suppressWarnings` function has been used so that only **error** (and not **warning**) condition objects will be generated when problems are encountered. These will be handled by a function that i) prints the **error** object, ii) asks for and then uses `scan` to acquire a new (hopefully) valid file name and then iii) attempts to read using the new name with a recursive call to `readFile`. The `quiet` argument for `scan` is set to `TRUE` when reading the new file name to suppress the output it would otherwise have produced. In the context of the group1.txt example this gives the desired result: namely,

```
> A <- readFile("grop1.txt")
<simpleError in file(file, "r"): cannot open the connection>
[1] "Please enter a file name followed by a return"
1: "group1.txt"
2:
Read 24 items
```

Let us again consider the `dimnames` example of the previous section. There the `dimnames` function was apparently used on a matrix `A` in two forms: to wit,

```
> dimnames(A)
```

and

```
> dimnames(A) <- list(c("C1", "C2"), c("R1", "R2", "R3"))
```

The first form merely returns the `list` object that contains the `dimnames` for `A` without altering the matrix object. In contrast, the second actually modifies `A` by changing its `dimnames` to the value on the right-hand side of the `<-` operator. This raises two questions: how can one function perform two quite different operations and how can a function that supposedly passes by value, as would be expected in a functional language, alter its arguments. The answer is that appearances have (possibly) fooled us. First, there are two functions at work: the `dimnames` function that returns the `dimnames` list and a *replacement* function `dimnames<-` that assigns values to the list elements. The replacement "version" of dimnames then reformulates its operation so that the altered matrix is the output, rather than input, of a function call thereby preserving the functional language paradigm.

The underlying syntax for the replacement form of `dimnames` that we used is

```
> A <- "dimnames<-"(A, value = list(c("C1", "C2"),
+ c("R1", "R2", "R3")))
```

That is, this is the basic R translation of the command. The quotes are necessary to keep the R interpreter from attempting to parse the <- symbol.

The `dimnames` function provides a model for how to write replacement functions. The requirements are that

a) the function name (in quotes) must end with <-,

b) the final argument of the replacement function must be named `value` and

c) the replacement function must return the object that it modifies.

The succeeding code uses this approach to create a function that replaces a vector by a trimmed version where only values from a specified middle percentage of the data are retained.

```
"trim<-" <- function(x, value){
    x <- x[(quantile(x, value) <= x & x <= quantile(x, 1 - value))]
    x
}
```

The R function `quantile` is used to compute the 100(value) and 100(1 - value) percentiles from the input data vector x. Then, the subset of values that fall between these two percentiles is extracted using the subsetting operator with a logical vector argument.

An example of using `trim` is given below.

```
> set.seed(123)
> x <- rnorm(10)
> mean(x, trim = .1)
[1] 0.03703159
> trim(x) <- .1
> x
[1] -0.56047565 -0.23017749  1.55870831  0.07050839  0.12928774  0.46091621
[7] -0.68685285 -0.44566197
> mean(x)
[1] 0.03703159
```

A pseudo-random sample of 10 standard normals is generated and the 10% trimmed mean is computed using the R mean function with `trim` argument equal to .1. Our `trim` replacement function is then used to trim off 10% of the values from each end of the data; i.e., only the middle 80% of the data are retained. An application of the `mean` function now produces the same value as for the 10% trimmed version of `mean`.

The R language includes a function `system.time` that can be employed for run-time calculations much like the `clock` function in C++. One calls `system.time` by providing it with the expression to be evaluated as its argument. The output is an object of class `proc_time` whose first three members are total user CPU time, system CPU time of the current R process and elapsed time since the process was started. In terms of our `meanComp` function from Listing 5.2 we can produce results such as

```
> set.seed(123)
> A <- matrix(rnorm(5*10^5), 10^5, 5)
> system.time(meanComp(A))
   user  system elapsed
  0.132   0.018   0.150
```

The `meanComp` function was applied to five samples of 100,000 observations from a standard normal distribution and the analysis was performed in .15 seconds.

As noted in Section 5.2, syntactically valid names in R consist of letters, digits, periods or underline characters and must start with a letter or period that is not followed by a digit. Apart from these conditions, any name is fair game. This allows for the possibility of using a pre-existing name for an R constant or function. The result is that the new function's name will mask that of the corresponding R function. All is not lost, however, as the masking takes place only in the current R environment. The hidden function can still be accessed using the scope resolution operator from the previous section with the name of the package where the masked function resides. Suppose, for example, that a spelling challenged statistician creates a version of R's built-in `sign` function that returns the sign of its numeric argument such as.

```
> sin <- function(x) ifelse(x < 0, -1, 1)
```

A test of the new function produces

```
> sin(3); sin(-3); sin(pi)
[1] 1
[1] -1
[1] 1
```

(Note the use of a semicolon here to place more than one expression on a command line.) The output reveals that the function is operating "correctly" and also that we have lost the ability to directly use R's built-in sine function. The sine version of `sin` resides in the `base` package for R and must now be called using the namespace resolution operator with

```
> base::sin(pi)
[1] 1.224647e-16
```

After examining a dictionary and the *Introduction to R* manual, our statistician might decide to remove his/her redundant function from the workspace. This can be done by applying `rm` to the name that needs to be removed: e.g.,

```
> rm(sin)
> sin(pi)
[1] 1.224647e-16
```

In some instances it is useful to suppress the printing of output. The `invisible` function provides a means to create this effect. One merely applies `invisible` to the object or objects that will be returned by a function; the output will then be returned but not printed. The function below represents another attempt at recreating R's `sign` function.

```
f <- function(x){
  if(x < 0) return(-1)
  if(x > 0) return(1)
  0
}
```

Output from using this function might look like

```
> f(-1); f(1); f(0)
[1] -1
[1] 1
[1] 0
```

In contrast, a reworking of f as

```
f <- function(x){
  if(x < 0) return(invisible(-1))
  if( x > 0) return(1)
  invisible(0)
}
```

uses `invisible` in strategic locations to produce

```
> f(-1); f(1); f(0)
[1] 1
> 2*f(-1)
[1] -2
```

There are at least three ways to physically create the code for an R function. The first is to use the command line. We have already seen that in this setting R gives a continuation "+" prompt until you have entered a syntactically complete expression (e.g., a complete function definition including the closing braces). The second is to use the dump and source functions discussed in Section 5.2. The third method is to use the edit command which connects to text editors that are available on the operating system.

The fact that R has the same basic looping and conditional statements as C++ means that it is possible to write C++ type code in R. This is generally not the best option as there are many built-in R functions to perform standard (and nonstandard) tasks that can provide more compact and faster code. Somewhat more generally, developing procedural-type code (such as from C/C++) in R runs counter to the functional nature of the language wherein one seeks (existing, if possible) functions that can be used in composition to solve a problem of interest. Illustrations of this will be seen in the next two sections and in the exercises.

5.5 Avoiding loops in R

Many of the functions in R have internal workings that rely on C or Fortran code. The loops in such lower-level languages are generally faster than explicit R loops and, consequently, more efficient code can often be produced by replacing explicit R loops with expedient calls of existing R functions. To do so it is necessary to know

a) the right functions to select and

b) how to apply them.

There is no easy fix for part a) in that R is a rich language that comes equipped with many functions. There are numerous texts about R and R packages whose study can increase one's R function "vocabulary". Examination of the source code for R functions can also be a useful learning experience. There is somewhat more that can be said about part b) and this section explores that issue through discussion of the apply family of functions and the outer and vectorize functions.

The apply function is discussed in Section 1 of Appendix B. It applies a specified function to the rows or columns of an array. There are also lapply and sapply functions that operate on lists. These two functions have the same basic prototype: e.g.,

```
lapply(X, FUN)
```

with X being a list object and FUN the function that is to be applied to its components. The difference is in the output; lapply always returns a list while sapply will return a vector or matrix object if possible. Results from an R session illustrate this difference.

```
> A <- list(1:10, 20:30)
> lapply(A, FUN = sum)
[[1]]
[1] 55

[[2]]
[1] 275

> sapply(A, FUN = sum)
[1]  55 275
```

The R sum function that was used here is discussed in Section 1 of Appendix B.

When the elements of a list are themselves lists, a recursive apply function rapply can be used for which an abbreviated description is

```
rapply(object, f)
```

The `object` argument is a list of lists and `f` is the function to be applied. The use of `rapply` is demonstrated below.

```
> A <- list(list(1:10, 20:30), list(list(40:50, 60:70), 80:90),
+   100:110)
> rapply(A, f = sum)
[1]    55   275   495   715   935  1155
```

Note that the action of `rapply` on a non-list object is just to apply `f` (or `sum` in this case) to the object while it continues to recursively work on any list of lists it encounters inside its `object` argument.

The mean comparison example of the previous section is one instance where `apply` and `sapply` can be used to produce more efficient code than a loop implementation. With this in mind we wrote a function to carry out pairwise mean comparisons.

Listing 5.3 *meanCompVec.r*

```
#meanCompVec.r
meanCompVec <- function(A){
  indexList <- list()

  #fill indexList with all the index pairs
  count <- 0
  for(i in 1:(ncol(A) - 1)){
    for(j in (i + 1):ncol(A))
      indexList[[count + j - i]] <- c(i, j)
    count <- count + ncol(A) - i
  }
  #compute the statistics for the columns
  meanVec <- apply(A, MARGIN = 2, FUN = mean, na.rm = TRUE)
  varVec <- apply(A, MARGIN = 2, FUN = var, na.rm = TRUE)
  nVec <- apply(A, MARGIN = 2, FUN =
         function(e) length(e) - sum(is.na(e)))

  #now compute the t statistics
  sapply(indexList, FUN = function(e){
    n1 <- nVec[e[1]]
    n2 <- nVec[e[2]]
    mDiff <- meanVec[e[1]] - meanVec[e[2]]
    n <- n1 + n2 - 2
    sdFactor <- sqrt(n1^{-1} + n2^{-1})
    sPooled <- sqrt(((n1 - 1)*(varVec[e[1]]) +
        (n2 - 1)*(varVec[e[2]]))/n)
    mDiff/(sPooled*sdFactor)
  } )
}
```

The first step in Listing 5.3 is to create a vector of list objects `indexList` that holds the index pairs for the columns of `A` that will be used to carry out the comparisons. For example, when the data matrix `A` has five columns the `indexList` array looks like

```
> list.tree(indexList)
 indexList = list 10 (392 bytes)
.  [[1]] = integer 2= 1 2
.  [[2]] = integer 2= 1 3
```

```
.  [[3]]  = integer  2= 1 4
.  [[4]]  = integer  2= 1 5
.  [[5]]  = integer  2= 2 3
.  [[6]]  = integer  2= 2 4
.  [[7]]  = integer  2= 2 5
.  [[8]]  = integer  2= 3 4
.  [[9]]  = integer  2= 3 5
.  [[10]] = integer  2= 4 5
```

The `list.tree` function used here performs pretty printing for lists. It is available through the Hmisc package that, as its name suggests, contains an eclectic collection of useful functions for data analysis, graphics, etc.

The next step in the `meanCompVec` function is to compute the means, variances and sample sizes for the columns of the data matrix. In all three cases this is accomplished with `apply`. In the first two instances, `apply` is used on the columns of `A` with the existing R functions `mean` and `var`. To get the sample sizes we supply our own function that computes the number of nonmissing entries for each column of `A` in much the same manner as in Listing 5.2. The function is applied to the columns of `A`.

Once all the means, variances and sample sizes are available, `sapply` is used to perform the standard calculations for pooled variance t-tests. Our choice for `FUN` uses the elements of `indexList` directly to obtain the indices to use in the mean comparisons.

Another useful function for avoiding explicit R loops is `outer`. The name derives from its application to the computation of outer products of vectors or matrices. Given vectors a and b of dimensions n and m, respectively, their outer product is defined as the $n \times m$ matrix whose (i,j)th entry is the product of the ith entry of a and the jth entry of b. The outer product of two vectors in R is written as `a%o%b`. For example, the 10 by 10 multiplication table may be generated by invoking the standard outer product operator

```
1:10%o%1:10
```

In fact, `%o%` is just a wrapper for the more general function `outer` that has prototype

```
outer(X, Y, FUN)
```

with X and Y typically vector or array objects and FUN some specified function. The effect of `outer` is to carry out an analog of `%o%` on the X and Y objects wherein multiplication is replaced by the operation specified in FUN. Thus, the output of `outer` is a matrix of dimension `length(X)` $\times$ `length(Y)` having elements of the form `FUN(X[i], Y[j])`. In particular, choosing `X = Y = c(1:10)` and `FUN = "*"` reproduces the multiplication table that was previously obtained via `%o%`.

In Listing 5.3 an array of index pairs for the columns of the data matrix was created that was then used to determine which means were to be compared. An alternative spin on our mean comparison example demonstrates how `outer` provides a means of bypassing explicit construction of such an index array.

Listing 5.4 *meanCompOuter.r*

```
#meanCompOuter.r
meanCompOuter<- function(A){

  #compute the statistics for the columns
  meanVec <- apply(A, MARGIN = 2, FUN = mean, na.rm = TRUE)
  varVec  <- apply(A, MARGIN = 2, FUN = var, na.rm = TRUE)
  nVec    <- apply(A, MARGIN = 2, FUN =
          function(e) length(e) - sum(is.na(e)))
```

```
#now compute the t statistics
outer(1:ncol(A), 1:ncol(A), FUN = function(i, j){
  n1 <- nVec[i]
  n2 <- nVec[j]
  mDiff <- meanVec[i] - meanVec[j]
  n <- n1 + n2 - 2
  sdFactor <- sqrt(n1^{-1} + n2^{-1})
  sPooled <- sqrt(((n1 - 1)*(varVec[i]) +
      (n2 - 1)*(varVec[j]))/n)
  mDiff/(sPooled*sdFactor)
}) [upper.tri(matrix(, ncol(A), ncol(A)))]
}
```

The vectors of sample means, variances and sample sizes are constructed exactly as in Listing 5.3. The difference is that the indices for the columns that will be used in the comparisons are provided by all the pairs of values from the first two arguments to `outer`. Since both of these are the integer arrays `1:ncol(A)`, all possible pairwise comparisons will be conducted. This is wasteful in terms of computational effort and also returns an array that contains redundant entries. The latter problem is easily solved using the `upper.tri` function that takes a matrix argument and returns an array of logical values of the same dimension with above diagonal entries set to TRUE and all other entries set to FALSE. The `lower.tri` function performs a similar operation where only the below diagonal entries evaluate to TRUE. When the output from `upper.tri` is given as an argument to the subsetting operator applied to a matrix of the same dimension, the outcome will be an array consisting of just the upper diagonal elements of that matrix.

To illustrate how `upper.tri` works, consider the output below that was produced with our `meanCompOuter` function.

```
> set.seed(123)
> A <- matrix(rnorm(5*100000), 100000, 5)
> meanCompOuter(A)
[1]  -0.9460884   0.3518758   1.2972025   0.0913167   1.0365320  -0.2603067
[7]   1.5338896   2.4765512   1.1820542   1.4414838
```

Mean comparisons were performed on the columns of a 100000×5 matrix of pseudo-random numbers generated from the standard normal distribution. The output corresponds to the TRUE entries in the matrix

```
> upper.tri(matrix(, ncol(A), ncol(A)))
        [,1]    [,2]    [,3]    [,4]    [,5]
[1,]  FALSE    TRUE    TRUE    TRUE    TRUE
[2,]  FALSE   FALSE    TRUE    TRUE    TRUE
[3,]  FALSE   FALSE   FALSE    TRUE    TRUE
[4,]  FALSE   FALSE   FALSE   FALSE    TRUE
[5,]  FALSE   FALSE   FALSE   FALSE   FALSE
```

When processing this array the [] operator works by columns. Thus, for example, the difference between the means of the second and fifth columns could be viewed as statistically significant with an approximate upper-tail p-value of

```
> 1 - pnorm(2.4765512)
[1] 0.00663293
```

The only hitch in application of `outer` is that FUN must be ready to take two vectors as arguments and carry out the necessary pairwise operations. That is, FUN must be "vectorized". The basic R arithmetic operators $(+, -, *$ and $/)$ have been implemented in a vectorized manner. Thus,

```
exp(outer(log(1:10), log(1:10), "+"))
```

will produce once more the 10 by 10 multiplication table. On the other hand, calling

```
outer(1:5, 1:5, max)
```

results in

```
Error in dim(robj) <- c(dX, dY) :
    dims [product 25] do not match the length of object [1]
```

because `max`, given multiple arguments, simply iteratively computes the maximum of all entries of all the arguments and returns a single value.

One way to vectorize a function such as `max` is to use R's function `mapply` that has prototype

```
mapply(FUN, ...)
```

The action produced by `mapply` is to apply `FUN` to the first elements of each argument given in the place of the ellipsis (i.e., the ...), then to the second elements of each argument, etc. In particular, for our `max` example this produces

```
> outer(1:5, 1:5, FUN = function(i, j) mapply(max, i, j))
      [,1] [,2] [,3] [,4] [,5]
[1,]    1    2    3    4    5
[2,]    2    2    3    4    5
[3,]    3    3    3    4    5
[4,]    4    4    4    4    5
[5,]    5    5    5    5    5
```

Thus, `max(i, j)` with i fixed at 1 was applied to every component of the array 1:5, then the process was repeated with i fixed at 2, etc. We should note that in this instance there is a more direct route that can achieve the same goal using the existing R function `pmax` (and an analogous `pmin` for computing minima) that computes the "parallel maxima" of the input values.

The function `Vectorize` provides a way of creating a new function that will behave as if `mapply` were being used. It has the somewhat simplified (by us) form

```
Vectorize(FUN, vectorizeArgs)
```

where `FUN` is the name of the function to be vectorized and `vectorizeArg` is a character vector containing the names of the arguments that will be vectorized. The default value for `vectorizeArg` is all the arguments to `FUN`. To conclude this section, we will use this approach to create a function that performs mean comparisons using R's `t.test` function.

An abbreviated prototype for `t.test` looks like

```
t.test(x, y = NULL, var.equal = FALSE)
```

For two-sample problems both the x and y numeric vector arguments can be supplied to obtain a t-statistic for comparing their respective means. If the `var.equal` argument is set to `TRUE`, the output component `statistic` will be the value of a pooled t-test statistic such as the ones we have been using in this and the previous section.

The R session fragment shown below creates a wrapper function for the pooled variance version of the `t.test` function that applies it to the columns of a matrix A that are indicated by its arguments. This function is vectorized and then used in `outer` for the analysis of the same data matrix as in the previous example with the results

```
> set.seed(123)
> A <- matrix(rnorm(5*100000), 100000, 5)
> pooledT <- function(i, j) t.test(A[, i], A[, j],
```

```
+ var.equal = TRUE)$statistic
> vecPooledT <- Vectorize(pooledT)
> outer(1:5, 1:5, vecPooledT)[upper.tri(matrix(, ncol(A), ncol(A)))]
 [1] -0.9460884  0.3518758  1.2972025  0.0913167  1.0365320 -0.2603067
 [7]  1.5338896  2.4765512  1.1820542  1.4414838
```

Given that we now have so many ways to carry out paired comparisons, it would seem worthwhile to assess their relative performance in terms of computing time. One such comparison is provided by

```
> set.seed(123)
> A <- matrix(rnorm(5*10^5), 10^5, 5)
> nreps <- 1000
> timeVec <- rep(0, 4)
> for(i in 1:nreps){
+ timeVec[1] <- timeVec[1] + system.time(meanComp(A))[3]
+ timeVec[2] <- timeVec[2] + system.time(meanCompVec(A))[3]
+ timeVec[3] <- timeVec[3] + system.time(meanCompOuter(A))[3]
+ timeVec[4] <- timeVec[3] + system.time(outer(1:5, 1:5,
+ vecPooledT)[upper.tri(matrix(, ncol(A), ncol(A)))])[3]
+ }
> timeVec/nreps
[1] 0.154999 0.047227 0.046631 0.046870
```

The three functions `meanComp`, `meanCompVec` and `meanCompOuter` along with the vectorized pooled t-test function `vecPooledT` were all applied 1000 times to the same data set and their respective elapsed computation times were accumulated. The resulting average times suggest there is little reason to prefer any of the three methods created in this section over the other. However, they all carried out their task over three times faster than the function `meanComp` that uses explicit R loops.

5.6 An example

In this section we will work our way through an example that demonstrates code development in the R environment. The objectives are twofold: i) creation of an artificial data set that will provide the basis for examples in Chapters 6 and 10 and ii) construction of the tools we will need to analyze the data that is created.

The scenario is that of a fictional game called Guess5 where a lottery organization draws five balls from a set of 40 without replacement. People who play the game purchase tickets that specify five numbers between 1 and 40 and win various prizes depending on how many of the numbers they choose match those drawn by the lottery.

The physical drawings for Guess5 are conducted using two different drawing machines referred to as A and B and 10 different ball sets numbered 1 to 10. Drawings are conducted twice a week on Monday and Thursday. The drawing machine and ball set that are used for a particular draw are chosen randomly with equiprobable selection prior to draw time. Before the actual draw a series of nine diagnostic test draws are conducted to check for possible problems with the selected machine or ball set.

With this background, we now want to write an R function that will generate data that could have come from drawings for the Guess5 game. For this purpose it is first necessary to decide on what information to record for each draw, a format for storing the data and then on a mechanism for generating the data that will be stored. In terms of the first issue some of the information we might want to retain would be the date of the drawing, whether or not the drawing was on a Monday or Thursday, the machine and ball sets that were used for the draw and the actual drawing results. By retaining information about the drawing machines and ball sets it will be possible to investigate the performance of the individual components of the drawing equipment. Similarly, including the date and day of the week for a draw will enable us to address questions related to game history.

As the data "arrives" it will be collected in a white-space delimited format with the date, machine name, ball set number and day of the week (M or T for Monday or Thursday, respectively) being given on the first line of the record for a given draw. The next nine lines will contain the results of the preliminary test draws with the eleventh (and final) line representing the actual game drawing results. This sequential accumulation perspective suggests that the data be generated in a `for` loop. Given what we learned in the previous section, this is a bit of a step backward into the C++ mindset. A little thought leads us to the conclusion that, since the end result is an artificial data set, we are free to produce it in any way we wish provided it has the same probabilistic structure as would be expected from a sequence of game draws. In this respect, the machine selection, ball set selection and ball selections are all assumed to be independent which allows us to generate these outcomes separately and then combine them to produce the actual data set. This is the direction in which our code development will proceed.

The function below employs the perspective of the previous section to produce the desired simulated data for the Guess5 game.

Listing 5.5 guess5.r

```
#guess5.r
guess5 <- function(n, fOut, drawDate = "1952-1-17"){
  if(weekdays(as.Date(drawDate)) == "Monday")
    drawDate <- as.Date(drawDate) +
        c(0, cumsum(rep(c(3, 4), ceiling((n - 1)/2),
        length.out = n - 1)))
  else
    drawDate <- as.Date(drawDate) +
        c(0, cumsum(rep(c(4, 3), ceiling((n - 1)/2),
        length.out = n - 1)))

  Machine <- sample(c("A", "B"), n, replace = TRUE)
  Set <- sample(1:10, n, replace = TRUE)
  Day <- substr(weekdays(drawDate), start = 1, stop = 1)
  info <- cbind(as.character(drawDate), Machine, Set, Day)
  A <- replicate(n*10, sample.int(40, 5))
  for(i in 1:n) {
    cat(info[i, ], "\n", file = fOut, append = TRUE)
    write(A[, (10*i - 9):(10*i)], file = fOut, ncolumns = 5,
        append = TRUE)
  }
}
```

The function `guess5` has three arguments:

a) `n`, the number of draws that are to be generated,

b) `fOut`, the name of the file where the data will be written and

c) `drawDate`, the day, month and year that the drawing sequence is to begin.

Perhaps the first thing to notice about this listing is the use of a default for the date argument. From this you can see that the assignment of defaults works much the same as in C++. Although, in R there are no rules about the ordering of default-specified and nondefault-specified arguments in an argument list.

The next new feature in Listing 5.5 is the coercion or casting of `drawDate` to a `Date` object.

The `Date` class in R allows for operations involving addition, subtraction and certain logical relation operators (e.g., $<, >, ==$) that make it possible to manipulate dates in useful ways. Our initial use of the `Date` class is to determine the day of the week for the input value of `drawDate`. For this purpose, the character string `drawDate` is first coerced to a `Date` object using the `asDate` function. Then, the `weekdays` function is applied to the output to ascertain if `drawDate` corresponds to a Monday or Thursday. If the starting day is Thursday, then the next draw will not occur for four days as compared to a three day interval for a starting date that falls on a Monday. An `if/else` block is employed to deal with the two different possibilities.

Once the day of the week has been determined, the overloaded `+` operator for the `Date` class is used to augment the value of `drawDate` for the other drawings that will be included in the data set. The effect of adding an integer to a `Date` object is to produce a new `Date` object that is advanced by the same number of days as the value of the integer. R vectorized addition applies here as well; adding a vector of integers to a `Date` object will produce a vector of `Date` objects that have been moved forward by the number of days in the integer vector.

For our particular case, the number of days that will pass between draws is either three or four and this must accumulate over the course of the drawing process. So, in the case of, e.g., a Monday start date, we first add 0 to ensure that the start data is the same as that specified in `drawDate`. Then, the object `as.Date(drawDate)` is advanced according to the vector

```
cumsum(rep(c(3, 4), ceiling((n - 1)/2), length.out = n - 1))
```

There are several steps involved in explaining this code segment and we will go through them one-by-one. First, the `rep` function discussed in Section 5.1 is applied to the vector (3, 4). The number of replications is taken to be the smallest integer that is at least half of the sample size `n` less 1. If the sample size is odd this gives the desired result. For even sample sizes, only one of the two elements from the last replications of (2, 3) is needed. To appropriately truncate the output from `rep` in this case an additional argument `length.out` for `rep` has been set to n - 1. As a result, the vector returned from `rep` will be no longer than n - 1 elements. The idea is illustrated below where `rep` was used in this context with and without specification of the `length.out` argument for n = 5 and n = 6.

```
> rep(c(3, 4), ceiling(4/2), length.out = 4)
[1] 3 4 3 4
> rep(c(3, 4), ceiling(5/2))
[1] 3 4 3 4 3 4
> rep(c(3, 4), ceiling(5/2), length.out = 5)
[1] 3 4 3 4 3
```

As seen from the previous example, `rep` will return a vector consisting of the alternating integers 3 and 4. These tell us how many days will pass between each draw. To translate that into actual dates, the increments must be taken cumulatively; i.e., if a drawing occurs on a Monday then 3 days pass before the Thursday draw, $3 + 4 = 7$ days pass before the next Monday draw, 10 days pass before the subsequent Thursday draw, etc. As suggested by this progression, the way to augment the initial value of `as.Date(drawDate)` is to add on the cumulative (partial) sums of the days between the n draws that will make up the data. The `cumsum` function produces such partial sums; given a `numeric` vector v, `cumsum(v)` is a vector of the same length as v with its ith entry equal to the sum of the first i elements of v. A combination of `cumsum` and `rep` then produces the requisite vector of `Date` objects: e.g.,

```
> weekdays(as.Date("1952-01-17"))
[1] "Thursday"
> as.Date("1952-01-17") + cumsum(rep(c(4, 3), ceiling(5/2),
+ length.out = 5))
[1] "1952-01-21" "1952-01-24" "1952-01-28" "1952-01-31" "1952-02-04"
```

Once the draw dates have been created in Listing 5.5, the next step is to generate the vectors of machine and ball set selections. Again, not having to actually mimic the drawing process leaves us free to generate these two arrays directly using the `sample` function from Section 3 of Appendix B with the populations consisting of the characters "A" and "B" and the integers from 1 to 10, respectively.

The day for each draw is implicitly given by the date and can therefore be extracted using the `weekdays` function. Rather than store the entire name for a day, only the characters "M" and "T" are retained. They are picked off from the string that holds the day's name with the `substr` function. The `start` and `stop` arguments for this function define the position of the beginning and ending character to be used in creating the substring. Thus, setting both `start` and `stop` as 1 will return the first character in the string.

The vectors `drawDate`, `Machine`, `Set` and `Day` are combined into an n-row array called `info` using the `cbind` function. The rows of this array must then be paired with the actual drawing data. This latter aspect of the simulation is accomplished by a combination of the `replicate` and `sample` functions: namely,

```
A <- replicate(n*10, sample.int(40, 5))
```

The `replicate` function is a member of the `apply` family that can be used for repeated evaluation of an expression. Its prototype has the form

```
replicate(n, expr)
```

where `expr` is the expression to be evaluated and `n` is the requested number of evaluations. For our application, `n` is 10 times the total number of drawings: 9 test draws and 1 game draw occur every time the game is played. The actual numbers for the balls that are drawn are produced using the function `sample.int`. This is a specialization of the `sample` function that has the basic form

```
sample.int(n, size, replace)
```

This version of `sample` assumes that the population being sampled is the integers from 1 to its first argument `n`. The `replace` argument defaults to **FALSE** as required for our case.

The only remaining task is writing the contents of `info` and `A` into an output file. The two arrays must be meshed correctly with each row of `info` being paired with 10 columns of `A`. This process is carried out directly with a `for` loop. Execution of our `guess5` function produces results such as

```
> set.seed(123)
> guess5(100, "guess5.txt")
NULL
> system("head -n 22 guess5.txt")
1952-01-17 A 6 T
10 38 23 20 15
36 15 11 7 37
20 10 9 25 2
29 14 16 31 34
12 38 28 26 2
16 19 22 26 33
25 17 21 3 10
16 8 32 6 29
```

```
22 26 7 24 12
29 16 37 36 27
1952-01-21 B 4 M
11 9 23 10 20
32 7 16 18 40
38 35 26 36 19
24 14 39 1 19
35 1 3 7 28
30 38 18 3 24
31 6 16 9 3
16 3 9 39 25
12 4 3 33 28
33 39 4 38 29
```

The presence of NULL in this output is a bit of a curiosity. The last value returned from an R function will be the result of the last expression evaluation. The return value from cat (and, hence, write) will be seen as NULL and that will be printed to standard output when the function ends. This can be suppressed using the invisible function (Exercise 5.13).

We will develop a framework for dealing with Guess5 data that allows us to look at subsets of the drawing data corresponding to different machines, ball sets, etc., in the next chapter. For the present, let us focus on the somewhat more direct problem of analyzing the data we have stored in guess5.txt in its entirety. From the standpoint of players of the Guess5 game the question of interest is whether the selection of ball numbers is uniform. By this we mean that every one of the $\binom{40}{5} = 658008$ possible combination of five balls chosen from a set of 40 without replacement has the same chance of occurrence on any given draw. A test of this hypothesis using, e.g., a chi-square statistic would require a collective history of millions of draws to be of value thereby precluding that option from practical consideration (Exercise 5.12). Instead, it is possible to carry out tests of hypotheses about various marginal distributions whose rejection would imply rejection of the overall uniform selection hypothesis (Joe 1993). The simplest of these concerns the individual ball selection probabilities which should be 5/40 per draw. A chi-square test can be used for this purpose and that is the approach that will be implemented.

The first step in the analysis of data that has been collected from Guess5 draws would be to import it into R in a form that is suitable for the planned analysis. If the aim is to calculate a chi-square test statistic using all the draws, only the drawing results are needed; that is, the information about drawing equipment and dates is extraneous.

For the sake of this example let us now adopt the perspective that the guess5.txt file represents collected data from Guess5 drawings that is the focus of our analysis. As the equipment/date information is not needed, one way to import guess5.txt is to simply read all the drawn numbers into a long vector while filtering out the information about date, ball set, machine and day of the week. This can be accomplished with a for loop that uses the skip argument for the scan function.

Assume that the number of draws in guess5.txt (i.e., 100) is known. By using scan to read from the file with skip = 1 and nlines, the number of lines to read, set equal to ten, the first line of the file (with all the date and equipment details) will be skipped and only the results for the 10 draws (nine tests and one game draw) will be imported. This takes care of the data for the first drawing in the file. The location for the next application of scan has now moved to the first line of data for the second draw which, again, contains date and equipment information. This line can now be skipped and the next 10 lines read exactly as before. The process can be repeated as many times as there are draws in order to bring in all the data. The code segment below implements this approach and checks the length of the vector that was created to hold the data.

```
> drawData <- NULL
> for(i in 1:100)
+ drawData <- c(drawData, scan("guess5.txt", skip = 1, nlines = 10,
+   quiet = TRUE))
> length(drawData)
[1] 5000
```

Note the use of the NULL object to initialize the drawData array. Analogous results would be obtained using a class constructor for either the vector or matrix class with an empty argument list: e.g., drawData <- vector(). Failure to initialize drawData would have caused an error the first time the concatenation operator was used in the loop because at that point its drawData argument would not exist.

The fact that guess5.txt contains a mix of character and numeric data and rows of unequal length is what precludes the use of read.table for importing the file. Our for loop formulation is perhaps the most obvious solution (at least to a C/C++ programmer). Alternatively,

```
> drawData <- as.integer(unlist(strsplit(readLines("guess5.txt")
+ [-seq(1, 1090, 11)], split = " ")))
```

avoids explicit looping and, instead, relies on the inherent features of R. To examine this approach, let us work our way out from the interior of the expression. As noted in Section 5.2, the readLines function brings in the content of a line in a file as an array of character strings. Thus, the effect of

```
readLines("guess5.txt")
```

is to bring in a string vector with 1100 rows corresponding to the 1100 ($= 11 \times 100$) rows in the guess5.txt file. Now seq(1, 1090, 11) produces the integer collection $\{1, 11, \ldots, 1090\}$ that correspond to the rows of guess5.txt where the date and other nonnumeric drawing information reside. Prepending – to these indices removes them from the array created with readLines. So, at the completion of the

```
readLines("guess5.txt")[-seq(1, 1090, 11)]
```

command, an array containing 10000 ($= 1100 - 100$) strings has been created with each string containing the five integers produced by a drawing for the Guess5 game. The remainder of the import command uses strsplit, unlist and as.integer exactly as in Section 5.2 to convert the strings to their integer form.

Yet another approach for reading in the guess5.txt file[‡] is to use

```
> drawData <- scan(textConnection(readLines("guess5.txt")
+ [-seq(1, 1090, 11)]), what = integer(0))
Read 5000 items
```

The initial import of data via readLines is the same as before. The difference begins with the use of the textConnection function that enables an object containing character strings to be read by a function that reads from a connection (as discussed in Section 5.2). In this case, the function that expects the connection is scan. The application of textConnection makes the output of readLines appear to scan as a text file of 5000 numbers which can be read in the usual way.

Now we have three ways to import the guess5.txt information. Which one works the best? The system.time function from Section 5.4 provides one means of comparison. Applying it to the three different import schemes produces

---

[‡] This idea is what motivated the strsplit based approach. It was kindly suggested by one of this book's reviewers.

```
> system.time(for(i in 1:100)
+ drawData <- c(drawData, scan("guess5.txt", skip = 1, nlines = 10,
+ quiet = TRUE)))
   user  system elapsed
  0.011   0.003   0.014
> system.time(drawData <- as.integer(unlist(strsplit(
+ readLines("guess5.txt")[-seq(1, 1090, 11)], split = " "))))
   user  system elapsed
  0.005   0.000   0.005
> system.time(scan(textConnection(readLines(
+ "guess5.txt")[-seq(1, 1090, 11)]), what = integer(), quiet = TRUE))
   user  system elapsed
  0.008   0.000   0.007
```

The loop approach is clearly inferior to the other two options in this case. A (much) more definitive evaluation would require us to repeat the computations in a manner similar to what we used for the different mean comparison functions in the last section (Exercise 5.18).

Assume now that the draw information recorded in guess5.txt has been imported into the numeric object drawData. This data can now be used to carry out a chi-square test of the uniform selection hypothesis. The first step is to count up the frequency of occurrence of each of the balls with numbers from 1 to 40. The function that accomplishes this in R is table. An application of this function to drawData produces

```
> Obs <- table(drawData)
> Obs[1:3]
drawData
  1   2   3
140 133 131
> sum(Obs)
[1] 5000
```

From this it can be seen, for example, that ball number 1 was drawn 140 times in the 5000 (i.e., 5 balls per draw multiplied by 10 draws per drawing times 100 drawings) draws that are represented in drawData. This can be compared to an expected frequency of $5000/40 = 125$. Note that summing up all the observed frequencies must give the total number of draws and this was checked and verified in the last session command.

It remains only to compute the chi-square statistic. The usual formula for this is

$$\sum_{i=1}^{K} \frac{(\text{Obs}[i] - \text{E}[i])^2}{\text{E}[i]}, \tag{5.1}$$

where the Obs[i] are the observed frequencies of the K population elements or categories and the E[i] are the expected frequencies for each category under the null hypothesis. Our focus is on equally likely categories; i.e., on any given draw from the population, the chance an object will be from any of the categories is $1/K$.

Our previous formula for the chi-square statistic is appropriate for drawing with replacement but must be altered for situations where drawings are conducted without replacement. Specifically, if n independent draws are conducted where k elements are selected without replacement, Joe (1993) shows that

$$\chi^2 = \frac{K-1}{K-k} \sum_{i=1}^{K} \frac{\left(\text{Obs}[i] - n\frac{k}{K}\right)^2}{n\frac{k}{K}}$$

will, for large values of n, have an approximate chi-square distribution with $K - 1$ degrees-of-freedom when outcomes are being selected with equal probability. Note that the total number of elements that are selected is $N = nk$. Thus, an equivalent formula is

$$\chi^2 = \frac{K-1}{K-k} \sum_{i=1}^{K} \frac{(\text{Obs}[i] - N/K)^2}{N/K}. \tag{5.2}$$

In the case of data from the Guess5 game $K = 40$ and $k = 5$.

The R function `chisq.test` computes the chi-square statistic in (5.1). We have wrapped it in the `hyperChi` function in Listing 5.6 to automatically produce the statistic in (5.2).

Listing 5.6 *hyperChi.r*

```
#hyperChi.r
hyperChi <-function (O, K, k){
    chistat <- (K - 1)*chisq.test(O)[[1]]/(K - k)
    names(chistat) <- NULL
    list(chistat, 1 - pchisq(chistat, K - 1))
}
```

The `chisq.test` function takes the observed frequencies as its first argument and has a vector argument p that specifies the probabilities of occurrence for observations corresponding to each row of this frequency vector. The p argument defaults to equally likely categories which suffices for our case. Our wrapper function has two additional arguments K and k that represent the size of the population and the number of draws without replacement, respectively. The `chisq.test` function returns a list whose first component is the value of the chi-square statistic. The list subsetting operator `[[]]` is used to isolate the test value which is then inflated appropriately according to the values of K and k. The `chisq.test` function has left the result with the name "X-squared" attached that we would prefer not to have appear in our output. The (replacement version of the) `names` function is used to suppress this behavior by setting the name to `NULL`. Finally, the output from the function is returned as a list whose first component is the chi-square statistic and the second is its associated upper-tail p-value.

An application of `hyperChi` to the Guess5 drawing data produces

```
> hyperChi(Obs, 40, 5)
[[1]]
[1] 27.77691

[[2]]
[1] 0.9100113
```

The p-value of .91 leads us to the expected conclusion that the data does not exhibit a significant departure from uniform selection of the ball numbers.

5.7 Using C/C++ code in R

The R "foreign language" interface allows computational routines to be coded in, e.g., C or C++, and then imported into R as a shared library. In this section we will explore and illustrate how this can be done.

One situation where it is common to use a combination of R and C/C++ programs is in the creation of R packages. Descriptions of this process are given in the *Writing R Extensions* manual and, e.g., in Chapter 7 of Gentleman (2009). The development here is of a much smaller scale. Our perspective is that of someone who has written a relatively small and focused set of C++ code designed for a specific task of their own personal interest. In some cases this type of program development can benefit from a pairing with R. For example, one may have developed a new estimation method that has been coded in C++.

Simulation experiments to assess the properties of the method could be managed in R and the data analysis tools that R provides could then be used directly for summary and analysis of the simulation results. There are also many R functions that become available through the R API. These are the same as the ones discussed in Section 4.9 in our treatment of the R Standalone Math Library that includes, e.g., methods for random number generation. The code development process can be simplified in many instances by making use of this resource.

There are two basic interfaces between R and C/C++: .C and .Call. Both of these functions pass pointers to R objects to C/C++ code. The difference is that the former cannot make use of R internal data structures while the latter one can. The use of .C is therefore less involved and is likely to be the preferred option for our particular setting. Thus, we focus on .C here. An in-depth description of using .Call and, more generally, programming with R data structures is contained in the *R Internals* manual and Chapters 5–6 of the *Writing R Extensions* manual. Chapter 6 of Gentleman (2009) provides a nice introduction to the topic.

Let us begin by importing both C and C++ implementations of simple "Hello world!" programs into R. C code for this purpose might look like

```
//hello.c
#include "R.h"

void hello(int* n){
  int i;
  for(i = 0; i < *n; i++)
    Rprintf("Hi ya'll %d times!\n", i);
  *n += 1;
}
```

After a little work we will be able to call this program from inside an R session where it will print "Hello y'all!" as many times as desired.

The simple hello.c program has some distinguishing features relative to what was seen in Chapter 3. First, there is no main function. The presence of one in either C or C++ programs will create problems in compiling the shared libraries that are needed for using such code in R. For creating a shared library that can be called from R, the function that serves the purpose of main[§] in C/C++ code must

1. have void as a return type and

2. have arguments that are pointers.

Both of these qualities are present in the hello.c listing. This program will print out the greeting phrase a number of times determined by the pointer to int that is the argument for the hello function. The phrase is then printed to the R console using the Rprintf function that is made available through the R.h header file. Finally, *n is augmented by one and the program terminates.

The Rprintf function is an R analog of the printf function that is used to produce formatted output in the C language. The first argument to Rprintf is a character string (enclosed in quotes) with % placeholders for the values of each of the argument variables. The % placeholder is augmented by a character that designates the storage type for its corresponding arguments. This is accomplished through syntax of the form %?, where ? can be

    c for a character variable,

[§] The name main can still be used. However, it must have return type void and that will generally produce a warning message from the compiler.

d for an integer variable,

e for a floating-point variable in scientific notation,

f for a floating-point variable and

s for printing a character string.

Carriage returns are created by placing a \n inside the character string argument.

The initial argument to Rprintf in our program is "Hello y'all %d times!\n". The %d informs the compiler that the next argument is an integer variable and that the printing format is to be tailored to a variable of that type. The \n character at the end of the character string produces a new line after printing out the message.

To compile the C code we enter

```
$ R CMD SHLIB hello.c
```

on the Unix shell command line. This creates a shared library hello.so that can now be imported into R using the dyn.load utility. Specifically, this is accomplished inside R with

```
> dyn.load("hello.so")
> is.loaded("hello")
[1] TRUE
```

The is.loaded function checks that the shared library has been successfully loaded.

Once a shared C/C++ library has been imported, the underlying program can be executed using the .C function. For our example the result might appear something like

```
> .C("hello", as.integer(3))
Hi ya'll 0 times!
Hi ya'll 1 times!
Hi ya'll 2 times!
[[1]]
[1] 4
```

The list information provided at the end of this output is what has been returned by the .C function. It represents the values of the arguments (i.e., the values stored in the memory locations for the pointers that were passed in the function call) that were returned from the hello program. In particular the output shows that the last line of the C program had the effect of advancing the value of n (from the initial value assigned in R) by one unit. Also, note that the use of the as.integer function for passing the integer 3 into the C program is necessary.

A similar program can be developed in C++ as we now demonstrate. First, we will have a class X with a header file X.hh (the hh file extension is necessary) that contains the code

```
//X.hh
#ifndef X_HH
#define X_HH
class X{
  int n;
public:
  X(int N);
  ~X();
};
#endif
```

The class constructor and destructor are defined in X.cc (again, the cc file extension is necessary) as

```
//X.cc
#include <R.h>
#include <Rmath.h>
#include "X.hh"

X::X(int N) {
  n = N;
  for(int i = 0; i < n; i++)
    Rprintf("Hello world %d times!\n", i);
  Rprintf("Normal %f \n", pnorm(1.96, 0, 1, 1, 0));
}

X::~X() {
  for(int i = 0; i < n; i++)
    Rprintf("Goodbye world %d times!\n", i);
}
```

A driver program shown below uses an object of the X class.

```
//useX.cc
#include "X.hh"

extern "C" {

  void func(int* n){
    X x(*n);
  }
}
```

This last program represents the conduit between R and our C++ code. It will be compiled and linked with C libraries using the C compiler gcc. To allow for this mixing of C and C++ code, it needs to be enclosed in an extern "C" { } "wrapper" that encapsulates the function between curly braces. Beyond that, the intent of our program is to call both the constructor and destructor for the X class and thereby have the corresponding greeting and farewell messages printed in R.

To compile the X class program use

```
$ R CMD SHLIB X.cc useX.cc
```

to create the shared object file X.so. A specific name could have been given to the library other than X.so using the -o compiler option. In absence of a specified name, the name of the first file (without extension) is used to name the shared library. Thus, switching X.cc and useX.cc on the command line would produce the shared library useX.so.

Now an R session can be started and the library can be loaded with

```
> dyn.load("X.so")
> is.loaded("func")
[1] TRUE
```

The response from is.loaded indicates that the library has been loaded successfully. Note that R refers to it by the name of the function that interfaces with R, or func in this case, rather than the library name. The linked program can now be executed using the .C function as before to produce, e.g.,

```
> .C("func", as.integer(3))
Hello world 0 times!
Hello world 1 times!
Hello world 2 times!
```

```
Normal 0.975002
Goodbye world 0 times!
Goodbye world 1 times!
Goodbye world 2 times!
[[1]]
[1] 3
> dyn.unload("X.so")
```

The final line of code illustrates how to unload a library that has been linked to an R session.

We noted earlier that R has an API that allows functions from the R math library to be used in our C++ code. Our X class used the R function `pnorm` for the normal cdf to illustrate this feature.

The R API becomes available by including the Rmath.h header file that contains the prototypes for the available functions. A technique for deciphering the meaning of the arguments in these prototypes was described in Section 4.9. Briefly, one matches the arguments in the prototype to the ones given for that function in its corresponding R help page. In the case of the `pnorm` function this approach revealed that

a) the first argument is the evaluation point for the cdf,

b) the second and third arguments are the mean and standard deviation, respectively and

c) choosing 1 and 0 for the fourth and fifth arguments will produce return values that are lower tail probabilities.

The numeric output produced by using `.C` to call `func` indicates that `pnorm` has been used correctly.

To use a random number generator from the R API there are two additional functions that must be used: `GetRNGstate` and `PutRNGstate`. They retrieve the current value of the seed from R and then return the new seed value after the generator has been used. As a result, `GetRNGstate` must be called before invoking the generator and `PutRNGstate` is called once the generator is no longer required. The idea is illustrated by

```
//rngEx.cc
#include <R.h>
#include <Rmath.h>

extern "C" {

  void rngFunc(int* n, double* mu, double* sig, double* x)
  {
    GetRNGstate();
    for(int i = 0; i < *n; i++)
      x[i] = rnorm(*mu, *sig);
    PutRNGstate();
  }
}
```

The function `rngFunc` calls the R `rnorm` function for generating pseudo-random numbers from a normal distribution. Its arguments for the mean and standard deviation are passed in as pointers to `double` with a pointer to `double`, x, being provided to store the numbers that are generated. Note that the memory for x must be allocated in R and failure to do so or a failure to allocate enough memory will produce unpredictable results. The actual random numbers are generated in a `for` loop that is sandwiched between the requisite calls to `GetRNGstate` and `PutRNGstate`.

The rngEx.cc code is compiled with

```
$ R CMD SHLIB rngEx.cc
```

and imported into R with

```
> dyn.load("rngEx.so")
```

Then, an illustration of the use of **rngFunc** is

```
> x <- vector("double", 3) #allocate memory in R
> set.seed(123)
> result <- .C("rngFunc", as.integer(3), as.double(1), as.double(2),
+ x)
> result[[4]]
[1] -0.1209513  0.5396450  4.1174166
> .C("rngFunc", as.integer(3), as.double(1), as.double(2), x)[[4]]
[1] 1.141017 1.258575 4.430130
> set.seed(123)
> .C("rngFunc", as.integer(3), as.double(1), as.double(2), x)[[4]]
[1] -0.1209513  0.5396450  4.1174166
```

The aim was to generate three random numbers and, as a result, memory was allocated in R for a three-component, numeric **vector** object x. The seed was then set and **rngFunc** was called using the .C interface function. The output was stored in the **list** object **result** whose fourth entry is the array x that was passed in as the fourth argument to .C. The **rngFunc** function was called again producing three new random numbers. The seed was then reset to its initial value and the resulting output from **rngFunc** is seen to agree with our first set of numbers.

The .C interface can be used to import any of the C++ code that has been developed so far into R. As an example, let us consider how this can be accomplished using our Fishman-Moore random number generator class from Chapter 4. First, the FM.cpp and FM.h files from Chapter 4 must be renamed as FM.cc and FM.hh, respectively. Next, an interface program must be created. For this particular case something like

```
//fmR.cc
#include "R.h"
#include "FM.hh"

extern "C" {

  void rngFM(int* N, int* SEED){
    int n = *N;
    unsigned long seed = (unsigned long)(*SEED);
    FM* pFM = new FM();
    double* pu = new double[n];
    pFM->ranGen(seed, n, pu);
    Rprintf("Your random numbers are\n");
    for(int i = 0; i < n; i++)
      Rprintf("%f \n", pu[i]);

    delete pFM;
    delete[] pu;
  }
}
```

will suffice. Then, the interface program file fmR.cc and the FM.cc file are compiled and linked into a shared library fmR.so as in

```
$ R CMD SHLIB fmR.cc FM.cc
```

The final step is to import the fmR.so library into the R environment using the **dyn.load** function in the form

```
> dyn.load("fmR.so")
```

The random number generator program can now be executed using the .C function. It is a bit less cumbersome to "wrap" rngFM in an R function that takes care of actually calling and passing the arguments to .C. A function that serves this purpose is

```
rngFM <- function(N, seed){
    .C("rngFM", as.integer(N), as.integer(seed))
}
```

An example of output from rngFM is

```
> rngFM(5, 123)
Your random numbers are
0.104714
0.723072
0.156521
0.328550
0.635906
[[1]]
[1] 5

[[2]]
[1] 123
```

Further examples of using C++ code in R will be given in Chapters 7 and 8.

5.8 Exercises

5.1. Write an R function that will produce a three-number summary as described in Exercise 3.41 for each of the columns of an array. Perform your calculations in two ways:

a) Apply sort to the individual columns of the array.

b) Work only with the full array and use order to rearrange all the rows of the matrix simultaneously.

Which approach is faster? How does the performance of the two methods compare to using the R quantile function discussed in Section 3 of Appendix B?

5.2. Write expressions for for loops in R with the properties that

a) the loop index proceeds over the index set of positive integers from 1 to n in steps of size k for a specified integer k or

b) the loop index proceeds backwards through the integers 1 to n (i.e., it starts at n and finishes when the index is smaller than 1) in steps of size k for a specified integer k.

5.3. The collection of built-in constants in R includes LETTERS and letters that contain the 26 letters of the Roman alphabet in upper- and lower-case forms, respectively. Create an R function that will produce a random sentence containing a specified number of "words" (in the sense of being a sequence of letters). Your function should print the sentence upon completion of its task and produce output that adheres to the guidelines given below.

a) The first word in the sentence should begin with an upper case letter.

b) After the first word of a sentence, no upper-case letter should appear.

c) The sentence should end with a period, question mark or exclamation point with the specific choice of punctuation being unpredictable.

d) Other punctuation should be allowed to appear (at random) in the body of the sentence.

e) Word lengths should be allowed to vary randomly between one letter and some specified (by the user) upper bound for their size.

5.4. Write two R functions: one that takes a function as an input argument and another that returns a function as output.

5.5. Use the R tryCatch function to create a function that will allow you to recover from an error where scan encounters character data while reading from a file that was presumed to contain only numeric information.

5.6. Develop an R function that takes an array and a logical expression concerning the array elements as arguments and returns the matrix elements that satisfy the logical criteria.

5.7. Write an R function that takes an integer argument n and then uses a for loop to generate n random uniforms using the runif function. Evaluate the performance of your function using the system.time function. For example, compute the difference in the length of time it takes your function to return 10 million numbers versus a direct call to runif.

5.8. Let f be a function on the interval $[0, 1]$ with $f(0)f(1) \leq 0$. An n-step bisection algorithm for finding a zero for f is (Exercise 8.20)

Algorithm 5.1 n-step bisection algorithm

$a = 0, b = 1$
for $j = 1$ to n **do**
  $m := (a + b)/2$
  if $f(a)f(m) \leq 0$ **then**
    $a := a, b := m$
  else
    $a := m, b := b$
  end if
end for
return $(a + b)/2$

Write an R function that will apply this algorithm to a specified function that is provided as its argument.

5.9. The eval function provides a means of evaluating an expression in R. As an illustration, consider the succeeding code segment

```
> e <- expression (A[ ,2] == 4)
> E <- A[eval(e),]
```

The parse function in R will parse a text string (or file) of R code supplied as its text (respectively, file) argument (without evaluating it) and return an expression object. Write an alternative to source for importing code into R from a file that uses the parse function in combination with eval.

5.10. The R function for solving a linear equation system is solve. In particular, solve(A) returns the inverse of an invertible matrix A. Write an R function that takes a matrix as input and then

a) checks to see if the matrix is square and returns an error message if this is not the case,

b) if the matrix is square, uses solve to (attempt to) compute its inverse and, assuming the matrix is nonsingular, returns the inverse of the matrix and the determinant (via the R function det) and trace of the inverse matrix and

c) uses the tryCatch function to deal with any errors produced in part b).

5.11. Refer to Exercise 3.13. Carry out the calculations for this exercise in R.

5.12. Use the R choose function that computes binomial coefficients to determine the probabilities for all the possible outcomes (e.g., match five, match four, etc.) for the Guess5 game.

If a minimum of five observations are required per cell/group for carrying out a chi-square test for uniformity, how many observations would be needed to test for uniform selection over every possible five number combination that can occur in the Guess5 game?

5.13. Use the `invisible` function to suppress the `NULL` output from the `guess5` function of Section 5.6.

5.14. Develop a C++ program that will read in the Guess5 data from Section 5.6 and count the frequency of occurrence of the ball numbers over the entire collection of draws.

5.15. Numbers are drawn five times with replacement from the integers 0 to 9. Use R to calculate the probability distribution for the sum of the numbers that are drawn.

5.16. Consider a lottery game where `k` balls are selected from balls numbered 1 to `n1` without replacement and an additional (single) bonus ball is drawn independently from a set of balls numbered 1 to `n2`. Players win various prizes depending on how many of the balls they match from the `k` balls the lottery draws from the 1 to `n1` range and the lottery's bonus ball selection.

a) Use the R `choose` function to create an R function that will compute the chance of winning for all possible outcomes for the game.

b) What are the chances for all the outcomes when `k = 5, n1 = 40` and `n2 = 20`?

5.17. Suppose that the Guess5 game from Section 5.6 adds a bonus ball where at every draw an additional independent draw is made of a single ball from a set of balls numbered 1 to 20. There are five ball sets and two machines that are used for the bonus drawings.

a) Create a program similar to the one is Listing 5.5 that will generate a set of artificial data for this game.

b) Use your program from part a) to create a data file containing the results of 100 simulated draws of the new Guess5 game.

c) Import the data from the file you created into R in a format that will allow you to analyze the data using a chi-square test. Note that there are two separate tests that must be performed corresponding to the 5 of 40 drawings and those involving the bonus ball.

d) Compute appropriate chi-square statistics to determine if the data you generated suggests a departure from uniform selection in your generation methodology.

5.18. Compare the run-times for the three ways of reading in the guess5.txt file in Section 5.6 using replicate runs of each program similar to the approach used to compare the `meanComp`, `meanCompVec`, `meanCompOuter` and `vecPooledT` functions in Section 5.5.

5.19. A person buys four computer-generated tickets for playing the Guess5 game. That is, the player purchases four sets of five numbers selected from the integers 1 to 40 without replacement that have been produced by an assumedly fair random number generation algorithm of some sort. Use R to calculate the probability distribution for the number of unique numbers on the four tickets.

5.20. Refer to Exercise 3.10. Write an R program that will perform the same calculations as the C++ program that you created for part a) of that exercise. Compare its computation speed to that of the C++ implementation.

5.21. Modify the class `X` from Section 5.7 to where the constructor and destructor will write out "Hello World!" and "Goodbye World!" a total of `n1` and `n2` times, respectively, for user-supplied values of `n1` and `n2`. Import the resulting program into R as a shared library and verify that it behaves as expected.

5.22. Import the C++ code for using the Cauchy density to generate pseudo-random normals (Listings 4.8–4.9) in Section 4.6 into R and package it in an R function that will generate a specified number of random deviates from a normal distribution with user-specified means and variances not all of which are required to be the same.

5.23. Another use of `outer` is to generate regular grids of function values. For example, a simple plot function might take as input a function of two variables `f`, lower and upper bounds in both the `x` and `y` coordinates and a granularity parameter `eps`. Then, it would evaluate `f` at a grid whose edges are at the given minimum and maximum values of `x` and `y` and of granularity `eps`. Assuming the values of `xmin`, `xmax`, `ymin`, `ymax` and `eps` are defined, write a one line expression using `outer` that returns a matrix containing the values of `f` at a grid spaced by `eps` and bounded by the given `x` and `y` bounds.

5.24. Create an R program that will produce an artificial data set from an $r \times k$ factorial experiment with factors A (at levels $A_1, \ldots, A_r$) and B (at levels $B_1, \ldots, B_k$) for user-specified values of r and k. The function should allow for any specified number n of observations per factor level combination, all of which come from normal distributions with unit variance. Let μ_{ij} be the mean for the responses that receive the treatment combination of A_i and B_j for $i = 1, \ldots, r$ and $j = 1, \ldots, k$. Then, the means should have the form

$$\mu_{ij} = \delta_1 i, \ i = 2, \ldots, r, \ j = 1, \ldots, k,$$

and

$$\mu_{1j} = \delta_2 j, \ j = 1, \ldots, k,$$

for specified values of δ_1, δ_2. Use `rbind` and `cbind` to return the generated data as either an $rk \times n$ or $n \times rk$ matrix.

5.25. Use `apply` to create an R function that will analyze the result of a balanced $r \times k$ factorial experiment and return the usual F statistics for testing significance of the main effects and interaction. [Note: This facility is built into R in the `aov` and `lm` functions. The `anova` function takes, e.g, an `lm` or `aov` object as its argument and returns a corresponding analysis of variance table.]

5.26. Use your programs from Exercises 5.24 and 5.25 to carry out a simple power study. Specifically, take $r = 2, k = 3, n = 100$ and work over a grid of values for the parameters δ_1 and δ_2 in Exercise 5.24 that starts with $\delta_1 = \delta_2 = 0$. For each point on the grid, generate 1000 data sets with the corresponding values for the parameters and, for each of these data sets, apply your function from Exercise 5.25 to determine if the null hypothesis of no treatment effect was rejected. Present the final result as a three-dimensional plot (using, e.g., the R `persp` function or the `cloud` function from the R lattice package) of the proportion of times the null hypothesis was rejected in the 1000 samples for each value of δ_1 and δ_2 in your grid. The grid points should extend sufficiently far that they produce empirical powers of one at the edge of the grid.

5.27. Implement the FM2 random number generator of Section 4.4 as an R function. Compare the running time of this program with the C++ version from Section 4.4 and the one in Section 5.4 that uses the `.C` interface.

5.28. Let $x = (x_1, \ldots, x_n)^T$ and $x = (y_1, \ldots, y_n)^T$ be two vectors of real numbers. The lag l cross-correlation between x and y is

$$r(l) = \frac{\sum_{t=r+1}^{n}(x_t - \bar{x})(y_{t-r} - \bar{y})}{s_x s_y}$$

with s_x and s_y the sample variances for the x and y vectors and $\bar{x}$ and $\bar{y}$ their corresponding means. Write R and C++ code that uses the `.C` interface to compute the cross-correlations for two `numeric` R objects and an integer array of lag values.

5.29. An R `matrix` object with `nRows` rows and `nCols` columns must be passed into C++ as a pointer to a vector of `nRows*nCols` elements if we use the `.C` interface. The `matrix` elements are passed in column major order. Write a C++ program that takes two matrices constructed in R, transforms them to objects of the `Matrix` class from Section 3.9 and Appendix D and adds them using the overloaded addition operator for that class.

5.30. The Fibonacci numbers are defined by $F_0 = F_1 = 1$ and

$$F_j = F_{j-1} + F_{j-2}, j = 2, \ldots.$$

a) Write a function in R that computes the Fibonacci numbers using recursion in the sense of Exercises 3.7-3.10.

b) Use the function created in part a) of Exercise 3.9 to created a shared library for import into R. Then, write an R function that will perform the same operations as the function created in part a) except that it employs the shared library using the `.C` interface.

c) Use the `system.time` function to compare the run-times for the two approaches for calculating the Fibonacci numbers in parts a) and b).

5.31. Revise the mean comparison functions from Listings 5.3–5.4 so that they carry out protected mean comparisons using an experiment-wise level obtained via the Bonferroni inequality.

5.32. An alternative to controlling the experiment-wise error rate is to control the false discover rate (FDR). The FDR is essentially the expected number of rejections when the null hypothesis is true. A procedure for controlling the FDR has been developed by Benjamini and Hochberg (1995) and Benjamini and Yekutieli (2001). Suppose m hypothesis tests have been conducted that produced p-values $p_1, \ldots, p_m$. Then, the Benjamini/Hochberg/Yekutieli method bounds the FDR by a specified value $\alpha \in (0, 1)$. The calculations proceed according to Algorithm 5.2.

Algorithm 5.2 Benjamini-Hochberg-Yekutieli FDR control method

Arrange $p_1, \ldots, p_m$ in numerically ascending order as $p_{(1)} \leq \cdots \leq p_{(m)}$
$q = \alpha / \sum_{j=1}^{m} \frac{1}{j}$
$k = \max \left\{ 1 \leq i \leq m : p_{(i)} \leq q\,(i/m) \right\}$
if k exists **then**
    Reject the null hypotheses corresponding to $p_{(1)}, \ldots, p_{(k)}$
else
    Reject nothing
end if

Revise the mean comparison functions from Listings 5.3–5.4 so that they carry out protected mean comparison using the Benjamini-Hochberg-Yekutieli method. [Note: The Benjamini-Hochberg and other methods that adjust p-values are available through the R function `p.adjust`. There are also several R packages that deal with FDR control. These include fdrtool, multtest, nFDR and qvalue.]

5.33. Apply the Bonferroni and Benjamini/Hochberg/Yekutieli methodology to 1000 data sets generated from your function in Exercise 5.24 using $r = 2, k = 3, n = 100$ and $\delta_1 = \delta_2 = 0$. Compare the results from this with what is obtained using unprotected comparisons.

5.34. The bootstrap is a resampling method that can be used to approximate the distribution of a statistic. Let $X_1, \ldots, X_n$ be a random sample from some unknown distribution and let $T(X_1, \ldots, X_n)$ be a corresponding statistic. Sampling from $X_1, \ldots, X_n$ is then carried out with replacement to obtain B bootstrap samples $X_{1b}^*, \ldots, X_{nb}^*, b = 1, \ldots, B$. These samples produce the values $T_b^* = T(X_{1b}^*, \ldots, X_{nb}^*), b = 1, \ldots, B$ that lead to

$$F_T^*(t) = \frac{1}{B} \sum_{b=1}^{B} I(T_b^* \leq t)$$

with $I(A)$ the indicator function for the set A. The function F_T^* provides an estimator of the true distribution function for T that has various consistency and related asymptotic properties (e.g., Davidson and Hinkley 1997).

The R boot package provides functions that can be used for bootstrapping. In particular, the boot function has the simplified prototype

```
boot(data, statistic, R)
```

The data argument is the array of data that is to be used in the inferential process while statistic is the function that will be applied to the data. The function that is specified for statistic must have two arguments: a data array and an array of indices that specifies the subset of the data that will be used to calculate the statistic. The R parameter is the number of bootstrap samples that was denoted by B in the above developments.

As an example, the function med0 in the listing below can be used to provide an approximation to the null distribution of the sample median under the hypothesis that the median is zero.

```
med0 <- function(y, indices){
  x <- y - med(y)
  median(x[indices])
}
```

The centering of the y array is to so that the sample will behave like data from the null model when resampling takes place. An illustration of how this might be used is provided by the following.

```
> set.seed(123)
> y <- runif(20, -1, 1)
> median(y)
[1] 0.0795405
> bootOut <- boot(y, med0, R = 1000)
> sort(bootOut$t)[c(25, 975)]
[1] -0.3426429   0.5917820
}
```

In this case 20 observations are generated from a uniform distribution on the interval $[-1, 1]$ from which 1000 bootstrap samples are taken. The values of the median that are obtained in resampling are returned as the t member of the boot class object bootOut. These values are sorted and their 2.5 and 97.5 percentiles are obtained. The sample median falls between these bounds and, accordingly, the null hypothesis of a zero population median is not rejected.

a) Perform a power study to see how the bootstrap median test performs against alternatives to the zero median model. For this purpose i) consider alternative distributions that are uniform on $[-1-\delta, 1-\delta]$ for $\delta \geq 0$, ii) test for the one-sided alternative that the median is less than zero and iii) use .05 as the level for the test. A grid of values should be used for δ with the choice of $\delta = 0$ corresponding to "powers" that represent the level of the test. In this regard you should replicate the experiment sufficiently many times that departures as small as .01 from the nominal .05 level can be detected with 95% confidence.

b) Repeat part a) using the Student's t-statistic for the hypothesis that the population mean is 0 and compare the results with the bootstrap approach.

Chapter 6

Creating classes and methods in R

6.1 Introduction

The previous chapter focused on developing applications that worked with existing R classes. This chapter describes how new R classes and methods can be created for dealing with data analysis problems that require a more customized approach.

It is possible to create classes in R with member elements that in most ways are consistent with the C++ paradigm. However, the primary focus in R is on functions rather than classes. To expand on this last statement, suppose ObjA is an object from a C++ class A. A method func for class A objects would be invoked via syntax resembling ObjA.func apart from any arguments that would be needed for the method. If ObjB was an object from another class B, an expression of the form ObjB.func need not have any meaning and, even if it did, there is no need for the method func to perform similar or even related operations for objects from class A or B. The R version of OOP essentially reverses the role of function and object as compared to the C++ treatment. Classes do not have member functions. Instead, functions are a class in their own right and specific functions have "methods" that provide implementation of the function for different types of objects. Thus, if func is an R function, it may have method functions with exactly the same name that would be applicable to and perform appropriately on the objects ObjA and ObjB from classes A and B, respectively. The correct version or method function of func for an object (or objects) supplied as its argument (or arguments) is determined by the class of the argument (or arguments) through a process termed *method dispatch*.

The functions in R can be (and often are) generic in a sense that is analogous to the template function from C++ discussed in Section 3.11. A template functions func is not tied to any particular class and, at least in terms of syntax, expressions of the form func(ObjA) and func(ObjB) for objects ObjA and ObjB from entirely different classes A and B are conceptually valid, if possibly not particularly meaningful from a practical perspective. When meanings can be ascribed to func(ObjA) and func(ObjB), the general spirit that underlies template function creation would lead us to expect that the specializations of func to class A and B objects will perform similar operations on the objects of each class. For example, template functions named print or plot would be expected to perform analogous procedures across objects from different classes. With this in mind, generic functions in R might be viewed as a collection of template functions that carry out similar operations on a variety of R class objects.

There are two class systems in R that are often referred to as S3 (or "old style") classes (e.g., Chapter 4 of Venables and Ripley 2000 and Chapter 3 of Gentleman 2009) and S4 (or "new style") classes (e.g., Chapters 7–8 of Chambers 1998, Chapter 5 of Venables and Ripley 2000 and Chapter 3 of Gentleman 2009). The S3 class concept is essentially a naming convention that is employed, for example, in method dispatch. In contrast, the S4 class system allows for the creation of objects with member elements and other features that would be expected from an object-oriented language. Accordingly, our attention will be focused on the S4 framework which represents the subject of the next section.

6.2 Creating a new class

Suppose that we wish to construct a new class named **regr** for dealing with a linear regression situation where an $n \times 1$ response vector y is to be related to p independent variables whose values are contained in the columns of an $n \times p$ matrix X. The analysis will produce, e.g., a vector b of least-squares coefficient estimators. Of course R already provides the facility for this through the linear models (i.e., **lm**) class and this would be the superior option in most cases of practical interest. The goal of the present exercise is not to produce a class that can compete with the **lm** class or something of much practical utility. Instead, the objective is to illustrate the steps involved in class creation in a relatively simple setting. A more detailed and useful illustration of the construction of classes and methods is the subject of Section 6.4.

To create a new class in R one uses the **setClass** and **representation** functions. For our **regr** class this might take the form

```
> setClass("regr", representation(X = "matrix", y = "numeric",
+ b = "numeric", intercept = "logical"))
[1] "regr"
```

This command creates a class with four members, **X**, **y**, **b** and **intercept** that are objects from the existing R classes **matrix** (for the X matrix), **numeric** (for the response vector y and estimated coefficient vector b) and **logical**. By using the **numeric** class for y and b we have stated that they will be represented in R by double precision **vector** objects. The Boolean variable **intercept** will be used to specify whether or not an intercept term is to be included in the linear regression fit to the data.

The **getClass** command provides a way to check that a class has been created correctly. For example, an application of **getClass** to our **regr** class returns

```
> getClass("regr")
Class "regr" [in ".GlobalEnv"]

Slots:

Name:        X          y          b intercept
Class:    matrix    numeric    numeric    logical
```

This output tells us that the class **regr** has four member elements which R refers to as being *slots*. The names of the elements (**X**, **y**, **b** and **intercept**) as well as their respective classes (**matrix**, **numeric** and **logical**) are also listed.

The command **removeClass("className")** will remove an existing class **className** from the current R session as demonstrated in the next listing.

```
> isClass("regr")
[1] TRUE
> removeClass("regr")
[1] TRUE
> isClass("regr")
[1] FALSE
```

Removal of the class was preceded and followed with an application of the **isClass** command that checks to see if its argument is a formally defined class. The **FALSE** return value from its second application tells us that the **regr** class is no longer defined.

An object from a class can be created with the **new** operator. In particular, using **new** with the **regr** class gives

```
> regrObj <- new("regr")
> regrObj
```

```
An object of class "regr"
Slot "X":
<0 x 0 matrix>

Slot "y":
numeric(0)

Slot "b":
numeric(0)

Slot "intercept":
logical(0)
```

This demonstrates that a "blank" regr object has been constructed where the class members/slots have been assigned the R default objects for their respective classes.

There are instances where the defaults that new supplies for a given class (or the way they will propagate into classes that are constructed from user-defined classes) may not be satisfactory. In the case of C++ a cure for such problems is the inclusion of default arguments to a constructor which is an option in R as well. But, R provides another means to accomplish this through use of the prototype function when the class is first created via the setClass function. For example, in our regr class the X matrix could be specified as defaulting to a 1×1 "matrix" with a single 0 element, the y and b slots could be set as 0 and the value of intercept could be set to TRUE. This latter choice will make the inclusion of a constant term the default fitting behavior. This can all be accomplished with

```
> setClass("regr", representation(X = "matrix", y = "numeric",
+ b = "numeric", intercept = "logical"), prototype(X =
+ matrix(0, 1, 1), y = 0, b = 0, intercept = TRUE))
[1] "regr"
> regrObj <- new("regr")
> regrObj
An object of class "regr"
Slot "X":
     [,1]
[1,]    0

Slot "y":
[1] 0

Slot "b":
[1] 0

Slot "intercept":
[1] TRUE
```

The output shows that a regr object regrObj created by using the new operator with no arguments produces an object with the specified default entries for the X, y, b and intercept slots.

Let us now create an regr object using a specific data set. For this purpose we will use a portion of the mtcars data frame that contains performance measures and other characteristics for 32 cars reported in a 1974 issue of *Motor Trend* magazine. The data set is loaded into the R environment with the command data(mtcars). The y slot of our regr object will then be filled with the first column from the mtcars data frame that corresponds to miles per gallon and the X slot will be composed of columns 3, 6 and 7 that give the displacement (in cubic inches), weight and quarter mile speed of the cars. To accomplish this a matrix is first created that contains the information to be used in the regr object: i.e.,

```
> A <- as.matrix(mtcars[,c(1, 3, 6, 7)])
```

The desired regr object could then be obtained from

```
> regrCar <- new("regr", X = A[, 2:4], y = A[, 1])
```

The slots of an object can be accessed using the @ operator or the function slot. Thus, if Obj is an object with a slot having the name s, Obj@s or slot(Obj, "s") will return the contents of that particular slot. The code listing below illustrates how @ and slot can be used to fill and access the b slot for the regrCar object that was created from the mtcars data.

```
> XTX <- t(cbind(1, regrCar@X))%*%cbind(1, regrCar@X)
> regrCar@b <- solve(XTX, drop(t(cbind(1, regrCar@X))%*%regrCar@y))
> regrCar@b
                      disp              wt            qsec
19.7775575655  -0.0001278962  -5.0344097167   0.9266492353
> slot(regrCar, "b")[1]

19.77756
```

The value that goes in the b slot is the least-squares estimator for the coefficients in the regression of y on X. To create this vector the X slot of regrCar is accessed and, since intercept is set to TRUE, a column of unit elements is added to the matrix so that the regression fit will include an intercept term. The resulting $X^T X$ matrix is evaluated, stored as XTX and the vector of regression coefficients is calculated using the R function solve. This function will be discussed in more detail in the next chapter. For now it is enough to know that a call to solve of the form solve(A, v) will return the solution b of the system Ab = v provided that A is a nonsingular matrix. In this particular instance, solve is applied with A = XTX, v = t(cbind(1, regrCar@X))%*%regrCar@y and the resulting solution is assigned to the b slot of regrCar. A point of interest here is the use of the drop function on the t(cbind(1, regrCar@X))%*%regrCar@y array that represents the right-hand side of the linear system being handled by solve. The problem here is that the multiplication of regrCar@y by the matrix object t(cbind(1, regrCar@X)) promotes the outcome to matrix status which conflicts with the numeric class designation that has been given to the b slot. One approach would be to simply coerce the outcome into the right form with the as.numeric function. The R help page for %*% suggests instead using the drop function that will delete the extra dimension from an array that has only one column or row.

Using the new operator directly for the construction of every object of a particular class is generally too tedious. An alternative approach is to provide class constructors that can be tailored to particular situations that frequently arise in practice. As in C++, constructor functions will generally have the same name as their class and involve arguments that are required to create a class object in some particular situation of interest. Constructor functions are created in the same way as any function in R. For example, a constructor for the regr class might look like

```
regr <- function(X, y, Intercept = TRUE){
  regrObj <- new("regr", X = as.matrix(X), y = as.numeric(y),
    intercept = Intercept)
  if(Intercept){
    XTX <- t(cbind(1, regrObj@X))%*%cbind(1, regrObj@X)
    regrObj@b <- solve(XTX, drop(t(cbind(1, regrObj@X))%*%regrObj@y))
  }
```

```
  else{
    XTX <- t(regrObj@X)%*%regrObj@X
    regrObj@b <- solve(XTX, drop(t(regrObj@X)%*%regrObj@y))
  }
  regrObj
}
```

Notice that a safeguard has been built into the constructor wherein an attempt is made to coerce X and y into their appropriate classes. This comes into play when, for example, a data frame object is inadvertently supplied for the X argument. Also, the behavior of making inclusion of an intercept term the default has been enforced by assigning intercept the default value of TRUE in the constructor's argument list.

Using the regr class constructor with the mtcar data will now produce results such as

```
> regrCar <- regr(A[,2:4], A[,1])
> regrCar@b
                  disp             wt            qsec
19.7775575655 -0.0001278962 -5.0344097167  0.9266492353
> regrCar <- regr(A[,2:4], A[,1], FALSE)
> regrCar@b
       disp            wt          qsec
 0.01519605  -5.99044330   2.00209915
```

Fits with and without a constant term have been obtained. To determine which is preferable the regr class could be expanded to include measures of lack-of-fit (Exercise 6.2).

It is usually a good practice to provide checks that an object has been constructed in the proper way. This will be taken care of automatically for slots that are from prespecified classes like matrix. So, in the case of the regr class, a matrix must be supplied for the X slot when calling new or an error will occur. But, there are no further checks carried out to ensure any other form of consistency or that the calculations to be carried out inside the constructor are actually possible.

One feature that would be expected for an object of type regr is that both the X and y arrays should have the same number of rows. The nrow function for class matrix and the length function for class numeric allow us to carry out comparisons of this nature. Also, in order for solve to work, the matrix X (or X augmented by a unit column vector) must have full column rank. To check this latter condition the R function qr can be used. This function returns a list with a component named rank giving the rank of the matrix supplied as the argument. We will return to the qr function in Section 7.5.

To make sure that the input data is appropriate for creation of an object from a user-created class one can use the setValidity function. The code listed below employs this function to check the row length of the X and y slots as well as the appropriate rank conditions for X.

```
setValidity("regr", function(object){
  if(nrow(object@X) != length(object@y))
    return("The number of rows for X and y differ")
  if(object@intercept){
    if(qr(cbind(1, object@X))$rank < (ncol(object@X) + 1))
      return("The matrix of predictors is singular")
  }
  else{
    if(qr(object@X)$rank < ncol(object@X))
      return("The matrix of predictors is singular")
  }
  return(TRUE)
})
```

The output below demonstrates that objects can no longer be constructed unless they satisfy the requisite validity conditions.

```
> regrObj <- regr(matrix(1, 10, 10), rnorm(3))
Error in validObject(.Object) :
  invalid class "regr" object: The number of rows for X and y differ
> regrObj <- regr(matrix(1, 10, 10), rnorm(10))
Error in validObject(.Object) :
  invalid class "regr" object: The matrix of predictors is singular
> regrObj <- regr(matrix(1, 10, 10), rnorm(10), FALSE)
Error in validObject(.Object) :
  invalid class "regr" object: The matrix of predictors is singular
```

One of the important polymorphic features of an object-oriented language is the ability to have inherited classes that extend a base class in various directions of interest. The inheritance mechanism in R is fairly straightforward requiring only the specification of an additional argument `contains` for the `setClass` function. For our purposes, `contains` will simply be a character string that gives the name of the parent class. The more general case of multiple inheritance will not be considered here.

To illustrate inheritance in R, suppose that it is of interest to extend the `regr` class by including an additional member corresponding to the vector of fitted values. Again, there are already tools for this purpose available in R and, in any case, the more direct approach would be to simply include the fit as a slot in the original `regr` class. For the purpose of illustration we will ignore these issues and proceed as if there is a valid reason for distinguishing between `regr` type objects that do or do not contain the fit to the data.

Let us call our new class `regrFit`. The class is created similarly to the `regr` class using

```
> setClass("regrFit", contains = "regr", representation(yHat
+ = "numeric"), prototype(yHat = 0))
[1] "regrFit"
```

This differs from our creation of the `regr` class through the presence of the `contains` argument in `setClass` that has been set to `regr` thereby indicating that `regrFit` is a derived class for `regr`. Objects of the `regrFit` class will automatically inherit all the slots from the `regr` class: namely, X, y, b and `intercept`. It is only necessary to define the new slot for the fitted values that has been named yHat.

The next step is to create a class constructor such as

```
regrFit <- function(X, y, intercept = TRUE){
  fitObj <- regr(X, y, intercept)
  fitObj <- as(fitObj, "regrFit")
  if(fitObj@intercept)
    fitObj@yHat <- drop(cbind(1, fitObj@X)%*%fitObj@b)
  else
    fitObj@yHat <- drop(fitObj@X%*%fitObj@b)

  fitObj
}
```

The arguments for the constructor are the same as those for the `regr` class. Now, there is no reason to start from scratch as we already know how to create `regr` objects. To make use of this fact a `regr` object is created first with the `regr` class constructor and then coerced or promoted to the `regrFit` class using the `as` function. In this case the general binary version of `as` is being used that has the form

```
as(object, className)
```

with `object` the R object that is to be coerced into an object of data type `className`. When used in this particular context, the effect of `as` is to map all the values for the slots of its first argument into the corresponding ones of a new object of the derived class whose name is supplied in its second argument. This leaves us with the remaining task of filling in the slots that are unique to the derived class: namely, the fitted values stored in the `yHat` slot in this instance. An application of the `regrFit` class constructor to the `mtcars` data will now lead to results such as

```
> regrCarFit <- regrFit(A[,2:4], A[,1])
> regrCarFit@y[1:3]
[1] 21.0 21.0 22.8
> regrCarFit@yHat[1:3]
   Mazda RX4 Mazda RX4 Wag     Datsun 710
    21.81959        21.05474       25.32886
```

6.3 Generic methods

As compared to the C++ setting, where functions are linked to a particular class and accessible only through a class object, functions in R may extend across multiple classes. This allows us to write a function that operates on objects from a given class and subsequently extend that function to work on and produce different types of output for objects of some other class. In this way generic or "template" functions are created with the same name that take arguments from different classes. This section presents details on how such functions are created in the R language. Since R comes with a number of predefined generic functions, there are basically two problems that must be addressed: how to extend one of the predefined generic functions to a new class and how to define a brand new generic function.

As noted at the beginning of the chapter there are two ways to create generic functions in R: the S3 and S4 systems. The syntax for creating and extending functions under the two systems is quite different. Thus, the first step in extending an existing function is to ascertain if it is of the S3 or S4 variety. One way to make such a determination is through use of the `isS4` or `isGeneric` functions that returns `TRUE` when applied to the names of S4 generic functions. In the case of the functions `show` and `coefficients` discussed in Appendix B this produces

```
> isGeneric("show")
[1] TRUE
> isGeneric("coefficients")
[1] FALSE
```

which tells us that `show` is S4 and `coefficients` is S3. An alternative and more informative strategy is to look at a function's "listing" that is returned by entering its names on the command line. In the case of `show` and `coefficients` this generates the output

```
> show
standardGeneric for "show" defined from package "methods"

function (object)
standardGeneric("show")
<environment: 0x1009212d8>
Methods may be defined for arguments: object
Use  showMethods("show")  for currently available ones.
(This generic function excludes non-simple inheritance; see ?setIs)

> coefficients
function (object, ...)
```

```
UseMethod("coef")
<environment: namespace:stats>
```

The presence of standardGeneric in the description of show designates it as an S4 function while useMethod signifies that coefficients is S3. The additional information obtained from the listings is that show has a single argument object while coefficients has arguments object and the catchall ellipsis. New functions that are written to extend show and coefficients to regr objects should have matching arguments. Functions that provide such extensions are called *method functions*.

To extend an existing S3 function func to a new class myClass one need only create a function named func.myClass. For example, consider the S3 coefficients function that currently has the method functions

```
> methods(coefficients)
[1] coef.Arima*    coef.aov*       coef.default* coef.listof*   coef.nls*

    Non-visible functions are asterisked
Warning message:
In methods(coefficients) :
  generic function 'coefficients' dispatches methods for generic 'coef'
```

The warning message tells us that methods for coefficients will be dispatched on calls to functions that have coef. in the beginning of their names. Thus, a version of coefficients that will work for regr objects is simply

```
coef.regr <- function(object){
  object@b
}
```

which now serves as an accessor function for the vector of regression coefficients. With coef.regr in place an application of methods now produces

```
> methods(coefficients)
[1] coef.Arima*    coef.aov*       coef.default* coef.listof*   coef.nls*
[6] coef.regr

    Non-visible functions are asterisked
Warning message:
In methods(coefficients) :
  generic function 'coefficients' dispatches methods for generic 'coef'
```

revealing that our new function has been successfully installed in the family of method functions for coefficients. An application of coef to the regr object regrCar from the previous section gives the expected result: viz,

```
> coefficients(regrCar)
                       disp               wt              qsec
19.7775575655  -0.0001278962  -5.0344097167   0.9266492353
> coefficients(regrCarFit)
                       disp               wt              qsec
19.7775575655  -0.0001278962  -5.0344097167   0.9266492353
```

The polymorphic behavior of coef is illustrated here as well: i.e., the coef function works equally well when applied to the regrCarFit object from the regrFit class that is derived from regr.

The presumption was that we were dealing with an existing S3 method. To create a new S3 method func one uses

```
> func <- function(x) UseMethod("func")
```

Method functions for func can now be added in the same manner as with the coefficients function. For example,

```
> func.regr <- function(x) print("func for regr")
> func(regrCar)
[1] "func for regr"
```

Let us now turn to the case of S4 generic functions. To see the currently available generic functions enter `getGenerics()` on the R command line. New method functions for the functions whose names appear in the resulting list may be created using the `setMethod` function. For functions that are not currently present, a new generic function must be created by first using `setGeneric` and then invoking `setMethod`. Simplified prototypes for `setGeneric` and `setMethod` are

```
setGeneric(name, definition)
```

and

```
setMethod(name, signature, definition)
```

with `name` a character string that represents the name of the function, `definition` a function specification and `signature` a character string array that defines data types for the non-ellipsis arguments of the generic function specified in `name`.

To illustrate the use of `setGeneric` and `setMethod` we will initially focus on a simple "toy" example that will demonstrate the ideas with a minimum of technical detail. Specifically, suppose that a new two-argument function `func` is to be introduced into our R session. This can be accomplished with the command

```
> setGeneric("func", function(object1, object2, ...)
+ standardGeneric("func"))
[1] "func"
```

This establishes `func` as a generic function having three arguments: `object1`, `object2` and the ellipsis. The presence of `standardGeneric` in the body of the function indicates that this incarnation of `func` will serve as the vehicle for methods dispatch. In this sense it represents a hub or "traffic control center" that directs function calls to the appropriate method function. It performs this operation based on the non-ellipsis arguments or `object1` and `object2` in this instance. To remove a generic function from the workspace use `removeGeneric`: e.g.,

```
> removeGeneric("func")
[1] TRUE
```

This will remove `func` as well as any method functions it might have. If only a particular method function should be removed, `removeMethod` can be used.

To see how method dispatch works consider the introduction of two new method functions for `func` via

```
> setMethod("func", signature("numeric", "numeric"),
+ function(object1, object2) object1*object2)
[1] "func"
> setMethod("func", signature = c(object1 = "character",
+ object2 = "numeric"), function(object1, object2)
+ print(paste(object1, object2)))
[1] "func"
```

The `signature` arguments for the two method functions have been specified using two different but equally valid ways; the first case employs the `signature` function that returns a named list of the classes that will be matched in order to the arguments of the hub function while the second just specifies the argument character array directly. The signatures that have been given for the two method functions tell us that the first one will work for cases where both `object1` and `object2` are from class `numeric` while the second is intended for cases where `object1` is of type `character` and `object2` is `numeric`. The first method function will compute the product of its two arguments while the second merely prints out their values. An application of `func` will now produce results such as

```
> func(2, 3)
[1] 6
> func("Hi ya'll!", 3)
[1] "Hi ya'll! 3"
> func(as.character(2), 3)
[1] "2 3"
> func("Hi", "ya'll!")
Error in function (classes, fdef, mtable)    :
  unable to find an inherited method for function "func",
  for signature "character", "character"
```

Thus, the hub method for `func` matches the data types for the arguments that are supplied to `func` with its method function signatures to determine which (if any) are appropriate and then invokes the most appropriate one to "answer" the call to `func`. If there are no method functions with signatures that can match the given arguments, an error message is generated.

Method dispatch cannot be based on additional arguments that come in from the ellipsis direction. This is illustrated by

```
> func(2, 3)
[1] 6
> setMethod("func",   signature("numeric", "numeric"),
+ function(object1, object2, object3) object1*object2*object3)
[1] "func"
> func(2, 3)
Error in object1 * object2 * object3 : 'object3' is missing
> func(2, 3, 4)
[1] 24
```

Initially the existing `func` method function with two numeric arguments is used. Then, a three-argument method function with `numeric` data types for `object1` and `object2` is introduced that is seen to replace the two-argument version. The conclusion is that only the types of the non-ellipsis arguments can serve as a means to distinguish between different method functions.

It is possible to have missing function arguments by specifying them to be of data type `missing`. In order for functions with `missing` arguments to be called, one of three things needs to be true:

- they are explicitly missing; e.g., a comma separated empty argument is present,

- the non-missing arguments are specified by name or

- only trailing arguments from the function's argument list are missing.

The opposite effect from using `missing` is achieved by specifying an argument as data type `ANY`. In this latter case an object from any class can be provided as the argument's value. An illustration of these ideas is provided in the R session output below.

```
> setMethod("func",   signature( "missing", "numeric"),
+ function(object1 ,object2) print("missing and numeric"))
[1] "func"
> setMethod("func",   signature("ANY", "numeric"),
+ function(object1, object2) print("ANY and numeric"))
[1] "func"
> func("a")
```

```
Error in function (classes, fdef, mtable)  :
  unable to find an inherited method for function "func",
  for signature "character", "missing"
> func("a", 2)
[1] "a 2"
> func(2)
Error in function (classes, fdef, mtable)  :
  unable to find an inherited method for function "func",
  for signature "numeric", "missing"
> func(, 2)
[1] "missing and numeric"
> func(object2 = 2)
[1] "missing and numeric"
> setMethod("func",  signature("ANY", "missing"),
+ function(object1, object2) print("ANY and missing"))
[1] "func"
> func("a")
[1] "ANY and missing"
> func(2)
[1] "ANY and missing"
```

Two additional method functions are introduced initially with `numeric` second arguments and first arguments that are `missing` and `ANY`. Neither of the functions (nor any of the previous method functions for `func`) have signatures that will match a call to `func` with a single first argument (and implicitly `missing` second argument) whether its value is from class `character` or `numeric`. The combination of a `character` and `numeric` argument results in a call to the hub `func` that will match two of the method function signatures: our previous one with a `character` and `numeric` argument and the new one with arguments of type `ANY` and `numeric`. The `character` and `numeric` arguments for the former method function are more specific and, in that sense, conform more closely (in fact, exactly) with the argument types that were supplied to the function. Therefore, this is the function that is evaluated. If the first argument is explicitly omitted as in the `func(, 2)` statement, the `func` method function with the `missing` argument will be dispatched for `numeric object2` data types; implicit omission of a leading argument will not work and generates an error message. The other option is to simply assign a `numeric` value for `object2` as demonstrated with the `func(object2 = 2)` statement. Finally, a `func` method function with `ANY` as the class for `object1` and `missing` as the data type for `object2` will respond to cases where only a single first argument is supplied to `func`; since the data type of the first argument is provided, the hub version of `func` has enough information to look for an appropriate method function whose signature will match the first argument (or more generally, the types of the arguments that have been supplied to the function) with other arguments (or just `object2` in this instance) that are designated as `missing`.

Arguments need not even be explicitly specified when one defines a method function; those that are not specified are implicitly assigned data type `ANY`. For example,

```
> setMethod("func",  signature("ANY", "ANY"),
+ function(object1, object2) print("ANY and ANY"))
[1] "func"
```

and

```
> setMethod("func",  signature("ANY"),
+ function(object1) print("just ANY"))
[1] "func"
```

are functions with the same signature.

Let us now apply some of what we have learned about creating generic methods to our `regr` and `regrFit` classes of the previous section. Our first illustration will use the existing S4 function `show` that can be used either explicitly or implicitly to print out information about an object. It is called implicitly when an object's name is entered on the command line.

At present, a default version of `show` would be applied to `regr` objects that employs appropriate versions of `show` for the objects that occupy its slot. For large data sets this is likely much more output than necessary since, for example, the entire X matrix and response vector would be exhibited. As a result, it is probably worthwhile to extend `show` to `regr` objects in a way that produces more succinct and interpretable output.

To create a new method function for `show` the arguments must be known for its parent or hub generic function that will be used for method function dispatch. To discover this information the function name can be entered on the command line as demonstrated at the beginning of the section. If there are R manual pages for the function these will also provide argument details. Still another option is use of the `args` function as in

```
> args(show)
function (object)
NULL
```

This tells us that `show` has a single argument named `object`. With this knowledge in hand, a (very) terse `show` method function for `regr` objects is

```
show.regr <- function(object){
  cat("An object of class regr with", ncol(object@X),
      "predictors", "\n")
}
```

that uses the `cat` function as described in Section 5.2 for printing output. This is then installed as a new method function using

```
> setMethod("show", signature("regr"), show.regr)
[1] "show"
```

Then,

```
> regrCar
An object of class regr with 3 predictors
> regrCarFit
An object of class regr with 3 predictors
```

which demonstrates that the new version of `show` is now being used for `regr` objects. As `regrFit` is a derived class of `regr`, this version of `show` will also work for `regrFit` objects.

In some cases it is of interest to perform a transformation on the dependent variable when conducting regression analysis. One option that is often used is a power transformation. Numeric objects are raised to a power in R with the ^ operator. A check with `getGenerics` reveals that ^ is an S4 function in the base package that is loaded whenever an R session is started. An application of `args` produces

```
> args("^")
function (e1, e2)
```

Thus, new method functions may be defined that have the arguments e1 and e2. A new method function that will produce an `regr` object with a power transformed dependent variable is seen to be

```
> setMethod("^", signature("regr", "numeric"), function(e1, e2)
+ regr(e1@X, (e1@y)^e2))
[1] "^"
```

This can now be used on either `regr` or `regrFit` objects as demonstrated by

```
> newRegrCar <- regrCar^(.5)
> newRegrCar@b
                 disp            wt           qsec
 4.5053469796 -0.0002690319 -0.5360063280  0.0961389814
> newRegrCarFit <- regrCarFit^(.5)
> newRegrCarFit@b
                 disp            wt           qsec
 4.5053469796 -0.0002690319 -0.5360063280  0.0961389814
> newRegrCarFit@yHat[1]
Error: no slot of name "yHat" for this object of class "regr"
```

This output reveals a problem. Even though the ^ operator will work with `regrFit` objects, the final product is not all that might be desired since the yHat slot is deleted. An alternative version of ^ that resolves this issue is obtained from

```
> setMethod("^", signature("regr", "numeric"), function(e1, e2) {
+ if(.hasSlot(e1, "yHat")) regrFit(e1@X, (e1@y)^e2)
+ else regr(e1@X, (e1@y)^e2)
+ })
[1] "^"
```

This uses inheritance along with the `.hasSlot` function to distinguish between `regr` and `regrFit` objects to allow the operator to work on both the base and derived class. With this new version of ^ in place the previous calculations will now produce the expected results: e.g.,

```
> newRegrCarFit <- regrCarFit^(.5)
> newRegrCarFit@yHat[1]
Mazda RX4
 4.640413
```

There is a bit more to the ^ operator than has been mentioned here. It is part of a *group generic* called **Arith** that includes other arithmetic operators such as +, -, *, etc. Group generic functions are collections of functions that perform related operations and have comparable mathematical structure in terms of, e.g., the number of arguments. With group generic functions it is possible to simultaneously extend all the operators in the group to work with objects of a given class. We will illustrate this in the next section using the **Compare** group generic.

The @ operator and `slot` function provide us with a means of accessing the slots of an object. However, as was true for the C++ setting, it is generally preferable to write accessor functions that simplify code and are named after a mnemonic of the slot of interest. For example, we might create an accessor function for the coefficient vector of an `regr` object with

```
> setGeneric("bHat", function(object) standardGeneric("bHat"))
[1] "bHat"
> setMethod("bHat", signature = "regr", function(object) object@b)
[1] "bHat"
```

The method will work equally well with `regr` and `regrFit` objects as seen from

```
> bHat(regrCar)
                disp              wt            qsec
19.7775575655  -0.0001278962  -5.0344097167  0.9266492353
> bHat(regrCarFit)
                disp              wt            qsec
19.7775575655  -0.0001278962  -5.0344097167  0.9266492353
```

An examination of the regression coefficients from our continuing mtcars example suggests that the b slot of `regrCar` could be simplified to $(20, 0, -5, 1)$. Thus, we might try to perform the operation

```
> bHat(regrCar) <- c(20, 0, -5, 1)
Error in bHat(regrCar) <- c(20, 0, -5, 1) :
    could not find function "bHat<-"
```

This error message is a bit confusing as we just wrote and successfully used our bHat accessor function. The problem is that we are not using bHat here but are, instead, attempting to use its replacement version bHat<- (see Section 5.4) that indeed has not been defined for the regr class. This oversight is easy to correct using

```
> setGeneric("bHat<-", function(x, value) standardGeneric("bHat<-"))
[1] "bHat<-"
```

We then specify the form of the replacement method function using

```
> setReplaceMethod("bHat", "regr", function(x, value)
+ {x@b <- value; x})
[1] "bHat<-"
```

Then, the alternative version of the coefficient vector can be installed with

```
> bHat(regrCar) <- c(20, 0, -5, 1)
> bHat(regrCar)
[1] 20  0 -5  1
```

The creation of accessor functions for the other slots of `regr` and `regrFit` objects is the subject of Exercise 6.3.

6.4 An example

To conclude the discussion in this chapter we will consider a particular data analysis problem corresponding to the fictional Guess5 lottery game discussed in Section 5.6. This game involved the selection of five (imaginary) balls from a set of forty (imaginary) balls numbered from 1 to 40. Drawings for the game were conducted twice a week on Monday and Thursday using two (imaginary) drawing machines (i.e., machines A and B) and 10 (imaginary) ball sets (i.e., sets 1 to 10). Both the machine and ball set for a drawing were selected at random and then used to carry out nine preliminary test draws before the actual game drawing.

In Section 5.6 an artificial data set was created that contained the results of 100 Guess5 draws. The data was stored in the file guess5.txt. The first draw in the file looks like

```
> system("head -n 11 guess5.txt")
1952-01-17 A 6 T
10 38 23 20 15
36 15 11 7 37
20 10 9 25 2
29 14 16 31 34
12 38 28 26 2
16 19 22 26 33
```

```
25  17  21  3  10
16  8  32  6  29
22  26  7  24  12
29  16  37  36  27
```

The results for the other draws in the file are in the same format. Consequently, every draw will produce 11 lines of text with the first line containing the date, the machine, the set and the day of the week (M or T for Monday and Thursday) for the draw. Let us now assume that the drawing results from the Guess5 game data will all be stored in the format shown above. If regular analysis is to be conducted on this data, time and energy can be conserved by creating a data analysis structure that is tailored to dealing with data of this type. A natural way to accomplish this is by building Guess5 related classes and functions and that is the approach that will be taken here.

The first step in the design process is to consider what classes would be useful and suitable for Guess5 data. In this respect the smallest data "unit" might be viewed as being the results of a particular drawing. It is important to be able to access the information about each individual draw for historical purposes as well as diagnostic analysis related to unusual drawing events that may occur at various points in time. This suggests that each draw needs to have its own "identity" and, with that in mind, a natural starting point would be the creation of a Guess5 class to hold the information about a single Guess5 draw.

In creating a Guess5 class we need to consider what information to retain about the draw results. The balls can exhibit structural problems from wear and tear and machines can malfunction. Thus, it may be necessary to look at subsets of the data that correspond to different ball sets and machines. Also, the date of the draw provides a unique identifier that can aid in sorting and locating data in a file suggesting that information of this nature should be included. These considerations lead to a "wish list" for properties that might be built into Guess5 class objects:

a) A Guess5 object should contain the results of a single night's draw.

b) Class members should include identifiers for the drawing machine and ball set that were used to carry out the draw. Other possible class members might represent information on the date and day of the week for the draw.

c) The actual drawing results (i.e., the numbers for each of the balls that are drawn in the Guess5 draw and pretest experiments) should be available in a form that is easy to manipulate for data analysis purposes: e.g., as a matrix.

An R class that adheres to our design guidelines for Guess5 drawing results might be established with something like

```
> setClass("g5", representation(drawDate = "Date",
+ mName = "character", sNo = "integer", Day = "character",
+ data="matrix"))
[1] "g5
```

This creates a class called g5 with class members for the date of the draw (drawDate), the machine name (mName), the ball set number (sNo) and the day of the week (Day). All these members derive from the native R classes character, integer and Date. We worked with the Date class previously in Section 5.6. The actual drawing results will be stored in the slot data as a matrix object.

The next step is to create a constructor for the class. There are various ways this could be accomplished depending on the way the drawing data can be accessed. Let us suppose here that the data come in the form of a 54-element character array: the first four array entries provide the date, machine name, ball set number and day of the week, respectively, and the next 50 give the drawing results. A constructor that will work for that scenario is

```
g5 <- function(a){
  drawdate <- as.Date(a[1])
  mname <- a[2]
  sno <- as.integer(a[3])
  day <- a[4]
  A <- matrix(as.integer(a[5:54]), 10, 5, byrow = TRUE)
  g5Obj <- new("g5", drawDate = drawdate, mName = mname, sNo = sno,
        Day = day, data = A)
  g5Obj
}
```

The first four elements of the input array a are used to set the values that will be used to fill the drawDate, mName, sNo and Day slots for the g5 object. Then, the next 50 elements are converted from character to integer using the as.integer function and reshaped into a 10 × 5 matrix. This information is used in the basic constructor function new to create a new class object.

An application of our g5 constructor to data from the guess5.txt file produced

```
> a <- scan("guess5.txt", what = character(0), n = 108)
Read 108 items
> g5Obj1 <- g5(a[1:54]); g5Obj2 <- g5(a[55:108])
> g5Obj1
An object of class "g5"
Slot "drawDate":
[1] "1952-01-17"

Slot "mName":
[1] "A"

Slot "sNo":
[1] 6

Slot "Day":
[1] "T"

Slot "data":
      [,1] [,2] [,3] [,4] [,5]
 [1,]   10   38   23   20   15
 [2,]   36   15   11    7   37
 [3,]   20   10    9   25    2
 [4,]   29   14   16   31   34
 [5,]   12   38   28   26    2
 [6,]   16   19   22   26   33
 [7,]   25   17   21    3   10
 [8,]   16    8   32    6   29
 [9,]   22   26    7   24   12
[10,]   29   16   37   36   27

> g5Obj2@drawDate
[1] "1952-01-21"
```

The scan function is used to import the information about the first two draws or first 108 elements from the guess5.txt file. This character array is then processed with the g5 class constructor to produce two g5 class objects. Examination of the objects indicates that they agree with the drawing data created in Section 5.6.

A validity check function should be installed to ensure that the constructor has produced a valid g5 object. A function that will work for that purpose is

```
validG5 <- function(object){
 if(object@mName != "A" & object@mName != "B")
   return(paste("Incorrect machine name for date", object@drawDate))
 if(object@sNo < 1 | object@sNo > 10)
   return(paste("Incorrect set number for date", object@date))
 if(object@Day != "M" & object@Day != "T")
   return(paste("Incorrect day for date", object@date))
 if(any(object@data < 1 | object@data > 40))
    return(paste("Incorrect ball number for date", object@date))
  return(TRUE)
}
```

This can be set as the function to check for incorrect g5 member input data through the command

```
   setValidity("g5", validG5)
```

The effect will now be to reject situations where the ball set, machine or day entries do not coincide with those that are appropriate for the game. Similarly a check is performed to see that all the ball numbers are between 1 and 40. The calculations in this latter case are carried out using the any function that takes a logical array as its argument and returns TRUE if any of its elements evaluate as TRUE. If there are any ball number entries with values that fall outside the 1 to 40 range, any will return TRUE and the date of the draw where an error has been found will be printed to help in locating the problem and correcting it in the text file containing the data.

Now that we have created the g5 class some of the R generic functions can be extended to work with g5 objects. One function of interest is the S3 function summary. We could use the S3 system discussed in the previous section to define a version of summary that would be applicable to the g5 class. There is another option: namely, make summary an S4 generic function with

```
> setGeneric("summary")
[1] "summary"
> summary
standardGeneric for "summary" defined from package "base"

function (object, ...)
standardGeneric("summary")
<environment: 0x102839c40>
Methods may be defined for arguments: object
Use  showMethods("summary")  for currently available ones.
```

The last part of the output confirms that summary is now an S4 function with the same arguments, object and the ellipsis, as its S3 progenitor. A method function that might serve our purpose is then installed with

```
setMethod("summary", "g5", function(object){
  cat(paste("Drawings for", object@drawDate, "used Machine",
  object@mName, "and Ball Set", object@sNo, "\n"))
  cat("Drawing results with pretests", "\n")
  print(object@data)
  }
)
```

An application of summary to the g5 object g5Obj1 will now produce

```
> summary(g5Obj1)
Drawings for 1952-01-17 used Machine A and Ball Set 6
Drawing results with pretests
         [,1] [,2] [,3] [,4] [,5]
   [1,]    10   38   23   20   15
   [2,]    36   15   11    7   37
   [3,]    20   10    9   25    2
   [4,]    29   14   16   31   34
   [5,]    12   38   28   26    2
   [6,]    16   19   22   26   33
   [7,]    25   17   21    3   10
   [8,]    16    8   32    6   29
   [9,]    22   26    7   24   12
  [10,]    29   16   37   36   27
```

Eventually we will create a function that can sort a collection of g5 objects. The critical ingredient in accomplishing this is having order relationships that allow us to compare one g5 object to another. For our purposes the ordering will be in terms of the drawDate slot that will produce a chronologically sorted arrangement. Rather than define ==, <=, etc., individually it is possible to define all such operators simultaneously using a slight modification of our previous approach for defining method functions. The difference is that now the method functions will be defined for a group or collection of functions named Compare in this instance. Other group generics include those with names Arith, Logic, Math and Summary as described on the R help page for S4groupGeneric. For Compare this help page describes the functions in the group as having the basic form

```
Compare(e1, e2)
```

Thus, they are all binary functions with two arguments e1 and e2. All of them can now be extended to the g5 class using

```
> setMethod("Compare", signature(e1 = "g5", e2 = "g5"),
+ function(e1, e2) callGeneric(e1@drawDate, e2@drawDate))
[1] "Compare"
```

The callGeneric that appears here is for method dispatch; it will call the current generic function that is appropriate for the Date class objects e1@drawDate and e2@drawDate in this instance. This will be sufficient because all the functions in the Compare group generic have method functions that apply to the Date class as indicated in the R help page for Ops.Date. A test of our comparison operations gives results such as

```
> g5Obj1 > g5Obj2
[1] FALSE
> g5Obj1 < g5Obj2
[1] TRUE
> g5Obj1 == g5Obj2
[1] FALSE
> g5Obj1 <= g5Obj2
[1] TRUE
```

The fact that the date for g5Obj1 (i.e., 1952–01–17) is earlier that the date for g5Obj2 (i.e., 1952–01–21) suggests that the comparison operators are working correctly.

So far the emphasis has been on constructing an object corresponding to a single Guess5 drawing. In practice a file of drawing data will contain information from multiple draws as in the guess5.txt file. Thus, a system is needed for bundling together a collection of g5 objects into a framework that is easy to use in a data analysis context. There are many ways to accomplish this. The approach that will be developed here is to process the raw

Guess5 data from a text file and then use our g5 constructor to create g5 objects that will
be stored in a list structure. To implement this idea another class called g5List is created
via the command

```
> setClass("g5List", representation(objList = "list", nDraws =
+ "integer", nSet = "integer", nMach = "integer"))
[1] "g5List"
```

The essential ingredient of this class is the objList slot that will hold the list of g5 objects.
Additional class members have been added to hold information about i) the number of
times each machine has been used (the nMach slot that will hold a two-component integer
vector), ii) the number of times each ball set was used (the nSet slot that will hold a
10-component integer vector) and iii) the number of draws nDraws or, equivalently, the
number of g5 objects that are stored in the list objList.

The next step is to create a constructor for the g5List class. The ensuing listing accomplishes this and we will spend some time discussing the details of this function.

```
g5List<-function(fileName){
  A <- readG5(fileName)
  ndraws <-length(A)/54
  objlist <- lapply(1:ndraws, FUN =
        function(i) g5(A[(54*i - 53):(54*i)]))
  temp <- sapply(objlist, FUN = function(g5obj) g5obj@mName)
  tempNMach <- sum(temp == "A")
  nmach <- c(tempNMach, ndraws - tempNMach)
  temp <- sapply(objlist, FUN = function(g5obj) g5obj@sNo)
  nset <- integer(10)
  for(i in 1:10) nset[i] <-  sum(temp == i)
  g5ListObj <- new("g5List", objList = objlist, nDraws =
          as.integer(ndraws), nSet = nset,
          nMach = as.integer(nmach))
}
```

The first action in the g5List constructor is to import the drawing data from the file whose
name fileName is supplied as its argument. The data is brought in as a character array via
the function readG5 that takes the form

```
readG5 <- function(fileName){
  out <- tryCatch(unlist(strsplit(readLines(fileName), split = " ")),
  warning = function(e){print(e)
  print("Please enter a file name followed by a return")
  newName <- scan(, what = character(0), quiet = TRUE)
  readG5(newName)},
  error = function(e) print(e))
  out
}
```

This reads the information from a file in exactly the same way as in Section 5.6 except that
the use of scan is now managed by the tryCatch function that, similar to developments
in Section 5.4, allows for recovery from an error in the file name specification. Once the
data has been imported, the number of draws nDraws can be determined as the length of
the A array divided by 54. The contents of A are then transformed into a list of g5 objects
using the lapply function. The first argument to lapply is a vector of integers from 1 to
the number of draws. These values are passed to a function that then uses the appropriate
subsets of the A character array to call the g5 class constructor for each draw in the data
set. The output from lapply is a list which is what is desired in this instance.

There are two members of the g5Array class that remain to be determined: the vector nMach that gives the number of times the two machines have appeared and the vector nSet that contains the number of times that each of the 10 ball sets have been used. To evaluate the first of these sapply is used on the list of g5 objects in a way that will extract the information in the mName slot for every element in the list. By using sapply (rather than lapply) the output will be returned as a character vector temp. The action of (temp == "A") is to return a vector of Boolean variables of the same length as temp so that applying sum to its components will return the number of times that "A" appeared in temp. The number of times the "B" machine was used is the difference between ndraws and the number of times that the "A" machine was employed.

The idea behind finding the number of occurrences for each ball set is similar to the calculations that were done concerning the two machines. First a vector of integers giving the ball set number for each draw is extracted from the list of g5 objects using sapply. Then, a for loop is used to count the number of TRUE values in each of the vectors of logical variables (temp == 1), ..., (temp == 10). The final step puts all the information that has been accrued into the basic constructor function new and returns the resulting g5List object.

For convenience of access to the elements of a g5List object we should extend the list sub-setting operator [[to work with the g5List class. The most natural way to accomplish this is by having g5ListObj[[i]] correspond to g5ListObj@objList[[i]] for a g5List object g5ListObj. The [[operator is an S4 function whose prototype from the corresponding R help page takes the form

```
x[[i, j, ..., exact = TRUE]]
```

The R help page also tells us that the method is based on internal code which is built into the R interpreter with the result that methods may only be dispatched on the x argument. So, by taking x = "g5List" an indexing method function can be introduced with

```
> setMethod("[[", signature(x = "g5List"), function(x, i, j, ...,
+ exact = TRUE) x@objList[[i]])
[1] "[["
```

It seems natural to also define a replacement version of [[at this point which is readily accomplished with

```
> setReplaceMethod("[[", signature(x = "g5List"),
+ function(x, i, j, ..., exact, value) {x@objList[[i]] <- value; x})
[1] "[[<-"
```

The effect is to replace a component of the list with a specified index with the g5 object on the right side of the <- symbol.

The list indexing operator is probably the most natural way to access the individual g5 objects in a g5List object. This alone is not sufficient in that it does not provide us with access to the other g5List class members nDraws, nMach and nSet. Of course, these can be obtained using the @ operator. But, a more aesthetically pleasing approach is provided by accessor functions created with the code

```
> setGeneric("nDraws", function(object) standardGeneric("nDraws"))
[1] "nDraws"
> setMethod("nDraws", signature = "g5List", function(object)
+ object@nDraws)
[1] "nDraws"
> setGeneric("nMach", function(object) standardGeneric("nMach"))
[1] "nMach"
> setMethod("nMach", signature = "g5List", function(object)
+ object@nMach)
```

```
[1] "nMach"
> setGeneric("nSet", function(object) standardGeneric("nSet"))
[1] "nSet"
> setMethod("nSet", signature = "g5List", function(object)
+ object@nSet)
[1] "nSet"
```

The constructor for the g5List class was applied to the data in our guess5.txt file. Some of the properties of the resulting g5List object are examined below.

```
> g5ListObj <- g5List("guess5.txt")
> nDraws(g5ListObj)
[1] 100
> nMach(g5ListObj)
[1] 53 47
> nSet(g5ListObj)
 [1]  4  8 12 14 12 12 12  7  8 11
> g5ListObj[[2]]
An object of class "g5"
Slot "drawDate":
[1] "1952-01-21"

Slot "mName":
[1] "B"

Slot "sNo":
[1] 4

Slot "Day":
[1] "M"

Slot "data":
      [,1] [,2] [,3] [,4] [,5]
 [1,]   11    9   23   10   20
 [2,]   32    7   16   18   40
 [3,]   38   35   26   36   19
 [4,]   24   14   39    1   19
 [5,]   35    1    3    7   28
 [6,]   30   38   18    3   24
 [7,]   31    6   16    9    3
 [8,]   16    3    9   39   25
 [9,]   12    4    3   33   28
[10,]   33   39    4   38   29

> temp <- g5ListObj[[2]]
> g5ListObj[[2]] <- g5ListObj[[1]]
> g5ListObj[[2]]@drawDate
[1] "1952-01-17"
> g5ListObj[[2]] <- temp
> g5ListObj[[2]]@drawDate
[1] "1952-01-21"
```

The sum of the values for nMach and nSet should be 100 as we find to be the case. A check with results from Section 5.6 also indicates that the second g5 object agrees with the one in the guess5.txt file. A test of our replacement version of [[successfully overwrites the information from the second draw with that from the first and then reverses the operation.

The task remains of actually analyzing the Guess5 data and, similar to what was done in Section 5.6, chi-square tests will be used for that purpose. Instead of looking at all the drawing information the idea is to now work with subsets of the data corresponding to different machines and ball sets. There are two obvious ways to proceed in creating the tools for this type of analysis: namely,

a) an analog of the `table` function (discussed in Section 5.6) could be created for working with g5List objects that produces results that can be used directly with the `chisq.test` function (also discussed in Section 5.6) or

b) the entire analysis could be packaged in a suitable version of the `chisq.test` function.

Our approach will take the second path. The other option is the topic of Exercise 6.13.

The first step is to make `chisq.test` an S4 generic function. This is accomplished with

```
> setGeneric("chisq.test")
[1] "chisq.test"
> chisq.test
standardGeneric for "chisq.test" defined from package "stats"

function (x, y = NULL, correct = TRUE, p = rep(1/length(x), length(x)),
    rescale.p = FALSE, simulate.p.value = FALSE, B = 2000)
standardGeneric("chisq.test")
<environment: 0x100a89e98>
Methods may be defined for arguments: x, y, correct, p, rescale.p,
                simulate.p.value, B
Use  showMethods("chisq.test")  for currently available ones.
```

The summary information about `chisq.test` states that new method functions may be created that use the arguments x, y, correct, p, rescale.p, simulate.p.value and B. A new feature is that the y, correct, p, rescale.p, simulate.p.value and B have all been given default values in the generic function's specification.

The question is which arguments to use in our new method function and how to use them. In some respects there is no reason to prefer one over the other (Exercise 6.12). If we want the name of the argument we choose to have some connection to its purpose, correct, rescale.p, simulate.p.value and B are not compelling options. The x argument in the original `chisq.test` method is the data to be analyzed and usually a vector or matrix of counts. When applied in the form

```
    chisq.test(x, y)
```

the vectors x and y provide pairs of values on two categorical variables for the observations in the data. This is not precisely what we have for our problem. But, a bit of imagination suggests taking x as a g5List object and y as a list of machines and ball sets to be used in the analysis.

The listing below gives one possible set of code that could be used to create a new g5List method function for `chisq.test`.

```
chisq.test.g5List <-function (x, y = list(machRange = c("A", "B"),
    setRange = 1:10), correct, p, rescale.p, simulate.p.value, B){
    objlist <- x@objList
    tempM <- sapply(objlist, FUN = function(g5obj) g5obj@mName)
    tempS <- sapply(objlist, FUN = function(g5obj) g5obj@sNo)
    tempM <- is.element(tempM, y[[1]])
    tempS <- is.element(tempS, y[[2]])
    indices <- as.logical(tempM*tempS)
    newList <- objlist[indices]
    ndraws <- length(newList)
    temp <- as.numeric(sapply(1:ndraws, FUN =
            function(i) newList[[i]]@data))
    chistat = 39*chisq.test(x = table(temp))[[1]]/35
```

```
    names(chistat) <- NULL
    list(chistat, 1 - pchisq(chistat, 39))
}
```

As discussed above, the x argument is a g5List object and y is a list with the machine or machines to use in the comparison as its first component and the numbers for the ball sets as its second component. A default value has been provided for the y argument that will result in the analysis of the entire data set: i.e., a chi-square statistic will be calculated for the draws produced using both machines and all 10 ball sets. The R help page for setMethods states that in order for a default argument to be used in a method function "the generic function must have some default expression for the same argument". Since, as noted above, this is true for the y argument in the chisq.test generic function, the specified default list can be used provided we furnish a a suitable signature when we define the method function.

The initial step in the chisq.test.g5List function is to extract the list of g5 objects from the x argument. Then, similar to the developments in the g5List constructor, sapply is used to obtain a vector of characters and a vector of integers that give the machines and ball sets for each of the g5 objects in the data. The is.element function is applied to these vectors using the y[[1]] and y[[2]] list components. The result is two vectors of logical variables, tempM and tempS, whose entries are TRUE or FALSE depending or whether or not the element's value lies in the set designated by the machine component y[[1]] and set component y[[2]] of the y list. Multiplication of two logical variables produces 0 (if either of the two variables is FALSE) or 1 (if both variables are TRUE). Thus, the product of tempM and tempS will have all 0 entries except for those g5 objects whose mName and sNo slots contained values that were in the intersection of y[[1]] and y[[2]]. The as.logical function returns these values to logical TRUE and FALSE form. The subsetting feature for R lists has the consequence that when this vector of logical variables is passed to the [] operator, the output will be a sublist corresponding to the elements of the list for which the index evaluates as TRUE. From this point the development proceeds along the lines of Section 5.6 except that the analysis is working with a subset, rather than the entire collection of data.

We can now establish chisq.test.g5List as a method function for chisq.test with

```
> setMethod("chisq.test", signature(x = "g5List", y = "ANY"),
+ definition = chisq.test.g5List)
[1] "chisq.test"
```

Our choice for the signature perhaps requires a bit of discussion. As a rule, our calls to chisq.test will involve two specified arguments with the first being a g5List object and the second an R list object. However, for a call to use the default argument that was supplied for the method function, the second argument will need to be missing. Thus, we need a signature that will match both (x = "g5List", y = "list") and (x = "g5List", y = "missing"). An option that will work is (x = "g5List", y = "ANY").

Examples of how the new chisq.test method function works are given by the output below.

```
> chisq.test(g5ListObj, list(machRange = "A", setRange = 1:10))
[[1]]
[1] 39.88722

[[2]]
[1] 0.4304843

> chisq.test(g5ListObj, list(machRange = c("A", "B"), setRange = 10))
[[1]]
[1] 39.02026
```

```
[[2]]
[1] 0.4689672

> chisq.test(g5ListObj)
[[1]]
[1] 27.77691

[[2]]
[1] 0.9100113
```

The analysis has produced results for subsets of the data that isolate the performance of machine "A" and ball set 10 alone. The final application of `chisq.test` uses the default y argument that entails analysis of the entire data set. The chi-square statistic produced in this latter case agrees with the answer that was previously obtained in Section 5.6.

Finally, let us consider how to create a method for sorting the data inside a `g5List` object. The ordering will be in terms of draw dates with the goal being to arrange the g5 objects in the `objList` slot in chronologically descending order. The comparison operators are already in place and the only remaining obstacle is the creation of code to implement a sorting routine.

Lists in R are basically arrays that can hold objects of general types. Thus, techniques for sorting arrays can be employed to solve our problem. Chapter 9 discusses some efficient routines for that purpose. For now we will use a simple, inefficient, method called *selection sort* that can be used to sort an array in place. The idea is straightforward; one finds the minimum value in the array, swaps it with the first entry, repeats the operation on the array composed of all but the first element, etc.

The driving force behind selection sort is a method that finds the smallest value in an array. A function that will serve this purpose for our setting is

```
minIndex <- function(x){
  minIndex <- 1
  minObj <- x[[1]]
  for(i in 1:length(x))
    if(x[[i]] < minObj){
      minIndex <- i
      minObj <- x[[i]]
    }
  list(minIndex, minObj)
}
```

The input argument x is assumed to be a list of objects that can be ordered by the comparison operator `<`. This is used in conjunction with a `for` loop to move through the list and locate the "smallest" (vis-à-vis the definition of `<`) object.

The function that actually sorts the list is

```
sortList <- function(x){
  y = NULL
  if(length(x) > 1){
    temp <- minIndex(x)
    x[temp[[1]]]<- x[[1]]
    x[[1]] <- temp[[2]]
    y <- sortList(x[2:length(x)])
  }
  y = c(y, x[[1]])
  y
}
```

The sorting process uses recursion (cf. Exercise 3.7); i.e., each time the minimum element is removed from the list `sortList` calls itself again. This process continues until the list is of size one. At that point the last two lines of the function are realized for the first time after which the minimum elements from the previous iterations are sequentially collected in the new list y.

There is another way to create an S4 generic function: add a method function to any existing function. A simple example that illustrates the basic idea is

```
> func <- function(x, y) x*y
> setMethod("func", signature(y = "integer"), function(x, y) y)
Creating a generic function from function "func"
[1] "func"
> func(3, 3); func(3, as.integer(3));
[1] 9
[1] 3
```

We can use this last approach to create an S4 generic function from the existing S3 function `sort`. It has arguments

```
> args(sort)
function (x, decreasing = FALSE, ...)
NULL
```

A new S4 generic function that adheres to this format can then be created with

```
> setMethod("sort", signature("g5List"), function(x,
+ decreasing = FALSE, ...) {newObject <- new("g5List",
+ objList = sortList(x@objList), nDraws = nDraws(x),
+ nSet = nSet(x), nMach = nMach(x))
+ newObject
+ })
Creating a new generic function for "sort" in ".GlobalEnv"
[1] "sort"
```

The listing below gives some illustrations of using the new `sort` method function.

```
> as.Date(sapply(g5ListObj@objList, function(e) as.Date(e@drawDate)),
+ origin = "1970-01-01")[1:5]
[1] "1952-01-17" "1952-01-21" "1952-01-24" "1952-01-28" "1952-01-31"
> as.Date(sapply(sort(g5ListObj)@objList, function(e)
+ as.Date(e@drawDate)), origin = "1970-01-01")[1:5]
[1] "1952-12-29" "1952-12-25" "1952-12-22" "1952-12-18" "1952-12-15"
> set.seed(123)
> newg5ListObj <- new("g5List", objList = sample(g5ListObj@objList),
+ nDraws = nDraws(g5ListObj), nSet = nSet(g5ListObj),
+ nMach = nMach(g5ListObj))
> as.Date(sapply(sort(newg5ListObj)@objList, function(e)
+ as.Date(e@drawDate)), origin = "1970-01-01")[1:5]
[1] "1952-12-29" "1952-12-25" "1952-12-22" "1952-12-18" "1952-12-15"
```

The original `g5List` object `g5ListObj` holds g5 objects arranged in chronologically ascending order. Our `sort` method function successfully reverses this order. It is also applied to a `g5List` object that is obtained from `g5ListObj` by randomly shuffling its contents with similar results.

6.5 Exercises

6.1. Add a method function to the S3 generic function `residuals` that will work with `regr` class objects.

6.2. Expand the `regr` class from Section 6.2 to include

a) both residual and regression sums of squares,

b) an F-statistic for testing the significance of the regression,

c) a p-value for your F-statistic from part b) and

d) the value of the coefficient of determination.

Using this extended class decide whether or not a constant term should be used in creating an `regr` object from the mtcars data in Section 6.2.

6.3. Write code for accessor functions to go with your class from Exercise 6.2.

6.4. Make the constructor for your `regr` class of Exercise 6.2 an S4 generic function. Then, add two method functions:

a) the original `regr` constructor from Exercise 6.2 and

b) a constructor that allows the `b` slot to be a specified vector.

6.5. Create a method for `plot` that can be used with objects from the `regr` and `regrFit` classes. Exploit the ... argument for plot by including additional arguments to your `plot` method that will produce plots of residuals or fits against the independent variables.

6.6. Write a method for the `show` function that will distinguish between `regr` and `regrFit` objects.

6.7. Extend the `Math` group generic functions to produce `regr` and `regrFit` objects that use transformations of the response vector. Then, evaluate the performance of the log transform for the mtcars data example of Sections 6.2–6.3.

6.8. Create a validity check function that can be used for g5List objects.

6.9. Write a method function for `print` that will work with g5List objects.

6.10. Create a `g5List` constructor that can be used to append or insert another new g5 object into an existing `g5List` object. [Hint: Begin with a generic `g5List` constructor and then create method functions that work with file input or existing `g5List` objects.]

6.11. Create new generic functions and associated method functions that provide accessor and assignment functions for the slots of the g5 class.

6.12. Rewrite the `chisq.test.g5List` function using one of the other (than y) arguments for the S4 generic version of `chisq.test`.

6.13. Extend the `table` function to work with `g5List` objects in a way that will produce frequency counts for the drawing results corresponding to different subsets of the machines and ball sets used to produce the data.

6.14. Write a method function for + that adds the ball frequencies for g5 objects. Then, use this addition operator to obtain an alternative version of the `chisq.test.g5List` function in Section 6.4.

6.15. Write code that implements a method function for `sort` with g5List objects that is based on the bubblesort method described in Algorithm 3.1.

6.16. Write code for a sorting method for `g5List` objects that relies on the R functions `unlist` and `order` while not directly using the comparison operators.

6.17. Consider the game in Exercise 5.17.

a) Create a class for individual draws.

b) Define generic methods for `summary`, `show` and `print` that will work with objects from your class in part a).

c) Create accessor and replacement functions for the class data members.

d) Develop another class that will allow for manipulation and storage of a collection of objects from the class in part a).

e) Develop a method function for the S4 version of `chisq.test` that will work with objects from the class developed in part d).

f) Analyze the data that was generated is Exercise 5.17 using your method from part e).

6.18. Rework Exercise 6.17 using inheritance from the g5 and g5List classes of Section 6.4.

6.19. Create a class that can be used for working with complex numbers. In addition,

a) extend `show`, `summary`, `print` and `[` to work with objects from your complex number class,

b) extend the arithmetic operators $*, +, -$ and $/$ to objects of the complex number class,

c) create a `sort` method function that will arrange complex number objects in numerically ascending or descending order as determined by their complex modulus or norm and

d) develop a version of the S3 generic `plot` function that will apply to complex number objects.

[Note: The resident R class for complex numbers is named `complex`. You should avoid using this name for your class.]

6.20. The R `array` class has the constructor

```
array(data, dim, dimnames)
```

with `data` a `vector` or `list` object, `dim` a vector of integers specifying the size of each dimension and `dimnames` the associated dimension names. For example,

```
> array(rnorm(27), c(3, 3, 3))
, , 1

            [,1]        [,2]        [,3]
[1,]   0.3741209  -0.3761964   0.4679258
[2,]  -2.6802921  -0.6930957  -0.7583162
[3,]  -0.4875758  -1.4664360   1.5395929


, , 2

            [,1]        [,2]        [,3]
[1,]  1.0333163  -0.3666477   1.1908588
[2,]  2.3681655  -0.2561474   0.3799085
[3,]  0.2180131   0.4169775  -1.1094986


, , 3

            [,1]        [,2]        [,3]
[1,]  -0.3157936  -1.9503897  -0.2257388
[2,]   0.5651924  -0.9226202  -0.0743885
[3,]   1.2550420  -0.3217654   1.0233531
```

creates a $3 \times 3 \times 3$ array of real numbers while

```
> A <- array(list(matrix(rnorm(2), 2, 2), 1:5, c("a", "b"),
+ expression(1 + x)), c(2, 2))
> A[1, 1]
[[1]]
            [,1]        [,2]
[1,]  -0.9755330  -0.9755330
[2,]   0.2602563   0.2602563
```

```
> x = 2; eval(A[2, 2][[1]]);
[1] 3
```

creates a 2 × 2 array of disparate R objects. Notice our use of the `eval` function from Exercise 5.9 to evaluate the `expression` object in the array.

Use the `array` class to create a class that can be used for working with three-dimensional numeric arrays. Also,

a) extend `show`, `summary` and `print` to work with objects from your array class and

b) extend the arithmetic operators ∗, + and − to objects of the array class.

6.21. Create a class that can be used for working with character strings. In addition,

a) extend `show`, `summary` and [to work with objects from your string class,

b) extend the arithmetic operator + to objects of the string class where its action is concatenation,

c) provide an S4 method function for `match` that applies to string class objects and

d) provide an S4 method function for the `print` function that is appropriate for your string class.

6.22. Create a class that can be used for working with polynomials.

a) Extend `show`, `summary`, `print` and [to work with objects from the polynomial class.

b) Extend the arithmetic operators +, − and ∗ to objects of the polynomial class.

c) Provide a method function for `plot` that will work with a polynomial class object.

d) Use the R function `polyroot` to create a method function for the `solve` function that will work for your polynomial class.

Venables and Ripley (2000)

Chapter 7

Numerical linear algebra

7.1 Introduction

Numerical techniques for solving linear algebra problems represent a dominant theme in scientific computing. In statistics, linear systems arise from the fitting of linear models and even iterative methods for fitting nonlinear models. Multivariate analysis includes methods such as factor analysis, principal components analysis and canonical correlation all of which attempt to reduce the dimensionality of data through construction of optimal linear combinations of collections of variables. This, in turn, leads to eigenvalue-eigenvector problems that must be addressed numerically.

The next section contains an introductory treatment of techniques for solving linear equation systems that covers Gaussian elimination and Cholesky factorization for both full and banded systems. Sections 7.3 and 7.4 deal with eigenvalue-eigenvector and singular-value decompositions of matrices. Least-squares problems and the QR decomposition are briefly addressed in Section 7.5. The chapter concludes with an introduction to the open source *Template Numerical Toolkit* collection of C++ routines for numerical linear algebra.

Throughout this chapter we will be developing new methods for the class `Matrix` that was introduced in Section 3.9. Complete listings of the header and source code files for the `Matrix` and `Vector` classes are contained in Appendix D.

7.2 Solving linear equations

Consider the problem of finding a solution (or solutions) b to the system

$$Ab = c, \tag{7.1}$$

where A is a specified $n \times n$ real matrix and c is a specified real vector. There are numerous numerical methods that can be employed for this purpose including variants of the Gaussian elimination technique discussed in Section 7.2.2.

Linear equations arise in statistics through, e.g., the normal equations for estimation in regression and linear models that have the form

$$X^T X b = X^T y$$

with X an $n \times p$ matrix for $n \geq p$, y an $n \times 1$ response vector and b a $p \times 1$ solution vector of estimated regression coefficients. This is a special case of (7.1) where the matrix $A = X^T X$ is positive-semidefinite. When X has rank p, $X^T X$ is positive-definite which allows us to employ more specialized procedures, such as the Cholesky decomposition treated in Section 7.2.3, that explicitly use this property.

A standard tactic for solving (7.1) is via factorization to produce a lower- or upper-triangular system. The reason for this is that triangular systems are easy to solve as explained in the next section.

7.2.1 Solving triangular systems

The basic premise behind many solution methods for linear equations involves triangularization where (7.1) is transformed into an equivalent system with (the transformed version of) A being upper-triangular. To see why such triangular systems are of interest, suppose that $U = \{u_{ij}\}_{i,j=1,n}$ is upper-triangular with $u_{ij} = 0, i > j$. Then, assuming all divisions are well defined, the system $Ub = c$ can be solved via $b_n = c_n/u_{nn}$, $b_{n-1} = (c_{n-1} - u_{(n-1)n}b_n)/u_{(n-1)(n-1)}$, etc., with the general step being

$$b_t = \left(c_t - \sum_{j=t+1}^{n} u_{tj}b_j \right) / u_{tt}$$

for $t = n - 1, \ldots, 1$. This is called a *back-solving* algorithm.

The computational effort involved in back-solving is on the order of n^2 floating-point operations or flops. Floating-point operations include addition, subtraction, multiplication and division. To see how we arrive at the n^2 figure observe that solving for b_n requires one division while solving for b_t involves $n - t$ multiplications, $n - t$ subtractions and one division for $t = n - 1, \ldots, 1$. Thus, the total effort is

$$1 + \sum_{t=n-1}^{1} \{2(n - t) + 1\} = n^2.$$

This type of operation count is often referred to as the *complexity* of the algorithm and that term is frequently employed along with O notation to describe the order of an algorithm's computational effort. Thus, in this case the back-solving algorithm has complexity $O(n^2)$.

Back-solving is easily implemented in an iterative form that is amenable for use with a computer as illustrated by the following new member function for our class `Matrix` from Chapter 3.

```
Vector Matrix::backward(const Vector& RHS) const {
  double temp;//temporary storage

  Vector b(nRows, 0.);//solution vector

  //initialize the recursion
  if(pA[nRows - 1][nRows - 1] != 0)
    b.pA[nRows - 1] = RHS.pA[nRows - 1]/pA[nRows - 1][nRows - 1];
  else{
    cout << "Singular system!" << endl;
    exit(1);
  }

  //now work through the remaining rows
  for(int i = (nRows - 2); i >= 0; i--){
    if(pA[i][i] != 0){
      temp = RHS.pA[i];
      for(int k = (i + 1); k < nRows; k++)
        temp -= b.pA[k]*pA[i][k];
      b.pA[i] = temp/pA[i][i];
    }
    else{
      cout << "Singular system!" << endl;
      exit(1);
    }
  }
  return b;
}
```

Notice that checking is done at each step of the outside loop to guard against division by zero. When a diagonal entry of a square, upper-triangular matrix is zero this means that the matrix is singular. In that case, there is no unique solution to the linear system. The calling `Matrix` object as well as the right-hand side `Vector` object should not be altered by `backward`. To account for this `backward` has been made a `const` member function and the method's argument is passed in as a `const` reference.

Lower-triangular systems can be solved similarly to the upper-triangular case. Specifically, suppose the system is $Lb = c$ with L a lower-triangular matrix having $l_{ij} = 0$, $j > i$. Then, $b_1 = c_1/l_{11}$, $b_2 = (c_2 - l_{21}b_1)/l_{22}$ and

$$b_t = (c_t - \sum_{j=1}^{t-1} l_{tj}b_j)/l_{tt}$$

for $t = 2, \ldots, n$. The member function `forward` for class `Matrix` that implements this recursion is given in Appendix D.

In general the process of solving (7.1) can be formulated as one of constructing an LU decomposition of A wherein $A = LU$ for L and U lower- and upper-triangular matrices, respectively.[*] Given such a decomposition for A, b can be computed using forward and backward substitution. Specifically, $Lz = c$ is first solved by forward substitution and then b is calculated from $Ub = z$ via the backward recursion.

7.2.2 Gaussian elimination

The results in the last section provide the means for solving any full rank, upper-triangular system. More generally, this entails that a solution can be obtained for any linear system that can be transformed to upper-triangular form through a series of row operations: i.e., by left multiplication of A and c by appropriate matrices. This can be accomplished with iterated Gauss transformations as will now be described.

Let $x = (x_1, \ldots, x_n)^T$ be a vector with $x_k \neq 0$ and define the Gauss transformation matrix

$$T_k(x) = \begin{bmatrix} 1 & \cdots 0 & 0 & \cdots & 0 \\ \vdots & \ddots & \vdots & \vdots & \vdots & \vdots \\ 0 & \cdots & 1 & 0 & \cdots & 0 \\ 0 & \cdots & -\frac{x_{k+1}}{x_k} & 1 & \cdots & 0 \\ \vdots & \vdots & \vdots & \vdots & \ddots & \vdots \\ 0 & \cdots & -\frac{x_n}{x_k} & 0 & \cdots & 1 \end{bmatrix}.$$

Then, one may check that $T_k(x)x = (x_1, \ldots, x_k, 0, \ldots, 0)^T$. With this in mind, the system (7.1) can be transformed to upper-triangular via the Gauss sweep process described in Algorithm 7.1. The algorithm begins by initializing a work matrix G as the matrix A for the linear system. Similarly, a right-hand side work vector h is initialized to c. Gauss transforms are then applied sequentially to the vector and work matrix until the system is upper-triangular. At that point the solution is available by back-solving.

---

[*] The LU decomposition of a matrix A exists provided that all the principal minors of A are nonzero. A more general factorization of A as LUP with L and U lower- and upper-triangular and P a permutation matrix always exists and reduces to the LU decomposition when the latter exists. An algorithm for computing an LUP decomposition can be found in Chapter 28 of Cormen, et al. (2001).

Algorithm 7.1 Gauss sweep algorithm

$G = [g_1, \ldots, g_n] = A, h = c$
for $k = 1, \ldots, (n-1)$ **do**
  $h := T_k(g_k)h$
  $G := T_k(g_k)G$
end for
Back-solve $Gb = h$
return b

An R function that was created to produce a Gauss transform for a given vector of input and row designation has the form

```
gTran <- function(x, k){
  n <- length(x)
  T <- diag(1, n, n)
  e <- vector(mode = "numeric", length = n)
  e[k] = 1
  temp <- vector(mode = "numeric", length = n)
  temp[(k + 1):n] <- -x[(k + 1):n]/x[k]
  T<-T + temp%*%t(e)
  T
}
```

This function creates the Gauss transform matrix $T_k(x)$ using the fact that $T_k(x) = I + v_k e_k^T$ with I the $n \times n$ identity matrix, $v_k = (0, \ldots, 1, -x_{k+1}/x_k, \ldots, -x_n/x_k)^T$ and e_k an $n \times 1$ vector of all zeros except for a 1 as its kth component. The R function `diag` is used to create the identity matrix. Somewhat more generally, `diag(a, n, n)` with a a scalar will create a square diagonal matrix with all the diagonal entries equal to `a`.

We will now use `gTran` to illustrate Algorithm 7.1 by working through a specific example in R. First, a matrix A and right-hand side c are generated via

```
> set.seed(123)
> A <- matrix(rnorm(25), 5, 5)
> A <- t(A)%*%A
> A
           [,1]       [,2]       [,3]       [,4]       [,5]
[1,]  2.8183730 -3.1452537 -0.2082627 -4.1931724 -1.0827824
[2,] -3.1452537  5.4246530  1.9299179  5.5110075  0.1453005
[3,] -0.2082627  1.9299179  2.1096700  1.9187222 -1.5299804
[4,] -4.1931724  5.5110075  1.9187222  8.0239285 -0.2145681
[5,] -1.0827824  0.1453005 -1.5299804 -0.2145681  3.1624022
> cVector <- rnorm(5)
> cVector
[1] -1.6866933  0.8377870  0.1533731 -1.1381369  1.2538149
```

Here A and the right-hand-side vector `cVector`[†] have been created using pseudo-random numbers from a standard normal distribution. If the values in A truly represented a sample from the standard normal distribution and they were stored in infinite precision, A would be of full rank with probability 1. In practice there is only a (very) remote chance that a matrix created in this fashion could be singular.

The first step in the recursion is

```
> ATemp <- A
> cTemp <- cVector
```

[†] Using `cVector` rather than `c` avoids masking R's concatenate function `c`.

```
> cTemp <- gTran(ATemp[, 1], 1)%*%cTemp
> ATemp <- gTran(ATemp[, 1], 1)%*%ATemp
```

This produces a transformed version of the A matrix of the form

```
> ATemp
           [,1]        [,2]        [,3]        [,4]        [,5]
[1,]   2.818373  -3.1452537  -0.2082627  -4.1931724  -1.082782
[2,]   0.000000   1.9146064   1.6975005   0.8315023  -1.063065
[3,]   0.000000   1.6975005   2.0942805   1.6088692  -1.609992
[4,]   0.000000   0.8315023   1.6088692   1.7853309  -1.825531
[5,]   0.000000  -1.0630653  -1.6099922  -1.8255307   2.746411
```

Succeeding steps proceed as

```
> cTemp <- gTran(ATemp[,2],2)%*%cTemp
> ATemp <- gTran(ATemp[,2],2)%*%ATemp
> cTemp <- gTran(ATemp[,3],3)%*%cTemp
> ATemp <- gTran(ATemp[,3],3)%*%ATemp
> cTemp <- gTran(ATemp[,4],4)%*%cTemp
> ATemp <- gTran(ATemp[,4],4)%*%ATemp
```

with the final outcome being

```
> ATemp
           [,1]        [,2]        [,3]        [,4]        [,5]
[1,]   2.818373  -3.145254  -0.2082627  -4.1931724  -1.0827824
[2,]   0.000000   1.914606   1.6975005   0.8315023  -1.0630653
[3,]   0.000000   0.000000   0.5892672   0.8716547  -0.6674727
[4,]   0.000000   0.000000   0.0000000   0.1348469  -0.3765101
[5,]   0.000000   0.000000   0.0000000   0.0000000   0.3488329
> cTemp
            [,1]
[1,]   -1.6866933
[2,]   -1.0445324
[3,]    0.9548237
[4,]   -4.6063557
[5,]  -11.7541565
```

This upper-triangular system can now be solved using the R backsolve function with the result

```
> b <- backsolve(ATemp, cTemp)
> b
          [,1]
[1,]  -303.89287
[2,]   -99.34466
[3,]   153.15147
[4,]  -128.24258
[5,]   -33.69567
```

The R language comes equipped with three basic functions for solving linear systems: backsolve, forwardsolve and solve. Commands of the form backsolve(U, cVector) and forwardsolve(L, cVector) provide solutions to the upper- and lower-triangular systems Ub = cVector and Lb = cVector, respectively. For a general, nonsingular, square matrix A, solve(A, cVector) returns the solution of Ab = cVector. The computations are carried out using a routine from the LAPACK numerical linear algebra package that performs Gaussian elimination. If solve is now applied to the original matrix A and right-hand side vector cVector from our example, we obtain

```
> solve(A, cVector)
[1] -303.89287  -99.34466  153.15147 -128.24258  -33.69567
```

which agrees with the result from the Gauss transform iteration. In addition to solving linear systems, when no right-hand side is specified for solve (i.e., the command is solve(A)), it returns the inverse of the matrix supplied as its argument.

The example serves to illustrate a case where the Gauss transform approach works. However, there is more to be realized here through examination of the transform matrices that were used in the recursion. This will help us see how to code up the algorithm in an efficient way.

The first Gauss transform matrix is

```
> ATemp <- A
> gTran(ATemp[,1],1)
          [,1] [,2] [,3] [,4] [,5]
[1,] 1.00000000    0    0    0    0
[2,] 1.11598205    1    0    0    0
[3,] 0.07389464    0    1    0    0
[4,] 1.48779897    0    0    1    0
[5,] 0.38418706    0    0    0    1
```

that produced the transformed matrix

```
> ATemp <- gTran(ATemp[,1],1)%*%ATemp
> ATemp
              [,1]        [,2]        [,3]       [,4]       [,5]
[1,] 2.818373e+00 -3.1452537 -0.2082627 -4.1931724 -1.082782
[2,] 0.000000e+00  1.9146064  1.6975005  0.8315023 -1.063065
[3,] 2.775558e-17  1.6975005  2.0942805  1.6088692 -1.609992
[4,] 0.000000e+00  0.8315023  1.6088692  1.7853309 -1.825531
[5,] 0.000000e+00 -1.0630653 -1.6099922 -1.8255307  2.746411
```

The next transform matrix is

```
> gTran(ATemp[,2],2)
     [,1]        [,2] [,3] [,4] [,5]
[1,]    1  0.0000000    0    0    0
[2,]    0  1.0000000    0    0    0
[3,]    0 -0.8866055    1    0    0
[4,]    0 -0.4342941    0    1    0
[5,]    0  0.5552396    0    0    1
```

that leads to

```
> ATemp <- gTran(ATemp[,2],2)%*%ATemp
> ATemp
              [,1]       [,2]       [,3]       [,4]       [,5]
[1,] 2.818373e+00 -3.145254 -0.2082627 -4.1931724 -1.0827824
[2,] 0.000000e+00  1.914606  1.6975005  0.8315023 -1.0630653
[3,] 2.775558e-17  0.000000  0.5892672  0.8716547 -0.6674727
[4,] 0.000000e+00  0.000000  0.8716547  1.4242143 -1.3638477
[5,] 0.000000e+00  0.000000 -0.6674727 -1.3638477  2.1561552
```

The pattern that underlies the Gauss sweep algorithm now becomes clear. On the kth step of the recursion, operations are performed only on the columns whose indices exceed that of the one being targeted by the transform: i.e., columns with indices $k + 1$ or larger. For these columns it is only necessary to work with elements below the kth row. The effect is that of working on progressively smaller submatrices of $A = \{a_{ij}\}_{i,j=1:n}$ as the calculations proceed from right to left in the sweeping process: the first submatrix is $\{a_{ij}\}_{i,j=2:n}$, the second is the transformed version of $\{a_{ij}\}_{i,j=3:n}$, etc.

The new method for class `Matrix` given below uses the ideas that have just been discussed.

```
Vector Matrix::gauss(const Vector& RHS) const {
  double multiplier = 0;
  Matrix G(nRows, nRows, pA);
  Vector h(nRows, RHS.pA);

  for(int j = 0; j < nCols; j++){//column to be swept

    for(int jj = j + 1; jj < nCols; jj++){//current operation column

      if(G[j][j] == 0){
        cout << "Oops! Division by 0!" << endl;
        exit(1);
      }

      for(int i = j + 1; i < nRows; i++){//work down rows
        multiplier = G[i][j]/G[j][j];
        G.pA[i][jj] = G.pA[i][jj] - multiplier*G.pA[j][jj];
      }
    }

    for(int i = j + 1; i < nRows; i++){//do the same to the RHS
      multiplier = G[i][j]/G[j][j];
      h.pA[i] = h.pA[i] - multiplier*h.pA[j];
    }
  }

  //now backsolve
  Vector b = G.backward(h);
  return b;
}
```

The roles of `ATemp` and `cTemp` in our R calculations are occupied by `G` and `h` in this listing. These two quantities are initialized using the (calling) matrix for the left-hand side and the right-hand side vector that is supplied as an argument to the method. The algorithm then proceeds across the columns of `G` working only with row elements below the diagonal while also carrying out the same calculations on `h`. At the end of the recursion `G` is upper-triangular and the system is solved using the `backward` method from the previous section. One may now check that an application of this C++ routine to the matrix and vector created in R for our example produces the same solution.

The sweep algorithm requires $O(n^3)$ flops (Exercise 7.3). Problems can arise with this approach when one of the diagonal elements of A or one of its Gauss transforms is small. This can be avoided by implementing a suitable row interchange strategy such as *partial pivoting* that ensures that no multiplier can exceed one in magnitude (Exercise 7.2).

7.2.3 Cholesky decomposition

When the matrix A in our linear equation system is positive-definite, it can be transformed to upper-triangular form using the Cholesky decomposition. While this method can exhibit some numerical inaccuracies when the matrix in question is near singular, it is still useful today and occupies an important role in statistics.

To appreciate the basic premise behind the Cholesky algorithm, suppose it is possible to write $A = \{a_{ij}\} = LL^T$ for

$$
L = \begin{bmatrix}
l_{11} & 0 & 0 & \cdots & 0 \\
l_{21} & l_{22} & 0 & \cdots & 0 \\
l_{31} & l_{32} & l_{33} & \cdots & 0 \\
\vdots & \vdots & \vdots & \ddots & \vdots \\
l_{n1} & l_{n2} & l_{n3} & \cdots & l_{nn}
\end{bmatrix}.
$$

But, if that is the case, the system $A = LL^T$ can be solved directly for the elements l_{ij} of L in terms of the elements of A. This approach produces

$$
\begin{aligned}
l_{11} &= \sqrt{a_{11}} \\
l_{21} &= a_{21}/l_{11} \\
&\vdots \\
l_{n1} &= a_{n1}/l_{11}
\end{aligned}
$$

which leads to

$$
\begin{aligned}
l_{22} &= \sqrt{a_{22} - l_{21}^2} \\
l_{32} &= (a_{32} - l_{31}l_{21})/l_{22} \\
&\vdots \\
l_{n2} &= (a_{n2} - l_{n1}l_{21})/l_{22}.
\end{aligned}
$$

The general or jth step is

$$
l_{jj} = \sqrt{a_{jj} - \sum_{k=1}^{j-1} l_{jk}^2} \tag{7.2}
$$

$$
l_{ij} = \left(a_{ij} - \sum_{k=1}^{j-1} l_{ik}l_{jk} \right) / l_{jj}, \quad i = j+1, \ldots, n. \tag{7.3}
$$

The Cholesky recursion can be summarized as in Algorithm 7.2.

Algorithm 7.2 Cholesky algorithm

$l_{11} = \sqrt{a_{11}}, l_{i1} = a_{i1}/l_{11}, i = 2, \ldots, n$
for $j = 2, \ldots, n-1$ **do**
   $l_{jj} = \sqrt{a_{jj} - \sum_{k=1}^{j-1} l_{jk}^2}$
   $l_{ij} = \left(a_{ij} - \sum_{k=1}^{j-1} l_{ik}l_{jk} \right) / l_{jj}, \quad i = j+1, \ldots, n$
end for
$l_{nn} = \sqrt{a_{nn} - \sum_{k=1}^{n-1} l_{nk}^2}$
Forward-solve $Lh = c$ for h
Back-solve $L^T b = h$ for b
return b

The algorithm uses (7.2)–(7.3) to factorize A. Then, given L, the system $Ab = c$ is solved as noted in Section 7.2.1 by forward-substitution and then back-solving.

Like Gaussian elimination the Cholesky algorithm has complexity on the order of n^3. However, one can show that it requires only half the effort of Gaussian elimination (Exercise 7.3).

The details required for implementing Algorithm 7.2 have been collected into another new method for class `Matrix`.

```cpp
Vector Matrix::cholesky(const Vector& RHS) const {
  double temp = 0;
  Matrix G(nRows, nCols, 0.);
  for(int j = 0; j < nCols; j++){//proceed by columns
    if(pA[j][j] == 0){
      cout << "Singular system!" << endl;
      exit(1);
    }
    temp = pA[j][j];//starting value for diagonal element recursion

    if(j > 0)
      for(int k = 0; k < j; k++)
        temp -= G[j][k]*G[j][k];

    G.pA[j][j] = sqrt(temp);

    for(int i = (j + 1); i < nRows; i++){//now do the rest
      temp = pA[j][i];
      for(int k = 0; k < j; k++)
        temp -= G[i][k]*G[j][k];
      G.pA[i][j] = temp/G[j][j];
    }
  }
  cout << "Cholesky factor:" << endl;
  G.printMatrix();
  Vector h = G.forward(RHS);
  Vector b = G.trans().backward(h);
  return b;
}
```

The code implements the scheme of Algorithm 7.2 to compute the lower-triangular Cholesky factor and store it in a `Matrix` object G. In general there is no need to retain G, although the developments in Section 7.2.5 demonstrate this need not always be the case. Once G has been obtained it provides the matrix that is used in forward-solving while the transpose of G is used for back-solving.

An application of our Cholesky method to the matrix A constructed for the example in the previous section produces

```
Cholesky factor
    1.678801              0              0              0              0
   -1.873512       1.383693              0              0              0
   -0.1240544      1.22679      0.7676374              0              0
   -2.497718       0.6009298    1.135503       0.367215              0
   -0.6449736     -0.7682812   -0.8695156     -1.025312       0.5906208
```

and the solution vector

```
-303.8929
-99.34466
153.1515
-128.2426
-33.69567
```

The solution agrees with the one returned previously by Gaussian elimination.

To check the accuracy of the Cholesky factor and solution we can compare it to answers produced with the R chol function that performs a Cholesky factorization: namely,

```
> chol(A)
            [,1]        [,2]         [,3]          [,4]           [,5]
[1,]  1.678801   -1.873512   -0.1240544   -2.4977185    -0.6449736
[2,]  0.000000    1.383693    1.2267898    0.6009298    -0.7682812
[3,]  0.000000    0.000000    0.7676374    1.1355032    -0.8695156
[4,]  0.000000    0.000000    0.0000000    0.3672150    -1.0253122
[5,]  0.000000    0.000000    0.0000000    0.0000000     0.5906208
```

The two Cholesky factors agree apart from the fact that the R chol function returns the upper- rather than lower-triangular factor. Thus, to solve a system in R with the Cholesky factorization one uses forward-solving with the transpose of the upper-triangular factor and then back-solving with the factor returned from chol. In terms of the example this translates to

```
> U <- chol(A)
> h <- forwardsolve(t(U), cVector)
> backsolve(U, h)
[1]  -303.89287   -99.34466   153.15147   -128.24258   -33.69567
```

A second option is to use the actual inverse matrix computed via the Cholesky method. This is accomplished with the chol2inv function which takes the Cholesky factor obtained from chol as input. Again, for our example this produces

```
> chol2inv(chol(A))%*%cVector
           [,1]
[1,]  -303.89287
[2,]   -99.34466
[3,]   153.15147
[4,]  -128.24258
[5,]   -33.69567
```

7.2.4 Banded matrices

Of some interest is the case where the matrix A in a linear system has a band-limited structure. More precisely, a matrix $A = \{a_{ij}\}$ is said to be banded with bandwidth q if $a_{ij} = 0$ whenever $|i - j| > q$. For example, the matrix below is 6×6 and two-banded

$$
\begin{bmatrix}
a_{11} & a_{12} & a_{13} & 0 & 0 & 0 \\
a_{21} & a_{22} & a_{23} & a_{24} & 0 & 0 \\
a_{31} & a_{32} & a_{33} & a_{34} & a_{35} & 0 \\
0 & a_{42} & a_{43} & a_{44} & a_{45} & a_{46} \\
0 & 0 & a_{53} & a_{54} & a_{55} & a_{56} \\
0 & 0 & 0 & a_{64} & a_{65} & a_{66}
\end{bmatrix}
$$

In the case of a positive-definite matrix A, it is easy to see that the matrix L in the Cholesky factorization must have the same lower bandwidth as A (Exercise 7.6). Thus, the order of computation can be substantially reduced by restricting attention to only the nonzero matrix elements.

Suppose, initially, that we are not at the boundary; that is, for a given index i, $i \pm q$ are valid row and column indices. Then, by retracing our previous arguments in the nonbanded scenario it can be seen that (excluding the trivial case of $q = 0$) for $t = 0, \ldots, q$

$$a_{(i+t)i} = \begin{bmatrix} 0, \ldots, l_{(i+t)(i+t-q)}, \ldots, l_{(i+t)i}, \ldots, l_{(i+t)(i+t)}, 0, \ldots, 0 \end{bmatrix} \begin{bmatrix} 0 \\ \vdots \\ l_{i(i-q)} \\ \vdots \\ l_{i(i+t-q)} \\ \vdots \\ l_{ii} \\ 0 \\ \vdots \\ 0 \end{bmatrix}.$$

This gives a modified version of (7.2)–(7.3) wherein

$$l_{ii} = \sqrt{a_{ii} - \sum_{k=i-q}^{i-1} l_{ik}^2} \tag{7.4}$$

$$l_{(i+t)i} = \left(a_{(i+t)i} - \sum_{k=i+t-q}^{i-1} l_{(i+t)k} l_{ik} \right) / l_{ii}, \; t = 1, \ldots, q. \tag{7.5}$$

The only issue that now remains is dealing with the boundaries. After initializing the first row of L exactly as for the nonbanded case, we can make certain that the boundaries are handled correctly by computing all sums in (7.4)–(7.5) from the bound $\max(1, i+t-q)$ and restricting t to the range from 0 to $\min(q, n-i)$.

To maintain computational efficiency, both the forward and backward substitution steps must explicitly use the band structure. This means beginning, as before, by solving $Lz = c$ with $z_1 = c_1/l_{11}$ and

$$z_j = \left(c_j - \sum_{k=\max(1, j-q)}^{j-1} l_{jk} z_k \right) / l_{jj}, j = 2, \ldots, n, \tag{7.6}$$

and then back-solving $L^T b = z$ via $b_n = z_n/l_{nn}$ and

$$b_j = \left(z_j - \sum_{k=j+1}^{\min(n, j+q)} l_{kj} b_k \right) / l_{jj}, j = n-1, \ldots, 1. \tag{7.7}$$

The claim now is that all of this work can be accomplished in only $O(n)$ flops. To see that this is true, observe that there are only $O(nq)$ nonzero elements in A and L. Each element of L can be computed in order q flops and, hence, all the nonzero components of L can be obtained in a total of order nq^2 flops. Finally, the backward and forward substitution steps are also of order nq which verifies the claim.

The development thus far has ignored storage considerations for the left-hand side matrix in the linear system. In general, all n^2 memory locations will be needed to hold this matrix. In the case of band structure this can be reduced to $n(2q + 1)$. The implications of this fact will be explored in this section from an R perspective and again in Section 7.2.6 from the C++ viewpoint. It should also be noted that symmetric matrices, banded or otherwise, allow for savings in terms of memory. In general, only $n(n + 1)/2$ elements of an $n \times n$ symmetric matrix need to be stored and, similarly, a symmetric q-banded matrix can be stored in $n(q + 1)$ memory locations without any loss of information. Memory efficient storage of symmetric matrices is the topic of Exercise 7.21.

The C++ code below implements the banded Cholesky algorithm for a given matrix with a specified bandwidth. The companion methods **bandBack** and **bandFor** for backward and forward solution of banded systems are given in Appendix D.

```
Vector Matrix::bandChol(const Vector& RHS, Matrix& G,
        int bWidth) const {
  double temp = 0;
  int low, up;

  for(int j = 0; j < nCols; j++){
    if(pA[j][j] == 0){
      cout << "Singular system!" << endl;
      exit(1);
    }
    temp = pA[j][j];

    low = std::max(0, j - bWidth);

    for(int k = low; k < j; k++)
      temp -= G[j][k]*G[j][k];

    G.pA[j][j] = sqrt(temp);

    up = std::min(j + bWidth, nRows - 1);

    for(int i = (j + 1); i <= up; i++){
      temp = pA[j][i];
      for(int k = low; k < j; k++)
        temp -= G[j][k]*G[i][k];
      G.pA[i][j] = temp/G[j][j];
    }
  }
  Vector h = G.bandFor(RHS, bWidth);
  Vector b = G.trans().bandBack(h, bWidth);
  return b;
}
```

In contrast to the nonbanded Cholesky method from the previous section, our banded Cholesky code explicitly returns the Cholesky factor to the user in a **Matrix** object that is passed in (by reference) from the calling program. The reason for this stems from an application for this method that will arise in the next section.

The **bandChol** method uses the C++ **min** and **max** functions that become available by including the **algorithm** header. Thus, we also need to add

```
#include <algorithm>
```

to the beginning of the file that contains the method definitions for class **Matrix**.

The R package **SparseM** provides access to tools for computing with sparse matrices. The routines in this package work on matrices that are stored in a compressed sparse row (csr) format that essentially adapts to the structure of the matrix. A **matrix.csr** class has been created to efficiently store sparse matrix objects with method functions for **t**, **%*%**, **backsolve**, **chol** and **solve** that apply to objects of this class. The library can be installed from the CRAN website via

```
> install.packages("SparseM")
```

and then loaded with

```
> library(SparseM)
```

The slots for objects of class `matrix.csr` can be determined by the `getClass` function. The result is

```
> getClass("matrix.csr")
Class "matrix.csr" [package "SparseM"]

Slots:

Name:          ra          ja          ia dimension
Class:    numeric     integer     integer     integer
```

The `ra` slot is a one-dimensional array that holds the nonzero elements of the matrix and `dimension` is a two-dimensional array that gives the number of rows and columns of the matrix as its first and second component, respectively. The `ja` and `ia` slots specify where the elements of `ra` are located in the matrix. First, `ja` is a one-dimensional array of the same length as `ra` with entries that give the column location of the corresponding entry in `ra`. The `ia` slot consists of a one-dimensional array with length equal to the number of rows plus 1. The integer elements of `ia` are the indices of the elements in `ra` where new rows should be started. The last entry is where a new row would start if one were to be appended to the array: i.e., it is the number of nonzero entries plus 1.

To illustrate the use of the **SparseM** package we first created a positive-definite two-banded matrix with

```
> set.seed(123)
> A <- matrix(rnorm(36), 6, 6)
> check <- function(i, j) {
+ if(abs(i - j) > 1) A[i, j] = 0
+ else A[i, j] = A[i, j]
+ }
> vCheck <- Vectorize(check)
> A <- outer(1:6, 1:6, vCheck)
> A <- t(A)%*%A
> A
             [,1]        [,2]        [,3]        [,4]        [,5]        [,6]
[1,]   0.36711463  0.03285631 -0.02547667  0.0000000  0.00000000  0.00000000
[2,]   0.03285631  2.28459052  0.24176066  0.7334378  0.00000000  0.00000000
[3,]  -0.02547667  0.24176066  3.51426859  0.2040381  0.27406444  0.00000000
[4,]   0.00000000  0.73343776  0.20403810  2.2404457  1.13430207 -0.84294584
[5,]   0.00000000  0.00000000  0.27406444  1.1343021  2.89093086 -0.07164435
[6,]   0.00000000  0.00000000  0.00000000 -0.8429458 -0.07164435  1.14922087
```

A 6×6 matrix is created first using random numbers from the standard normal distribution. To convert it to a banded matrix the `Vectorize` function from Section 5.5 has been used to vectorize a function that will replace those elements outside a unit band length with 0 entries. This vectorized function is used with `outer` to actually carry out the restructuring. A positive-definite two-banded matrix is then produced from the product of the transpose of the banded matrix with itself. The `as.matrix.csr` function can now be used to convert the `matrix` object A that we created into an object of class `matrix.csr`: i.e.,

```
> A.csr <- as.matrix.csr(A)
```

The slots of `A.csr` are

```
> A.csr@ra
 [1]   0.36711463   0.03285631  -0.02547667   0.03285631   2.28459052   0.24176066
 [7]   0.73343776  -0.02547667   0.24176066   3.51426859   0.20403810   0.27406444
[13]   0.73343776   0.20403810   2.24044566   1.13430207  -0.84294584   0.27406444
[19]   1.13430207   2.89093086  -0.07164435  -0.84294584  -0.07164435   1.14922087
```

```
> A.csr@ja
 [1] 1 2 3 1 2 3 4 1 2 3 4 5 2 3 4 5 6 3 4 5 6 4 5 6
> A.csr@ia
 [1]   1   4   8 13 18 22 25
> A.csr@dimension
 [1] 6 6
```

From this we see that `A.csr` has the 24 nonzero elements of `A` in its `ra` slot. The `ja` slot tells us that the first three nonzero elements of `A` are in columns 1, 2 and 3, the next four nonzero elements are in columns 1, 2, 3 and 4, the next five nonzero elements have column indices one through five, etc. Finally, the components of `A.csr@ja` tell us that the first row has elements 1 through 3 of `A.csr@ra`, the second row has elements 4 through 7, the third row has elements 8 through 12, etc.

As noted previously, a method for the `solve` function has been provided for `matrix.csr` objects. The solution is carried out via an efficient, sparse matrix, Cholesky algorithm which entails that in this case `solve` is intended for use only with positive-definite systems. To illustrate the idea the following calculations were performed.

```
> set.seed(456)
> rhs
[1] -1.3435214  0.6217756  0.8008747 -1.3888924 -0.7143569 -0.3240611
> b <- solve(A.csr, rhs)
> b
[1] -3.7187735  0.8250369  0.2124663 -1.6256003  0.3345638 -1.4534930
```

One may check that the same result is obtained by applying `solve` to the noncompressed version of `A`.

7.2.5 An application: linear smoothing splines

One application of banded Cholesky methods arises in the computation of linear smoothing splines that are used for nonparametric regression and scatter plot smoothing. The problem involves bivariate data (x_i, y_i), $i = 1, \ldots, n$, with $0 \le x_1 < \cdots < x_n \le 1$. A linear smoothing spline fit f_λ to this data is obtained through minimization of the criterion

$$\sum_{i=1}^{n}(y_i - f(x_i))^2 + \lambda \int_0^1 (f')^2(x)dx, \quad \lambda > 0, \tag{7.8}$$

over all continuous functions f with square-integrable derivatives f'. It is known (see, e.g., Wahba 1990) that $f_\lambda(\cdot)$ is unique and can be expressed as a natural linear spline with knots or join points at $x_1, \ldots, x_n$. More precisely, $f_\lambda(\cdot)$ is a piecewise linear function whose derivative has jump discontinuities at $x_1, \ldots, x_n$ and vanishes outside of $[x_1, x_n]$. The so-called *smoothing parameter* λ in (7.8) governs the smoothness or wiggliness of the fit to the data. As λ grows large f_λ tends to the average of the y_i while as $\lambda \to 0$ the fitted function interpolates the y_i; that is, $\lim_{\lambda \to 0} f_\lambda(x_i) = y_i$, $i = 1, \ldots, n$.

Examples of linear smoothing spline fits to an artificial data set are shown in Figure 7.1 for three different values of λ. The data in the plot was generated using R from the model

$$y_i = 10x_i + 10 \exp\{-15(x_i - .5)^2\} + e_i, \ i = 1, \ldots, 100,$$

for uniformly spaced x_i and e_i from a normal distribution with mean zero and standard deviation 2. Specifically, the data was created with

```
> set.seed(123)
> x <- 1:100/101
> y <- 10*x + 10*exp(-15*(x - .5)^2) + rnorm(100, 0, 2)
```

The true response mean function, which can be viewed as the target for the smoothing spline estimator f_λ, has been overlaid on the plot. Examination of the three fits to the data indicates that the values of $\lambda = .01$ and $\lambda = 100$ are too small and too large, respectively. The $\lambda = .1$ choice seems to provide a fit that is more satisfactory from a visual perspective. The question of how to make objective comparisons of smoothing parameter values will be addressed shortly.

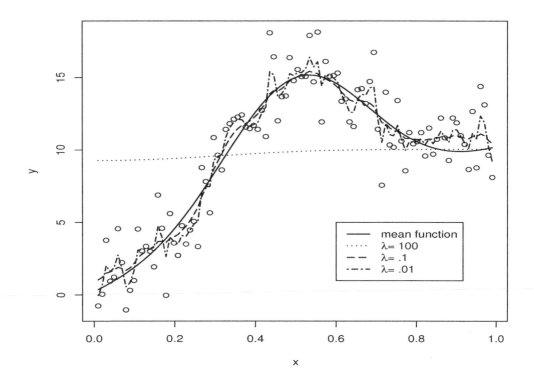

Figure 7.1 *Linear smoothing splines with different choices for* λ

The set of natural linear splines with knots at $x_1, \ldots, x_n$ is a linear space with dimension n. As a result, it follows that $f_\lambda = \sum_{j=1}^{n} b_j B_j$ with $B_1, \ldots, B_n$ some suitable set of linear natural spline basis functions. Substituting this representation into the estimation criterion produces the normal equations

$$(B^T B + \lambda \Omega)b = B^T y \tag{7.9}$$

for $b = (b_1, \ldots, b_n)^T$ with

$$B = \{B_j(x_i)\}_{i,j=1,n},$$
$$\Omega = \left\{ \int_0^1 B_i'(x) B_j'(x) dx \right\}_{i,j=1,n}$$

and $y = (y_1, \ldots, y_n)^T$ the response vector. The resulting vector of fitted values then has the form

$$f_\lambda = (f_\lambda(x_1), \ldots, f_\lambda(x_n))^T = H(\lambda)y$$

for

$$H(\lambda) = B(B^T B + \lambda \Omega)^{-1} B^T. \tag{7.10}$$

One route to efficient computations is through the use of a local support basis for the natural linear splines such as

$$
B_1(x) = \begin{cases} 1, & x \leq x_1, \\ (x_2 - x)/(x_2 - x_1), & x_1 < x < x_2, \\ 0, & x \geq x_2, \end{cases}
$$

$$
B_j(x) = \begin{cases} 0, & x \leq x_{j-1}, \\ (x - x_{j-1})/(x_j - x_{j-1}), & x_{j-1} < x < x_j, \\ (x_{j+1} - x)/(x_{j+1} - x_j), & x_j \leq x < x_{j+1}, \\ 0, & x \geq x_{j+1}, \end{cases}
$$

for $j = 2, \ldots, n - 1$, and

$$
B_n(x) = \begin{cases} 0, & x \leq x_{n-1}, \\ (x - x_{n-1})/(x_n - x_{n-1}), & x_{n-1} < x < x_n, \\ 1, & x \geq x_n. \end{cases}
$$

Using this basis in our normal equation system produces

$$
[I + \lambda \Omega] f_\lambda = y \tag{7.11}
$$

with $\Omega = \{\omega_{ij}\}$ a symmetric one-banded matrix having diagonal elements

$$
\omega_{11} = \frac{1}{x_2 - x_1}, \quad \omega_{ii} = \frac{1}{x_{i+1} - x_i} + \frac{1}{x_i - x_{i-1}}, i = 2, \ldots, n - 1, \quad \omega_{nn} = \frac{1}{x_n - x_{n-1}} \tag{7.12}
$$

and above diagonal entries

$$
\omega_{i(i+1)} = -\frac{1}{x_{i+1} - x_i}, \quad i = 1, \ldots, n - 1. \tag{7.13}
$$

One consequence of (7.12)–(7.13) is that this particular form of the linear smoothing spline normal equation system can be solved in $O(n)$ operations using the banded Cholesky algorithm from the previous section.

The discussion to this point has failed to address the issue of how the smoothing parameter λ should be selected in practice. For this purpose data driven smoothing parameter selection techniques are frequently employed. One such method relies on minimization of the generalized cross validation criterion

$$
\text{GCV}(\lambda) = \frac{n^{-1} \sum_{i=1}^{n} (y_i - f_\lambda(x_i))^2}{(1 - \text{tr} H(\lambda)/n)^2} \tag{7.14}
$$

with tr denoting the matrix trace; i.e., $\text{tr} H(\lambda) = \sum_{i=1}^{n} h_i(\lambda)$ with $h_i(\lambda), i = 1, \ldots, n$, the diagonal elements of the matrix in (7.10). The minimizer of the GCV criterion is known to have a variety of optimality properties and tends to estimate the mean-squared error optimal choice for λ in large samples (e.g., Eubank 1999). There is no closed form for the minimizer which entails that the minimization process must be carried out numerically using optimization procedures such as those developed in the next chapter. Thus, for now our attention will be restricted to the computation of the GCV criterion for a given value of λ.

A brute force, order n^2, approach for computing the diagonal elements of $H(\lambda)$ is to simply solve the systems

$$
[I + \lambda \Omega] b_j = e_j, \quad j = 1, \ldots, n,
$$

for e_j as before the jth column of the n-dimensional identity matrix that consists of all zeros except for a one in the jth row. Then, $h_j(\lambda)$ is the jth component of b_j. It turns out that all n diagonal elements $h_1(\lambda), \ldots, h_n(\lambda)$ can actually be obtained in $O(n)$ operations through careful use of the available band structure. Specifically, if $H(\lambda)$ has (banded) Cholesky factorization $L = \{l_{ij}\}$, it can be shown (Exercise 7.13) that the $h_j(\lambda)$ evolve according to the recursion

$$h_n(\lambda) \;=\; 1/l_{nn}^2 \tag{7.15}$$

$$h_j(\lambda) \;=\; \left[1 + l_{(j+)j}^2 h_{j+1}(\lambda)\right]/l_{jj}^2, \; j = n-1, \ldots, 1. \tag{7.16}$$

The fitting of linear smoothing splines has been implemented as a C++ class. The header file for class Linss appears in Listing 7.1.

Listing 7.1 linss.h

```
//linss.h
#ifndef LINSS_H
#define LINSS_H

#include "vector.h"
#include "matrix.h"

class Linss{
  int n;
  Vector x, y;
  Matrix lhsMat(double lam) const;

public:
  Linss(const Vector& X, const Vector& Y);
  Vector smooth(double lam, double* GCV) const;

};
#endif
```

The class members are two Vector objects that contain the values for the response and independent variable and an integer variable that represents the number of observations. The method smooth will return the linear smoothing spline fitted values as a Vector object for a specified value of the smoothing parameter. The GCV criterion is also evaluated and returned in a pointer that is passed in from the calling program. As the Linss class requires the Vector and Matrix class by value, the header files for both classes must be included in the Linss class header file.

The method definitions for class Linss take the form

Listing 7.2 linss.cpp

```
//linss.cpp
#include "linss.h"

Linss::Linss(const Vector& X, const Vector& Y){
  x = Vector(X);
  y = Vector(Y);
  n = x.getnRows();
}


Vector Linss::smooth(double lam, double* GCV) const {
  Matrix G(n, n, 0.);
  Matrix H = lhsMat(lam);
  Vector fit = H.bandChol(y, G, 1);
```

```
//compute the GCV criterion
double temp = 1/(G[n - 1][n - 1]*G[n - 1][n - 1]);
double tr = temp;
*GCV = (y[n - 1] - fit[n - 1])*(y[n - 1] - fit[n - 1]);
for(int i = (n - 2); i >= 0; i--){
  temp = (1 + G[i + 1][i]*G[i + 1][i]*temp)/(G[i][i]*G[i][i]);
  *GCV += (y[i] - fit[i])*(y[i] - fit[i]);
  tr += temp;
}
*GCV /= (1. - tr/((double)n))*(1. - tr/((double)n))*(double)n;

return fit;
}

Matrix Linss::lhsMat(double lam) const {
  double** pH = new double*[n];
  pH[0] = new double[n];
  pH[n - 1] = new double[n];
  pH[0][0] = 1. + (lam/(x[1] - x[0]));
  pH[0][1] = -lam/(x[1] - x[0]);
  pH[n - 1][n - 2] = -lam/(x[n - 1] - x[n - 2]);
  pH[n - 1][n - 1] = 1. + (lam/(x[n - 1] - x[n - 2]));
  for(int i = 1; i < (n - 1); i++){
    pH[i] = new double[n];
    pH[i][i - 1] = -lam/(x[i] - x[i - 1]);
    pH[i][i] = 1 + (lam/(x[i] - x[i - 1])) + (lam/(x[i + 1] - x[i]));
    pH[i][i + 1] = -lam/(x[i + 1] - x[i]);
  }
  Matrix H(n, n, pH);

  for(int i = 0; i < n; i++)
    delete[] pH[i];
  delete[] pH;

  return H;
}
```

Rather than clutter up the operative method smooth for the class, the task of creating the matrix $(I + \lambda\Omega)$ in (7.11) has been relegated to a method lhsMat. This method is only for internal use and has been designated as private. Notice that the memory that is allocated in lhsMat is explicitly released. Had this not been done it would have become inaccessible once lhsMat returned control to smooth. This would have created a memory leak that could become problematic if the calling program were to evaluate fits over a number of different smoothing parameter values as would be the case for an optimization method. The diagonal elements of $(I + \lambda\Omega)^{-1}$ are obtained via (7.15)–(7.16). Rather than store them, their sum is merely accumulated and the individual values are overwritten on each step.

The data in Figure 7.1 were written to the file linssDat.txt. Smoothing splines were then fitted to the data using the simple driver program

Listing 7.3 *linssDriver.cpp*

```
//linssDriver.cpp
#include <cstdlib>
#include <iostream>
#include <fstream>
```

```
#include "vector.h"
#include "linss.h"

int main(int argc, char* argv[]){
  double lam = atof(argv[1]);
  int n = atoi(argv[2]);
  std::ifstream inFile(argv[3]);
  double* py = new double[n];
  double* px = new double[n];

  for(int i = 0; i < n; i++){
    inFile >> px[i];
    inFile >> py[i];
  }
  inFile.close();
  Vector y(n, py);
  Vector x(n, px);
  Linss S(x, y);
  double gcv = 0.;
  Vector fit = S.smooth(lam, &gcv);
  std::cout << "The value of GCV at lambda = "
            << lam << " is " << gcv << std::endl;

  return 0;
}
```

This program takes the value of the smoothing parameter, the number of observations and the name for the file containing the data as command line arguments. It then reads in the data and creates the two Vector objects that are needed for the Linss class constructor. Once the Linss object has been created its smooth method is invoked and the value of the GCV criterion for the specified value of λ is written to standard output. For the values of λ that were used with our example data set, the executable linssDriver for the linear smoothing spline code produced the results

```
$ ./linssDriver .01 100 linssDat.txt
The value of GCV at lambda = 0.01 is 4.21511
$ ./linssDriver .1 100 linssDat.txt
The value of GCV at lambda = 0.1 is 3.7224
$ ./linssDriver 100 100 linssDat.txt
The value of GCV at lambda = 100 is 21.6474
```

The values of the GCV criterion coincide with our original visual impression that $\lambda = .1$ gave the better fit.

Smoothing data is an inherently visual procedure. Thus, it would be useful to be able to interactively view plots that correspond to different values of λ or possibly have certain data points removed from a fit. If R is to serve as our graphical engine, the current set-up would require us to import fits into R and, in that sense, does not have the desired interactive quality. A more effective approach is to import the C++ linear smoothing spline code into R as a shared library in a way that can be accessed directly from R.

Recall from Section 5.7 that to create a shared library all files have to end with a .cc or .hh. Therefore, the first step is to replace the .cpp and .h file extensions on the files for class Linss, Matrix and Vector with .cc and .hh, respectively. Then, the following driver program was created to provide the interface to R.

Listing 7.4 *linssDriver.cc*

```
//linssDriver.cc
#include "R.h"
#include "vector.hh"
#include "linss.hh"

extern "C" {

  void linss(int* n, double* lam, double* px, double* py,
      double* pfit, double* GCV){

    Vector y(*n, py);
    Vector x(*n, px);

    Linss S(x, y);
    Vector fit = S.smooth(*lam, GCV);
    for(int i = 0; i < *n; i++)
      pfit[i] = fit[i];
  }
}
```

The arguments to the linss function are pointers to variables that correspond to the number of responses (n), the value of the smoothing parameter to use in computing the fit (lam), the vector of values for the independent variable (px), the response vector (py), a vector (pfit) being passed in from R that will be filled with the fitted values and an empty memory location that will hold the value of the GCV criterion on return to the R environment.

The shared library linssDriver.so is created with

```
$ R CMD SHLIB linssDriver.cc linss.cc matrix.cc vector.cc
```

Then, from inside R the library is loaded with

```
> dyn.load("linssDriver.so")
```

Linear smoothing splines may now be fit to data in R by using the imported library with the .C function. Rather than do this directly, we will hide the mechanics in the wrapper function

Listing 7.5 *linss.r*

```
#linss.r
function(x, y, lam){
  fit <- .C("linss", as.integer(length(x)), as.double(lam),
      as.double(x), as.double(y), vector("numeric", length(x)), 0.)
  list(fit[[5]], fit[[6]])
}
```

In calling linss existing values will be used for x and y that originate in R. The numeric vector that will hold the fitted values and the variable that will hold the value of the GCV criterion need to be initialized inside of R as well. Upon completion of its task the C++ program returns control to the .C function that packages the outcome in the form of a list containing the return values for all the variables that were passed to the C++ program. For this particular case the result is a list of length six with the fitted values and the value of the GCV criterion being the fifth and sixth components, respectively. These two quantities are what the user will want and, accordingly, our wrapper program returns them in a two component list.

Section 8.3.2 of the next chapter develops an automated procedure for selecting a good value of λ in the GCV sense. A crude approach that can be used at this juncture is to carry out a grid search over values of the smoothing parameter and use this to locate at least a region (or regions) where the better choices for λ can be found. An R function that carries out such a search is

```
gcvSearch <- function(x, y, low, up, nEvals){
  gcVals <- seq(low, up, length.out = nEvals)
  cbind(gcVals, sapply(gcVals, function(lLam)
        linss(x, y, 10^lLam)[[2]]))
}
```

The gcvSearch function takes as input the data vectors x and y along with lower and upper bounds (low and up, respectively) for the search and size of the grid (nEvals). The actual grid is then constructed using the R seq function. A slight twist here is that the search is carried out on a log scale with the grid representing a partition of powers of 10. The logarithmic scale is appropriate because it is known that the GCV criterion tends to behave like a power of the smoothing parameter near its minimum. The sapply function is used in conjunction with our linss wrapper function to evaluate the GCV criterion across the specified grid.

The gcvSearch function was applied to the data in Figure 7.1 using

```
> gcVals <- gcvSearch(x, y, -2, 0., 100)
```

This carried out a search over a 100 point grid for values of λ that are evenly spaced in log10 scale between -2 and 0. A plot of the result is shown in Figure 7.2. This suggests that the minimizer lies in the interval $[10^{-1}, 10^{-1/2}]$. We will return to this observation in Section 8.3.2.

7.2.6 Banded matrices via inheritance

So far banded matrices have been treated as if they were full matrices (i.e., no pattern of zero elements) in terms of how they are stored, multiplied, transposed, etc., with efficient solution of linear systems being the only real use that has been made of the band structure. This is wasteful from a storage perspective. After all, if the matrix is symmetric with bandwidth q, there are at most q + 1 unique, nonzero entries per row. This means that if the matrix has nRows rows, all the nonzero elements can be held in, e.g., nRows*(q + 1) memory locations rather than in the nRows*nRows locations that would usually be required. The special structure of band matrices could be built directly into class Matrix by adding new constructors and methods to the class. Another avenue that will be explored in this section is to make use of the C++ inheritance mechanism.

Inheritance is a means of formalizing an "is a" relationship between two classes; such a relationship exists between classes A and B if every object from class B has all the qualities (i.e., members and methods) of a class A object. Thus, every B object can be viewed as an A object in the sense that it will have all the functionality provided by the A class. The relationship is established syntactically in the class B declaration as

```
class B : public A
```

which indicates that class B is a class derived from the base class A. Derived classes automatically contain the elements and methods of the base class. The degree of access that is allowed to the base class components is determined by the type of inheritance that is specified in the derived class declaration. The most common one is public and that is the only one that will be considered here.

Under public inheritance, the private members/methods of the base class cannot be accessed directly by a derived class object. The public portion of the base class is available for use with the exception of

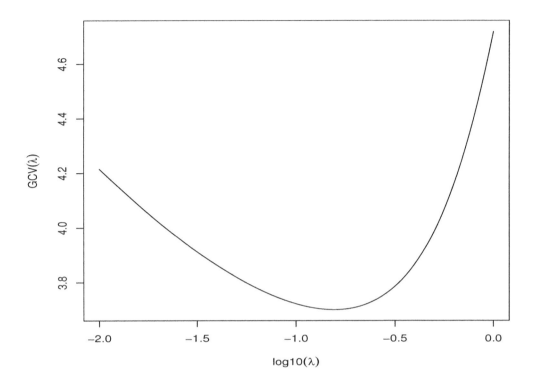

Figure 7.2 *GCV criterion over* $\log 10(\lambda) \in [-2, 0]$

a) constructors,

b) destructors,

c) the = or assignment operator and

d) friends.

The reason for a)–c) is that a derived class may have additional members that must be created, destroyed or assigned values. Friend status indicates a relationship between classes and, as such, there is no a priori reason to expect it to be inherited in very much the same sense as with people where friends of a child need not be friends of a parent and conversely.

Listing 7.6 provides a simple example of most of the inheritance syntax and features that we will use in the remainder of this section.

Listing 7.6 *inheritEx.cpp*

```
//inheritEx.cpp
#include <iostream>

using std::cout; using std::endl;

struct A{
  int iA;
  A(int ia){iA = ia;}
  void output(){f();}
```

```
   void f(){cout << iA << endl;}                                          11
   //virtual void f(){cout << iA << endl;}                               12
};                                                                        13
                                                                          14
struct B : public A {                                                     15
  int iB;                                                                 16
  B(int ia, int ib):A(ia){                                               17
    iB = ib;                                                              18
  }                                                                       19
  void f(){cout << iA << " " << iB << endl;}                             20
};                                                                        21
                                                                          22
int main(){                                                               23
  A aObj(2);                                                              24
  B bObj(3, 5);                                                           25
  A* pA;                                                                  26
  pA = &aObj;                                                             27
  pA->f();                                                                28
  pA = &bObj;                                                             29
  pA->f();                                                                30
  bObj.output();                                                          31
  return 0;                                                               32
}                                                                         33
```

Here struct A has a single class member iA of type int and two member functions: the function f that writes the value of iA to standard output and a function output that serves as a wrapper for f. The derived struct/class B has an int member iB while also inheriting iA from class A. The value of iA is set on line 17 of Listing 7.6 where the class B constructor employs an initializer list that prefaces the constructor's definition with a colon and a call to a base class A constructor. We discussed initializer lists in Section 3.8.3 in the context of const class members. In the inheritance setting initializer lists serve the purpose of forcing the creation of the members of the inherited base class object prior to the construction of any that are unique to the derived class. This is essentially the only way derived objects can be initialized.[‡]

Class B in Listing 7.6 also has a function named f and the behavior of this function with B class objects is what we want to investigate. To see what can occur, objects from both the A and B structs are created in main and a pointer pA is declared for class A objects. The fact that class B is derived from class A means that pA will point to either class A or class B objects and this feature is used to call the function f using a pointer to objects of both classes. Finally, the object from class B calls the output function that it inherits from class A. The results obtained from compilation and execution of Listing 7.6 are

```
2
3
3
```

This indicates that the pA pointer is only using the A class version of the f function regardless of whether the calling object is from class A or B. The inherited output function performs similarly in that it uses the A class version of f when called by a B class object. A more desirable outcome would be polymorphic behavior wherein the call to output as well as use of the A class pointer would adapt to the object involved in the function call. This is accomplished by commenting out line 11 and uncommenting line 12 of Listing 7.6. A new keyword *virtual* now comes into play whose use allows functions in the derived class to

---

[‡] If an explicit base class constructor is not specified the base class default constructor is called.

override functions in the base class that have the same name, return type and arguments. The outcome from running our program is now

```
2
3 5
3 5
```

Consequently, the pointer and output function are both adapting to produce the appropriate behavior for A or B class objects. One may check that similar results are obtained if references to A class objects are used instead of an A class pointer in Listing 7.6 (Exercise 7.16).

Let us now consider the creation of a class pdBand to represent positive-definite banded matrices. The "is-a" relationship that exists here makes it natural to develop the pdBand class as a derived class of class Matrix. The benefit of this approach is that class Matrix already embodies many of the methods that would be needed for the pdBand class and these (public) features would become part of the pdBand class automatically under an inheritance relationship. As a result, design of the pdBand class can instead focus on just those modifications and enhancements that are appropriate for matrices that are positive-definite with banding.

To begin consider a pared down version of class Matrix with the header file

Listing 7.7 *simpleMat.h*

```cpp
//simpleMat.h
#ifndef MATRIX_H
#define MATRIX_H
#include <iostream>
#include "vector.h"

class Matrix{
  int nRows, nCols;
  double** pA;
  Vector gauss(const Vector& RHS) const;

 public:
  //constructors and destructors
  Matrix(int nrows = 0, int ncols = 0, double a = 0.);
  Matrix(int nrows, int ncols, const double* const* pa);
  Matrix(const Matrix& A);
  virtual ~Matrix();

  //overloaded operators
  Matrix& operator=(const Matrix& B);
  const double* operator[](int i) const {return pA[i];}

  //solution of a linear system
  virtual Vector solve(const Vector& RHS) const{
    std::cout << "Solution via gauss from class Matrix" << std::endl;
    return gauss(RHS);
  }

  //utilities and accessors
  void printMatrix() const;
  int getNRows() const {return nRows;}
  int getNCols() const {return nCols;}
};
#endif
```

This class has the usual pointer-to-pointer and diagonal matrix (default) constructors, a copy constructor, an explicit destructor to release the memory in the private member pA and overloaded assignment and indexing operators. Accessor functions are provided to give access to the private members nRows and nCols.

The principal difference between the Matrix class in Listing 7.7 and previous incarnations is how we handle the solution of linear equations. The gauss method has been wrapped in a virtual function solve. The intention is to use a band-limited formulation for solve in the pdBand class that will be derived from class Matrix.

The header file for the pdBand class is given in the listing below.

Listing 7.8 pdBand.h

```
//pdBand.h                                                                    1
#ifndef PDBAND_H                                                             2
#define PDBAND_H                                                             3
#include <iostream>                                                          4
#include "simpleMatrix.h";                                                   5
#include "vector.h"                                                          6
                                                                            7
class pdBand : public Matrix{                                                8
   int q;//the bandwidth                                                     9
   Vector bandBack(const Vector& RHS) const;                                10
   Vector bandForward(const Vector& RHS) const;                             11
   Vector bandChol(const Vector& RHS) const;                               12
                                                                           13
 public:                                                                    14
                                                                           15
   pdBand(int nrows = 0, int Q = -1, double a = 0.);                        16
   pdBand(int nrows, int Q, const double* const* pa);                      17
   pdBand(const pdBand& A);                                                18
                                                                           19
   ~pdBand(){};                                                            20
                                                                           21
   pdBand& operator=(const pdBand& B);                                     22
                                                                           23
   Vector solve(const Vector& RHS) const {                                 24
     std::cout << "Solution via banded Cholesky from class pdBand"         25
               << std::endl;                                               26
     return bandChol(RHS);                                                 27
   }                                                                       28
                                                                           29
   Matrix bandToFull() const;                                             30
};                                                                        31
#endif                                                                    32
```

The class declaration for pdBand begins on line 8 of Listing 7.8. It takes the usual form except that is has been augmented by : public Matrix to indicate that pdBand is a derived class that inherits the public members and member functions from its parent class Matrix. This entails that a pdBand object can use getNRows, getNCols, the indexing operator, etc., just like an object from class matrix without having to include them in the class declaration. In addition to the members that pdBand inherits from class Matrix is has its own unique member q that will hold the bandwidth of the matrix. With this in mind, it is probably useful at this point to detail the plan for storing the matrix elements in a pdBand object.

If A is an $n \times n$ symmetric banded matrix with bandwidth q and elements $a_{ij}, i, j = 1, \ldots, n$, it is only necessary to store its diagonal and above diagonal elements. This means that A's contents can be placed in an $n \times (q+1)$ matrix of the form

$$\tilde{A} = \begin{bmatrix} a_{11} & a_{12} & a_{13} & \cdots & a_{1q} & a_{1(q+1)} \\ a_{22} & a_{23} & a_{24} & \cdots & a_{2(q+1)} & a_{3(q+2)} \\ \vdots & \vdots & \vdots & \vdots & & \\ a_{(n-q)(n-q)} & a_{(n-q)(n-q+1)} & a_{(n-q)(n-q+2)} & \cdots & a_{(n-q)(n-1)} & a_{(n-q)n} \\ a_{(n-q+1)(n-q+1)} & a_{(n-q+1)(n-q+2)} & \cdots & a_{(n-q+1)(n-1)} & a_{(n-q+1)n} & 0 \\ a_{(n-q+2)(n-q+2)} & \cdots & a_{(n-q+2)(n-1)} & a_{(n-q+2)n} & 0 & 0 \\ \vdots & \vdots & \vdots & \vdots & \vdots & \vdots \\ a_{nn} & 0 & 0 & \cdots & 0 & 0 \end{bmatrix}$$

There is a useful relationship between the elements of A and those in $\tilde{A} = \{\tilde{a}_{ij}\}$ in that

$$a_{ij} = \tilde{a}_{i(j-i+1)}, \ j \leq \min(i+q, n), \tag{7.17}$$

with $\tilde{a}_{ij}$ the (i,j)th element of $\tilde{A}$ (cf. page 220 of Björck 1996). This translates into $a_{ij} = \tilde{a}_{i(j-i)}$ for 0-offset indexing. Relation (7.17) will be exploited to translate routines from Section 7.2.4 to work with matrices that are in this compressed storage mode.

As a specific example of a matrix in band storage mode we will use the two-banded matrix A that was produced in R in Section 7.2.4. This matrix is converted to our band storage mode with

```
> Atilde <- matrix(0, 6, 3)
> for(i in 1:4)
+ Atilde[i,] <- A[i, i:(i + 2)]
> Atilde[5, 1:2] <- A[5, 5:6]
> Atilde[6, 1] <- A[6, 6]
```

with the result that

```
> Atilde
           [,1]        [,2]        [,3]
[1,]  0.3671146  0.03285631 -0.02547667
[2,]  2.2845905  0.24176066  0.73343776
[3,]  3.5142686  0.20403810  0.27406444
[4,]  2.2404457  1.13430207 -0.84294584
[5,]  2.8909309 -0.07164435  0.00000000
[6,]  1.1492209  0.00000000  0.00000000
```

As mentioned above, it is necessary to define constructors (including a copy constructor), a destructor and the assignment operator in a derived class. The definitions for the three constructors for class pdBand take the form

```
pdBand::pdBand(int nrows, int Q, double a)
  : Matrix(nrows, Q + 1, a) {
  q = Q;
}


pdBand::pdBand(int nrows, int Q, const double* const* pa)
  : Matrix(nrows, Q + 1, pa) {
  q = Q;
}


pdBand::pdBand(const pdBand& A)
  : Matrix(A) {
  q = A.q;
}
```

The approach is the same in all three cases: an initializer list calls the appropriate class Matrix constructor and the body of the constructor sets the value of the bandwidth parameter that is unique to the pdBand class. The utility of the diagonal matrix constructor for the pdBand class is limited to its default role. In this regard, it uses a default value of -1 for q. Since the number of columns is q + 1, this has the effect of producing an inherited Matrix object with both nRows and nCols set as zero and, hence, a null pointer for its pointer-to-pointer member.

The destructor in class Matrix was designated as virtual. Declaring the base class destructor as virtual is generally viewed as a good practice when developing an inheritance relationship. Even though destructors are not inherited, the base class destructor will still be called when an object goes out of scope. The default behavior for destruction of a derived class object is to proceed in reverse order through the inheritance chain moving downward through the hierarchy of children classes until the base class is reached. For example, in the case of a pdBand object this means that the pdBand destructor will be called prior to that of the base class Matrix. In most instances this will work fine. However, as seen from Listing 7.6, it is possible to have a pointer for a base class point to a derived class object. In such cases, only the base class destructor would be called when the pointer goes out of scope. If memory was dynamically allocated in the derived class the result would be a memory leak. By making the base class destructor virtual the behavior of the destructor will adapt to the type of the object being pointed to rather than the type of the pointer. For the case of a pdBand object, the only new data member is the integer q which needs no special treatment. Thus, the body of the destructor for the pdBand class on line 21 of Listing 7.8 is left empty.

Our implementation of the pdBand assignment operator takes the form

```
pdBand& pdBand::operator=(const pdBand& A){

   if(this == &A)//avoid self assignment
      return *this;
   this->Matrix::operator=(A);
   q = A.q;

   return *this;
}
```

The development is similar to what transpires with our constructors in that the base class version of = is used to assign the base class members of the pdBand object with the pdBand part of assignment taking care of any nonshared class members or q in this case. The syntax this->Matrix::operator=(A) means that the pdBand object that resides on the left-hand side of the = operator (i.e., the one that owns the pointer this) is invoking the (public) Matrix function = to assign values to the class members nRows, nCols and pA. Actually, the use of this in this context is not necessary. Its use here helps to clarify who is actually calling the base class = operator.

A new utility method bandToFull has been added to the pdBand class that converts a matrix from band to full storage form. Its definition is

```
Matrix pdBand::bandToFull() const{
  int nrows = this->getNRows();
  double** pTemp = new double*[nrows];
  for(int i = 0; i < nrows; i++){
    pTemp[i] = new double[nrows];
    for(int j = i; j < nrows; j++){
      if(j <= std::min(nrows, i + q))
        pTemp[i][j] = (*this)[i][j - i];
      else pTemp[i][j] = 0.;
```

```
      }
   }
   for(int i = 1; i < nrows; i ++)
      for(int j = 0; j < i; j++)
         pTemp[i][j] = pTemp[j][i];
   Matrix A(nrows, nrows, pTemp);
   for(int i = 0; i < nrows; i++)
      delete[] pTemp[i];
   delete[] pTemp;
   return A;
}
```

This function allocates new memory for a `Matrix` object and then fills in the diagonal and above diagonal elements using (7.17). The below diagonal entries are obtained from symmetry. The full pointer-to-pointer configuration is passed to the `Matrix` class constructor to obtain a `Matrix` object. Our use for this function arises in a subsequent example. It also has other uses that include development of an overloaded + operator for addition of a `pdBand` and `Matrix` object (Exercise 7.20).

The `bandToFull` method employs the syntax `(*this)[i][j]` to access the elements of the `pA` pointer in a `pdBand` object. This can be read as follows. Because `this` is a pointer to the `pdBand` object making the call to `bandToFull`, `*this` is the `pdBand` object itself which means it can use the `[]` operator from the `Matrix` class for (read) access of the pointer `pA[i]` and, hence, of the stored values in `pA[i][j]`. The use of `this` becomes necessary as a result of `pA` being a private member of class `Matrix` that cannot be accessed directly by a `pdBand` object. Another way to deal with this is to make `pA` `protected` rather than `private` in class `Matrix` via the syntax

```
protected:
   double** pA;
```

Once this change is made, the occurrence of expressions like `(*this)[i][j]` in this and other listings can be replaced by `pA[i][j]`.

One aim of this inheritance exercise is to create a generic `solve` method that will adapt to the form of the calling `Matrix` object; i.e., a method that will adapt to objects from derived `Matrix` classes. The method that was used for `solve` in class `Matrix` was Gauss elimination. For a positive-definite banded matrix a much more efficient approach is to use a banded Cholesky recursion and that is the way `solve` has been formulated in the `pdBand` class.

To use a banded Cholesky it is necessary to have methods for forward and backward substitution that employ banded structure as do `bandBack` and `bandFor` in Appendix D. There is a little more to do in this case as the Cholesky recursion will also produce Cholesky factors that are held in a compressed storage format. In this regard, a method for back-solving that will work with an upper-triangular matrix in band storage mode is

```
Vector pdBand::bandBack(const Vector& RHS) const {
   double temp;//temporary storage
   int up;
   int nrows = this->getNRows();
   Vector b(nrows, 0.);//solution vector

   //initialize recursion on last row of right-hand side
   if((*this)[nrows - 1][0] != 0)
      b.pA[nrows - 1] = RHS.pA[nrows - 1]/(*this)[nrows - 1][0];
   else{
      cout << "Singular system!" << endl;
```

```
      exit(1);
   }

   //now work through the remaining rows
   for(int i = (nrows - 2); i >= 0; i--){
     if((*this)[i][0] != 0){
        temp = RHS.pA[i];
        up = std::min(nrows, q + 1);
        for(int k = 1; k < up; k++)
           temp -= b.pA[i+k]*(*this)[i][k];
        b.pA[i] = temp/(*this)[i][0];
     }
     else{
        cout << "Singular system!" << endl;
        exit(1);
     }
   }
   return b;
}
```

This is a straightforward modification of the **bandBack** method in Appendix D. There is an important change to the **Vector** class that is necessary to make this and subsequent methods work: namely, the **pdBand** class must be added as a **friend** to class **Vector**. This allows us to write statements such as **b.pA[i]** for a **Vector** object **b** without producing compilation errors. As **friend** status is not inherited from class **Matrix**,

```
   friend class pdBand;
```

must be explicitly placed in the **Vector** class header.

The Cholesky recursion that we will employ will produce an upper-triangular banded matrix that will be held in band storage mode. This will work fine for the final back-solving step in the Cholesky method. However, the first step requires us to forward solve a lower-triangular system using the transpose of the upper-triangular matrix that is being held in a compressed storage format. Symmetry considerations entail that all the information that is needed for forward solving is available from the compressed, upper-triangular, array. The only question is how to extract it. The solution is provided by observing that, due to symmetry and relation (7.17), formula (7.6) can be expressed as

$$
b_j = \left(c_j - \sum_{k=\max(1,j-q)}^{j-1} a_{jk}b_k \right) / a_{jj}
$$

$$
= \left(c_j - \sum_{k=\max(1,j-q)}^{j-1} \tilde{a}_{j(k-j+1)}b_k \right) / \tilde{a}_{j0}
$$

with the b_j and c_j representing the elements of the solution vector and the right-hand-side vector, respectively. The next listing uses this formula to produce the code for our forward solution method.

```
Vector pdBand::bandForward(const Vector& RHS) const {
   double temp;//temporary storage
   int low;
   int nrows = this->getNRows();
   Vector b(nrows, 0.);//solution vector

   //initialize recursion on first row of right-hand side
```

```
if(*this[0][0] != 0){
  b.pA[0] = RHS.pA[0]/(*this)[0][0];
}
else{
  cout << "Singular system!" << endl;
  exit(1);
}

//now work through the remaining rows
for(int i = 1; i < nrows; i++){
  if((*this)[i][0] != 0){
    temp = RHS.pA[i];
    low = std::max(0, i - q);
    for(int k = low; k < i; k++)
      temp -= b.pA[k]*(*this)[k][i - k];
    b.pA[i] = temp/(*this)[i][0];
  }
  else{
    cout << "Singular system!" << endl;
    exit(1);
  }
}
return b;
}
```

The pieces are now in place to solve a system provided the Cholesky factorization is available. The last step is therefore to translate the banded Cholesky algorithm from Section 7.2.4 into one that will work in the band storage setting. That is essentially what will be done. But, the previous Cholesky algorithms have produced lower-triangular factors. It is more convenient for band storage to have the upper-triangular factor. Rather than compute the lower-triangular matrix in band storage and transpose the result, the (0-offset) upper-triangular matrix $\tilde{U} = \{\tilde{u}_{i,j}\}_{i=0:(n-1),j=0:q}$ in band storage can be computed directly from $\tilde{A}$, the (0-offset) version of A in band storage, via

$$\tilde{u}_{i0} = \left(\tilde{a}_{i0} - \sum_{k=\max(0,i-q)}^{i-1} \tilde{u}_{k(i-k)}^2\right)^{1/2} \tag{7.18}$$

for $i = 0,\ldots,n-1$ and for $j = (i+1),\ldots,\min(i+q,n-1)$

$$\tilde{u}_{i(j-i)} = \left(\tilde{a}_{i(j-i)} - \sum_{k=\max(0,j-q)}^{i-1} \tilde{u}_{k(i-k)}\tilde{u}_{k(j-k)}\right)/\tilde{u}_{i0}. \tag{7.19}$$

The listing below is a translation of these relations.

```
//solve a banded system via Cholesky factorization
Vector pdBand::bandChol(const Vector& RHS) const{
  double temp = 0;
  int low, up;
  int nrows = this->getNRows();

  //temporary storage
  double** pG = new double*[nrows];
  for(int i = 0; i < nrows; i++){
    pG[i] = new double[q + 1];
    for(int j = 0; j < q + 1; j++){
```

```
      pG[i][j] = (*this)[i][j];
    }
  }
  for(int i = 0; i < nrows; i++){
    if((*this)[i][0] == 0){
      cout << "Singular system!" << endl;
      exit(1);
    }
    temp = (*this)[i][0];

    low = std::max(0, i - q);
    for(int k = low; k < i; k++)
      temp -= pG[k][i - k]*pG[k][i - k];

    pG[i][0] = sqrt(temp);

    up = std::min(i + q, nrows - 1);

    for(int j = (i + 1); j <= up; j++){
      temp = (*this)[i][j - i];
      low = std::max(0, j - q);
      for(int k = low; k < i; k++)
        temp -= pG[k][i - k]*pG[k][j - k];
      pG[i][j - i] = temp/pG[i][0];
    }
  }
  pdBand G(nrows, q, pG);
  Vector h = G.bandForward(RHS);
  Vector b = G.bandBack(h);

  for(int i = 0; i < nrows; i++)
    delete[] pG[i];
  delete[] pG;
  return b;
}
```

The following driver program was written to test the **pdBand** class.

```
//pdBandDriver.cpp
#include <iostream>
#include <fstream>
#include "simpleMat.h"
#include "vector.h"
#include "pdBand.h"

int main(int argc, char* argv[]){
  double* pc = new double[6];
  double** pA = new double*[6];

  std::ifstream fIn1, fIn2;
  fIn1.open(argv[1]); fIn2.open(argv[2]);

  for(int i = 0; i < 6; i++){
    pA[i] = new double[3];
    for(int j = 0; j < 3; j++){
      fIn1 >> pA[i][j];
    }
```

```
    fIn2 >> pc[i];
  }
  fIn1.close(); fIn2.close();

  pdBand A(6, 2, pA);
  Vector c(6, pc);
  Matrix* pMat = &A;
  pMat->solve(c).printVec();

  Matrix B = A.bandToFull();
  pMat = &B;
  pMat->solve(c).printVec();
  return 0;
}
```

This program begins by reading in the left-hand-side matrix and right-hand-side vector for a linear system from two different files. The corresponding pdBand and Vector objects are then created and a pointer to Matrix is initialized with the address of the pdBand object. This pointer is first used to call the solve method with the pdBand object. A new Matrix object is created next using the bandToFull method of the pdBand object. The Matrix pointer is assigned the address of this Matrix object and solve is called again.

The two-banded matrix that was created using R earlier in this section was used as input to the pdBandDriver.cpp program along with a right-hand-side vector produced by

```
> set.seed(456)
> rhs <- rnorm(6)
> rhs
[1] -1.3435214  0.6217756  0.8008747 -1.3888924 -0.7143569 -0.3240611
```

The results produced by our program are

```
Solution via banded Cholesky from class pdBand
-3.71877
0.825037
0.212466
-1.6256
0.334564
-1.45349
Solution via Gauss from class Matrix
-3.71877
0.825037
0.212466
-1.6256
0.334564
-1.45349
```

The output confirms that solve is exhibiting the desired polymorphic behavior.

7.3 Eigenvalues and eigenvectors

Eigenvalues play a fundamental role in statistics. This is particularly true in the case of multivariate analysis where eigenvalues of sample covariance matrices and related quantities provide the foundation for many tools that are used for statistical inference.

The problem that will be addressed in this section is the computation of eigenvalues corresponding to a real, positive-semidefinite (and, hence, symmetric), $p \times p$ matrix A. The rank of A is $r \leq p$ so that it admits an eigenvalue-eigenvector decomposition of the form

$$A = \sum_{j=1}^{r} \lambda_j v_j v_j^T \tag{7.20}$$

for positive eigenvalues $\lambda_1, \ldots, \lambda_r$ and associated eigenvectors $v_1, \ldots, v_r$. For simplicity it will be assumed that the λ_j are distinct in which case they can be arranged so that $\lambda_1 > \lambda_2 > \cdots > \lambda_r$.

The Euclidean norm of a p-vector $v = (v_1, \ldots, v_p)^T$ is

$$||v|| = \left\{ \sum_{j=1}^{p} v_j^2 \right\}^{1/2}$$

and the inner product of two p-vectors v and $x = (x_1, \ldots, x_p)^T$ is

$$\langle v, x \rangle = \sum_{j=1}^{p} v_j x_j.$$

The eigenvalue-eigenvector pairs $(\lambda_i, v_i), i = 1, \ldots, r$, for A then satisfy

$$Av_i = \lambda_i v_i \tag{7.21}$$

subject to

$$\langle v_i, v_j \rangle = \delta_{ij} \tag{7.22}$$

with δ_{ij} equal to 1 when $i = j$ and 0 otherwise. The combination of (7.21)–(7.22) gives

$$v_i^T A v_i = \langle v_i, A v_i \rangle = \lambda_i, \; i = 1, \ldots, r.$$

The approach that will now be described for computing eigenvalues is generally called the *power method*. The idea behind it is quite simple. The eigenvectors $v_1, \ldots, v_r$ are known to form an orthonormal basis for an r-dimensional subspace of $\mathbb{R}^p$. As a result, an additional $p - r$ vectors $v_{r+1}, \ldots, v_p$ can be found so that $v_1, \ldots, v_p$ provide an orthonormal basis for all of $\mathbb{R}^p$. Using this basis any p-vector v can be written as

$$v = \sum_{i=1}^{p} a_i v_i \tag{7.23}$$

with $a_i = \langle v, v_i \rangle$. Thus,

$$Av = \sum_{i=1}^{r} \lambda_i a_i v_i, \; A(Av) = A^2 v = \sum_{i=1}^{r} \lambda_i^2 a_i v_i, \ldots, A^k v = \sum_{i=1}^{r} \lambda_i^k a_i v_i, \ldots.$$

But, if $a_1 \neq 0$,

$$\sum_{i=1}^{r} \lambda_i^k a_i v_i = \lambda_1^k \left[a_1 v_1 + \sum_{i=2}^{r} (\lambda_i/\lambda_1)^k a_i v_i \right]$$

and

$$\left\| \sum_{i=2}^{r} (\lambda_i/\lambda_1)^k a_i v_i \right\| \leq ||v|| \left\{ \sum_{i=2}^{r} (\lambda_i/\lambda_1)^{2k} \right\}^{1/2}$$

$$\leq \sqrt{r-1} ||v|| (\lambda_2/\lambda_1)^k$$

since $\sum_{j=2}^{r} a_j^2 \leq ||v||^2$. The fact that $\lambda_2/\lambda_1 < 1$ has the consequence that the remainder term converges to 0 as k increases.

The procedure for computing λ_1 and v_1 that stems from the previous development is summarized in Algorithm 7.3 below. The algorithm depends on user-specified values of the tuning parameters δ and $itMax$ that represent the desired level of approximation and maximum number of iterations, respectively, as well as a starting vector v with $v^T v_1 \neq 0$.

Algorithm 7.3 Power method

$v^{(0)} = Av/\|Av\|, \lambda^{(0)} = (v^{(0)})^T Av^{(0)}, niter = 0, change = \infty$
while $change > \delta$ and $niter < itMax$ **do**
   $niter := niter + 1$
   $v^{(niter)} := Av^{(niter-1)}/\|Av^{(niter-1)}\|$
   $\lambda^{(niter)} := (v^{(niter)})^T Av^{(niter)}$
   $change = |\lambda^{(niter)} - \lambda^{(niter-1)}|$
end while
return $v^{(niter)}, \lambda^{(niter)}$

Algorithm 7.3 successively applies A to the vector v, renormalizes and evaluates the ratio of quadratic forms in A that provides the approximation to λ_1. The variable $change$ holds the changes in the approximations to λ_1 from successive iterations. If the magnitude of $change$ falls below the specified tolerance level δ, execution is terminated. The value of $change$ is initialized at ∞ so that the `while` loop will begin.

To illustrate how Algorithm 7.3 works we will use the positive-definite matrix A that was used for demonstrating the Gauss transform in Section 7.2 and will choose v as the vector

```
> v
[1] -1.6866933   0.8377870   0.1533731  -1.1381369   1.2538149
```

that served as the right-hand side for that example. The condition that is needed for v is that it has a nonzero inner product with the first eigenvector of A. Random selection seems like a good strategy to produce a v with this property.

The eigenvalues and eigenvectors for

```
> A
          [,1]         [,2]         [,3]         [,4]         [,5]
[1,]   2.8183730  -3.1452537  -0.2082627  -4.1931724  -1.0827824
[2,]  -3.1452537   5.4246530   1.9299179   5.5110075   0.1453005
[3,]  -0.2082627   1.9299179   2.1096700   1.9187222  -1.5299804
[4,]  -4.1931724   5.5110075   1.9187222   8.0239285  -0.2145681
[5,]  -1.0827824   0.1453005  -1.5299804  -0.2145681   3.1624022
```

can be obtained using the R `eigen` function. The results from `eigen` are returned as a list with the components `values` and `vectors` containing the eigenvalues and eigenvectors, respectively. For the current matrix of interest this gives

```
> eigen(A)
$values
[1] 15.199405715   4.577881146   1.344645025   0.413226152   0.003868673

$vectors
            [,1]         [,2]         [,3]         [,4]         [,5]
[1,]   0.385560633  -0.32572975   0.1754596   0.2661721   0.80225275
[2,]  -0.560721905  -0.02432660   0.6461808  -0.4437891   0.26551988
[3,]  -0.191942614  -0.50619678   0.4039680   0.6156238  -0.40588179
[4,]  -0.707157822  -0.02592964  -0.5284342   0.3267940   0.33647946
[5,]  -0.004448737   0.79774734   0.3305018   0.4964037   0.08905766
```

With this as a basis for comparison, let us see how the power method performs in terms of computing both $\lambda_1 = 15.199405715$ and its associated eigenvector.

The first step in the power iteration is

```
> v1 <- v
> v1 <- A%*%v1
> v1 <- v1/sqrt(drop(t(v1)%*%v1))
```

It turns out that `t(v1)%*%v1` is of the R class `matrix` and must be "demoted" to `numeric` using the `drop` function discussed in Section 6.2. The approximation to the first eigenvalue of A that results from this is

```
> drop(t(v1)%*%A%*%v1)
[1] 8.327123
```

Certainly this is not very satisfactory. So, another iteration appears to be needed which produces

```
> v1 <- A%*%v1
> v1 <- v1/sqrt(drop(t(v1)%*%v1))
> drop(t(v1)%*%A%*%v1)
[1] 13.82744
```

The next iteration gives

```
> v1 <- A%*%v1
> v1 <- v1/sqrt(drop(t(v1)%*%v1))
> drop(t(v1)%*%A%*%v1)
[1] 15.05940
```

The fourth, fifth and sixth iterations provide further improvements with

```
> v1 <- A%*%v1
> v1 <- v1/sqrt(drop(t(v1)%*%v1))
> drop(t(v1)%*%A%*%v1)
[1] 15.18656
> v1 <- A%*%v1
> v1 <- v1/sqrt(drop(t(v1)%*%v1))
> drop(t(v1)%*%A%*%v1)
[1] 15.19824
```

and

```
> v1 <- A%*%v1
> v1 <- v1/sqrt(drop(t(v1)%*%v1))
> drop(t(v1)%*%A%*%v1)
[1] 15.1993
```

We might be tempted to stop iterating at this point. However, the vector `v1` produced in the last iteration is

```
> v1
           [,1]
[1,]  -0.38658660
[2,]   0.56064349
[3,]   0.19034451
[4,]   0.70707149
[5,]   0.00696751
```

This represents the approximation to the first eigenvector v_1 of A that corresponds to our approximation for λ_1. As such it could stand further improvement. So, four more iterations are carried out to obtain

```
> drop(t(v1)%*%A%*%v1)
[1] 15.19941
> v1
            [,1]
[1,]  -0.38556909
[2,]   0.56072127
[3,]   0.19192947
[4,]   0.70715715
[5,]   0.00446946
```

This agrees to several decimals with the results from R apart from the difference in sign.

The example demonstrates that the power method can work in a particular case. More generally, Theorem 7.1 confirms that for k sufficiently large the value $\lambda^{(k)}$ and vector $v^{(k)}$ returned at the kth iteration of the power method algorithm provide numerical approximations to the largest eigenvalue of A and its associated eigenvector, respectively. Upper bounds for the size of the approximation errors for both the eigenvalues and eigenvectors are obtained as a by-product of the proof.

Theorem 7.1. *Let v be a p-vector with $\langle v, v_1 \rangle \neq 0$ and let $v^{(k)}, \lambda^{(k)}$ be the approximations to v_1, λ_1 that are returned on the kth step of the power method iteration. Then, as $k \to \infty$, $v^{(k)} \to -\pm v_1$ in Euclidean norm and $|\lambda^{(k)} - \lambda_1| \to 0$.*

Proof. Note that $v^{(1)} = Av/||Av||$ and

$$v^{(2)} = \frac{Av^{(1)}}{||Av^{(1)}||} = \frac{A^2v/||Av||}{||A^2v||/||Av||}.$$

Proceeding by induction one concludes that on the kth step of the recursion

$$v^{(k)} = A^k v/||A^k v||.$$

Since $\lambda^{(k)} = (v^{(k)})^T A v^{(k)}$, this gives

$$\lambda_1 - \lambda^{(k)} = \lambda_1 \left[1 - \frac{1 + \sum_{j=2}^r (\lambda_j/\lambda_1)^{2k+1} (a_j/a_1)^2}{1 + \sum_{j=2}^r (\lambda_j/\lambda_1)^{2k} (a_j/a_1)^2} \right]$$

which converges to 0 at the rate $(\lambda_2/\lambda_1)^{2k}$.

To establish convergence of $v^{(k)}$, first observe that there is a sign ambiguity in eigenvectors: i.e., both v_1 and $-v_1$ are eigenvectors for λ_1. The vectors $v^{(k)}$ will converge to one of these depending on the sign of a_1 in (7.23). The convergence is to v_1 if $a_1 > 0$ and to $-v_1$ otherwise. For specificity suppose that $a_1 > 0$ as the other case can be handled analogously. Then,

$$
\begin{aligned}
\langle v^{(k)}, v_1 \rangle &= \frac{\langle v_1, A^k v \rangle}{||A^k v||} \\
&= \frac{a_1 \lambda_1^k}{\left\{ a_1^2 \lambda_1^{2k} + \sum_{j=2}^r \lambda_j^{2k} a_j^2 \right\}^{1/2}} \\
&= \frac{1}{\left\{ 1 + \sum_{j=2}^r (\lambda_j/\lambda_1)^{2k} (a_j/a_1)^2 \right\}^{1/2}}.
\end{aligned}
$$

Thus, $||v^{(k)} - v_1||^2 = 2 - 2\langle v^{(k)}, v_1 \rangle$ converges to zero like $(\lambda_2/\lambda_1)^k$ and the theorem is proved. $\square$

The proof technique reveals that the sequence of approximations $\lambda^{(k)}$ returned from Algorithm 7.3 converges to λ_1 at the rate $(\lambda_2/\lambda_1)^{2k}$ while the rate for the corresponding eigenvectors is only $(\lambda_2/\lambda_1)^k$. A conclusion that might be drawn from this is that eigenvectors are harder to approximate numerically than eigenvalues.

Up to this point only the largest eigenvalue has been considered. But, the power method approach can be used for other eigenvalues as well. For example, let $\tilde{v}_1, \tilde{\lambda}_1$ be the approximations to v_1, λ_1 that are returned at the end of the power method recursion for the largest eigenvalue. Then, a new matrix $\tilde{A} = A - \tilde{\lambda}_1 \tilde{v}_1 \tilde{v}_1^T$ and a new vector $\tilde{v} = v - \langle v, \tilde{v}_1 \rangle \tilde{v}_1$ can be defined with the power method now being applied to $\tilde{A}$ and $\tilde{v}$ to obtain approximations to v_2 and λ_2.

To illustrate how the power method can be used for computing eigenvalues other than the largest, let us continue with the matrix A and initialization vector v from the previous example. Using the approximate eigenvector we found for A we make the adjustment

```
> A <- A - drop(t(v1)%*%A%*%v1)*v1%*%t(v1)
> v <- v - drop(t(v1)%*%v)*v1
```

Then, the recursion for the second eigenvalue and eigenvector begins with

```
> v2 <- A%*%v
> v2 <- v2/sqrt(drop(t(v2)%*%v2))
> drop(t(v2)%*%A%*%v2)
[1] 4.366477
```

and continues to the second step as

```
> v2 <- A%*%v2
> v2 <- v2/sqrt(drop(t(v2)%*%v2))
> drop(t(v2)%*%A%*%v2)
[1] 4.558759
```

Proceeding in this manner, the fifth iteration, for example, produces

```
> v2 <- A%*%v2
> v2 <- v2/sqrt(drop(t(v2)%*%v2))
> drop(t(v2)%*%A%*%v2)
[1] 4.577869
```

and

```
> v2
              [,1]
[1,]  -0.32535367
[2,]  -0.02311193
[3,]  -0.50542452
[4,]  -0.02702335
[5,]   0.79839001
```

The power method is straightforward to implement in C++. One way this can be accomplished is encoded in the `eigen` method that was added to our class `Matrix`.

```
Vector Matrix::eigen(int Nvals, Vector& v, Matrix& U, int itMax,
        double delta) const {
  double change, temp;
  int niter;

  Matrix ACopy(nRows, nCols, pA);//work copy of A
  Vector vCopy;//work copy of v
  Vector lambda(Nvals, 0.);
  for(int i = 0; i < Nvals; i++){
    change = std::numeric_limits<double>::infinity();
    niter = 0;
```

```
    vCopy = v;
    lambda.pA[i] = 0;

    while(change > delta && niter < itMax){
      vCopy = ACopy*vCopy;
      vCopy = (1./sqrt(vCopy.dotProd(vCopy)))*vCopy;
      lambda.pA[i] = vCopy.dotProd(ACopy*vCopy);
      if(niter == 0)
        temp = lambda.pA[i];
      else if(temp != 0.){
        change = fabs(1. - lambda.pA[i]/temp);
        temp = lambda.pA[i];
      }
      niter++;
    }
    U.vecToMat(i, vCopy);
    v -= (vCopy.dotProd(vCopy))*vCopy;
    ACopy -= lambda.pA[i]*(vCopy*vCopy);
  }
  return lambda;
}
```

The input to the `eigen` method includes the number of eigenvalues/eigenvectors that are to be computed (i.e., `Nvals`), a `Vector` object `v` that will be used to initialize the recursion and a `Matrix` object passed in by reference from the calling program that will be filled with `Nvals` eigenvectors. The final two arguments represent stopping criteria for the recursion. The actual number of iterations must be controlled in some fashion. The way this has been accomplished here is to halt the power method recursions if the number of iterations exceeds `itMax` or the absolute relative change in the computed eigenvalue from the previous step (i.e., `change`) is less than `delta`. Similar stopping rules are discussed in more detail in the next chapter. The absolute value function `fabs`, that becomes accessible via inclusion of the cmath header, was used to evaluate `change`.

In order to start the power method iterations a value for the tolerance variable `change` is needed at the beginning of the `while` loop. For this purpose `change` is set equal to the C++ representation of positive infinity using the C++ `numeric_limits` template class that requires the limits header file. A slight twist involves the `vecToMat` function detailed in Appendix D (cf. Exercise 3.21). It furnishes a solution to a problem that arises at the end of each `while` loop where a vector object that contains the approximate eigenvector needs to be inserted into the `Matrix` object U that was provided for eigenvector storage. The `dotProd` method that appears in the listing computes the inner product of two vectors and the outer product of two vectors is computed with an overloaded version of the `*` operator. Their definitions can also be found in Appendix D.

As a test for the C++ power method code the original matrix A and vector v from our running example were written into the file `test.txt` as a 5×6 array with v in the last column via

```
> set.seed(123)
> A <- matrix(rnorm(25), 5, 5)
> v <- rnorm(5)
> write.table(cbind(A, v), "test.txt", quote = FALSE,
+ row.names = FALSE, col.names = FALSE)
```

Then, the C++ power method function was applied to the data in `test.txt` with the number of iterations set at 35 and `delta` at 10^{-8}. This produced the results

```
Eigenvalues and vectors for A
15.1994
4.57788
1.34465
0.413226
0.00386867
```

-0.385561	-0.325727	0.17546	-0.266174	-0.802252
0.560722	-0.0243302	0.646182	0.443783	-0.265521
0.191941	-0.506198	0.403968	-0.615627	0.405883
0.707158	-0.0259358	-0.528435	-0.326789	-0.336479
0.00445062	0.797748	0.330497	-0.496407	-0.0890564

The eigenvalues and vectors are in general agreement with those that were computed in R.

Although the simplicity of the power method is appealing, it is not the method of choice for eigenvalue computations. Instead, an algorithm that employs the power method in conjunction with orthogonal tridiagonalization and the QR decomposition from Section 7.5 tends to give faster convergence (e.g., Golub and Van Loan 1996). The JAMA package discussed in Section 7.6 uses a version of this approach for computing the eigenvalues of a symmetric matrix in its Eigenvalue class.

7.4 Singular value decomposition

An extension of the eigenvalue-eigenvector decomposition for a symmetric real matrix leads to the *singular value decomposition*, or merely SVD, that is applicable to matrices that are neither symmetric nor square. The SVD appears in multivariate analysis in the context of canonical correlation analysis, for example, and also has applications in the solution of general least-squares problems.

Let A be a real $p \times m$ matrix of rank $r \leq \min(p, m)$. Then, AA^T is a symmetric positive-semidefinite matrix that can be written as

$$AA^T = \sum_{j=1}^{r} \lambda_j^2 u_j u_j^T$$

for eigenvalues $\lambda_1^2 \geq \lambda_2^2 \geq \cdots \geq \lambda_r^2 > 0$ and associated eigenvectors $u_1, \ldots, u_r$. Now, define

$$v_j = \frac{1}{\lambda_j} A^T u_j \tag{7.24}$$

for $j = 1, \ldots, r$ and observe that v_j is an eigenvector for $A^T A$ because

$$
\begin{aligned}
A^T A v_j &= \frac{1}{\lambda_j} A^T A A^T u_j \\
&= \lambda_j^2 \left(\frac{1}{\lambda_j} A^T u_j \right) \\
&= \lambda_j^2 v_j.
\end{aligned}
$$

Thus, AA^T and $A^T A$ have the same nonzero eigenvalues and their eigenvectors are related by (7.24). Additionally.

$$
\begin{aligned}
u_j^T A v_i &= (A u_j)^T v_j \\
&= \lambda_j v_j^T v_i \\
&= \lambda_j \delta_{ij} \tag{7.25}
\end{aligned}
$$

with δ_{ij} either 1 or 0 depending on whether or not i is equal to j.

The (left) singular vectors $u_1, \ldots, u_r$ can be augmented by an additional $p - r$ orthonormal vectors $u_{p+1}, \ldots, u_p$ to create an orthonormal matrix $U = [u_1, \ldots, u_p]$. Similarly, an additional $m - r$ orthonormal vectors $v_{r+1}, \ldots, v_m$ can be constructed to create the orthonormal matrix $V = [v_1, \ldots, v_m]$ of (right) singular vectors. The vectors $u_{r+1}, \ldots, u_p$ and $v_{r+1}, \ldots, v_m$ necessarily correspond to 0 eigenvalues for AA^T and $A^T A$, respectively. As a result, (7.25) can be rewritten in the standard SVD form

$$U^T A V = \left[\begin{array}{cc} \Lambda & 0_{r \times (m-r)} \\ 0_{(p-r) \times r} & 0_{(p-r) \times (m-r)} \end{array} \right] \tag{7.26}$$

or, equivalently, as

$$A = U \left[\begin{array}{cc} \Lambda & 0_{r \times (m-r)} \\ 0_{(p-r) \times r} & 0_{(p-r) \times (m-r)} \end{array} \right] V^T \tag{7.27}$$

for

$$\Lambda = \text{diag}(\lambda_1, \ldots, \lambda_r). \tag{7.28}$$

In combination these developments prove our next theorem.

Theorem 7.2. *Let A be a $p \times m$ matrix of rank $r \leq \min(p, m)$. Then,*

$$A = \sum_{j=1}^{r} \lambda_j u_j v_j^T, \tag{7.29}$$

where $\lambda_1^2, \ldots, \lambda_r^2$ are the nonzero eigenvalues of AA^T and $A^T A$ and $u_j, v_j = \lambda_j^{-1} A^T u_j, j = 1, \ldots, r$ are the corresponding eigenvectors.

As a result of Theorem 7.2, an SVD can be produced using the computational scheme described in Algorithm 7.4.

Algorithm 7.4 SVD algorithm

---

if number of rows for $A \leq$ number of columns **then**
    compute eigenvalues λ_j^2 and eigenvectors $u_j, j = 1, \ldots, r$ of AA^T
    $v_j = \lambda_j^{-1} A^T u_j, \ j = 1, \ldots, r$
else
    compute eigenvalues λ_j^2 and eigenvectors $v_j, j = 1, \ldots, r$ of $A^T A$
    $u_j = \lambda_j^{-1} A v_j, \ j = 1, \ldots, r$
end if
return $\lambda_j, u_j, v_j, j = 1, \ldots, r$

To illustrate how this algorithm is applied we will again work with a specific example in the R environment. The matrix that will be used is

```
> set.seed(123)
> A <- matrix(rnorm(20),5,4)
> A
           [,1]        [,2]        [,3]        [,4]
[1,] -0.56047565   1.7150650   1.2240818   1.7869131
[2,] -0.23017749   0.4609162   0.3598138   0.4978505
[3,]  1.55870831  -1.2650612   0.4007715  -1.9666172
[4,]  0.07050839  -0.6868529   0.1106827   0.7013559
[5,]  0.12928774  -0.4456620  -0.5558411  -0.4727914
```

This is a 5×4 array. So, $A^T A$ has the smaller dimension and its eigenvalues and eigenvectors are the ones that will be computed. The power method recursion will be initialized with the vector composed of the first four elements of the initialization vector that was used for the eigenvalue example in the previous section: viz,

```
> v <- v[1:4]; v;
[1] -1.6866933  0.8377870  0.1533731 -1.1381369
```

To begin, take

```
> v1 <- t(A)%*%A%*%v
> v1 <- v1/sqrt(drop(t(v1)%*%v1))
> sqrt(drop(t(v1)%*%t(A)%*%A%*%v1))
[1] 3.727011
```

Then, four more iterations lead to

```
> sqrt(drop(t(v1)%*%t(A)%*%A%*%v1))
[1] 3.898612
> v1
          [,1]
[1,] -0.3852381
[2,]  0.5606925
[3,]  0.1924203
[4,]  0.7072411
```

and

```
> u1 <- A%*%v1/sqrt(drop(t(v1)%*%t(A)%*%A%*%v1)); u1;
           [,1]
[1,]  0.68661807
[2,]  0.19710637
[3,] -0.67294214
[4,]  0.02694537
[5,] -0.19007241
```

The other eigenvalues can also be computed using the power method as demonstrated in the previous section.

The R function svd that computes the SVD of a matrix provided as its argument can be used to assess the accuracy of our power method iterations. The output from svd is returned as a list; the first component of the list (named d) holds the singular values while the second and third components (named v and u) contain the right and left singular vectors. An application of svd to the matrix used in our example produces

```
> svd(A)
$d
[1] 3.8986117 1.4848168 0.9795070 0.1154849

$u
            [,1]        [,2]        [,3]        [,4]
[1,] -0.68663120 -0.5941114 -0.02638716  0.41766365
[2,] -0.19710967 -0.1488653 -0.01012230 -0.49508266
[3,]  0.67292599 -0.7277940 -0.08942548  0.06879205
[4,] -0.02694425  0.1033245 -0.99312907  0.04115105
[5,]  0.19007890  0.2907073  0.06999283  0.75764403

$v
            [,1]        [,2]        [,3]        [,4]
[1,]  0.3852091 -0.4864560 -0.1870774  0.7615619
[2,] -0.5607034 -0.2474217  0.7290891  0.3046691
[3,] -0.1924690 -0.8234246 -0.2252240 -0.4839440
[4,] -0.7072350  0.1552902 -0.6186318  0.3049563
```

Thus, the power method approach would also seem to be effective in this example.

The method SVD below was added to our class Matrix.

```
Vector Matrix::SVD(int Nvals, Vector& v, Matrix& U, Matrix& V,
         int itMax, double delta) const {
  Vector temp;
  Matrix ACopy(nRows, nCols, pA);

  if(nRows > nCols){//work with A^TA
    temp = (ACopy.trans()*ACopy).eigen(Nvals, v, V, itMax, delta);
    Matrix Temp(Nvals, Nvals, 0.);
    for(int i = 0; i < Nvals; i++)
      Temp.pA[i][i] = 1./sqrt(temp[i]);
    U = ACopy*V*Temp;
  }
  else{//work with AA^T
    temp = (ACopy*ACopy.trans()).eigen(Nvals, v, U, itMax, delta);
    Matrix Temp(Nvals, Nvals, 0.);
    for(int i = 0; i < Nvals; i++)
      Temp.pA[i][i] = 1./sqrt(temp[i]);
    V = ACopy.trans()*U*Temp;
  }
  Vector lambda(Nvals, 0.);
  for(int i = 0; i < Nvals; i++)
    lambda.pA[i] = sqrt(temp[i]);
  return lambda;
}
```

The arguments for the SVD method are the same as those for eigen except that now two Matrix objects must be supplied to store both the right and left singular vectors. Beyond that, the program simply carries out Algorithm 7.4 directly with the assistance of our previously developed eigen method.

An application of the SVD method to the matrix used in the R calculations with 35 iterations and delta $= 10^{-8}$ produced

```
Singular values for A
3.89861
1.48482
0.979507
0.115485
```

```
Left singular vectors
     0.686629    -0.594217    -0.0263068    -0.417663
     0.197109    -0.148896    -0.0101026     0.495083
    -0.672928    -0.727694    -0.0892991    -0.0687902
     0.0269445    0.103291    -0.993146     -0.0411305
    -0.190078     0.290738     0.0699508    -0.757645
```

```
Right singular vectors
    -0.385213    -0.486442    -0.187025    -0.761561
     0.560702    -0.247421     0.729112    -0.304671
     0.192462    -0.823446    -0.22514      0.483945
     0.707236     0.155221    -0.618651    -0.304955
```

Again, these results are in general agreement with those obtained from R.

As was true for computing eigenvalues and eigenvectors, the power method is not generally favored for SVD calculations. Beyond the issue of slow convergence there are also accuracy problems that arise from (unnecessary) round-off errors produced by working with $A^T A$ or

AA^T. The preferred SVD algorithm is described in Golub and Van Loan (1996) and Björck (1996) a version of which is used in the JAMA SVD class introduced in Section 7.6.

7.5 Least squares

Suppose that we have an $n \times 1$ vector y that is to be approximated by a linear combination of the columns of an $n \times p$ real matrix X. That is, the objective is to find a p-vector b such that Xb provides a good "estimator" for y in some sense. In statistics the vector y typically represents observed values for a random variable and the columns of X correspond to observed or preset values of p "predictor" variables. For problems of this nature one can expect $n > p$ and that is the case that will be addressed here.

The fact that $n > p$ has the consequence that the system

$$Xb = y \tag{7.30}$$

is over-determined and generally has no solution. Instead, a least-squares solution can be sought by attempting to minimize $||y - Xb||$ over all $b \in \mathbb{R}^p$ with b then chosen as the minimizer with smallest Euclidean norm. This process turns out to be equivalent to looking for a minimum norm solution to the normal equations

$$X^T X b = X^T y$$

which puts us back in the framework of Section 7.3 with $A = X^T X$ and $c = X^T y$ provided that $X^T X$ is nonsingular or, equivalently, X has rank p. More generally, for analysis of variance and other problems arising from the general linear model the rank p assumption may not hold and some additional work is needed.

At this juncture it is useful to introduce the concept of a generalized inverse for a matrix as part of our linear algebra repertoire. There are actually several ways to define such quantities. But, for our purposes is suffices to only consider the *Moore-Penrose generalized inverse* (e.g., Ben-Israel and Greville 2003). For an $n \times p$ matrix X, its Moore-Penrose generalized inverse is the unique matrix $X^\dagger$ that satisfies the *Penrose equations*

$$
\begin{align}
XX^\dagger X &= X \tag{7.31}\\
X^\dagger X X^\dagger &= X^\dagger \tag{7.32}\\
(XX^\dagger)^T &= XX^\dagger \tag{7.33}\\
(X^\dagger X)^T &= X^\dagger X. \tag{7.34}
\end{align}
$$

The next theorem gives the connection between the Moore-Penrose inverse and the least-squares solution of an over-determined linear system. Its proof can be found in Ben-Israel and Greville (2003, Chapter 3).

Theorem 7.3. *Let X be an $n \times p$ matrix of rank r with $n > p \geq r$. The vector*

$$b = X^\dagger y \tag{7.35}$$

is the minimizer of $||y - X\theta||$ that has minimal norm in that

$$||b|| = \inf \left\{ ||\theta|| : X^T X \theta = X^T y \right\}. \tag{7.36}$$

The SVD can be used to provide a computable characterization of the solution in (7.35). To see this, let $X = UDV^T$ be the singular-value decomposition of X with $U = [u_1, \ldots, u_n]$ and $V = [v_1, \ldots, v_p]$, respectively, $n \times n$ and $p \times p$ orthogonal matrices and $D = \{d_{ij}\}$ an $n \times p$ matrix whose only nonzero entries are the square roots of the nonzero eigenvalues $\lambda_i^2, i = 1, \ldots, r$ of $X^T X$. We now claim that

$$X^\dagger = VD^\dagger U^T \tag{7.37}$$

for

$$D^\dagger = \left[\begin{array}{cc} \text{diag}(1/\lambda_1, \ldots, 1/\lambda_r) & 0_{r \times (n-r)} \\ 0_{(p-r) \times r} & 0_{(p-r) \times (n-r)} \end{array} \right]. \tag{7.38}$$

The proof (Exercise 7.23) consists of showing that the Penrose equations are satisfied with this choice for $X^\dagger$. The implication of (7.37)–(7.38) is that b in (7.35) has the form

$$\begin{aligned} b &= VD^\dagger U^T y \\ &= \sum_{j=1}^{r} \lambda_j^{-1} a_j v_j \end{aligned}$$

with $a_j = u_j^T y, j = 1, \ldots, r$. As a result, b can be evaluated once the SVD for X is available.

To give an example of how the SVD can be used in a least-squares context, consider finding the least-squares solution of $Xb = y$ with X given by

```
> X
     [,1] [,2] [,3]
[1,]    1    1    0
[2,]    1    1    0
[3,]    1    1    0
[4,]    1    0    1
[5,]    1    0    1
[6,]    1    0    1
```

and the response vector y created in R via

```
> set.seed(123)
> y <- rnorm(6)
```

This choice for the X matrix corresponds to a two-sample mean comparison problem under the model

$$y_{ij} = \mu + \tau_j + e_{ij}, \ i = 1, \ldots, 6, j = 1, 2,$$

for treatment effects τ_1 and τ_2 and random errors $e_{ij}, i = 1, \ldots, 6, j = 1, 2$. With this formulation the X matrix has rank two with the consequence that $X^T X$ is singular.

An application of formulas (7.35) and (7.24)–(7.25) in conjunction with the SVD method for class Matrix (with parameters itMax and delta set at 35 and 10^{-8}, respectively) produces

```
Moore-Penrose inverse of A
    0.11111      0.11111      0.11111     0.111112     0.111112     0.111112
    0.22222      0.22222      0.22222    -0.111112    -0.111112    -0.111112
   -0.11111     -0.11111     -0.11111     0.222224     0.222224     0.222224
Least square solution
0.298103
-0.0420867
0.34019
```

One may check these calculations with R using the ginv function from the MASS package that computes the Moore-Penrose inverse of a matrix that is given as its argument (Exercise 7.26).

Assume now that X has rank p. In that instance another avenue to computing a least-squares solution of $Xb = c$ is through a QR decomposition of X obtained via Gram-Schmidt orthogonalization. First, the general idea is to factorize X as

$$X = QR \tag{7.39}$$

with $Q = [q_1, \ldots, q_p]$ an $n \times p$ orthogonal matrix and R a $p \times p$ nonsingular, upper-triangular matrix. To create the Q matrix the columns of X can be orthonormalized using the procedure we now describe.

Let $x_1, \ldots, x_p$ be linearly independent, n-vector columns of X. The Gram-Schmidt algorithm creates an orthonormal basis $q_1, \ldots, q_p$ for the linear manifold spanned by $x_1, \ldots, x_p$.

Algorithm 7.5 Gram-Schmidt algorithm

$q_1 = x_1/\|x_1\|$
for $i = 2, \ldots, p$ **do**
$\quad q_i = x_i - \sum_{j=1}^{i-1} \left(x_i^T q_j \right) q_j$
$\quad q_i := q_i/\|q_i\|$
end for

A proof that $q_1, \ldots, q_p$ are orthonormal and provide a basis for the linear span of $x_1, \ldots, x_p$ can be obtained by induction (Exercise 7.31). This algorithm can exhibit numerical instability. But, it is easily modified to resolve such problems (Exercise 7.32).

Since the vectors $q_1, \ldots, q_p$ returned from the Gram-Schmidt algorithm are an orthonormal basis for the columns of X it must be that $X = QR$ with R a $p \times p$ matrix with elements $r_{ij} = q_i^T x_j$ the coefficients of x_j in its representation under the $q_1, \ldots, q_p$ basis. The order in which the q_i have been constructed guarantees that R is upper-triangular. As a result, the existence has been established of a factorization of the form (7.39). In addition,

$$X^T X = R^T Q^T Q R$$
$$= R^T R.$$

But, R is upper-triangular and the Cholesky factorization is unique. Therefore, it must be that R^T is the lower-triangular Cholesky factor for the matrix $X^T X$.

Using our QR decomposition the least-squares normal equations $X^T X b = X^T y$ become

$$R^T R b = R^T Q^T y.$$

The fact that R is nonsingular has the consequence that this identity is the same as

$$Rb = Q^T y \tag{7.40}$$

which can be solved directly using back substitution.

The modified Gram-Schmidt method from Exercise 7.32 was implemented in C++ (see Appendix D) and then used to solve the least-squares problem corresponding to the matrix used for the SVD example in the previous section with the right-hand-side vector that was used for our Gaussian elimination example in Section 7.2.2. This produced

```
The Q matrix
  -0.3338547      0.7874465      0.2822091      0.4340472
  -0.1371083      0.1474621      0.2109068     -0.4703158
   0.9284652      0.3428718      0.1241733      0.01466849
   0.04199925    -0.4395243      0.8533937      0.2759971
   0.07701195    -0.2178078     -0.363561       0.7169494
The R matrix
   1.678801       -1.873512     -0.1240544     -2.497718
          0        1.383693      1.22679        0.6009298
          0               0      0.7676374      1.135503
          0               0              0      0.367215
The product QR
  -0.5604756       1.715065      1.224082       1.786913
  -0.2301775       0.4609162     0.3598138      0.4978505
   1.558708       -1.265061      0.4007715     -1.966617
   0.07050839     -0.6868529     0.1106827      0.7013559
   0.1292877      -0.445662     -0.5558411     -0.4727914
```

```
Least-squares solution
-1.789488
0.01582229
-0.05267274
-1.468022
```

The product of the Q and R matrices agrees with the original X matrix through all digits shown in the output.

The R function for carrying out a QR decomposition is `qr`. Given a `matrix` object argument, it returns an object of class `qr` that contains the matrices from the decomposition in a compressed storage format. An application of the functions `qr.Q` and `qr.R` to a `qr` object will restructure its associated Q and R matrices into their standard forms. The least-squares solution of `Xb = y` with X and y both `matrix` objects is obtained from `qr.solve(qr(X), y)`. One may check that the results from our C++ QR code for the previous example agree with those obtained from R (Exercise 7.30).

7.6 The Template Numerical Toolkit

Our particular implementation of the classes `Matrix` and `Vector` should be viewed as merely a first step in the creation of classes that can be employed in statistical computing applications. In particular, the algorithms for solving linear systems, finding eigenvalues-eigenvectors, etc., need further refinement before they can be seriously considered for day-to-day usage. In this regard, a thorough development would require a detour from our current path into a detailed study of numerical matrix methods along the lines of that in Golub and Van Loan (1996). In lieu of fine tuning the `Matrix` and `Vector` classes another option is to simply employ existing packages that provide similar features for matrix/vector manipulation with better supporting routines for numerical methods. There are several alternatives that can be used for this purpose including the newmat library, Lapack++ and the Template Numerical Toolkit (TNT) in conjunction with JAMA/C++ that provides C++ translations of routines in Lapack that use the C++ template feature. We conclude this chapter by giving a brief introduction to TNT.

The TNT and JAMA libraries can be downloaded from `http://math.nist.gov/tnt`. The download will produce all the necessary header files. The fact that the classes are templates also means that the actual source code is contained in the headers. Consequently, there are no libraries to compile and everything that is needed for compilation is accomplished with include directives. In particular, by including the file tnt.h access is obtained to all the basic matrix and vector functions that include indexing and elementary numerical operations such as addition and multiplication.

The TNT matrix and vector classes are named `Array2D` and `Array1D`, respectively. There are a variety of constructors that include specifications such as

```
Array2D();
Array2D(int m, int n,  T* a);
Array2D(int m, int n,  const T& a);
```

for two-dimensional arrays or matrices and

```
Array1D();
Array1D(int n,  T* a);
Array1D(int n,  const T& a);
```

for one-dimensional arrays or vectors. The default constructors create null arrays. The other constructors employ the template format discussed in Section 3.11 with T being a generic place holder for a specific data type whose designation can be postponed until the array object is actually constructed. The integers m and n are the row and column dimensions of the two-dimensional arrays while **n** designates the number of rows for the one-dimensional case. The constructors with `const T&` a arguments return arrays with every entry set equal to the value **a**. As was true for our `Matrix` and `Vector` classes these constructors can be used for creation of arrays with all zero entries. The one-dimensional array constructor with the pointer argument acts like the constructor that was used for our class `Vector` in Section 3.9 in that it creates the `Array1D` object using a pointer (or array). The two-dimensional case involving a pointer is somewhat more involved. Here a one-dimensional array of dimension **m*n** must be given to the constructor with the contents of the underlying array being arranged in row major form.

In Listing 7.9 below the contents of a matrix are read from a file and subjected to various manipulations using tools from the TNT package. The scope resolution operator is employed to specify references to names in the TNT namespace.

Listing 7.9 *tntEx1.cpp*

```
//tntEx1.cpp
#include <iostream>
#include <fstream>
#include "tnt.h"

using std::cout; using std::endl;

int main(int argc, char** argv){
  int m = atoi(argv[2]); int n = atoi(argv[3]);
  double* pc = new double[n];
  double* pA = new double[m*n];
  std::ifstream fIn;
  fIn.open(argv[1]);

  for(int i = 0; i < m; i++){
    for(int j = 0; j < n; j++)
      fIn >> pA[n*i + j];//row major form
    fIn >> pc[i];
  }
  fIn.close();
  TNT::Array1D<double> c(n, pc);
  cout << "The vector c" << endl;
  cout << c << endl;
  TNT::Array2D<double> A(m, n, pA);
  cout << "The matrix A" << endl;
  cout << A << endl;

  TNT::Array2D<double> B(m, n);
  for(int i = 0; i < m; i++){
    for(int j = 0; j < n; j++)
      B[i][j] = (double)(i + 1)*(j + 1);
  }
  cout << "The matrix B" << endl;
  cout << B << endl;
  cout << "The sum of A and B" << endl;
  cout << A + B << endl;
  cout << "The product of A and B" << endl;
  cout << matmult(A, B) << endl;
  return 0;
}
```

This program uses the one-dimensional pointer based constructors to create both a one- and two-dimensional array. In accessing the file information that will fill the arrays it is assumed that the storage format progresses across the n columns and then down the m rows with the elements of the one-dimensional array residing in the last column. The pointers that provide access to the data are then used to create the **Array2D** object A and **Array1D** object c that are printed out using overloaded output insertion operators.

Listing 7.9 illustrates the use of an overloaded + operator for matrix addition. There are also overloaded versions of =, -, +=, -=, *, *=, / and /=. The *, *=, / and /= operators carry out their respective computations on an element-wise basis. In particular, this means that * does not produce matrix multiplication. This is instead obtained from the **matmult** function as illustrated in the program. It should be noted that both the **Array1D** and **Array2D** copy constructors create shallow copies that are essentially just an alias for the original; that is, altering the copy changes the original. To obtain an independent copy it is necessary to use the **copy** method for the class. For example.

```
Array2D B = A.copy();
```

produces an **Array2D** object B whose memory is independent of that for object A.

To illustrate the use of Listing 7.9, we will use the 5×5 array and five-element right-hand-side vector that were created with R and used with the Gauss transform example in Section 7.2.2. For the purposes of this example the two arrays reside in the file array1.txt. An application of our program to this data produces

```
$ g++ -Wall tntEx1.cpp -o tntEx1
$ ./tntEx1 array1.txt 5 5
The vector c
5
-1.68669
0.837787
0.153373
-1.13814
1.25381

The matrix A
5 5
2.81837 -3.14525 -0.208263 -4.19317 -1.08278
-3.14525 5.42465 1.92992 5.51101 0.1453
-0.208263 1.92992 2.10967 1.91872 -1.52998
-4.19317 5.51101 1.91872 8.02393 -0.214568
-1.08278 0.1453 -1.52998 -0.214568 3.1624

The matrix B
5 5
1 2 3 4 5
2 4 6 8 10
3 6 9 12 15
4 8 12 16 20
5 10 15 20 25

The sum of A and B
5 5
3.81837 -1.14525 2.79174 -0.193172 3.91722
-1.14525 9.42465 7.92992 13.511 10.1453
2.79174 7.92992 11.1097 13.9187 13.47
-0.193172 13.511 13.9187 24.0239 19.7854
3.91722 10.1453 13.47 19.7854 28.1624
```

```
The product of A and B
5 5
-26.2835 -52.567 -78.8506 -105.134 -131.418
36.2643 72.5287 108.793 145.057 181.322
10.0056 20.0111 30.0167 40.0223 50.0279
43.6079 87.2158 130.824 174.432 218.039
9.57162 19.1432 28.7148 38.2865 47.8581
```

As one consequence of this output it can be seen that the output insertion operator applied to `Array1D` and `Array2D` objects prints out the dimensions as well as the elements of the array.

The TNT package can be augmented with the JAMA package to provide the facility for solving linear systems and carrying out eigenvalue and singular-value decompositions. JAMA contains the template classes `Cholesky`, `Eigenvalue`, `QR` and `SVD` for carrying out Cholesky and QR matrix factorizations and eigenvalue or singular value decompositions. All the classes have constructors that take `Array2D` objects as arguments and return objects that can access class methods that correspond to various numerical routines. The `Cholesky` and `QR` classes have `solve` methods that take `Array1D` objects as a right-hand side for a linear system and return an `Array1D` object solution. The `Eigenvalue` class has the methods `getV` and `getRealEigenvalues` that return the eigenvectors and eigenvalues of a real symmetric matrix in `Array2D` and `Array1D` objects that are provided as its arguments. The analogous functions for the SVD class are `getU`, `GetV` and `getSingularValues` that return the left and right singular vectors in `Array2D` objects and the singular values in an `Array1D` object with all objects being passed into the functions by reference.

A program that uses the JAMA package is given in the next listing. The idea is that two arrays will be read in from text files. The first array is comprised of a square $n \times n$ array with an n-vector right-hand side as the last column in the file. This first array and vector will be used to demonstrate the solution of a linear system and an eigenvalue-eigenvector decomposition using the JAMA `Cholesky` and `Eigenvalue` classes. A second, nonsquare array will also be imported and used to exemplify using the SVD class.

Listing 7.10 *tntEx2.cpp*

```cpp
//tntEx2.cpp
#include <iostream>
#include <fstream>
#include <algorithm>
#include "tnt.h"
#include "jama_cholesky.h"
#include "jama_eig.h"
#include "jama_svd.h"

using std::cout; using std::endl;

int main(int argc, char** argv){
   int m = atoi(argv[1]); int n = atoi(argv[2]);
   double* pA1 = new double[n*n];
   double* pc = new double[n];
   double* pA2 = new double[n*m];
   std::ifstream fIn1, fIn2;
   fIn1.open(argv[3]); fIn2.open(argv[4]);

   for(int i = 0; i < n; i++){
     for(int j = 0; j < n; j++)
       fIn1 >> pA1[n*i + j];//row major form
```

```
      fIn1 >> pc[i];
   }
for(int i = 0; i < n; i++)
   for(int j = 0; j < m; j++)
      fIn2 >> pA2[m*i + j];

   fIn1.close(); fIn2.close();
   TNT::Array2D<double> A1(n, n, pA1);
   cout << "The matrix A1" << endl;
   cout << A1 << endl;
   TNT::Array1D<double> c(n, pc);
   cout << "The right-hand side" << endl;
   cout << c << endl;
   JAMA::Cholesky<double> Ch(A1);
   cout << "The solution by Cholesky" << endl;
   cout << Ch.solve(c) << endl;

   JAMA::Eigenvalue<double> A1Eig(A1);
   TNT::Array2D<double> V1(n, n);
   TNT::Array1D<double> lambda(n);
   A1Eig.getRealEigenvalues(lambda);
   A1Eig.getV(V1);
   cout << "The eigenvalues for A1" << endl;
   cout << lambda << endl;
   cout << "The eigenvectors for A1" << endl;
   cout << V1 << endl;

   TNT::Array2D<double> A2(n, m, pA2);
   cout << "The matrix A2" << endl;
   cout << A2 << endl;

   int dMin = min(m, n);
   JAMA::SVD<double> A2Svd(A2);
   TNT::Array2D<double> V2(m, m);
   TNT::Array2D<double> U(n, n);
   TNT::Array1D<double> s(dMin);
   A2Svd.getSingularValues(s); A2Svd.getU(U); A2Svd.getV(V2);

   cout << "The singular values for A2" << endl;
   cout << s << endl;
   cout << "The left singular vectors for A2" << endl;
   cout << U << endl;
   cout << "The right singular vectors for A2" << endl;
   cout << V2 << endl;
   return 0;
}
```

First observe that the header files for the relevant JAMA package classes have been included: namely, jama_cholesky.h, jama_eig.h and jama_svd.h for classes Cholesky, Eigenvalue and SVD, respectively. The solution of our linear system is obtained as a direct application of the solve function using a Cholesky class object. For the eigenvalue and singular-value decompositions different constructors were employed for the TNT Array1D and Array2D classes that require specification of only the array dimensions. Blank arrays created in this manner are passed as arguments to the functions that return the eigenvalues, eigenvectors, singular values and singular vectors.

Listing 7.10 was compiled and applied to the 5×5 array and right-hand-side vector in the file `array1.txt` that was used to demonstrate Listing 7.9. The 5×4 array that was created in Section 7.4 to illustrate our C++ SVD code was stored in the file array2.txt and used for illustrating the SVD calculations. The results produced by application of the program to these two files are

```
$ g++ -Wall tntEx2.cpp -o tntEx2
$ ./tntEx2 4 5 array1.txt array2.txt
The matrix A1
5 5
2.81837 -3.14525 -0.208263 -4.19317 -1.08278
-3.14525 5.42465 1.92992 5.51101 0.1453
-0.208263 1.92992 2.10967 1.91872 -1.52998
-4.19317 5.51101 1.91872 8.02393 -0.214568
-1.08278 0.1453 -1.52998 -0.214568 3.1624

The right-hand side
5
-1.68669
0.837787
0.153373
-1.13814
1.25381

The solution by Cholesky
5
-303.893
-99.3447
153.151
-128.243
-33.6957

The eigenvalues for A1
5
0.00386867
0.413226
1.34465
4.57788
15.1994

The eigenvectors for A1
5 5
-0.802253 0.266172 0.17546 0.32573 -0.385561
-0.26552 -0.443789 0.646181 0.0243266 0.560722
0.405882 0.615624 0.403968 0.506197 0.191943
-0.336479 0.326794 -0.528434 0.0259296 0.707158
-0.0890577 0.496404 0.330502 -0.797747 0.00444874

The matrix A2
5 4
-0.560476 1.71506 1.22408 1.78691
-0.230177 0.460916 0.359814 0.49785
1.55871 -1.26506 0.400771 -1.96662
0.0705084 -0.686853 0.110683 0.701356
0.129288 -0.445662 -0.555841 -0.472791

The singular values for A2
4
3.89861
1.48482
0.979507
0.115485
```

```
The left singular vectors for A2
5 4
-0.686631 -0.594111 0.0263872 -0.417664
-0.19711 -0.148865 0.0101223 0.495083
0.672926 -0.727794 0.0894255 -0.0687921
-0.0269443 0.103324 0.993129 -0.0411511
0.190079 0.290707 -0.0699928 -0.757644

The right singular vectors for A2
4 4
0.385209 -0.486456 0.187077 -0.761562
-0.560703 -0.247422 -0.729089 -0.304669
-0.192469 -0.823425 0.225224 0.483944
-0.707235 0.15529 0.618632 -0.304956
```

Reference to the results that were obtained in Sections 7.2.3, 7.3 and 7.4 reveals that the JAMA functions have worked as expected.

7.7 Exercises

7.1. Develop a method for class **Matrix** that will create a Gauss transform matrix corresponding to a given input vector. Verify that an application of the method to the matrix **A** and right-hand-side vector **c** from Section 7.2.2 gives the correct solution to the equation system.

7.2. Gaussian elimination can experience accuracy problems when the sweep process encounters a (relatively) small diagonal entry. One approach to avoiding such difficulties is to employ partial pivoting. The idea is that at the kth step of the recursion rows are swapped so that the array entry with the largest magnitude on or below the diagonal occupies the diagonal position. As an example, consider the matrix

$$\begin{bmatrix} 6 & 3 & 2 \\ 2 & 12 & 7 \\ 9 & 20 & 4 \end{bmatrix}.$$

The largest element in the first column is 9 and the first and third rows are swapped to obtain the matrix

$$\begin{bmatrix} 9 & 20 & 4 \\ 2 & 12 & 7 \\ 6 & 3 & 2 \end{bmatrix}.$$

After application of the Gauss transform for the first column the transformed matrix is

$$\begin{bmatrix} 9 & 20 & 4 \\ 0 & 7.555556 & 6.11111 \\ 0 & -10.333333 & -.666667 \end{bmatrix}.$$

The next target column is the second for which the value in the third row is again largest in magnitude. So, the third and second rows are switched and the Gauss transform is again applied to give

$$\begin{bmatrix} 9 & 20 & 4 \\ 0 & -10.333333 & -.666667 \\ 0 & 0 & 5.623656 \end{bmatrix}$$

as the final outcome. With this example in mind,

a) create a method for class **Matrix** that will swap two specified rows of a **Matrix** object.

b) use your method from part a) to implement a method for class **Matrix** that solves a nonsingular linear system by Gaussian elimination with partial pivoting.

7.3. Show that the Gaussian elimination and Cholesky methods involve $2n^3/3$ and $n^2/3$ flops, respectively, to convert an $n \times n$ system to upper-triangular form.

7.4. Derive an algorithm for computing the inverse of the lower-triangular Cholesky factor. Then, use this to create a method for class `Matrix` that will compute the inverse of a positive-definite matrix.

7.5. Develop a method for class `Matrix` that will perform an efficient Gauss sweep for a banded matrix system.

7.6. Let A be an $n \times n$ positive-definite matrix with bandwidth q and let LL^T be its Cholesky factorization. Show that L is banded below the diagonal with lower bandwidth q.

7.7. Write an R function that combines the `chol`, `backsolve` and `forwardsolve` functions to solve a positive-definite linear system via the Cholesky method. Use the `system.time` function to compare the performance of your algorithm to using the `chol2inv` function for this purpose.

7.8. Consider the linear system $Ab = c$ with $A = \{a_{ij}\}_{i,j=1:n}$ an $n \times n$ matrix, $b = (b_1, \ldots, b_n)^T$ and $c = (c_1, \ldots, c_n)^T$. An iterative approach to solving the system is Algorithm 7.6 that computes an approximate solution according to a specified number of iterations $itMax$ and relative change in the approximation δ.

Algorithm 7.6 Gauss-Seidel algorithm

---

    select an initial guess $b^{(0)} = (b_1^{(0)}, \ldots, b_n^{(0)})^T$ for b
    $change = \infty, niter = 0$
    while $change > \delta$ and $niter < itMax$ **do**
        for $i = 1, \ldots, n$ **do**
            $b_i^{(niter)} = \left(c_i - \sum_{j=1}^{\min(1,i-1)} a_{ij} b_j^{(niter)} - \sum_{j=\min(i+1,n)}^{n} a_{ij} b_j^{(niter-1)} \right) / a_{ii}$
        end for
        $change = \|b^{(niter)} - b^{(niter-1)}\| / \|b^{(niter-1)}\|$
        $niter = niter + 1$
    end while

Create a method for class `Matrix` that will perform a Gauss-Seidel iteration and compare its performance with other solution methods in the class.

7.9. Let Z have a p-variate normal distribution with a mean vector of all zero elements and identity variance-covariance matrix.

a) Let Σ be a positive-definite matrix with Cholesky factorization LL^T. Prove that $\mu + LZ$ is p-variate normal with mean vector μ and variance-covariance matrix Σ.

b) Create a random number generation class that will generate vectors from a p-variate normal distribution corresponding to some user-specified mean vector and variance-covariance matrix.

7.10. Modify the methods in class `Matrix` that can be used for solving linear systems to allow for multiple right-hand sides.

7.11. Show that the number of nonzero elements in an $n \times n$ banded matrix with bandwidth q is $(n - q)(2q + 1) + q^2$.

7.12. Verify the form of the linear smoothing spline normal equations (7.9). Then, show that they take the form (7.11)–(7.13) under the local support, linear smoothing spline basis of Section 7.2.5.

7.13. Explicitly carry out the forward and backward substitution steps that are needed to establish (7.15)–(7.16). Also, show that the entire computational effort is of order n.

7.14. Develop R code for fitting the linear smoothing spline from Section 7.2.5 using the `SparseM` package.

7.15. Linear smoothing splines have a boundary bias that can reduce the efficiency of the estimator near the edges of the interval $[0, 1]$. The problem is easily corrected by including appropriate polynomial functions in the fit (e.g., Eubank and Kim 1998). Specifically, let

$$q_0(x) = x - .5x^2, \quad q_1(x) = .5x^2.$$

Then, a boundary corrected estimator is obtained by minimizing

$$\sum_{i=1}^{n} \left(y_i - a_0 q_0(x_i) - a_1 q_1(x_i) - \sum_{j=1}^{n} b_j B(x_i) \right)^2 + \lambda b^T \Omega b,$$

with respect to $a = (a_0, a_1)^T$ and $b = (b_1, \ldots, b_n)^T$ for $B_1, \ldots, B_n$ the local support basis functions from Section 7.2.5 and Ω the matrix defined in (7.12)–(7.13).

a) Derive a closed form expression for the fitted values at points $x_1, \ldots, x_n$ for the boundary corrected estimator.

b) Let $Q = \{q_j(x_i)\}_{i=1:n, j=0:1}$ be the $n \times 2$ matrix containing the values of q_0 and q_1 at the x_i. Show that the fitted values from part a) can be expressed as $\tilde{f}_\lambda = f_\lambda + \tilde{Q} a_\lambda$ with f_λ the fitted values for the noncorrected estimator obtained from (7.11),

$$\tilde{Q} = [I - (I + \lambda\Omega)^{-1}]Q$$

and a_λ is the solution of

$$\tilde{Q}^T \tilde{Q} a = \tilde{Q} \tilde{y}$$

for $\tilde{y} = [I - (I + \lambda\Omega)^{-1}]y$.

c) Use your result from part b) to develop an order n algorithm for computing $\tilde{f}_\lambda$ and the GCV criterion.

d) Create a C++ class that implements your algorithm from part c).

e) Generate data from the model

$$y_i = \mu(x_i) + e_i, \quad i = 1, \ldots, 100,$$

with the e_i being uncorrelated normal random variables having mean 0 and standard deviation .05, $x_i = (2i - 1)/200, i = 1, \ldots, 100$ and

$$\mu(x) = \begin{cases} 16x^2(1 - x)^2, & 0 < x < .7, \\ 16x^2(1 - x)^2 + 64(x - .7)^3, & x \geq .7. \end{cases}$$

Fit this data using both the boundary corrected and non-boundary-corrected linear smoothing spline estimator for various values of λ and compare the results.

7.16. Rework Listing 7.6 to use references rather than pointers.

7.17. Verify identities (7.18)–(7.19).

7.18. Add code to the **Matrix** and/or **pdBand** classes that will produce an operable diagonal matrix constructor for **pdBand** objects.

7.19. Suppose that a matrix is symmetric and banded but not positive-definite. Develop a derived class from a class **Matrix** such as the one in Listing 7.7 that will solve linear systems for such a matrix via a **solve** method based on the banded Gaussian elimination approach of Exercise 7.5.

7.20. Add a **virtual +** operator to the **Matrix** class in Listing 7.7. Then, use the **bandToFull** method from class **pdBand** to extend the operator to addition of a **pdBand** and **Matrix** object.

7.21. Let $A = \{a_{ij}\}_{i,j=1:n}$ be a symmetric $n \times n$ matrix. Then, the unique entries in A can be stored in a one-dimensional array $\tilde{a}$ of length $n(n + 1)/2$ using the relationship

$$\tilde{a}_{g(i,j)} = a_{ij}, i \geq j,$$

with the index mapping function g defined by

$$g(i, j) = (j - 1)n - j(j - 1)/2 + i.$$

a) Create a derived class `SyMatrix` from class `Matrix` in Listing 7.7 that deals with symmetric matrices and stores them efficiently. Provide your class with the requisite constructors, destructor and assignment operator.

b) Provide a `virtual solve` method that uses Gaussian elimination to solve a linear system.

7.22. Let $X_k = [x_1, \ldots, x_k]^T$ be a $k \times p$ matrix. If $(X_m^T X_m)^{-1}$ for $m \geq p$ is available it is possible to compute $(X_{m+1}^T X_{m+1})^{-1}$ using the rank-one update

$$(X_{m+1}^T X_{m+1})^{-1} = (X_m^T X_m)^{-1} \left(I - \frac{x_{m+1} x_{m+1}^T}{1 - x_{m+1}^T (X_m^T X_m)^{-1} x_{m+1}} \right) (X_m^T X_m)^{-1}.$$

Suppose now that observations $y_1, y_2, \ldots$, arise in a temporal sequence and at each point $t > 2p$ we wish to compute the coefficients that minimize

$$\sum_{i=p+1}^{t} \left(y_i - b_0 - \sum_{j=1}^{p} b_j y_{i-j} \right)^2$$

as a function of $b_0, \ldots, b_p$.

a) Develop an order n algorithm that uses rank-one updates to sequentially compute the least-squares coefficient estimators.

b) Implement your algorithm in C++.

c) Use the R `arima.sim` function (see, e.g., Section 10.4.1) to simulate data from an autoregressive model and use that to test your code from part b).

7.23. Show that $X^\dagger$ defined in (7.37)–(7.38) is the Moore-Penrose generalized inverse of X.

7.24. Consider the linear system $Ab = c$ with A and c determined from

```
> set.seed(123)
> A <- matrix(rnorm(18), 6, 3)
> v <- matrix(rnorm(6), 6, 1)
```

Compute a least-squares solution for this system using the `svd` function from R.

7.25. Expand on Exercise 7.24 and use the `svd` function in R to create a function that will compute the generalized inverse of a matrix.

7.26. Write C++ code that will allow you to verify the form of the least-squares solution for $Xb = y$ in Section 7.5 that was obtained with the X matrix

```
> X
     [,1] [,2] [,3]
[1,]    1    1    0
[2,]    1    1    0
[3,]    1    1    0
[4,]    1    0    1
[5,]    1    0    1
[6,]    1    0    1
```

and response vector y determined from

```
> set.seed(123)
> y <- rnorm(6)
```

Compare the results from your code to that returned by the R functions `svd` and `ginv` (from the MASS package).

7.27. Provide a method for class `Matrix` that will compute the Moore-Penrose inverse of a matrix. Use this method to obtain a least-squares solution for the linear system $Ab = c$ with A and c obtained from R as in Exercise 7.24.

7.28. Add the following functions to the TNT package.

a) A function to transpose a matrix.

b) An overloaded version of $*$ that performs matrix multiplication.

c) An overloaded version of $*$ that performs matrix-vector multiplication.

7.29. Create `Matrix` and `Vector` template classes that rely on the constructors and methods from the TNT package. The idea is to create wrapper classes for the TNT and JAMA classes that add both functionality and ease of access.

7.30. Use the R functions `qr` and `qr.solve` to check the C++ results for the least-squares example of Section 7.5.

7.31. Show that the Gram-Schmidt process in Algorithm 7.5 produces an orthogonal basis for the linear space spanned by the vectors to which it is applied.

7.32. Let $x_1, \ldots, x_p$ be linearly independent n-vectors for $n \geq p$. A numerically stable modification of the Gram-Schmidt procedure is given in Algorithm 7.7 below.

Algorithm 7.7 Modified Gram-Schmidt algorithm

$q_1 = x_1$
$r_{11} = \|q_1\|$
$q_1 = q_1/r_{11}, r_{1j} = q_1^T x_j, j = 2, \ldots, p$
$x_j^{(1)} = x_j - r_{1j}q_1, j = 2, \ldots, p$
for $i = 2, \ldots, p - 1$ **do**
   $q_i = x_i^{(i)}$
   $r_{ii} = \|q_i\|$
   $q_i = q_i/r_{ii}, r_{ij} = q_i^T x_j, j = i + 1, \ldots, p$
   $x_j^{(i)} = x_j - r_{ij}q_i, j = i + 1, \ldots, p$
end for
$q_p = x_p^{(p-1)}, r_{pp} = \|q_p\|$
$q_p = q_p/\|q_p\|$

---

a) Show that the vectors $q_1, \ldots, q_p$ produced by the modified Gram-Schmidt Algorithm 7.7 provide an orthonormal basis for the linear subspace spanned by $x_1, \ldots, x_p$.

b) Let $X = [x_1, \ldots, x_p]$. Show that $X = QR$, where $Q = [q_1, \ldots, q_p]$ and R is an upper-triangular matrix with nonzero elements $r_{ij}, j \geq i$.

c) Use part b) as a basis for code to implement a method for class `Matrix` that carries out a QR factorization of a matrix object. There is an outer product form for the changes that are made to the columns of X that can make the formulation easier to encode.

d) Using the matrix `X` and right-hand-side vector `v` from Exercise 7.24, compare the least-squares solutions obtained from the C++ code developed in part c) to that obtained from the TNT/JAMA packages (using jama_qr.h) and from R via the function `solve.qr`.

7.33. Extend the R `sqrt` function to apply to positive-definite matrices. If A is such a matrix, `sqrt` should return the unique, symmetric, square matrix $A^{1/2}$ with the property that $A^{1/2}A^{1/2} = A$.

Chapter 8

Numerical optimization

8.1 Introduction

Statistical estimation methods generally require maximizing or minimizing some sort of performance criterion. For example, in the case of nonlinear regression one may estimate the parameters θ in a regression function $g(\cdot; \theta)$ via least-squares: i.e., by minimizing

$$f(\theta) = \sum_{i=1}^{n} (y_i - g(x_i; \theta))^2$$

with respect to θ, where y_i, $i = 1, \ldots, n$, are responses that were observed at values $x_1, \ldots, x_n$ of a vector of predictor variables. As another instance, maximum likelihood estimation of a parameter vector θ involves maximizing the likelihood function $L(\theta)$ which is the joint probability density or mass function for the data evaluated at the observed sample values. In general, this is accomplished by equivalently minimizing $l(\theta) = -\ln L(\theta)$. As illustrated by the likelihood setting, all maximization problems can be made into minimization problems. Thus, it suffices to concentrate on numerical minimization.

The task of finding a global minimizer of an objective function is complicated by the possibility that there may be multiple local minima. On the other hand, any continuous, nonconstant, function will necessarily be convex or concave up (i.e., bowl shaped) in an area around a local minimum. So, in that sense, our focus can be directed toward the simpler problem of finding the minimum of a convex function. The minimization methods that will be considered in this chapter are designed for that purpose. The remaining task is then one of locating all the local minima. Techniques for accomplishing that goal will also be discussed.

Minimization of the GCV criterion for the linear smoothing spline in Section 7.2.5 represents a prototypical illustration of the type of optimization problems that will be of the most interest for our study. The basic framework is that there is a class (i.e., Linss) that has a method (i.e., GCV) and the aim is to find a minimizing argument or arguments of that method for a particular object from the class. The first step in this process is to give a minimization routine access to the object's member function. This means that the function to be minimized will originate outside the minimization class and some way of passing the function into the class will be needed. Consequently, this is the first topic that will be explored.

8.2 Function objects

The phrase *callback* refers to the passage of executable code as arguments to other programs. The discussion throughout this section will focus on the creation of a canonical callback function that typifies the basic features of the function passing problems that will be encountered throughout the remainder of the chapter. Specifically, we have a function f that takes a single argument and the objective is to pass this function to another object that will "call back" by evaluating it for some given value of its argument.

First, to provide some perspective on the problem, let us assume that the code is being written in R. In that case, the problem is no "problem" at all since functions are objects in this language. The point is illustrated by the two R functions in the listing below.

Listing 8.1 *func.r*

```
#func.r
func <- function(x){
  floor(x)
}


useFunc <- function(f, x){
  print(f(x))
}
```

The first function `func` is just a wrapper for the R function `floor` that returns the greatest integer that does not exceed its argument. The second function `useFunc` takes any function `f` as its first argument and evaluates it at the value of its second argument. Using these two functions in combination can produce results like

```
> useFunc(func, 2.5)
[1] 2
```

The development of an analogous version of Listing 8.1 in C++ is a more formidable task as C++ is missing the key ingredient from R that endows functions with object status. The solution is to simply correct this "deficiency" and devise a way to create function objects in C++. Of course, this is not nearly as easy as it might sound and the formulation of a satisfactory solution will lead us through some interesting new language territory.

In C++ function objects are often referred to as *functors*. In a broad sense this can be taken to mean objects that are created from a class with an overloaded version of (). A specific example of a functor is provided by the next listing.

Listing 8.2 *functorEx.cpp*

```
//functorEx.cpp
#include <iostream>
#include <cmath>

using std::cout; using std::endl;

struct Functor{
  int nCalls;
  Functor(){nCalls = 0;}

  int operator() (double x){
    nCalls++;
    cout << "Calling from ()" << endl;
    return floor(x);
  }
};

int main(){
  Functor f;
  cout << f(2.5) << endl;
  cout << f.nCalls << endl;
  return 0;
}
```

The struct `functor` in Listing 8.2 is similar to the R `func` in Listing 8.1 in that its purpose is to provide a construct that can hold the C++ `floor` function. The mechanism for accomplishing this is an overloaded () operator that takes a `double` argument and returns an `int`. The body of the operator merely evaluates `floor` at the operator's argument and returns the result.

Listing 8.2 reveals another property of functors that differentiates them from the functions they encapsulate: namely, they can have a state. As functors are classes/structs, they can also have member elements or data components. This allows them to, e.g., "remember" how they were last called or their last output. In terms of Listing 8.2 the function object `f` in `main` will retain a record of how many times it is used through the member variable `nCalls`. The default constructor for the struct initializes `nCalls` to 0. So, the program's output

```
Calling from ()
2
1
```

demonstrates that the initial state of `f` has been updated.

One criticism that can be directed at a functor such as the one in Listing 8.2 is that it can encapsulate only one function. For example, a new functor would have to be created to contain the ceiling function `ceil` that returns the greatest integer that is not smaller than its argument. These two functions are of a very similar nature in the sense that they both take a `double` argument and return an integer.[*] It would be nice to have a single functor class/struct that could hold either of these functions or, more generally, any function that takes a `double` argument and returns an `int`. One avenue for creating this type of functor object is provided by function pointers.

The syntax for creating an ordinary function pointer is

```
returnType (*pFunc)(arguments)
```

This produces a pointer `pFunc` that can point to any function with `returnType` as its return type and an argument list of the same length with data types that match, in order, those listed for `pFunc`. Similarly, the pointer `pFunc` can be passed to a function which has a function pointer as an argument with `returnType` as its return type and an argument list with length and data types matching the one for `pFunc`. The `pFunc` pointer can be used as if it were an ordinary function; writing `pFunc(argumentValues)` with `argumentValues` the specific values to be used for the arguments has the effect of dereferencing the pointer.

The listing below shows how a function pointer can be used to solve our simple function passing problem.

Listing 8.3 *pointToFunc.cpp*

```cpp
//pointToFunc.cpp
#include <iostream>
#include <cmath>

using std::cout; using std::endl;

struct Z{
  int func(double x){
    cout << "Calling from func" << endl;
    return ceil(x);
  }
```

---

[*] Actually, the C++ `floor` and `ceil` functions have `double`, rather than `int`, return types even though the values they return are integers. The `int` return type is instead enforced in the wrapper functions that have been created to hold them.

```cpp
};

struct Functor{
  int (*pF)(double x);

  Functor(int (*pf)(double x)){
    pF = pf;
  }

  int operator()(double x){
    cout << "Calling from ()" << endl;
    return pF(x);
  }
};

int func1(double x){
  cout << "Calling from func1" << endl;
  return floor(x);
}

int func2(double x){
  cout << "Calling from func2" << endl;
  return ceil(x);
}

int main(){
  Functor f1(&func1);
  cout << f1(2.5) << endl;
  Functor f2(&func2);
  cout << f2(2.5) << endl;
  Z z;
  //Functor f3(&z.func);
  //cout << f3(2.5) << endl;
  return 0;
}
```

The Functor struct in Listing 8.3 has a member that is a pointer to a function with a single double argument and an int return type. This pointer member is initialized via the struct's constructor and a callback is carried out via the overloaded () operator. The program also contains two functions, func1 and func2, that provide wrappers for the floor and ceil functions, respectively. Both wrapper functions have a single double argument and an int return type.

The main function in Listing 8.3 uses the addresses for func1 and func2 to create two functor objects that are used to access their respective functions. The output

```
Calling from ()
Calling from func1
2
Calling from ()
Calling from func2
3
```

illustrates the flexibility of our new Functor struct.

Listing 8.3 also contains a struct Z whose member method func is another wrapper for the ceil function. At the end of main an attempt is made to create a functor object using func.

This fails and removal of the comments from this block of code will result in a compilation error. To make this work it is necessary to use a slightly different approach.

To create a pointer to a member function for a class/struct X one uses syntax such as

```
returnType (X::*pFunc)(arguments)
```

The only real difference between this case and an ordinary function pointer is that the scope resolution operator is used to specify that the pointer is for functions/methods from a class/struct X. Then, just like a regular function pointer, pFunc can point to any member function of the class with the same return and argument types.

The code listing below provides an alternative spin on executing the tasks performed by the previous listings in this section. In this case a class Z has two member functions func1 and func2 that are wrappers for floor and ceil. A functor struct is created with two members: a pointer to a Z object and a pointer to a Z class method that has a double argument and int return type.

Listing 8.4 *pointToMemFunc.cpp*

```
//pointToMemFunc.cpp
#include <iostream>
#include <cmath>

using std::cout; using std::endl;

class Z{

public:
  int func1(double x){
    cout << "Calling from Z func1" << endl;
    return floor(x);
  }

 int func2(double x){
   cout << "Calling from Z func2" << endl;
   return ceil(x);
  }
};

struct Functor{
  Z* pZ;
  int (Z::*pF)(double x);

  Functor(Z* pz, int (Z::*pf)(double x)){
    pZ = pz; pF = pf;
  }

  int operator() (double x) const {
    return (pZ->*pF)(x);
  }
};

int main(){
  Z z;
  Functor f1(&z, &Z::func1);
  cout << f1(2.5) << endl;
  Functor f2(&z, &Z::func2);
```

```
    cout << f2(2.5) << endl;
    return 0;
}
```

In the `main` function of Listing 8.4 an object is created from the Z class whose address is passed to the `Functor` constructor along with the address of the `func1` method. The resulting `Functor` object is then used to call `func1`. As before, this is accomplished with an overloaded version of () in the `Functor` struct that makes the process resemble evaluation of an ordinary function in `main`. The syntax that appears in the body of the () operator seems somewhat complicated at first. But, `pZ->*pF` translates to i) dereference the pointer `pF` to obtain the actual Z method and ii) use the pointer `pZ` to access the method with the member access operator `->`. The same results would be produced using `*pZ.*pF` instead of `pZ->*pF`. Both `.*` and `->*` are called *member dereferencing operators*. The sole purpose of the functor struct is to evaluate a class method and it should not need nor be able to alter the class object that is used to invoke that method. The `const` designation for the () operator ensures that this is the case.

The output from Listing 8.4 is

```
Calling from Z func1
2
Calling from Z func2
3
```

Thus, `Functor` class objects have successfully encapsulated both of the member functions of the Z class.

Simple functors such as those in Listings 8.2 and 8.4 can be quite effective and will suffice for many purposes. Nonetheless, the `Functor` class in Listing 8.4 is still constrained to work only with the particular class Z. The next step is to allow for both Z and the member function to be arbitrary. Template classes provide the means to attain that end.

Proceeding as in Section 3.10, a "general purpose" functor might take the form

Listing 8.5 funcTemp.h

```
//funcTemp.h
template <class T> class Functor{

    int (T::*pF)(double);
    T* pT;

public:
    Functor(){}
    Functor(T* pt, int (T::*pf)(double)){pT = pt; pF = pf;}
    int operator() (double u) const {
        return (pT->*pF)(u);
    }
};
```

In essence the functor in Listing 8.5 is the same as the one in Listing 8.4. The only difference is that the class Z argument has been replaced with the generic "place-holder" T.

The following program was written to test our new `Functor` class.

Listing 8.6 functorDriver.cpp

```
//functorDriver.cpp
#include <iostream>
#include <cmath>
```

```
#include "funcTemp.h"

using std::cout; using std::endl;

struct Y{

  int func(double x){
    cout << "Calling from Y func" << endl;
    return ceil(x);
  }
};

class Z{

public:
  int func(double x){
    cout << "Calling from Z func" << endl;
    return floor(x);
  }
};

int main(){
  Y y; Z z;
  Functor<Y> f1(&y, &Y::func);
  cout << f1(2.5) << endl;
  Functor<Z> f2(&z, &Z::func);
  cout << f2(2.5) << endl;
  return 0;
}
```

Both Y and Z objects are created in the program and **Functor** class objects are used to call the **func** method from both classes. The only trick is that the template parameters must be specified (i.e., the specific choices for T must be given) when the **Functor** class constructor is called. The output from the program is

```
Calling from Y func
3
Calling from Z func
2
```

The complaint could now be lodged that our newest version of the **Functor** class is limited by the fact that it can handle only a member function with prespecified argument and return types of **double** and **int**. This restriction can also be removed using the template approach (Exercise 8.1). For now, the generality provided by the formulation in Listing 8.5 will be sufficient.

8.3 Golden section

In this section we develop our first optimization method. To begin, let us deal with a very simple case where a convex function f is defined on the integers $\mathbb{N}_m = \{1, \ldots, m\}$. The objective is to find the fastest algorithm in the sense that it takes the fewest evaluations ν_m of f on $\mathbb{N}_m$ to find its minimum. Our discussion of this problem stems from developments in Kiefer (1953) and Monahan (2001).

One may check that if an arbitrary choice is allowed for f, then $\nu_2 = 2$ and $\nu_3 = 3$. For $m = 4$ and $\mathbb{N}_4 = \{1, 2, 3, 4\}$, the minimizer can be found in $\nu_4 = 3$ evaluations if f is first evaluated at 2 and 3. If $f(2) < f(3)$ it is only necessary to evaluate $f(1)$ to decide between 2 and 1 for the minimizer. Similarly, if $f(3) < f(2)$ only $f(4)$ need be evaluated.

For $m = 7$ evaluation at 3 and 5 will eliminate either $\{5, 6, 7\}$ or $\{1, 2, 3\}$, reduce the problem to the situation with $m = 4$ and thereby allow for minimization in $\nu_7 = 4$ evaluations. Finally, for $m = 12$ evaluation at 5 and 8 eliminates five points (i.e., either $\{8, 9, 10, 11, 12\}$ or $\{1, 2, 3, 4, 5\}$) and puts the remaining evaluation in the ideal location for the best seven point search.

The pattern can now be seen to go like this. Define the Fibonacci numbers $F_0 = F_1 = 1$ and

$$F_j = F_{j-1} + F_{j-2}, \ j = 2, \ldots.$$

Then, for $m = F_k - 1$, evaluate f at F_{k-1} and F_{k-2} thereby eliminating F_{k-2} points and reducing the problem to one that is equivalent to having $m = F_{k-1} - 1$. The total number of evaluations is $\nu_{F_k - 1} = k - 1$.

Of course, situations involving minimization of a function over a grid whose size is determined by a Fibonacci number are rare. However, the results for this setting can be applied to deal with a case that actually is of some interest.

Suppose that $f(\theta)$ is a continuous function with argument θ taking values in the interval $[0, 1]$. Let us now consider doing a grid search over the m-point set

$$\left\{ 0, \frac{1}{m-1}, \ldots, \frac{m-2}{m-1}, 1 \right\}.$$

We want to see what transpires as the grid grows dense on $[0, 1]$. So, it suffices to take $m = F_k - 1$ and let $k \to \infty$.

The search is defined by the first two evaluation points which are $\frac{F_{k-2}}{F_k - 1}$ and $\frac{F_{k-1}}{F_k - 1}$. Thus, letting $r = \lim_{k \to \infty} \frac{F_{k-1}}{F_k}$, we see that

$$\lim_{k \to \infty} \frac{F_{k-2}}{F_k - 1} = \lim_{k \to \infty} \frac{F_{k-2}}{F_{k-2} + F_{k-1} - 1} = \frac{r}{r+1}$$

and

$$\lim_{k \to \infty} \frac{F_{k-1}}{F_k - 1} = \lim_{k \to \infty} \frac{F_{k-1}}{F_{k-2} + F_{k-1} - 1} = \frac{1}{r+1}$$

are the limiting points for evaluation of f. It only remains to determine r. This derives from the fact that

$$r = \lim_{k \to \infty} \frac{F_{k-1}}{F_k} = \lim_{k \to \infty} \frac{F_{k-1}}{F_{k-2} + F_{k-1}} = \frac{1}{1+r}.$$

Thus, r is the positive root of the quadratic equation $r^2 + r - 1 = 0$; i.e.,

$$r = \frac{-1 + \sqrt{4 + 1}}{2} \doteq .618.$$

Consequently, the two evaluation points are now seen to be $r^2 = 1 - r$ and r.

The development thus far suggests a simple dyadic minimization strategy. At the beginning $f(1 - r)$ is compared to $f(r)$. If $f(1 - r) < f(r)$, the minimizer must lie in the interval $[0, r]$ and it now suffices to search for the minimizer of $f(ru)$ with $u \in [0, 1]$. This entails evaluations at $u = 1 - r$ and $u = r$ as before or, equivalently, $f(r(1 - r))$ is compared to $f(r^2) = f(1 - r)$ with the latter value already available from the previous comparison.

On the other hand, if $f(1 - r) > f(r)$, the minimizer must lie in the interval $[1 - r, 1]$ and the search proceeds by looking for the minimizer of $f(1 - r + ur)$ over $u \in [0, 1]$. In this instance $f(1 - r + r(1 - r)) = f(1 - r^2) = f(r)$ has already been computed.

To describe the general case, suppose that f is defined on some interval $[a, b]$ that need not be $[0, 1]$. In that case $f(a + u(b - a))$ is a function on $[0, 1]$ and our minimization scheme tells us to evaluate it at $u = 1 - r$ and $u = r$. If $f(a + (1 - r)(b - a)) < f(a + r(b - a))$, the minimizer is in $[a, a+r(b-a)]$ and we consider $f(a+ur(b-a))$ as a function on $[0,1]$ that must be evaluated at $u = 1 - r$ and $u = r$. But, at $u = r$ the value of $f(a + r^2(b - a)) = f(a + (1 - r)(b - a))$ has already been computed. So, redefine b as $b := a+r(b-a)$, evaluate $f(a+(1-r)(b-a))$ and begin the minimization again with this "new" function on a new subinterval.

Alternatively, if $f(a+(1-r)(b-a)) > f(a+r(b-a))$, the minimizer is in $[a+(1-r)(b-a), b]$. We now work with $f(a + (1 - r)(b - a) + ur(b - a))$ on $[0, 1]$. At $u = 1 - r$ this is

$$f(a + (1 + r)(1 - r)(b - a)) = f(a + (1 - r^2)(b - a)) = f(a + r(b - a))$$

that has already been calculated. Thus, we now redefine a as $a := a+(1-r)(b-a)$, evaluate f at $a + r(b - a)$ and proceed to the next iteration.

In its current form our search procedure would, in theory, continue indefinitely. Thus, for implementation with a computer there must be safeguards that will eventually terminate the search. A typical stopping rule would conclude the search if some preset maximum number of iterations was exceeded or

$$|minTemp^{(k+1)} - minTemp^{(k)}| < \delta|minTemp^{(k)}|$$

with $minTemp^{(k)}$ our approximation to the actual minimizer on the kth iteration and δ a small constant. From Section 2.5 we know that for single and double precision one cannot detect relative differences that are much smaller than 10^{-7} or 10^{-16} which would seem to make them obvious choices for δ. However, these options are much too small for general purposes and instead $10^{-7/2}$ and 10^{-8} represent more attainable choices as will be explained below.

The search procedure that we have described is called the *golden section algorithm*. A formal pseudo-code summary of the method is provided in Algorithm 8.1 for user-specified values of δ and the maximum number of iterations $itMax$.

Algorithm 8.1 Golden section algorithm

$niter = 1, r = \frac{-1+\sqrt{5}}{2}, minTemp = b, change = \infty$
$low = a, up = b, f_{low} = f(a + (1 - r)(b - a)), f_{up} = f(a + r(b - a))$
while $|change| > \delta\,|minTempLast|$ and $niter < itMax$ **do**
   $minTempLast = minTemp$
   $minTemp = (low + up)/2$
   if $f_{low} < f_{up}$ **then**
      $up = low + r(up - low), f_{up} = f_{low}$
      $f_{low} = f(low + (1 - r)(up - low))$
   else
      $low = low + (1 - r)(up - low), f_{low} = f_{up}$
      $f_{up} = f(low + r(up - low))$
   end if
   $change = minTemp - minTempLast$
   $niter = niter + 1$
end while
return $minTemp$

A measure of the efficiency of an optimization procedure such as Algorithm 8.1 is its convergence rate. If θ_{min} is the value of θ that minimizes f and $\theta^{(k)}$ is the approximation to θ_{min} that is produced on the kth step of the search, the convergence rate for the algorithm is the (relative) rate at which $\varepsilon_k = \theta_{min} - \theta^{(k)}$ decays to 0 as $k \to \infty$. The convergence rate is linear if $|\varepsilon_{k+1}| \doteq C|\varepsilon_k|$ for some constant C when k is large. The golden section search provides one example of an algorithm that exhibits linear convergence.

The golden section algorithm will provide the first member function for the optimization class Optim that will be extended and modified in subsequent sections and in the exercises. The initial version of the class header file is

Listing 8.7 optim.h

```
//optim.h
#ifndef OPTIM_H
#define OPTIM_H

class Functor;//forward declaration

class Optim{
  int itMax;
  double delta;

public:

  Optim(int ItMax = 38, double delta = 0.);

  double golden(const Functor& f, double low, double up) const;
};
#endif
```

The Optim class has two data members: itMax and delta. The itMax and delta members are user-supplied stopping parameters and therefore initialized in the class constructor. Default values of 38 and 0 are specified for itMax and delta, respectively. So, calling the Optim class constructor with no arguments will constrain any golden section search to no more than 38 iterations.

The golden method in Listing 8.7 takes three arguments. The last two are the lower and upper bounds for the interval to be used in the search. The first argument is a reference to a Functor object that represents some encapsulation of the function to be minimized. As Functor objects are not used by value in the header, it suffices to merely give a forward reference to the Functor class. The Functor class header is then included in the optim.cpp file shown in Listing 8.8 where the Optim class methods are defined.

Listing 8.8 optim.cpp

```
//optim.cpp
#include <cmath>
#include <limits>
#include "optim.h"
#include "functor.h"

const double rup = (-1. + sqrt(5.))/2;
const double rlow = 1 - rup;

Optim::Optim(int ItMax, double Delta){
    itMax = ItMax;
    delta = Delta;
}

double Optim::golden(const Functor& f, double low, double up) const {
    int niter = 0;
    double minTempLast = low, minTemp = up;
    double flow = f(low + rlow*(up - low));
    double fup = f(low + rup*(up - low));
```

```
double change = std::numeric_limits<double>::infinity();

while(change > delta*fabs(minTempLast) && niter < itMax){
  minTempLast = minTemp;
  minTemp = (low + up)/2.;
  if(flow < fup){
    up = low + rup*(up - low), fup = flow;
    flow = f(low + rlow*(up - low));
  }
  else{
    low = low + rlow*(up - low), flow = fup;
    fup = f(low + rup*(up - low));
  }
  change = fabs(minTemp - minTempLast);
  niter++;
}
return minTemp;
}
```

First note the definition of the two constants at the beginning of the file that correspond to r and $1 - r$ in Algorithm 8.1. Seeing that they are specified outside the body of any method they have global file scope which makes them available to all the functions in the file. Perhaps of more importance for future applications of the method is that they will not need to be recomputed each time the **golden** method is called. The **const** designation also makes sure that their values cannot be inadvertently altered. The code for the **golden** method essentially proceeds as in Algorithm 8.1. A question does arise about how to implement the initialization step of $change = \infty$. This is accomplished through the C++ **numeric_limits** template class that allows us to set the program's **change** variable at the machine representation for positive infinity (cf. the **eigen** method for class **Matrix** in Section 7.3).

The **golden** function takes a functor reference as an argument. For the present purpose a functor formulation along the lines of Listing 8.3 will suffice. Specifically, we will use

```
//functor.h
#ifndef FUNCTOR_H
#define FUNCTOR_H

struct Functor{
  double (*pF)(double theta);

  Functor(double (*pf)(double theta)){
    pF = pf;
  }

  double operator()(double theta){
    return pF(theta);
  }
};
#endif
```

Since the class is so simple, it is given in its entirety in a header file. The include guards become essential in this case and strange things can occur if they are not present.

As a test, the **Optim** class was used with the simple function $f(\theta) = -\theta(1 - \theta)$ that has a global minimum of $-.25$ at $\theta_{min} = .5$. The minimization process was then managed by the following driver program.

```
//optDriver.cpp
#include <iostream>
#include <cstdlib>
#include <iomanip>
#include "optim.h"
#include "functor.h"

double fun(double theta){
  return (-theta*(1. - theta));
}

int main(int argc, char** argv){
  int itMax = atoi(argv[1]);
  double delta = atof(argv[2]);
  Functor f(&fun);
  Optim Opt = Optim(itMax, delta);
  double fMin = Opt.golden(f, 0., 1.);
  std::cout << "The approximate minimizer is "
            << std::setprecision(16) << fMin << std::endl;

  return 0;
}
```

This program takes the maximum number of iterations (itMax) and the lower bound for the relative change in the approximate minimizer (delta) as command line arguments and converts them from character to int and double using the atoi and atof utility functions (accessed by including the cstdlib header). The address of the function fun that is to be minimized is used to create a Functor object. Next the Optim class object is initialized using the user-supplied values of itMax and delta and the golden method is applied to the Functor object. The final step is to print a summary of the results to standard output. For reasons that will become clear momentarily, the output is to be printed to 16 decimal accuracy. For this purpose the C++ standard library function setprecision (see Section 2.5) is accessed via inclusion of the iomanip header.

To actually carry out the minimization process using the optDriver.cpp program it is necessary to provide values for the stopping parameters itMax and delta. As noted above, since the computations are being done in double precision there is no obvious reason to choose delta smaller than about 10^{-16}. The selection of itMax can be guided by similar considerations. Specifically, if the golden section search is over an interval $[a, b]$ then at each step of the recursion the length of the interval that remains to be searched is reduced in size by the factor r. Thus, after the kth iteration the current search interval will have length $r^k(b - a)$. For $a = 0$ and $b = 1$ this gives $k = -16/\log_{10}(r) \doteq 76$ as the number of iterations that will be required to produce changes in the minimizer that are no longer detectable in double precision.

The optim.cpp and optDriver.cpp programs were compiled and linked to create an executable titled opt. An application of the golden section method then produced

$ ./opt 76 .0000000000000001
The approximate minimizer is 0.5000000037252904

The performance of golden section here seems somewhat disappointing. Certainly, the search has placed us in the right general location for minimization of the target function. But, there are only eight decimals of accuracy and as many as 16 might have been expected. The culprit here is round-off error; but, not from the approximation of the minimizer per se. The problem is that the objective function $f(\theta) = -\theta(1 - \theta)$ behaves like $-\theta^2$ for θ near

.5. Consequently, in the neighborhood of the minimizer, there are really only eight digits of the argument that are effectively being used to obtain the double precision value of f. The only way to achieve greater accuracy is to increase precision in the program (Exercise 8.2).

The effect of round-off error on minimization that was seen in our example holds quite generally. To see this observe as in Press, et al. (2005) that for θ near the minimum θ_{min} of a twice differentiable objective function f

$$f(\theta) \doteq f(\theta_{min}) + \frac{f''(\theta_{min})}{2}(\theta_{min} - \theta)^2$$

or

$$|\theta_{min} - \theta| \doteq \sqrt{\frac{2(f(\theta) - f(\theta_{min}))}{|f''(\theta_{min})|}}$$

with f'' the second derivative of f. Thus, if the difference $f(\theta) - f(\theta_{min})$ can be evaluated to an accuracy of no more than 10^{-k}, the best that can be expected is that the minimizer can be approximated to within an order of $10^{-k/2}$. As a result, $10^{-7/2}$ and 10^{-8} rather than 10^{-7} and 10^{-16} represent obtainable relative error bounds for single and double precision calculations, respectively. In terms of the number of effective iterations in double precision for the golden section method on $[0, 1]$ this translates to an upper bound of about $-8/\log_{10}(r) \doteq 38$.

The function `optimize` in R conducts a search for a minimizer (or maximizer) using Brent's method (e.g., Press, et al. 2005). This algorithm is a modification of the golden section search that attempts to enhance its performance by mixing in parabolic extrapolation (Exercise 8.5) with the ordinary golden section partitioning. A simplified version of the prototype for `optimize` is

```
optimize(f = func, lower = a, upper = b, maximum = FALSE)
```

Here `func` is a function that has been defined in R and `a` and `b` are the lower and upper bounds for the interval to be searched. The logical parameter `maximum` has `False` as its default value and need not be specified for minimization. By setting `maximum = TRUE` the search will be conducted for a maximizer of `func`. The `optimize` function returns a two-component list containing the approximate minimizer and the value of the objective function at the approximate minimizer.

An application of `optimize` to $f(\theta) = -\theta(1 - \theta)$ produces

```
> func <- function(theta){-theta*(1 - theta)}
> optimize(func, lower = 0, upper = 1)
$minimum
[1] 0.5

$objective
[1] -0.25
```

The objective function is quadratic which entails that parabolic extrapolation is exact here and the minimizer can be produced in a single step without the need for sectioning (Exercise 8.5).

The golden section algorithm is one of several optimization methods that do not require the existence (or knowledge) of the objective function's derivative. Other options include the bisection (Exercise 8.20), regula falsi and secant methods discussed in, e.g., Conte and de Boor (1972), Press, et al. (2005) and Flowers (2006).

8.3.1 Dealing with multiple minima

Up until now the premise has been that the objective function was globally convex and, as a result, had a single, global minimizer. Situations such as this are rare in practice. A more

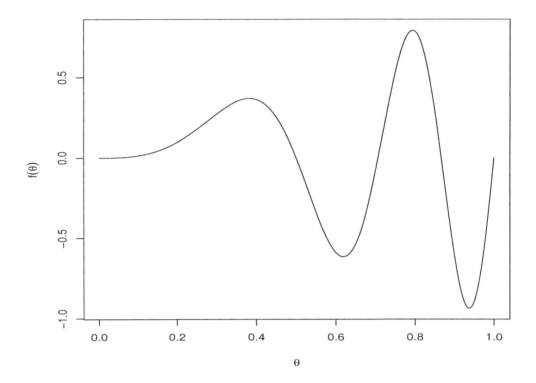

Figure 8.1 *The function* $f(\theta) = \theta \sin(4\pi\theta^2)$

realistic example is provided by the function in Figure 8.1 that has multiple local minima. An application of the Optim class golden method to this function results in

```
$ ./opt 38 .00000001
The approximate minimizer is 0.619058
```

A glance at Figure 8.1 reveals what has happened; the search has become trapped in the function's shallower middle trough and, as a result, has returned only a local minimizer.

In this particular case it is easy to see how to proceed by simply looking at the function. The solution is to start the golden section search in an interval that contains the minimizer such as [.9, 1]. But, for many practical applications the minimization process needs to be automated. It may be that the minimization step is just a small part of some larger computing endeavor and looking at the function is not an option. Simulation experiments fall into this category where, e.g., estimators are being computed via optimization of some random criterion function. Human intervention in the minimization process is simply not feasible in that context. It is also typically the case that θ is vector valued in which event visual location of a suitable search region tends to becomes problematic if there are more than two dimensions for the parameter space.

The only method that can be assured of locating a global minimum is a full search where the function to be minimized is evaluated at every value of its arguments. This approach can be implemented exactly if the arguments take values on a finite-dimensional set. In the case of continuous arguments, searches are carried out over a fine grid to find an approximate answer. This is generally too time consuming to be practical. Suppose, for example, that

θ is a p-vector and the minimization is carried out over a p-dimensional hyper-cube with the same number of grid points, n, being allocated to the range of each variable. Then, the search grid has size n^p. As a result, global searches in even three dimensions can be tasks that require a multiprocessor strategy and the search process can quickly pass beyond the capabilities of even modern super-computers as p increases.

A simple alternative to a global search is to partition the search interval into smaller subintervals and apply the golden section method (or some other minimization algorithm) to each of the subinterval. Of course, the success of this approach will depend on the partition and could fail if some of the subintervals contain more than one local minimum. Another similar, but distinctly different, option is to search using random subintervals. Two numbers u_1, u_2 are generated from the uniform distribution over the search interval $[a, b]$ and a golden section search is conducted over the interval $[\min(u_1, u_2), \max(u_1, u_2)]$. The process is repeated until some stopping criterion is met. This is the approach that will be considered here.

The random golden section search method in Listing 8.9 was added to class Optim.

Listing 8.9 Optim::ranGolden

```
double Optim::ranGolden(const Functor& f, double a, double b,
        int nSearch, unsigned long seed) const {
  double newMinTemp, minTemp = b;
  double up, low;

  //instantiate ranGen object
  ranGen RNG;
  RNG.setSeed(seed);
  //pointer for random uniforms
  double* pu = new double[2];

  //random search begins
  for(int i = 0; i < nSearch ; i++){
    //generate uniforms for the search interval
    RNG.ranUnif(2, pu);

    up = a + (b - a)*std::max(pu[0], pu[1]);
    low = a + (b - a)*std::min(pu[0],pu[1]);

    newMinTemp = golden(f, low, up);

    //compare local to global minimizer
    if(f(newMinTemp) < f(minTemp)){
      minTemp = newMinTemp;
    }
  }
  delete[] pu;
  return minTemp;
}
```

The WH random number generator from Section 4.5 is used here to generate random numbers from a uniform distribution over a search interval with a user-specified lower bound a and upper bound b. The WH algorithm is obtained from a class ranGen of random number generators that was developed in the same spirit as Exercise 4.33 and is given explicitly in Appendix E. An object RNG from the ranGen class is created and used to call the class method setSeed with a user-supplied initial seed. This method uses the input value seed in conjunction with the FM2 generator from Section 4.4 to create the three seeds that are needed to initialize the WH generator.

The `ranGen` object in Listing 8.9 is used to access the `ranUnif` class method and thereby produce two random uniforms from the interval $[0, 1]$ that are returned in the pointer `pu`. These two values are translated into a subinterval of the form `[low, up]` using the functions `min` and `max` that become available by inclusion of the algorithm header as in Section 7.2.4.

The ordinary golden section method is applied to the random search intervals bounded by `low = a + (b - a)*min(pu[0], pu[1])` and `up = a + (b - a)*max(pu[0], pu[1])` with `min(pu[0], pu[1])` and `max(pu[0], pu[1])` the smaller and larger of `pu[0]` and `pu[1]`, respectively. This process is repeated a specified number `nSearch` times with a check made on each iteration to see if the corresponding subinterval has produced a value that improved the approximation to the minimizer.

The `ranGolden` method for class `optim` was combined with a driver program that took the values of `itMax`, `nSearch`, `seed` and `delta`, in that order, as command line arguments. The resulting executable named `opt` was used to produce the results below.

```
$ ./opt 38 4 123 .00000001
The approximate minimizer is 0.619058
$ ./opt 38 5 123 .00000001
The approximate minimizer is 0.937337
```

The number of random search intervals is one of the "tuning" parameter for the algorithm. In general, the right choice for this quantity will depend on the complexity of the function and, of course, the choice for the generator's seed. In this instance it turns out that there is no further improvement to six decimals by choosing `nSearch` to be larger than 5.

It is also quite easy to implement the same type of random search in R with the `optimize` function. A program that accomplishes this is given in the next listing.

Listing 8.10 ranGolden.r

```
#ranGolden.r
ranGolden <- function(f, a, b, nSearch, seed){
  set.seed(seed)
  bounds <- matrix(runif(2*nSearch), 2, nSearch)
  minVec <- apply(bounds, MARGIN = 2,
    FUN = function(v){temp  = a + (b - a)*sort(v)
      optimize(f, lower = temp[1], upper = temp[2])[[2]]})
  location <- order(minVec)[1]
  temp  = a + (b - a)*sort(c(bounds[1, location],
    bounds[2, location]))
  optimize(f, lower = temp[1], upper = temp[2])
}
```

The `ranGolden` R function has the same basic arguments as its C++ analog: the objective function `f`, the lower and upper limits of the search interval `a` and `b`, the number of random searches to be performed and the value of the initial seed for the random number generator. The first line of code in the function's body sets the seed for the local R environment to the supplied value of `seed` using the R `set.seed` function. Next, an array named `bounds` of uniform random deviates is created with two rows and `nSearch` columns. The columns of this array will be used to construct the random subintervals of $[0, 1]$ to which `optimize` will be applied. The `apply` function manages the search. Its input argument is the `bounds` array. Our choice for the `FUN` argument to `apply` will work on the columns of `bounds`, sorting each pair of random uniforms and rescaling them to fall in the search interval between `a` and `b`. The `optimize` function is applied to this subinterval and only the second component from its output list is retained. The result is that the output from `apply` is a

vector `minVec` with `nSearch` elements that give all the local minimum function values that
have been found in the subinterval searches. The smallest value in this vector represents the
value of the objective function at the best approximation that has been found to the global
minimum. The index for this element is `order(minVec)[1]` which means it was produced by
using `optimize` on the interval determined by the `order(minVec)[1]` column of `bounds`.
Accordingly, `optimize` is used a final time on this subinterval and its output is what is
returned by the function.

An application of our R `ranGolden` function to the function in Figure 8.1 led to the
output

```
> ranGolden(func, 0, 1, 11, 123)
$minimum
[1] 0.619059

$objective
[1] -0.615748

> ranGolden(func, 0, 1, 12, 123)
$minimum
[1] 0.9373562

$objective
[1] -0.9363776
```

The `ranGolden` function finds the location of the minimum after 12 searches.

8.3.2 An application: linear smoothing splines revisited

Let us now return to the linear smoothing spline example of Section 7.3 and minimization
of the GCV criterion as a function of the smoothing parameter λ. A class `Linss` was created
there that could be used to fit linear smoothing splines to data. In particular, the class had
a method `smooth` with the prototype

```
Vector smooth(double lam, double* GCV) const;
```

This function returns the linear smoothing spline fit to a set of data for a given value of the
smoothing parameter furnished in the argument `lam`. The corresponding value of the GCV
criterion is also returned through a pointer that is passed in from the calling function.

Perhaps the first task that should be undertaken is creation of a function object that can
be used to manage the evaluation of the GCV criterion for `Linss` objects. For this purpose
a member function `gcv` with prototype

```
double gcv(double lLam) const;
```

was added to the `Linss` class. Its definition is

```
double Linss::gcv(double lLam) const {
    double GCV = 0;
    Vector fit = smooth(pow(10, lLam), &GCV);
    return GCV;
}
```

The `gcv` method is essentially an alternate way of packaging the `smooth` method that returns
only the value of the GCV criterion. One important difference is that the argument is
assumed to be in units of $\log_{10}(\lambda)$ that represents the preferred scale for optimization
purposes in this instance. Inside the method the C++ function `pow` (made available through
the cmath header) takes the value of $\log \lambda$, `lLam`, and transforms it back into smoothing
parameter units.

With the gcv method in place a functor class that is suitable for use in the optimization process must now be created. Our choice is essentially the one in Listing 8.5 apart from replacing int with double throughout the struct. Note that the file extension should also be changed to hh as we plan to use it in the creation of a shared library that will allow us to fit linear smoothing splines directly within an R session.

Since class Functor depends on an arbitrary class T and class Optim uses class Functor, we should make our optimization class a template class as well. A version of Optim that will serve this purpose has the declaration

```
template <class T> class Optim{
  int itMax;
  double delta;

public:

  Optim(int ItMax = 38, double Delta = 0.){
    itMax = ItMax; delta = Delta;
  }

  double golden(const Functor<T>& f, double low, double up) const;
  double gridSearch(const Functor<T>& f, double low, double up,
      int nGridPts) const;
};
```

In addition to the template formulation a new method gridSearch has been introduced. It takes the lower and upper bounds low and up of the search interval as input and then searches for the smallest value of the objective function corresponding to the Functor object over nGridPts evenly spaced evaluation points in the interval. The method definition is

```
template<class T> double Optim<T>::gridSearch(const Functor<T>& f,
      double low, double up, int nGridPts) const {
  double minTemp, fTemp, gridVal, range = (up - low);
  double minF = std::numeric_limits<double>::infinity();
  for(int i = 0; i < nGridPts; i++){
    gridVal = low + range*((double)i)/((double)nGridPts);
    fTemp = f(gridVal);
    if(fTemp < minF){
      minF = fTemp;
      minTemp = gridVal;
    }
  }
  return minTemp;
}
```

Note that Optim is now a template class with the consequence that all the method definitions must be included in the header file as well. Also, the other class methods will need to have similar syntax to modify their definition as that used for gridSearch; for example, the definition of the golden method must now begin with

```
template<class T> double Optim<T>::golden(const Functor<T>& f,
      double low, double up) const {
```

Minimization of the GCV criterion will be carried out in two phases. The Optim class gridSearch method will be used first. Once it has done its job the neighboring grid points for the approximate minimizer can be used as an interval for a golden section search to complete the minimization process. The entire task will be managed by an overloaded smooth method from the Linss class for which the declaration is

```
Vector smooth(double* lam, double* GCV, double a = -5, double b = 5,
       int nGridPts = 100) const;
```

When the `Optim` class object has completed its work the `GCV` and `lam` arguments will point to memory that contains the value for the approximate minimizing smoothing parameter and the corresponding value of the objective function. Additional arguments that provide the lower and upper search interval (for powers of 10) and the number of grid points to use in the search have been given suitable default values that will be effective in most instances. Thus, the user can simply call the `smooth` method and receive a "best" fit in the returned `Vector` object. There is a bit of subtlety here that merits a comment. The reason the compiler can distinguish between the second version of `smooth` with its default arguments and the first is due to the use of `double*` (as opposed to `double`) for the type of the `lam` argument of the newer version. The pointer type for `lam` is the correct choice in the second instance because `lam` is really an output rather than input entity.

The definition for the new `smooth` method is

```
Vector Linss::smooth(double* lam, double* GCV, double a, double b,
        int nGridPts) const {
  Functor<const Linss> F(this, &Linss::gcv);
  Optim<const Linss> Opt;
  //initial grid search
  double gcvMin = Opt.gridSearch(F, a, b, nGridPts);

  //golden section in the neighborhood of the minimizer
  double halfRange = (b - a)/((double)nGridPts);
  double low = gcvMin - halfRange;
  double up = gcvMin + halfRange;
  gcvMin = Opt.golden(F, low, up);
  *lam = pow(10, gcvMin);
  //now compute the fit at the GCV optimal lambda
  Vector fit = smooth(*lam, GCV);
  return fit;
}
```

The method begins by creating a functor for the object's gcv function using the `this` pointer and with template parameter set at T = `const Linss`. A class `Optim` object, also with T = `const Linss`, is then created and used with the gcv functor to apply the `gridSearch` method. If the default arguments are used, a 100 point grid search is carried out over the range $\lambda \in [10^{-5}, 10^5]$. This provides two points that bound the (possibly local) minimizer that are then passed on to the `Optim` class `golden` method. Finally, the value of λ returned from `golden` is used by the original, two argument, `smooth` method to obtain the fit to the data at the approximate minimizer that is returned to the calling program.

Some discussion may be helpful here and subsequently on how the `const` modifier was used in creating the functor objects in the `smooth` method. First, `smooth` should not alter the members of a `linss` object which is why it has been designated as a `const` class method. This has the effect of making the object's `this` pointer a `const` pointer to a `const linss` object. The template parameter must therefore be set as `const linss`. However, there is a bit more going on here. The optimization is applied to the `linss` class gcv function which will be called using the `this` pointer by an `optim` object. Seeing as only `const` functions can be called by `const` objects, it is necessary that gcv also be made a `const` class methods. Failure to do so will lead to a compilation error.

Upon compilation, the `Linss` class automated `smooth` method was applied to the data set created in Section 7.2.5 that is shown again in Figure 8.2. This produced a minimizing value of $\lambda_{opt} = .1559798$ and $GCV(\lambda_{opt}) = 3.700569$. In Section 7.2.5 the value $\lambda = .1$ was used to obtain a linear smoothing spline fit to this same data that produced a value of 3.7224 for the GCV function. Given the extra effort that went into obtaining the .1559798 value, the fact that it leads to a similar value for the GCV function as our ad hoc choice of $\lambda = .1$ raises the question of whether the additional labor has provided any visible improvement in the fit to the data. Plots are needed to address this type of question which means it is now time to return to R.

The most effective way to proceed is via creation of a shared library using, e.g., the interface program

```
//linssDriver.cc
#include "R.h"
#include "vector.hh"
#include "linss.hh"

extern "C" {

  void linss(int* n, double* lam, double* px, double* py,
        double* pfit, double* GCV){

    Vector y(*n, py);
    Vector x(*n, px);

    Linss S(x, y);
    Vector fit = S.smooth(GCV, lam);
    for(int i = 0; i < *n; i++)
      pfit[i] = fit[i];

    Rprintf("The value of GCV at lambdaOpt = %f is %f\n", *lam,
        *GCV);
    }
  }
```

This is a slightly modified version of Listing 7.4. The first difference is in function rather than form. For Listing 7.4 the `lam` pointer provided information being passed into the program whereas now it is a memory location that will be filled in with the GCV optimal smoothing parameter that is returned from the `smooth` method. The second difference is the use of the `Rprintf` function that was discussed in Section 5.7.

A shared library using all the required classes is compiled with

```
$ R CMD SHLIB linssDriver.cc linss.cc matrix.cc vector.cc
```

Once `linssDriver.so` is brought into R with the `dyn.load` function, it will be accessible through the name `linss` of the function that manages the transfer of information from and to the R session. As in Section 7.2.5 a wrapper function along the lines of Listing 7.5 can be created to facilitate its use. One possibility is

```
#linssGCV.r
linss <- function(x, y){
    fit <- .C("linss", as.integer(length(x)), 0., as.double(x),
        as.double(y), vector("numeric", length(x)), 0.)
    list(fit[[5]], fit[[6]])
  }
```

An application of this function to the data in Figure 8.2 produces

```
> linss(x, y)
The value of GCV at lambdaOpt = 0.155980 is 370.056890
```

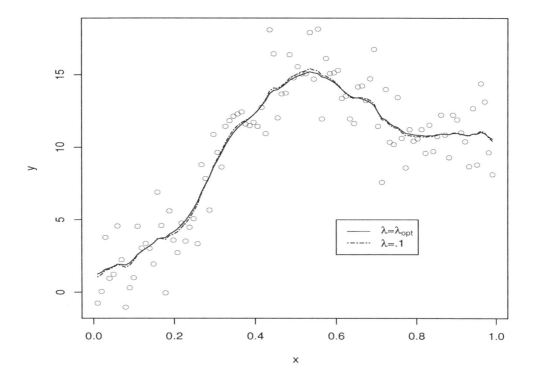

Figure 8.2 *Linear smoothing spline fits*

The resulting fit is shown in Figure 8.2 along with the one having $\lambda = .1$. As might have been suspected by the similarity of their two GCV values, there is little that distinguishes one from the other.

8.4 Newton's method

The golden section method is quite satisfactory in many settings. Its performance is nonetheless limited by the fact that no derivative information is utilized. In something like our GCV example of the previous section, golden section is likely the most reasonable approach. On the other hand, when f's first two derivatives are known (or can be approximated) a much faster rate of convergence to the minimizer can be obtained by using the *Newton* or *Newton-Raphson* method.

Assume that f has two continuous derivatives and is convex on an interval $[a, b]$ of the real line. In that case minimizing the function is equivalent to finding the unique point θ_{min} in $[a, b]$ where $f'(\theta_{min}) = 0$. An application of Taylor's theorem with exact remainder gives

$$f'(\theta_{min}) = 0 = f'(\theta) + f''(\theta^*)(\theta_{min} - \theta),$$

where θ^* is a point between θ and θ_{min}. This suggests that an initial guess $\theta^{(0)}$ for the minimizer can be updated using

$$\theta^{(1)} = \theta^{(0)} - \frac{f'(\theta^{(0)})}{f''(\theta^{(0)})}.$$

The process can be continued and, in general, movement from the kth to the $(k + 1)$st update is accomplished via

$$\theta^{(k+1)} = \theta^{(k)} - \frac{f'(\theta^{(k)})}{f''(\theta^{(k)})}.$$

Algorithm 8.2 summarizes the form of the Newton recursions for user-specified values of the tolerance parameter δ for the minimum allowable relative difference between successive approximations to the minimizer, the maximum number of iterations $itMax$ and an initial starting point $\theta^{(0)}$.

Algorithm 8.2 Newton-Raphson algorithm

$niter = 1, minTemp = \theta^{(0)}, minTempLast = b$
while $|minTempLast - minTemp| > \delta|minTempLast|$ and $niter < itMax$ **do**
$\quad minTempLast = minTemp$
$\quad minTemp = minTempLast - f'(minTempLast)/f''(minTempLast)$
$\quad niter = niter + 1$
end while
return $minTemp$

Not only does the Newton approach work but it actually produces quadratic rates of convergence near a minimum as compared to the linear rate achieved by the golden section method. To see this first define

$$\varepsilon_k = \theta_{min} - \theta^{(k)}.$$

Now note that if f has a third derivative f''' the condition $f'(\theta_{min}) = 0$ implies that

$$f'(\theta^{(k)}) + f''(\theta^{(k)}) \left(\theta_{min} - \theta^{(k)} \right) + \frac{f'''(\theta^*)}{2} \left(\theta_{min} - \theta^{(k)} \right)^2 = 0$$

or, equivalently, that

$$\theta_{min} - \left[\theta^{(k)} - \frac{f'(\theta^{(k)})}{f''(\theta^{(k)})} \right] = -\frac{f'''(\theta^*)}{2f''(\theta^{(k)})} \left(\theta_{min} - \theta^{(k)} \right)^2$$

with θ^* a point between θ_{min} and $\theta^{(k)}$. Thus,

$$|\varepsilon_{k+1}| = \varepsilon_k^2 \left| \frac{f'''(\theta^*)}{2f''(\theta^{(k)})} \right|$$

and our claim has been verified.

The values of the objective function's first two derivatives will be needed to implement the Newton method. Rather than spread them across different function objects they can instead be packaged collectively using a functor class such as the one in the next listing.

Listing 8.11 *functor.h (with derivatives)*

```
//functor.h
#ifndef FUNCTOR_H
#define FUNCTOR_H

struct Functor{
  double (*pF)(double theta);
  double (*pF1)(double theta);
  double (*pF2)(double theta);

  Functor(double (*pf)(double theta), double (*pf1)(double theta),
          double (*pf2)(double theta)){
```

```
   pF = pf; pF1 = pf1; pF2 = pf2;
 }

 Functor(double (*pf)(double theta)){
   pF = pf;
 }

 double operator()(double theta, int i) const {
   if(i == 1) return pF1(theta);
   return pF2(theta);
 }

 double operator()(double theta) const {
   return pF(theta);
 }
};
#endif
```

This class `Functor` uses what is probably the most direct approach and overloads the () operator with two arguments. The first corresponds to the optimization variable, as before, while the second is an integer that is combined with `if` statements to determine which of the two functions f' or f'' will be evaluated. Our previous one-argument version of () can live in harmony with the two-argument version and suffices for access to the function itself. An overloaded constructor that takes only one function pointer has been provided to handle cases where no derivative information is available. The fact that the one-argument version of () will work with an object created from either class constructor means that the Listing 8.11 functor can be used interchangeably with derivative and nonderivative based methods without changing our previous code that was created for the latter methods.

The prototype for the `newton` method for class `Optim` now takes the form

```
double newton(Functor& f, double a, double b, double start,
      int* ier) const;
```

The arguments for `newton` are a `Functor` object, the lower and upper terminals (a and b) of the interval that should contain the minimizer and a value `start` to be used as an initial "guess" for the minimizer's location. For a convex function, specification of a and b would be extraneous to the algorithm. But, when dealing with functions that have multiple minima the inclusion of a and b provides the means to check that the Newton method does not take us beyond the region of interest. In general, there is no reason that the Newton step $f'(\theta^{(k)})/f''(\theta^{(k)})$ should produce a new value $\theta^{(k+1)}$ that lies in the region of interest and the divisor in this ratio $f''(\theta^{(k)})$ can also evaluate to a floating-point 0. Both of these problems must be taken into account and that is the purpose of the `ier` pointer argument for `newton`. This variable will return an error code that indicates whether either of these two problems has been encountered.

The definition for the `newton` method is

```
double Optim::newton(const Functor& f, double a, double b,
      double start, int* ier) const {
  int niter = 0;
  *ier = 0;
  double minTemp = start, minTempLast = b;

  while(fabs(minTemp - minTempLast) > delta*fabs(minTempLast)
      && niter < itMax){
    minTempLast = minTemp;
```

```
    if(f(minTempLast, 2) != 0)
      minTemp = minTempLast - f(minTempLast, 1)/f(minTempLast, 2);
    else{
      *ier = 1;
      break;
    }
    if(minTemp > b || minTemp < a){
      *ier = 2;
      break;
    }
    niter++;
  }
  return minTemp;
}
```

The code in the method mimics the pseudo-code development in Algorithm 8.2. The actual Newton steps are computed using a `Functor` object with the second argument for the () operator being used to evaluate both the first and second derivatives of f. Additional if blocks have been included to guard against cases where f'' might be 0 or the Newton update takes us outside the interval over which f is being minimized. Such occurrences produce return values of 1 and 2, respectively, for the integer corresponding to `ier`. When either error occurs there is no reason to proceed further and the most expedient solution is likely to try a new value for `start`. Rather than stopping execution with the `exit` function, a `break` statement (e.g., Section 3.4) has been used. The current loop is terminated when `break` is encountered but the program will continue to run and end gracefully. In particular, this allows the value of `*ier` to be returned to the calling program for use in diagnosing the source of the difficulty.

The driver program in Listing 8.12 was written to apply the Newton method to the function $f(\theta) = -\theta(1 - \theta)$ from Section 8.3.

Listing 8.12 *optDriver.cpp*

```
//optDriver.cpp
#include <iostream>
#include <cstdlib>
#include "optim.h"
#include "functor.h"

double fun(double theta){
  return (-theta*(1. - theta));
}

double fun1(double theta){
  return (-1. + 2.*theta);
}

double fun2(double theta){
  return 2.;
}

int main(int argc, char** argv){

  int itMax = atoi(argv[1]);
  double delta = atof(argv[2]);
  double start = atof(argv[3]);
```

```
    Functor f(&fun, &fun1, &fun2);
    Optim Opt = Optim(itMax, delta);
    int ier = 0;
    double fMin = Opt.newton(f, 0., 1., start, &ier);
    std::cout << "The approximate minimizer is "
              << fMin << std::endl;
    std::cout << "The error code is " << ier << std::endl;

    return 0;
}
```

The functions fun, fun1 and fun2 are f and its first two derivatives. The addresses of these three functions are used to obtain the requisite Functor object. An Optim class object is created next using values for delta and itMax that are provided through command line input. Finally, the Optim object calls the newton class method using the Functor object and a starting value obtained from the command line. The resulting approximate minimizer and the value of the error code returned from newton are then written to standard output.

The optim.cpp and optimDriver.cpp files were compiled and linked to create the executable opt. Using this we obtained

```
$ ./opt 38 .00000001 .1
The approximate minimizer is 0.5
The error code is 0
$ ./opt 38 .00000001 .25
The approximate minimizer is 0.5
The error code is 0
$ ./opt 38 .00000001 .9
The approximate minimizer is 0.5
The error code is 0
```

It seems like every choice for the starting value works. A little checking reveals that iterations are always terminated after the niter variable inside the newton method reaches two. Of course, this is no real surprise because a function of this nature is an ideal case for the Newton algorithm. The same would occur for any other convex quadratic function; i.e., the newton method for class Optim would always return the exact (to machine precision) minimizer with niter $= 2$ (Exercise 8.14).

A criticism could be leveled against the way that errors are handled by our implementation of the Newton method. The current "solution" is to simply terminate the search when things go awry. This actually works fine if the method is applied to a sequence of random starting points as we did with the golden section algorithm. For general use there is no reason to stop computation. A new search can be started with the same parameters, apart from the starting value, using the C++ exception mechanism from Section 3.8.7. An alternative version of the newton method that employs this fix is shown in the listing below.

```
double Optim::newton(const Functor& f, double a, double b,
        double start) const {
  int niter = 0;
  double minTemp = start, minTempLast = b;

  while(fabs(minTemp - minTempLast) > delta*fabs(minTempLast)
        && niter < itMax){
    minTempLast = minTemp;
    try{
    if(f(minTempLast, 2) == 0)
       throw std::overflow_error("Division by zero: ");
    minTemp = minTempLast - f(minTempLast, 1)/f(minTempLast, 2);
```

```
    }

    catch(std::overflow_error e){
      std::cout << e.what() << std::endl;
      std::cout << "Enter another starting value (q to quit)"
                << std::endl;
      std::cin >> minTemp;
      minTempLast = b;
      continue;
    }

    try{
      if(minTemp > b || minTemp < a)
        throw std::out_of_range("Value outside [a, b]: ");
    }

    catch(std::out_of_range e){
      std::cout << e.what() << std::endl;

      std::cout << "Enter another starting value (q to quit)"
                << std::endl;
      std::cin >> minTemp;
      minTempLast = b;
      continue;
    }
    niter++;
  }
  return minTemp;
}
```

The basic search procedure is, of course, the same. The difference is that the user is allowed to input a new starting value if the second derivative evaluates as zero or if the search goes outside of the designated domain. The error objects that are thrown are from the classes overflow_error and out_of_range that require inclusion of the stdexcept header file.

We tried our alternative version of the newton method on the function in Figure 8.1 with the results shown below.

```
$ ./opt 38 .00000001 .1
The approximate minimizer is 3.58977e-13
$ ./opt 38 .00000001 .25
Value outside [a, b]:
Enter another starting value (q to quit)
.5
The approximate minimizer is 0.382298
$ ./opt 38 .00000001 .75
The approximate minimizer is 0.793737
$ ./opt 38 .00000001 .9
The approximate minimizer is 0.937337
```

The performance of the method is notably less impressive for this choice of an objective function and a little bit of everything has occurred here. A value for start of .1 causes the method to travel along the gentle downhill slope with increasingly smaller steps toward $\theta = 0$. The choice of start = .25 places the algorithm on the right-hand side of a local minimum and it takes an overly aggressive backward step (of about $-.28$) that moves it out of the search region and causes a prompt for a new starting value by the error handling catch block. The Newton method searches for a place where the derivative is zero and, as a consequence, is equally adept at finding a maximum as a minimum which is what transpired with the new choice of .5 for `start`. This value is to the left of an inflection point leading the method to climb up the hill to the smaller of the function's two peaks. The value `start` = .75 is also to the left of an inflection point except on the right-hand side of a trough. So, the Newton recursion goes down the slope to a local minimum. Finally, by taking `start` = .9 we land in the right position for locating the global minimum.

The previous example illustrates the way to use Newton's method for functions with multiple local minima: try lots of different starting values. A simple way to implement this idea is to proceed along the lines of Listing 8.9 and generate starting points randomly over the search interval. A program that accomplishes this is given in the listing below.

```
double Optim::ranNewton(Functor& f, double a, double b,
          int nSearch, unsigned long seed, int* nFail) const {
  double start, newMinTemp, minTemp = b;
  int ier = 0;
  *nFail = 0;

  //instantiate ranGen object
  ranGen RNG;
  RNG.setSeed(seed);
  //pointer for random uniforms
  double* pu = new double;

  //random search begins
  for(int i = 0; i < nSearch ; i++){
    //generate a uniform for the starting value
    RNG.ranUnif(1, pu);
    start = a + (b - a)*(*pu);
    newMinTemp = newton(f, a, b, start, &ier);

    //compare local to global minimizer
    if(f(newMinTemp, 0) < f(minTemp, 0) && ier == 0)
      minTemp = newMinTemp;

    else if(ier != 0)
      ++*nFail;
  }
  delete pu;
  return minTemp;
}
```

The premise behind the `ranNewton` method is essentially the same as that in Listing 8.9; `nSearch` random numbers are generated in the interval `a` to `b` that are used as starting values for Newton iterations. The method requires a function object and the limits of the search interval as input which it then passes on to our first version of the `newton` method that terminates the recursion when errors occur. It also requires a seed to start the random number generator that is obtained through a `ranGen` object as in Listing 8.9. The use of `break` statements in the `newton` method pays dividends here by allowing us to simply return to `ranNewton` when a starting value causes problems for `newton` rather than having the entire program terminate. A record is maintained through the variable `nFail` that tells us how many of the searches were ended before they were completed.

A driver program (stored in the file optDriverTrig.cpp whose name appears below) was written to apply `ranNewton` to the Figure 8.1 function. Its `main` function has the form

```
int main(int argc, char** argv){

  int itMax = atoi(argv[1]);
  int nSearch = atoi(argv[2]);
  int seed = atoi(argv[3]);
  double delta = atof(argv[4]);
  Functor f(&fun, &fun1, &fun2);
  Optim Opt = Optim(itMax, delta);
  int nFail = 0;

  double fMin = Opt.ranNewton(f, 0., 1., nSearch, seed, &nFail);
  std::cout << "The approximate minimizer is "
            << fMin << std::endl;
  std::cout << "The number of failures is " << nFail << std::endl;

  return 0;
}
```

The values of itMax, nSearch, delta and the starting seed for the random number generator are taken from command line input and the addresses of the function and its first two derivatives are used to create a Functor object. The ranNewton method is then applied with both the approximate minimizer and the number of starting values that caused difficulties being reported on completion of the optimization process.

The various programs were compiled and linked using the makefile ranNewton.mk that is shown in the listing below.

```
opt : optDriverTrig.o optim.o ranGen.o
        g++ -Wall optDriverTrig.o optim.o ranGen.o -o opt

ranGen.o : ranGen.cpp ranGen.h
        g++ -Wall -c ranGen.cpp

optim.o : optim.cpp optim.h ranGen.h functor.h
        g++ -Wall -c optim.cpp

optDriverTrig.o : optDriverTrig.cpp optim.h functor.h
        g++ -Wall -c optDriverTrig.cpp
```

This was employed to produce the results

```
$ make -f ranNewton.mk
g++ -Wall -c optDriverTrig.cpp
g++ -Wall -c optim.cpp
g++ -Wall -c ranGen.cpp
g++ -Wall optDriverTrig.o optim.o ranGen.o -o opt
$ ./opt 38 9 123 .00000001
The approximate minimizer is 0.619058
The number of failures is 3
$ ./opt 38 10 123 .00000001
The approximate minimizer is 0.937337
The number of failures is 3
```

The ranNewton method therefore requires five more searches to find the minimizer than did ranGolden. However, in this instance the Newton approach converges much faster to local minimizers usually using only three to six of the 38 allotted iterations while the golden section based search tends to use all 38 cycles.

There are (at least) two functions that carry out Newton type optimization in R: optim and `nlm`. Both of these functions perform minimization via quasi-Newton type algorithms that use finite difference approximations for the first and second derivatives that appear in the Newton updating step. The `optim` function provides access to several other minimization techniques including simulated annealing that will be covered in Section 8.6. It also allows for and enforces range constraints while `nlm` does not have that option. As a result, our focus will be on the use of `optim`.

A simplified version of the prototype for `optim` is

```
optim(par, fn, method = "L-BFGS-B", lower, upper)
```

The `par` argument is the starting value to initiate the recursion, `fn` is the function to be minimized and `lower` and `upper` are the bounds for the search interval that default to the entire real line in the univariate case. The method L-BFGS-B is a limited memory, quasi-Newton algorithm that employs the Broyden-Fletcher-Goldfarb-Shanno or BFGS updating scheme for second derivative approximations and allows for constraints in the form of bounds (see, e.g., Byrd, et al. 1995 and Zhu, et al. 1997). The limited memory aspect is not particularly relevant for our one-dimensional setting. In general, the `optim` function can be used for optimization in higher dimensions where such issues become more important.

An application of the `optim` function for minimization of $f(\theta) = \theta \sin(4\pi\theta^2)$ resulted in

```
> optim(.5, func, method = "L-BFGS-B", lower = 0, upper = 1)
$par
[1] 0.61905665

$value
[1] -0.61574799

$counts
function gradient
      17       17

$convergence
[1] 0

$message
[1] "CONVERGENCE: REL_REDUCTION_OF_F <= FACTR*EPSMCH"
```

A five component list is returned by `optim` whose first two components are the approximate minimizer (`par`) and the corresponding value of the objective function (`value`). The `counts` component indicates the number of evaluations that were required for the function (i.e., `func` in this case) and the number of approximations that were computed for the function's derivatives both of which are 17 in this instance. The value of `convergence` represents an error message with 0 indicating successful convergence. The last component `message` is a report of any additional information returned by the optimizing algorithm. In this instance, the optimizer has indicated that the algorithm terminated when the relative change in the objective function fell below a constant multiple `factr` of the machine epsilon or minimum detectable relative absolute rounding error that was discussed in Section 2.5. The value of `factr` can be specified as an argument to `optim` and defaults to 10^7 otherwise. The derivative of the objective function can also be supplied to `optim` as its `gr` argument.

Some further experimentation with `optim` leads to

```
> optim(.25, fn = func, method="L-BFGS-B", lower = 0, upper = 1)[1]
$par
[1] 0

> optim(.65, fn = func, method="L-BFGS-B", lower = 0, upper = 1)[1]
```

```
$par
[1]  0.6190567

> optim(.8, fn = func, method="L-BFGS-B", lower = 0, upper = 1)[1]
$par
[1]  0.9373372
```

So, `optim` will find the global minimizer if it is started in the right location. To make sure that this happens a scheme such as the one in Listing 8.10 is needed to automate the task of selecting starting values that will produce a global minimum. This is the topic of Exercise 8.6.

8.5 Maximum likelihood

The technique of maximum likelihood estimation leads to an optimization problem that is unique to statistics. First, let $X_1, \ldots, X_n$ be a collection of random variables with joint density (or probability mass function) $g(\cdot; \theta^*)$ for θ^* some fixed, unknown member of a set Θ that is a subset of the real line. Suppose now that the X_i have been observed in the sense that we have seen $X_1 = x_{1O}, \ldots, X_n = x_{nO}$ with the O subscript indicating observed values that are now fixed rather than random. The question is how to "best" extract the information about the parameter value θ^* from the observed data.

The sample likelihood is defined to be

$$L(\theta) = g(x_{1O}, \ldots, x_{nO}; \theta).$$

It is a function of θ alone because $x_{1O}, \ldots, x_{nO}$ are now fixed. The maximum likelihood estimator, or mle, of θ is then given by

$$\widehat{\theta} = \mathrm{argmax}_{\theta \in \Theta} L(\theta).$$

The motivation for using the mle as an estimator is most easily seen for the case where the components of X are discrete valued. In that instance

$$L(\theta) = \mathrm{Prob}(X_1 = x_{1O}, \ldots, X_n = x_{nO}; \theta);$$

i.e., $L(\theta)$ is the probability of seeing the observed sample values $x_{1O}, \ldots, x_{nO}$ for a given value θ of the parameter. This gives $\widehat{\theta}$ the interpretation of being the choice for θ that assigns the highest probability for the data that was actually collected.

Our treatment of maximum likelihood will deal with the special case where $X_1, \ldots, X_n$ represent a random sample in the sense of Definition 4.1. Let $f(\cdot; \theta)$ be the common (univariate) density or probability mass function for some value of θ in which case $g(x_1, \ldots, x_n; \theta) = \prod_{i=1}^n f(x_i; \theta)$. This makes it easier to work with the negative natural logarithm of the likelihood

$$\ell(\theta) = -\sum_{i=1}^n \ln f(x_{iO}; \theta)$$

for optimization purposes. Minimizing $\ell(\theta)$ is equivalent to maximizing $L(\theta)$ and this problem can be addressed using our work in the previous two sections. In order to accomplish that the likelihood function (and its derivatives if we are to use Newton's method) must first be constructed from the observed data and packaged in a form that is amenable for use with a version of our `Optim` class. The C++ way to approach this is through creation of a likelihood class.

The design that will be used for our likelihood class is similar in spirit to the approach that was used for class `Linss` in Section 8.3.2. Access to the data will be provided by the calling program. Then, a function object will be created that will be passed to an optimization class object. The end result will be the mle that will be returned as output.

To provide some variation in the development, we will have our likelihood class `Like` manage the input of data using a template input class `fileIn` that is similar to the one created in Exercise 3.33. The calling program will simply supply the name of the file that contains the data and furnish the functions that are to be used in the optimization process. For the Newton method the logarithm of the density and its first two derivatives will need to be passed in to a `Like` class object. The `Functor` class that will be used for this is

Listing 8.13 functorIn.h

```cpp
//functorIn.h
#ifndef FUNCTORIN_H
#define FUNCTORIN_H

struct FunctorIn{
  double (*pF )(double x, double theta);
  double (*pF1)(double x, double theta);
  double (*pF2)(double x, double theta);

  FunctorIn(){}
  FunctorIn(double (*pf)(double x, double theta),
       double (*pf1)(double x, double theta),
       double (*pf2)(double x, double theta)){
    pF = pf; pF1 = pf1; pF2 = pf2;
  }

  double operator()(double x, double theta, int i) const {
    if(i == 0) return pF(x, theta);
    if(i == 1) return pF1(x, theta);
    return pF2(x, theta);
  }
};
#endif
```

The primary difference between Listing 8.13 and Listing 8.11 is that `()` has three arguments rather than two: the variable to be used in optimization, `theta`, the index `i` that determines which function to evaluate and an `x` variable that corresponds to the observed sample values.

This header file for our likelihood class is given in Listing 8.14 below.

Listing 8.14 like.h

```cpp
//like.h
#ifndef LIKE_H
#define LIKE_H
#include "functorIn.h"

class Like{
  int n;
  FunctorIn f;
  double* px0;
  double l(double theta) const;
  double lp(double theta) const;
  double lpp(double theta) const;

public:

  Like(int N, char* fname, FunctorIn& F);
```

```
    double mle(double a, double b, double start, int itMax = 38,
        double delta = 0.) const;
};
#endif
```

The `Like` class has three members: the sample size `n`, a `FunctorIn` class object and a pointer to `double` that will hold the data values. The class constructor takes the sample size and a `FunctorIn` object as arguments as well as a C-style string (i.e., a null-terminated via "\0" character array or pointer to `char`) that contains the name of the file where the data is stored. The class method `mle` that manages the optimization work requires specification of the lower and upper bounds `a` and `b` for the parameter space interval and a starting value to use in initialization of Newton's method. The remaining two arguments `itMax` and `delta` that set the stopping criteria for the Newton iteration have been given default values and need not be specified. Recall that once an argument is given a default value all subsequent arguments in the function's signature must have defaults as well. This is why `itMax` and `delta` appear at the end of the argument list for the `mle` method.

The content of the file that contains the class `Like` method definitions is shown in the next listing.

<hr>

Listing 8.15 *like.cpp*

<hr>

```
//like.cpp
#include "like.h"
#include "functor.h"
#include "optim.h"
#include "fileIn.h"

Like::Like(int N, char* fname, FunctorIn& F){
  n = N; f = F;
  px0 = new double[n];
  fileIn<double> inObj(fname);
  inObj.read(n, px0);
}

double Like::l(double theta) const {
  double logL = 0;
  for(int i = 0; i < n; i++)
    logL += f(px0[i], theta, 0);
  return (-logL);
}

double Like::lp(double theta) const {
  double logLp = 0;
  for(int i = 0; i < n; i++)
    logLp += f(px0[i], theta, 1);
  return (-logLp);
}

double Like::lpp(double theta) const {
  double logLpp = 0;
  for(int i = 0; i < n; i++)
    logLpp += f(px0[i], theta, 2);
  return (-logLpp);
}
```

```
double Like::mle(double a, double b, double start, int itMax,
        double delta) const {
  int ier = 0;
  Functor<const Like> logL(this, &Like::l, &Like::lp, &Like::lpp);
  Optim<const Like> Opt(itMax, delta);
  double thetaHat = Opt.newton(logL, a, b, start, &ier);
  return thetaHat;
}
```

The class constructor allocates sufficient memory to hold the data array and then uses the input file name to create a `fileIn` object that is suitable for reading floating-point data in double precision. The `read` method for this particular incarnation of class `fileIn` looks like

```
template<class T> void fileIn<T>::read(int n, T* pData){
  std::ifstream inFile(fName);
  if(!inFile.is_open()){
    std::cout << "Error opening input file!" << std::endl;
    exit(1);
  }
  for(int i = 0; i < n; i++)
    inFile >> pData[i];

  inFile.close();
}
```

This method uses an `ifstream` object (cf. Section 3.10) to open the input file, reads in the data while placing it in the supplied memory location and then closes the file connection.

The functions l, lp and lpp in Listing 8.15 calculate the negative log-likelihood and its derivatives at a given value `theta` for the parameter. For this purpose they use the `FunctorIn` object that was supplied to the class constructor to sum across the observed sample values and directly evaluate ℓ, ℓ' and ℓ''. These functions were designated as `private` in the class header because they are primarily for internal use in this particular application.

The form of the `mle` method in Listing 8.15 is perhaps somewhat deceptive in its seeming simplicity. Certainly the idea is straightforward; a function object that encapsulates the l, lp and lpp methods (i.e., a `Functor` object from Listing 8.16 below) is constructed and is then passed out to an optimization object that uses Newton's algorithm to find the mle. However, similar to the smoothing spline example of Section 8.3.1, the fact that `mle` is a `const` method has the consequence that the template class parameter must be set as `const` like. Also, the functions l, lp and lpp must all be `const` member functions in order for them to be invoked by the `this` pointer that is passed to the `Optim` object.

The function objects in Listing 8.15 derive from

Listing 8.16 *functor.h (template with derivatives)*

```
//functor.h
#ifndef FUNCTOR_H
#define FUNCTOR_H

template <class T>
class Functor{
  double (T::*pF)(double);
  double (T::*pF1)(double);
  double (T::*pF2)(double);
  T* pT;
```

```
public:
  Functor(){}

  Functor(T* pt, double (T::*pf)(double)){
    pT = pt; pF = pf;
  }

  Functor(T* pt, double (T::*pf)(double), double (T::*pf1)(double),
      double (T::*pf2)(double)){
    pT = pt; pF = pf; pF1 = pf1; pF2 = pf2;
  }

  double operator() (double u) const {
    return (pT->*pF)(u);
  }

  double operator() (double u, int i) const {
    if(i == 1) return (pT->*pF1)(u);
    return (pT->*pF2)(u);
  }
};
#endif
```

This is a template version of the functor class in Listing 8.11. As it is a template class, our optimization class should be structured similarly. The template version of `Optim` described in Section 8.3.2 was adapted for the job by adding in a template version of the Newton method with prototype

```
template <class T> double Optim<T>::newton(Functor<T>& f, double a,
      double b, double start, int* ier)
```

The `Like` class will be used to obtain the mle for data from the exponential distribution with density

$$f(x) = \begin{cases} \theta \exp\{-\theta x\}, & x > 0, \\ 0, & \text{otherwise,} \end{cases}$$

for $\theta > 0$. This is a trivial case in the sense that the mle has a closed form and is found to be the reciprocal of the sample mean $\sum_{i=1}^{n} x_{iO}/n$. Its utility lies in the fact that it gives a test case where the answer is known.

The driver program for the exponential scenario is

```
//likeDriver.cpp
#include <cmath>
#include <iostream>
#include <cstdlib>
#include <limits>
#include "like.h"
#include "functorIn.h"

double f(double x, double theta){
  return (log(theta) - theta*x);
}

double fp(double x, double theta){
  return ((1./theta) - x);
}
```

```
double fpp(double x, double theta){
  return (-(1./(theta*theta)));
}

int main(int argc, char* argv[]){
  int n = atoi(argv[1]);
  FunctorIn F(&f, &fp, &fpp);
  Like expLike(n, argv[2], F);
  double up = std::numeric_limits<double>::infinity();
  double mle = expLike.mle(0., up, atof(argv[3]));
  std::cout << mle << std::endl;
  return 0;
}
```

The functions f, fp and fpp are the natural logarithm of the exponential density and its first two derivatives that are used to create the FunctorIn object in main. The sample size, file name (for the data) and a starting value for the recursion are obtained from command line arguments. For the exponential distribution the parameter space is the positive real line. Thus, a lower limit value of a = 0 is specified while the upper limit is set to the machine definition of infinity for a double using the numeric_limits template class.

The data to use in testing our Like class was generated in R and written to a file with

```
> set.seed(123)
> x0 <- rexp(100, 1)
> 1/mean(x0)
[1] 0.9562801
> write.table(x0,"edat.txt",quote = FALSE, row.names = FALSE,
+ col.names = FALSE)
```

from which the maximum likelihood estimator is seen to be .95628. A makefile like.mk was then created and used to compile our C++ code with the result being

```
$ make -f like.mk
g++ -Wall -c likeDriver.cpp
g++ -Wall -c like.cpp
g++ -Wall likeDriver.o like.o -o mle
$ ./mle 100 edat.txt .5
0.95628
```

The code seems to perform as expected. However, not all is perfect here and a poor choice for the starting value can lead the Newton method astray. Preliminary grid or random searches (Exercise 8.16) provide means to resolve such difficulties.

The R stats4 package includes the function mle that can be used for univariate and multivariate maximum likelihood problems. This function appears to be a wrapper for the optim function that allows some arguments to be fixed in the likelihood function. In practice this provides us with an ability to fix the data aspect of the likelihood and only optimize with respect to the parameter(s).

The prototype for mle appears like

```
mle(minuslogl, start = list(), method = "BFGS", fixed = list(), ...)
```

The minuslogl argument is the negative of the logarithm of the likelihood function that is to be minimized. The start parameter is a list of starting values for the minuslogl arguments specified by the parameter name. The fixed parameter is a similar list that designates arguments that are to be held fixed while assigning them specific values. The default optimization method is the quasi-Newton BFGS algorithm. But, any of the other methods available for optim can be used as with their arguments specified via the ellipsis.

To illustrate the use of `mle`, consider the function

```
lfunc <- function(theta, xbar, n){-n*log(theta) + n*xbar*theta}
```

This evaluates the negative log-likelihood for an exponential distribution given the sample size `n` and the sample mean `xbar`. For optimization purposes, the values of `n` and `xbar` must be held fixed and the `fixed` argument for `mle` can be used to handle this. The listing below demonstrates this with the same data that was used with our `like` class.

```
> set.seed(123)
> x0 <- rexp(100, 1)
> library("stats4")
> mle(lfunc, start = list(theta = .5), fixed = list(xbar = mean(x0),
+ n = length(x0)))

Call:
mle(minuslogl = lfunc,start = list(theta = 0.5),fixed = list(xbar = mean(x0),
    n = length(x0)))

Coefficients:
        theta           xbar                n
   0.95628036     1.04571871   100.00000000
```

8.6 Random search

Deterministic optimization methods such as the golden section and Newton's method provide practical alternatives to a global search strategy for finding the global minimizer of a function f. However, they are limited by their requirement of a good initial search location. The use of random number generators to find such locations can be quite effective. This raises the question of why use deterministic methods at all? Why not simply search the entire optimization region with a sequence of random numbers? There are many methods that do precisely this. An overview of this area is provided by Spall (2003).

The simplest strategy is a blind search method where one simulates in the region of interest until, e.g., a suitable approximation to the minimizer has been found. Spall (2003) shows that, under weak restrictions on f, the resulting sequence of approximate minimizers will converge to the global minimizer with probability one. The approach is summarized formally in Algorithm 8.3 that requires a choice for the maximum iteration count parameters $itMax$.

Algorithm 8.3 Blind random search

  $niter = 0$
  Generate $minTemp$ uniformly distributed on $[a, b]$
  while $niter < itMax$ **do**
     Generate U uniformly distributed on $[a, b]$
     if $f(U) < f(minTemp)$ **then**
        $minTemp = U$
     end if
     $niter = niter + 1$
  end while
  return $minTemp$

The `ranSearch` method below that was added to Listing 8.8 directly implements Algorithm 8.3 using a `ranGen` object from Appendix E.

```
double Optim::ranSearch(Functor& f, double a, double b,
         unsigned long seed) const {
  double minTemp, U;
  int niter = 0;
  ranGen RNG;
  RNG.setSeed(seed);
  double* pu = new double;
  RNG.ranUnif(1, pu);
  minTemp = a + (b - a)*(*pu);

 while(niter < itMax){
    RNG.ranUnif(1, pu);
    U = a + (b - a)*(*pu);
    if(f(U) < f(minTemp))
       minTemp = U;
    niter++;
  }
  delete pu;
  return minTemp;
}
```

Simulated annealing is a blind search variant that attempts to move more quickly toward the minimum using an accept-reject strategy. Values that decrease the objective function are always accepted; on the kth step a value θ that increases the objective function will be accepted with probability $p = \exp\left\{(f(\theta^{(k-1)}) - f(\theta))/T_k\right\}$, where $\theta^{(k-1)}$ is the approximate minimizer from the previous iteration and the T_k are a decreasing sequence of positive values called the *cooling schedule*. Since T_k is decreasing, values that increase the objective function will occur with decreasing probability while the accept-reject scheme allows for movement to escape local minima in the early stages of the search. Simulated annealing produces a sequence that will converge with probability one to the global minimizer assuming, e.g., continuity of the objective function (e.g., Bélisle 1992 and Spall 2003).

Simulated annealing relies on user choices for the cooling schedule, the maximum number of iterations ($itMax$) and a parameter d that governs the size of the step that can be taken away from the current approximation to the minimizer. A pseudo-code summary of the method is provided in Algorithm 8.4.

Algorithm 8.4 Simulated annealing algorithm

$niter = 0$, generate $minTemp$ uniformly on $[a, b]$
while $niter < itMax$ **do**
   Generate U uniformly on $[\max(minTemp - d, a), \min(minTemp + d, b)]$.
   if $f(U) < f(minTemp)$ **then**
     $minTemp = U$
   else
     $p = \exp\left\{(f(minTemp) - f(U))/T_{niter}\right\}$
     Generate V uniformly distributed on $[0, 1]$
     if $V < p$ **then**
       $minTemp = U$
     end if
   end if
   $niter = niter + 1$
end while
return $minTemp$

Our C++ implementation of simulated annealing for class `Optim` is given in the listing below.

```
double Optim::sAnn(Functor& f, double a, double b, double d,
        double rho, unsigned long seed) const {
  double minTemp, U, low, up, prob;
  int niter = 0;

  ranGen RNG;
  RNG.setSeed(seed);

  double* pu = new double;
  RNG.ranUnif(1, pu);

  //random search begins
  double T = 1.;
  while(niter < itMax){
    T *= rho;
    RNG.ranUnif(1, pu);
    low = std::max(minTemp - d, a);
    up = std::min(minTemp + d, b);
    U = low + (up - low)*(*pu);
    //always accept if the objective function is decreased
    if(f(U) < f(minTemp)){
      minTemp = U;
    }
    else{//the annealing step
      prob = exp((f(minTemp) - f(U))/T);
      RNG.ranUnif(1, pu);
      if(*pu < prob)
        minTemp = U;
    }
    niter++;
  }
  delete pu;
  return minTemp;
}
```

The geometric cooling schedule is used where $T_k = \rho T_{k-1}$ and $T_0 = 1$ for $0 < \rho < 1$. The value of ρ is yet another tuning parameter that is required as user input to the method. There are other popular cooling schedule options that include $T_k = 1/\log(k+1)$.

The `ranSearch` and `sAnn` methods were combined with our previous versions of `ranGolden` and `ranNewton` to obtain a new `Optim` class. This was compiled along with a driver program that contains the Figure 8.1 function to produce an executable called `opt`. The command line input in this case is, in order, the maximum number of iterations, the number of searches (for `ranGolden` and `ranNewton`), the random seed, the tolerance factor `delta`, the maximum allowed change in the approximate minimizer `d` for simulated annealing and the multiplier `rho` for the geometric cooling schedule. Some experimentation with the choice of these tuning parameters produced

```
$ ./opt 100 10 123 .00000001 .25 .99
The approximate minimizer from Newton is 0.93733736
The number of failures is 3
The approximate minimizer from Golden section is 0.93733735
The approximate minimizer from a blind search is 0.92800939
The approximate minimizer from simulated annealing is 0.59431479
```

```
$ ./opt 500 10 123 .00000001 .25 .99
The approximate minimizer from Newton is 0.93733736
The number of failures is 3
The approximate minimizer from Golden section is 0.93733735
The approximate minimizer from a blind search is 0.93680426
The approximate minimizer from simulated annealing is 0.93380905
$ ./opt 1000 10 123 .00000001 .25 .99
The approximate minimizer from Newton is 0.93733736
The number of failures is 3
The approximate minimizer from Golden section is 0.93733735
The approximate minimizer from a blind search is 0.93786242
The approximate minimizer from simulated annealing is 0.93715934
```

A "glass half full" view on this type of outcome would be to conclude that all four methods are capable of producing solutions that are in the neighborhood of the global minimizer. Further experience with this particular example suggests that simulated annealing is the most sensitive to the choice of its tuning parameters and, in that respect, the other three methods are more reliable. Part of this is undoubtedly due to the details of our particular implementation. However, in a general sense, it should be pointed out that simulated annealing is really more relevant for high-dimensional situations with possibly nonsmooth objective functions. In that context, blind search is not a viable option and the golden section concept quickly becomes intractable. Although Newton's method readily extends to the case of more than one variable, the objective function needs to be twice differentiable.

As noted in Section 8.4 the R optim function has a simulated annealing option that becomes available through the choice of method = "SANN". The cooling schedule that is used by optim is $T_k = \text{temp}/\ln((k-1) \bmod \text{tmax})\text{tmax} + e)$, where temp is the starting temperature for the cooling schedule (which defaults to 10) and tmax is the number of function evaluations that are allowed at each temperature (and also defaults to 10). The values of temp and tmax are set by the argument control for optim that must be given as a list that specifies the parameter values by name. Another general parameter for optim is maxit that sets the maximum number of iterations. It is also set using the control argument and defaults to 100 for the derivative-based methods and 10,000 for SANN.

To compare with our C++ code we should apply optim to the Figure 8.1 function. There is a bit of a problem in doing this because the SANN method does not allow for a specification of bounds for the search interval. Without such restrictions results such as

```
> set.seed(123)
> optim(par = .5, fn = func, method = "SANN")[1:2]
$par
[1] 11.52728

$value
[1] -11.51742
```

will be produced. Here optim was used to apply the simulated annealing method with a starting value of par = .5, func = function(theta){theta*sin(4*pi*theta^2)} and all other arguments left at their default values. As a result, optim attempts to search the whole real line and locates a point that is not in the desired optimization region.

In the case of something like simulated annealing that does not rely on continuity, the range problem is easy to fix: simply redefine func as

```
func <- function(theta){
  if(theta < 1 & theta > 0) f = theta*sin(4*pi*theta^2)
  else f = Inf
  f }
```

This version of `func` agrees with the original on $[0, 1]$ and evaluates at the R representation for positive infinity otherwise. Thus, values outside of $[0, 1]$ are effectively excluded from consideration. With this new version of `func` things go much more smoothly with `optim` returning

```
> set.seed(123)
> optim(par = .5, fn = func, method = "SANN")[1:2]
$par
[1]  0.9373296

$value
[1]  -0.9363776
```

The default parameter choices were used here and worked reasonably well in this instance. To see how sensitive the method is to the choices of `temp`, `tmax` and `maxit` a few other cases were considered giving results like

```
> set.seed(123)
> optim(par = .2, fn = func, method = "SANN",
+ control = list(maxit = 5, temp = 2, tmax = 1))[1]
$par
[1]  0.1829546

> set.seed(123)
> optim(par = .2, fn = func, method = "SANN",
+ control = list(maxit = 10, temp = 2, tmax = 1))[1]
$par
[1]  0.9542554

> set.seed(123)
> optim(par = .2, fn = func, method = "SANN",
+ control = list(maxit = 100, temp = 2))[1]
$par
[1]  0.9651814

> set.seed(123)
> optim(par = .2, fn = func, method = "SANN",
+ control = list(tmax = 250))[1]
$par
[1]  0.9373416
```

At least in the context of this example, the algorithm appears to behave robustly with respect to the choice of the tuning parameters. It placed us in the right area for locating the global minimizer even when the number of iterations, for example, has been set to be quite small.

8.7 Exercises

8.1. It is possible to have more than one template parameter. (See, e.g., Section 9.5.2 for an example.) In particular, syntax such as

```
template class<T1, T2, T3> class myClass
```

will produce a class with three template parameters. Using this fact, create a functor template class with the properties that

a) the class encapsulates a pointer to a member function from an arbitrary class with a single arbitrary argument and return type and

b) the class members are i) a pointer to an object from an arbitrary class and ii) a pointer to (any) one of its member functions.

Demonstrate that your functor class works as expected.

8.2. Let $f(\theta) = -\theta(1 - \theta)$ for $\theta \in [0, 1]$.

a) How many actual (not approximate) iterations does it take the golden section algorithm as implemented in the Optim class to reach the point where the approximate minimizer no longer changes its value.

b) Rework the golden section code that was applied to f so that 16 digits of accuracy can be achieved. [Hint: Evaluate f as a long double.]

8.3. Show that the golden section algorithm will produce a sequence of values that converge to the minimizer of f at a linear rate and determine the corresponding slope.

8.4. Write code that will allow you to replicate the C++ and R examples for the random interval golden section algorithm of Section 8.3.1.

8.5. Another minimization technique can be based on parabolic interpolation. Suppose that $f(\theta)$ is a continuous function on an interval $[a, b]$ and that $\theta_0 < \theta_1 < \theta_2$ are points in that interval such that $f(\theta_1) < f(\theta_0), f(\theta_2)$. Let $\tilde{f}(\theta) = c_0 + c_1\theta + c_2\theta^2$.

a) Find values for c_0, c_1, c_2 such that $\tilde{f}$ interpolates f at the θ_i: i.e., $\tilde{f}(\theta_i) = f(\theta_i), i = 0, 1, 2$.

b) Derive a formula for the minimizer of the quadratic function that was obtained in part a).

c) Construct an iterative algorithm based on your formula from part b) that can be used to find the minimizer of a convex function.

d) Implement the algorithm from part c) in C++ and add it to the Optim class from Section 8.3.

e) Apply your algorithm to $f(\theta) = -\theta(1 - \theta)$ with $a = 0$ and $b = 1$. How many steps does it take to produce the minimizer? Explain why this occurs.

f) Experiment with the code developed in part d) on the function $f(\theta) = \sin(2\pi\theta) \exp\{-\theta\}$ with $a = 1$ and $b = 2$. The minimizer is at $\theta_{min} = (2\pi)^{-1} \arctan(2\pi) + 1.5$.

g) Adapt the method created in part d) for use with a function having multiple minima as in Listing 8.9 and add the result to your Optim class.

Flowers (2006)

8.6. Create an R function that uses the optim function with the "L-BFGS-B" method option in conjunction with random placement of starting values to deal with functions having multiple minima.

8.7. Instead of the template functor class that was used in Section 8.3.2 consider using

```
class Functor{

  Linss* pS;
public:
  Functor(Linss* ps){pS = ps;}
  double operator() (lam) const {
    return (pS->gcv)(lam);
  }
};
```

which encapsulates the gcv method directly using a Linss object. Combine this functor class with the Optim class to obtain an alternative version of the Linss smooth method for minimization of the GCV criterion.

8.8. Create your own version of class `Linss` that reads data in from a file using your file input class from Exercise 3.33.

8.9. Bring the `Linss` class into R as a shared library and use the R `optimize` function (instead of class `Optim`) to find the smoothing parameter value that minimizes the GCV criterion.

8.10. Use your results from Exercise 8.9 to create a class in R that will hold linear smoothing spline objects with the smoothing parameters being either specified or selected by generalized cross validation. In addition,

a) Develop methods for `show`, `summary` and `print` that are applicable to linear smoothing spline objects.

b) Write accessor and replacement functions that will work with objects from the linear smoothing class.

c) Create a method function for `fitted` that can be used on linear smoothing spline objects.

d) Provide a method function for `plot` that will work with objects from the linear smoothing spline class.

8.11. Refer to Exercise 7.15.

a) Develop C++ code that will carry out the minimization of the GCV criterion for this estimator using a combination of grid search and golden section algorithms.

b) Apply your code from part a) to the data generated in part e) of Exercise 7.15 and compare the result with the non-boundary-corrected linear smoothing spline using its GCV optimal value for the smoothing parameter.

8.12. Bring the C++ boundary corrected linear smoothing spline class from Exercise 7.15 into R as a shared library and use the R `optimize` function (instead of class `Optim`) to find the smoothing parameter value that minimizes the GCV criterion.

8.13. Repeat Exercise 8.10 using the boundary corrected linear smoothing spline from Exercises 7.15, 8.11 and 8.12.

8.14. Let $f(\theta) = c_0 + c_1\theta + c_2\theta^2$ be convex and attain its minimum at θ_{min} in the interval (a, b). Show that for any starting point $\theta_0 \in (a, b)$, the Newton update will produce θ_{min}.

8.15. Let f be a function with three derivatives and define

$$\Delta f(\theta) = \frac{f(\theta + h) - f(\theta - h)}{2h}$$

$$\Delta^2 f(\theta) = \frac{f(\theta + h) - 2f(\theta) + f(\theta - h)}{h^2}$$

for $h > 0$.

a) Obtain expressions (in terms of h) for the errors in the approximation of f' and f'' by Δf and $\Delta^2 f$, respectively.

b) Use the formulas for Δf and $\Delta^2 f$ to create C++ code for numeric differentiation. Then, experiment with the functions from Figure 8.1 and part f) of Exercise 8.5, to assess the sensitivity of these approximations to the choice of h.

c) Use your results from parts a) and b) to create a version of the `Optim` class `newton` method that does not require the function's derivative. Experiment to determine how this new method fares relative to `newton`.

8.16. Expand the template `Optim` class used in Section 8.5 to include preliminary grid or random searches prior to use of the Newton algorithm.

8.17. Evaluate the performance of simulated annealing for minimization of $f(\theta) = \theta \sin(4\pi\theta^2)$ with the cooling schedule $T_k = 1/\log(k + 1), k = 1, \ldots$. Also, compare the geometric and logarithmic cooling schedules for minimization of the function in part f) of Exercise 8.5.

8.18. Use the simulated annealing option for the R `optim` function to minimize the GCV criterion for linear smoothing spline objects deriving from the class in Exercise 8.10. Compare its performance to the grid search/golden section scheme that was used in Section 8.3.2.

8.19. Use simulated annealing for minimization of the GCV criterion for the boundary-corrected linear smoothing spline objects that derive from the class created in Exercise 8.13.

8.20. Assume that f is a continuous convex function on a closed interval $[a, b]$ of the real line with $f(a)$ and $f(b)$ having opposite signs and that there is a single unknown root $x \in [a, b]$ for which $f(x) = 0$. The modified version of Algorithm 5.1 below can be used to locate the root. The values of δ and the maximum number of iterations $itMax$ are provided by user input.

Algorithm 8.5 Bisection algorithm

$low = a, up = temp = b$
while $up - low > \delta$ and $niter < itMax$ **do**
  $temp = (up + low)/2$
  if $f(low)f(temp) < 0$ **then**
    $up = temp$
  else
    $low = temp$
  end if
  $niter = niter + 1$
end while
return $temp$

---

a) Prove that this algorithm produces a sequence of approximations to the true root that converges linearly to the solution and give an expression for the corresponding slope.

b) If $a = 0, b = 1$ and $\delta = 10^{-8}$, show that the loop in Algorithm 8.5 will necessarily terminate after about 27 iterations in double precision.

c) Implement Algorithm 8.5 in C++ with the function f passed into the method using a functor.

d) Provide an alternative version of your answer to part c) that uses recursion in the sense of Exercises 3.7–3.10.

e) Compare your answers for parts c) and d) in terms of computation times.

8.21. Develop analogs of Algorithm 8.5 that derive from the golden section and Newton optimization methods and implement them in C++. Combine these methods with your code from Exercise 8.20 to form a root-finder class.

8.22. Develop an alternative version of Algorithm 8.5 that can be used to find the minimum of a convex function on an interval $[a, b]$. Implement the algorithm in C++ and add it to the `Optim` class from Section 8.3.

8.23. Modify the method created in Exercise 8.22 to work on functions having multiple minima as was done for the golden section algorithm in Listing 8.9 and add it to the `Optim` class.

8.24. Compare the performances of `ranGolden`, `ranNewton`, the random bisection search of Exercise 8.23 and the random parabolic extrapolation search of Exercise 8.5 for finding the global minimizer of the function from Figure 8.1. Your comparisons should be in terms of running times (measured via the `clock` function discussed in Section 3.8) and in terms of the total number of iterations.

8.25. Repeat Exercise 8.24 except with all the functions being written in R. For timing purposes you can use the `system.time` function discussed in Section 5.4.

8.26. Describe how golden section or bisection algorithms can be used to numerically evaluate the quantile function for a continuous distribution. Implement this approach as a quantile function class in C++ and evaluate its performance in the case of the exponential distribution (where the quantile function is known and given in Section 4.6) and for the normal distribution. There is no closed form for the normal cumulative distribution function and it must be evaluated numerically. The function `pnorm` from the R Standalone Math Library or the R API performs this approximation and can be used for this purpose in your code.

8.27. Repeat the code development part of Exercise 8.26 except with the `optimize` function in R.

8.28. The R function `uniroot` can be used to find the zeros of a function. Write an R function that uses `uniroot` to find a specified quantile of a cumulative distribution function. Assess the performance of your code with the exponential and normal distributions and compare it to the bisection and golden section algorithms of Exercises 8.26–8.27.

8.29. Develop a version of the C++ `Like` class in R. Then, construct a new method function for `mle` that can be used on objects from your R class. [Note: The `stats4` package contains a likelihood class `mle-class`. So, this name should be avoided.]

8.30. Create a set of code that will allow you to replicate the C++ and R maximum likelihood examples for the exponential distribution in Section 8.5.

8.31. A standard result concerning Fisher information is that (under regularity conditions) when θ^* is the true parameter value

$$\mathrm{E}\left[\left(\frac{\partial f(X;\theta^*)}{\partial \theta}\right)^2\right] = -\mathrm{E}\left[\frac{\partial^2 f(X;\theta^*)}{\partial^2 \theta}\right].$$

This suggests an alternative strategy wherein the second derivative of the likelihood function is replaced with

$$-\sum_{i=1}^{n}\left(\frac{\partial f(x_{iO};\theta)}{\partial \theta}\right)^2.$$

This approach is called Fisher scoring and has the advantage of circumventing the need for evaluation of a second derivative. Modify the `Like` class to incorporate a method that uses Fisher scoring and compare its performance to the `mle` method.

Chapter 9

Abstract data structures

9.1 Introduction

The developments in Chapters 5–8 focused on methods for computing various quantities of statistical interest. Very little thought was given to the data that was used in the illustrative examples beyond it being appropriate for use with the method and in a form that was readily accessible to the program or function that used it in calculations. In practice, neither of these properties are likely to obtain for any data analysis problem of substance. Instead, the data will need to be manipulated (e.g., ordered, subsetted, etc.) in order to carry out the desired analyses. Thus, the development of effective statistical computing methods must begin at the point of access to the data with the creation of algorithms to perform the requisite manipulations. The form of such algorithms is dictated by the way the data is stored. As a result, the design of data storage methods can be viewed as the statistical computing analog of experimental design; experiments must be run in a way that makes the factors of interest estimable and, similarly, data must be stored in a way that makes it possible to perform the proper analysis.

Until now our primary vehicle for storing data has been an array. Although arrays work quite well for many applications, there are situations that arise where they become unwieldy to the point of being impractical. For example, they are not ideal for dealing with dynamic data sets that must grow or shrink in a manner that requires frequent resizing of the storage container. Similarly, insertion of new elements or deletion of current ones from an array will require shifting of the other elements which can be prohibitively time consuming when such operations are commonplace. Fortunately, there are many alternatives to array storage that provide great flexibility in tailoring a particular storage method to work best with the data at hand.

In the next two chapters we explore the concept of data storage from the perspective of *abstract data structures* or *abstract data types* (ADTs, subsequently). In a general sense, an ADT represents a model for storing and organizing data in a way that simplifies the computations that are of interest and, in some cases, makes the desired computations possible. They generally come equipped with operations that let us efficiently examine, modify (e.g., insert or delete) and carry out computations (e.g., sorting) with the stored data. What distinguishes one ADT from another is not only the storage space it requires to hold the data (often referred to as *space complexity*), but also the associated efficiency, in terms of execution time, of the operations it provides under its particular storage paradigm (often referred to as *time complexity*). Although our discussions will pay some attention to the space and computation time issues, a more satisfactory treatment of computational complexity goes beyond what can be presented here. Detailed studies of this area are provided, for example, in Papadimitriou (1994) and Arora and Barak (2009).

The collection of all ADTs represents a data storage toolbox of sorts. In that respect the proper tool should be selected for the storage job at hand. For example, if one has data that needs to be maintained in a particular order while easily allowing element insertion, the *linked list* structure of Section 9.2.2 serves as a good storage option. For this type of ADT insertion and deletion operations involve a fixed time expense that is independent of the number of elements in the data set. Searching for specific elements is another matter and this type of operation requires an effort that grows linearly with the data size. For situations where searches are an important part of the data management process, the *chaining hash table* of Section 9.2.4 may provide a better option wherein searching and insertion/deletion tasks can both be carried out in nearly constant time. Comparisons of this nature extend across all ADTs and, as a result, selection of the proper "tool" from the ADT "toolbox" requires a familiarity with both the available types of ADTs as well as their respective strengths and weaknesses. This and the next chapter are aimed at initializing this familiarization process.

The present chapter gives a high level overview of ADTs. Our perspective here is much the same as for creating a class declaration in C++. With arrays as a case in point, the listing below is the declaration for a pared down version of our `Vector` class from Chapter 3 and Appendix D.

```
class Vector{
   double* pA;
   int nRows;

 public:

   Vector(int nrows, double const* pa);
   Vector(const Vector& v);
   ~Vector();
   Vector& operator=(const Vector& v);
   double operator[](int i) const {return pA[i];}
};
```

This declaration can be viewed as a contract that guarantees i) `Vector` objects can be created and copied, ii) a `Vector` object may be assigned to another `Vector` object and iii) (random) access (using `[]`) will be provided to the elements of the underlying array of doubles that is encapsulated by a `Vector` object.

There is nothing special about using doubles for array elements and, as in Exercise 3.31, a template approach can be used to allow them to be of type `int` or `char`, for example. Taking this one step further leads to `Vector` objects that serve as *containers* in the sense of being able to hold arbitrary, user-defined data types. But, even in this generic form the operations laid out in the `Vector` class declaration make sense and represent at least the minimum one would want from something that provides a generalization of a numeric array; i.e., a generic `Vector` container class should at least include functions that

a) create a new container object,

b) create a copy of an existing `Vector` object,

c) assign one `Vector` object to another and

d) allow (random) access to the objects in the container.

This list of operations represents a public interface for a `Vector` class in the same sense as does a class declaration.

Our treatment of ADTs in this chapter will be geared toward describing their public interfaces through abstract, somewhat axiomatic "statements" that indicate the way that data will be stored and the operations that will be expected from the structure. This will often be supplemented by *pseudo-code* implementations of the operations. In most cases, specific examples of simple C++ implementations will be provided that illustrate the concept with a (relative) minimum of technical clutter. The R language has certain ADT capabilities that will also be explored here.

Much of the code development in this chapter illustrates how to build ADT implementations from the ground up. The resulting structures are simplistic in nature and of conceptual, rather than practical, utility. Instead, there are excellent packages of existing code that can be used for applications that require ADTs and these will be the preferred option for most users. For C++ programmers, the first resort should be the Standard Template Library (STL) which provides implementations of basic data structures that include queues, priority queues, stacks and dictionaries. A much more comprehensive, yet nevertheless free, collection of C++ libraries is provided by BOOST. In Chapter 10, we examine some of the basic features of the STL.

9.2 ADT dictionary

The ADT called *dictionary* is a general data storage framework whose implementation leads to several useful data structures. As its name suggests, a dictionary ADT provides a mechanism for data storage that is modeled after the way a literary dictionary holds words; it is intended to contain a set of objects each of which has a unique name. The name is referred to as the *key* and represents the means by which a particular object can be distinguished from other objects in the collection. Each object will generally also have a data component that corresponds to the information that is to be stored. The lexicon analogy becomes complete by taking the key to correspond to a word and the data to correspond to information on the word's pronunciation, definition and usage.

The desired operations for an abstract dictionary implementation mimic those for the physical, linguistic variety. They are

- *makeDictionary*() which creates a new empty dictionary,

- *lookUp*(k) that returns the object with key k and

- *insert*(obj, k) whose purpose is to insert the object *obj* with key k into the dictionary.

It will be seen that various implementation of a dictionary will allow some of these operations to be performed in constant time (on average), while using space proportional to the number of objects that need to be stored.

There exist a number of ways to implement ADT dictionary. If the keys in a dictionary are a fixed set of integers or naturally associated with integer values, it may be possible to store the dictionary in a simple array. For example, suppose each key is an integer in $\{1, \ldots, n\}$. Then, all the dictionary objects can be placed in the array A with elements $A[1], \ldots, A[n]$. If the number of elements to be stored is not known a priori or will be subject to change, the array that holds the data will need to expand or shrink adaptively. This leads to the *dynamic array* ADT that represents the subject of the next section.

9.2.1 Dynamic arrays and quicksort

For this section it will be assumed that integer values are to be used for the dictionary keys. In that case one possible implementation of a dictionary is through the use of a dynamic array. This data structure stores objects of arbitrary type in a way that mimics how an ordinary vector stores numeric values. Unlike vectors of fixed length, the size of the array is allowed to grow (or even shrink) depending on the space/memory requirements for storing its content.

To illustrate the idea, the listing below provides the class declaration for a dynamic array class dArray in C++. To simplify the presentation, the class will produce objects that can hold data of type int rather than generic forms. This is less restrictive than it might seem because it is relatively simple to reformulate this particular class as a template to accommodate general data types (Exercise 9.37).

Listing 9.1 dArray.h

```
#ifndef DARRAY_H
#define DARRAY_H

class dArray{
  int* pA;
  int Size, containerSize;
  void pointerCheck(int* pa) const;
  void reSize();
  void swap(int i, int j);

 public:
  dArray() : pA(0), Size(0), containerSize(0){}
  dArray(int size, const int* pa);
  dArray(int size);

  dArray(const dArray& v);
  ~dArray();

  dArray& operator=(const dArray& v);
  int operator[](int k) const;

  void quickSort(int left, int right);
  void append(int value);
  void insert(int k, int value);
  int getSize() const {return Size;}
  int getContainerSize() const {return containerSize;}
};
#endif
```

The dArray class has three member elements: a pointer to int, pA, that holds the address of the beginning of a block of memory where the data in the dArray object is stored, a variable Size that gives the number of int objects that are currently held in storage and containerSize which represents the total number of storage locations. An overloaded [] operator and two accessor functions provide access to these class members.

The constructors for class dArray needs to allocate sufficient memory to hold the initial input data as well as allow some room for additional data that may arrive after the array's creation. There are various ways to accomplish this. A simple option might look something like

```
dArray::dArray(int size, const int* pa){
  Size = size;
  containerSize = 2*Size;
  pA = new int[containerSize];
  pointerCheck(pA);
  for(int i = 0; i < Size; i++)
    pA[i] = pa[i];
}
```

The input to this constructor is a pointer to int and an integer value size that gives the number of int objects that are stored (beginning) at the address held in the pointer. The value of size is used to initialize Size and pA after which the information in pa is transferred to the pA class member. Although only Size memory locations are needed to store the input data, memory is allocated for twice that many, or containerSize, objects and the data is copied into just the first Size of these locations. Prior to copying the success of the memory allocation step for pA is checked using a variant of the pointerCheck function seen previously in Section 3.9. Thus, the resulting dArray object will hold all the input data while being able to accommodate an additional Size objects as needed.

Another possibility is that one wants an "empty" array object that will then be filled by adding new elements. The constructor provided for this purpose is

```
dArray::dArray(int size) : Size(size){
  containerSize = 2*Size;
  pA = new int[containerSize];
  pointerCheck(pA);
}
```

The only parameter that must be specified in this instance is Size. Its value is taken as input and assigned via an initializer list (cf. Section 3.8.3). The memory is then obtained as in the previous constructor. A default constructor has also been given that uses an initializer list to set the Size and containerSize members to 0 and define pA as a null pointer.

If data is being continually added to a dynamic array at some juncture the array will become full and require expansion. The reSize function in Listing 9.1 provides the means to grow dArray objects. Its definition is

```
void dArray::reSize(){
  std::cout << "Resize was called" << std::endl;
  containerSize *= 2;
  int* temp = new int[containerSize];
  pointerCheck(temp);
  for(int i = 0; i < Size; i++)
    temp[i] = pA[i];

  delete[] pA;
  pA = temp;
}
```

As in the class constructor, the current array size containerSize is doubled by reSize. This is accomplished in four steps. First, the necessary memory is allocated. Then, the data is copied into the new memory locations, the current memory is released and the member pointer pA is reassigned to point to the new memory. The copying aspect of resizing is time consuming in that it requires an operation count that is proportional to Size. Thus, the frequency of resizing should be minimized. Our strategy for accomplishing this is to expand to twice the current capacity whenever growth is needed, thereby attempting to strike a balance between wasting memory and excessive copying. Note that the output statement that has been included in the reSize method is only for illustrative purposes in an example.

The dArray class involves dynamic memory allocation and therefore needs explicit definitions for i) a copy constructor, ii) a destructor and iii) the = operator. The prototypes for these methods are shown in Listing 9.1. Creation of appropriate definitions is the subject of Exercise 9.1.

The methods for component access and insertion in Listing 9.1 highlight the strengths and weaknesses of dynamic arrays. The overloaded [] operator gives the analog of the dictionary lookUp function. Its definition is

```
int dArray::operator[](int k) const {
  if(k > Size){
    std::cout << "This is not a valid index!" << std::endl;
    exit(1);
  }
  return pA[k];
}
```

Thus, (random) access to the contents of a known memory location can be obtained in constant time. The same is true of appending a new entry to the end of a dArray object, except for the case when the object is already full. Specifically, our definition for the append method is

```
void dArray::append(int value){
  if(Size < containerSize){
    pA[Size] = value;
    Size++;
  }
  else{
    reSize();
    pA[Size] = value;
    Size++;
  }
}
```

The location of the last object in the array is always available through the value of the member variable Size; i.e., it has index Size - 1. Thus, provided an open memory location exists, the append operation requires constant time. Otherwise, the reSize function must be called before the new int object can be stored.

While arrays are good for accessing and appending new elements, the insertion operation can be quite slow depending on where a new data object must be placed. Algorithm 9.1 describes the process for inserting a new entry x into the kth slot of an n-element array with existing entries $A[1], \ldots, A[m]$ for $m < n$.

Algorithm 9.1 $insert(k, a)$: Insert a new array entry

$tempLow = A[k]$
$A[k] = a$
for $j = k$ to $m - 1$ **do**
    $tempUp = A[j + 1]$
    $A[j + 1] = tempLow$
    $tempLow = tempUp$
end for
$A[m + 1] = tempLow$

The time needed to accomplish insertion is therefore dependent on the location where insertion is to take place, with the average time complexity being proportional to the number of elements in storage.

Our insert method for class dArray below uses a variation of Algorithm 9.1.

```
void dArray::insert(int k, int value){
  if(k > Size){
    std::cout << "This is not a valid index!" << std::endl;
    exit(1);
  }

  if(Size == containerSize)
    reSize();
  Size++;
  for(int j = Size; j > k; j--)
    pA[j] = pA[j - 1];
  pA[k] = value;
}
```

The first modification of Algorithm 9.1 is that `reSize` is called when the array becomes too full to accommodate another stored entry. The other is the way the elements of the array are shifted. First, the value of `Size` is augmented. This allows us to start at the slot with index `Size - 1` and work our way backward while shifting each entry forward by one array slot. An advantage of this approach is that it avoids the need for temporary storage.

The other side of insertion is deletion. It may be of interest to remove elements from a `dArray` object and, if the `Size` variable becomes sufficiently small after a number of deletions, resizing may be of value as a means of conserving memory. Implementation of these operations is the subject of Exercises 9.3–9.4.

The code below provides a test for our `dArray` class.

```
//dArrayDriver.cpp
#include <iostream>
#include "dArray.h"

int main(){
  int* pa;
  pa = new int[2];
  pa[0] = 2; pa[1] = 4;

  dArray v(2, pa);
  v.append(7); v.append(14);
  std::cout << v.getSize() << " " << v.getContainerSize()
            << std::endl;
  v.insert(1, 9);
  std::cout << v.getSize() << " " << v.getContainerSize()
            << std::endl;
  for(int i = 0; i < v.getSize(); i++)
    std::cout << v[i] << " ";
  std::cout << std::endl;

  return 0;
}
```

A `dArray` object is initialized with a pointer to `int` that corresponds to two integer objects. Thus, the initial `Size` value should be 2 and the initial value of `containerSize` should be 4. An additional pair of integers are appended to the array and the values of `Size` and `containerSize` are checked. An `int` is then inserted to force a call to the `reSize` method, after which `Size` and `containerSize` are checked again and the contents of the `dArray` object are printed to standard output. The program produces

```
4 4
Resize was called
5 8
2 9 4 7 14
```

It is often necessary to sort the elements in a numeric array to obtain, e.g., statistics of interest. We have seen examples of a sorting method in the form of the bubblesort Algorithm 3.1 and the selection sort method of Section 6.4. These sorting schemes are slow and require an order of n^2 operations on the average to sort an n-element array. A better sorting option is the *quicksort* algorithm that requires an average of only $O(n \log n)$ operations to order the array elements.

The idea behind quicksort is simple: recursively split the unsorted array into two parts that can be sorted independently. The easiest way to achieve this is to ensure that each value in the first part is no greater than any in the second part. This is perhaps best illustrated with a specific example. For that purpose consider the following unordered collection of integers.

$$15 \quad 1 \quad 18 \quad 28 \quad 27 \quad 19 \quad 49 \quad 38 \quad 6 \quad 2 \quad 40$$

The numbers at the right and left ends of this sequence will be called *left* and *right*, respectively. The first step in the sorting operation is to pick a value for a pivot that will be used to split the numbers into two parts. One possibility is to choose the value in the "middle" location or, in this case, the value in slot 6 of the array. This gives a pivot value of 19. Two new indices, *low* and *high*, are now introduced and initialized to *left* and *right*, respectively. Thus, at this point we have our array represented as

$$\underbrace{15}_{left=low} \quad 1 \quad 18 \quad 28 \quad 27 \quad 19 \quad 49 \quad 38 \quad 6 \quad 2 \quad \underbrace{40}_{right=high}$$

with the pivot value at 19.

The next step is to increment the *low* index sequentially until it corresponds to a value that is greater than the pivot. Similarly, the *high* index is decremented until it reaches an array entry that is less than the pivot. This results in

$$\underbrace{15}_{left} \quad 1 \quad 18 \quad \underbrace{28}_{low} \quad 27 \quad 19 \quad 49 \quad 38 \quad 6 \quad \underbrace{2}_{high} \quad \underbrace{40}_{right}$$

The array element indexed by *low* and the one indexed by *high* are swapped, *low* is incremented and *high* is decremented. If necessary, the process of incrementing and decrementing *low* and *high* would be repeated again until they reached elements that were larger and smaller than the pivot element, respectively. The initial increment/decrement is sufficient in this instance as it produced

$$\underbrace{15}_{left} \quad 1 \quad 18 \quad 2 \quad \underbrace{27}_{low} \quad 19 \quad 49 \quad 38 \quad \underbrace{6}_{high} \quad 28 \quad \underbrace{40}_{right}.$$

The *low* and *high* index elements are swapped again and the indices are incremented and decremented to give

$$\underbrace{15}_{left} \quad 1 \quad 18 \quad 2 \quad 6 \quad \underbrace{19}_{low} \quad \underbrace{49}_{high} \quad 38 \quad 27 \quad 28 \quad \underbrace{40}_{right}$$

Since *high* still exceeds the pivot it is decremented again at which point *low* and *high* coincide. A trivial swap of their two (identical) values is made and the indices are increased and decreased a final time. At this point they have crossed and the index point *low* (i.e., the seventh element of the array) has the property that i) all elements with indices smaller than *low* have corresponding values that are smaller than or equal to the pivot and ii) elements with indices of *low* or higher have values that are greater than or equal to the pivot. Thus, we can now independently apply the same process to each of the two subarrays

$$\underbrace{15}_{left=low} \quad 1 \quad 18 \quad 2 \quad 6 \quad \underbrace{19}_{right=high}$$

and

$$\underbrace{49}_{left=low} \quad 38 \quad 27 \quad 28 \quad \underbrace{40}_{right=high}.$$

If we continue to split arrays in this fashion, the end result will be eleven arrays with one entry each. All this work is done in place; i.e., the subarrays are not physically formed but rather swapping occurs inside the original parent array. The final outcome of the recursion will be an array with values arranged in numerically ascending order.

With our example as a guideline, Algorithm 9.2 provides a formulation of the quicksort procedure.

Algorithm 9.2 $quickSort(A, left, right)$: Sort array elements $A[left], \ldots, A[right]$

$low = left, pivotIndex = \lfloor (left + right)/2 \rfloor, pivot = A[pivotIndex], high = right$
while $low \leq high$ **do**
   while $pivot > A[low]$ **do**
      $low = low + 1$
   end while
   while $A[high] > pivot$ **do**
      $high = high - 1$
   end while
   if $low \leq high$ **then**
      $temp = A[low], A[low] = A[high], A[high] = temp$
      $low = low + 1, high = high - 1$
   end if
end while
if $low < right$ **then**
   $quickSort(A, low, right)$
end if
if $left < low - 1$ **then**
   $quickSort(A, left, low - 1)$
end if

The algorithm starts with low and $high$ as the left-most and right-most array indices, respectively, and takes the pivot to be the value of the entry with a "middle" array index. It then moves the low index across the array until it encounters an entry whose value is greater than or equal to the pivot succeeded by a decrease of the $high$ index until a value smaller than or equal to the pivot is located. If low is less than or equal to $high$ the array elements are swapped and low and $high$ are incremented and decremented. Otherwise, the subarray composed of those array elements with indices of low or larger will have entries whose values are all greater than or equal to the pivot. A recursive call can then be made to quickSort using this subarray as the array argument. Similarly, the values in the subarray constructed from elements with indices less than low will all be smaller than or equal to the pivot and the quickSort function can be applied to them as well.

The average running time for quicksort is proportional to $n \log n$ for an n-element array. To gain some intuition regarding why this is so, assume that $n = 2^k$ for some integer k and that the pivot index falls at the center of the array in each step of the recursion. Then, the first partitioning step will require order n operations to produce two arrays of size 2^{k-1}. These two arrays will, in turn, each need $O(2^{k-1})$ flops to produce two subarrays giving a total of four arrays of size 2^{k-2} which, again, will require a total of order n operations to advance to the next stage of the recursion. This remains true at every step in the recursion which means that the overall effort is on the order of $nk = n \log_2(n)$. It may be proved that, assuming the initial order of the elements in the array is random (regardless of what the elements' values really are), the expected number of comparisons in quicksort is $O(n \log n)$ (see Exercise 9.6 or Cormen, et al. 2003).

On the other hand, the worst case in terms of running time for quicksort occurs when the partitioning always produces a subarray of size one. If this happens, either the low or $high$ index will remain at $left$ or $right$, respectively. For an array of size n this gives rise to $n + 1$ comparisons: one comparison between the stationary index (either $left$ or $right$) and the pivot and the n comparisons that occur as the other index works its way down or up the array. Applying this to the subarrays that are produced by the sequential partitioning produces an effort on the order of $\sum_{i=1}^{n+1} i = (n + 2)(n + 1)/2$ operations; i.e., the method requires $O(n^2)$ operations. Note that this is far from an impossible event; it is exactly what happens if the pivot is selected from the beginning (or the end) of a sorted array. Somewhat paradoxically, we see that the hard cases for quicksort are those where the array elements are already ordered or partially ordered.

The methods swap and quickSort provide an implementation of the quicksort algorithm for our dArray class. The swap method is a utility that is used by the quickSort method. Thus, it was made a private member of the dArray class while quickSort was made public. The int variables in swap designate the elements that are to be swapped in the dArray object while the left and right arguments for quickSort specify the index subrange on which the algorithm will be applied.

The definition of the swap method is

```
void dArray::swap(int i, int j){
  int temp = pA[i]; pA[i] = pA[j]; pA[j] = temp;
}
```

Temporary storage is used to hold the value of one array element while it is overwritten after which the remaining component is replaced with the stored value. The implementation of the quicksort algorithm then takes the form

```
void dArray::quickSort(int left, int right){

  int pivotIndex = floor((left + right)/2);
  int pivot = pA[pivotIndex];
  int low = left, high = right;
  while(high >= low){
    while(pivot > pA[low])
      low++;
    while(pA[high] > pivot)
      high--;
    if(low <= high){
      swap(low, high);
      low++; high--;
    }
  }
  if(left < low - 1)
    quickSort(left, low - 1);
  if(low < right)
    quickSort(low, right);
}
```

The following main function was written to test our quickSort method.

```
int main(){
  int* pa = new int[11];
  pa[0] = 15; pa[1] = 1; pa[2] = 18; pa[3] = 28; pa[4] = 27;
  pa[5] = 19; pa[6] = 49; pa[7] = 38; pa[8] = 6; pa[9] = 2;
  pa[10] = 40;

  dArray dArrayObj(11, pa);
  dArrayObj.quickSort(0, 10);
  for(int i = 0; i < 10; i++)
    std::cout << dArrayObj[i] << " ";
  std::cout << std::endl;

  return 0;
}
```

The unordered set of integers used to illustrate the quicksort algorithm is placed into memory corresponding to a pointer to `int` that is then used to create a `dArray` object. This object calls its `quickSort` method and its contents are printed. This resulting output is

```
1  2  6  15  18  19  27  28  38  40  49
```

9.2.2 Linked lists and mergesort

Linked lists provide another approach to implementation of a dictionary. They arise, for example, in applications such as the storage of sparse matrices and polynomial arithmetic (Exercises 10.26–10.27). Like dynamic arrays, they also provide a foundation from which other useful data structures such as hash tables, queues and stacks can be built.

As we saw in the previous section, insertion of new elements or resizing operations are computationally expensive for array type structures. Linked lists provide one means of overcoming such problems. For these data types resizing is actually unnecessary, while insertion can be performed in constant time. As with most compromises, there is a downside which, in this case, comes in the form of slow performance for access of list elements.

A linked list data structure consists of a collection of objects that have both key and data components, as well as a member *next* which designates (e.g., by a pointer) the next element in the list. The components of the list are generally called *nodes* with the first node having the name *head* and the last node referred to as the *tail*. The *next* value of the tail of a linked list is the *NULL* value (e.g., a null pointer) indicating it has no successor. Lists of this particular type are generally called *singly linked* because the connection between nodes only goes in one direction; i.e., it is only possible to move toward the tail of the list by "stepping" through the *next* members of the node objects in the list. A simple extension is the doubly linked list, where the nodes have an additional member that points to the preceding list entry thereby making both forward and reverse movements possible.

Let us now take a look at the operations required to implement a linked list. A linked list can be built by "pushing" new nodes onto the front of the list as in

---

Algorithm 9.3 $push(a)$: Add a new node to a linked list

$a.next = head$
$head = a$

Here and subsequently the . notation represents a generic member selection operators; e.g., $a.next$ is the "value" of *next* for the a object. So, the value of *next* for the new node a is assigned the address of the current head node. Then, a becomes the head node and the list has grown by one object. Note that no matter how many nodes are currently in the list only two simple steps are needed to add a node. The operation can therefore be accomplished in constant time.

The look-up operation for a linked list is described in Algorithm 9.4.

Algorithm 9.4 $lookUp(key)$: Look-up in a linked list

$a := head$
while $a \neq NULL$ **do**
  if $a.key = key$ **then**
    return a
  end if
  $a := a.next$
end while
return $NULL$

The implementation is quite simple in that one iteratively steps through the list by moving from each node to its *next* member until either the end of the list is encountered or the designated key value is found. The algorithm returns the node that has the desired key value or $NULL$ if no node exists with that key. Note that potentially every node in the list must be examined to find the one with the targeted key value. Thus, this operation can require order n operations in an n-element list.

To insert a node to the right of (i.e., immediately after) a node with a given key value, we first look up the key to find where the new node should be placed. This also allows us to check if an element with the same key is already present which should trigger an error report. The idea is implemented in Algorithm 9.5.

Algorithm 9.5 $insert(a, key)$: Find the location and insert a node into a linked list

---

   if $lookUp(a.key) == NULL$ **then**
       $b := lookUp(key)$
       $a.next = b.next$
       $b.next = a$
   else
       return Error: duplicate key!
   end if

There is nothing special about insertion to the right. In fact, insertion to the left or between two specified keys may be of more interest in some applications (Exercise 9.12).

As was true for dynamic arrays, it is also of interest to be able to sort the elements in a list. The method of choice in this case is known as *mergesort*. Like quicksort, it uses a divide-and-conquer strategy although the details differ in how that is accomplished.

To illustrate the mergesort method, let us again consider sorting the integer array

$$15 \quad 1 \quad 18 \quad 28 \quad 27 \quad 19 \quad 49 \quad 38 \quad 6 \quad 2 \quad 40$$

that was used in Section 9.2.1. To begin, divide this array into the subarrays $(15, 1, 18, 28, 27)$ and $(19, 49, 38, 6, 2, 40)$. Let us focus on the second array which we split into two more three component arrays $(19, 49, 38)$ and $(6, 2, 40)$. Again, working with the second subarray, we arrive at a singleton 6 and $(2, 40)$ the latter of which is already sorted. The next step is to merge these last two "arrays". To accomplish this the larger of the two array elements, 40, is retained to start the new merged array. The pair $(6, 2)$ is now sorted and appended to 40 to give us $(2, 6, 40)$. By exactly the same process $(19, 49, 38)$ is ordered as $(19, 38, 49)$ and this must now be merged with the $(2, 6, 40)$ array. But, the basic idea remains the same; first, the largest value from the two arrays, 49, is retained to start the new array and the problem reduces to merging $(2, 6, 40)$ and $(19, 38)$. For this latter purpose we retain 40 and then must merge $(2, 6)$ and $(19, 38)$. Now 38 is retained and $(2, 6)$ is merged with 19. Working our way backward, $(2, 6, 19)$ becomes $(2, 6, 19, 38)$ that leads to $(2, 6, 19, 38, 40)$ and, finally, to $(2, 6, 9, 38, 40, 49)$. The algorithmic scheme is now hopefully clear and it should be apparent that taking this process to its logical conclusion will return a fully ordered array.

Our example illustrates that the real work in this sorting method is done by the recursive merging procedure. For the case of two arrays, $A1$ and $A2$, this can be described as in Algorithm 9.6.

Algorithm 9.6 $merge(A1, A2)$: Merge $A1$ and $A2$ of size n and m into A of size $n + m$

---

  if $A1[n] > A2[m]$ **then**
   $A[n + m] = A1[n]$
   $merge((A1[1], \ldots, A1[n - 1]), A2)$
  else
   $A[n + m] = A2[m]$
   $merge(A1, (A2[1], \ldots, A2[m - 1]))$
  end if
  return A

With the merge method in place, the mergesort algorithm for an array takes the form

---

Algorithm 9.7 $mergeSort(B)$: Sort an n-element array

  $mid = \lfloor n/2 \rfloor$
  if $mid = 0$ **then**
   return B
  end if
  $A1 = mergeSort((B[1], \ldots, B[mid]))$
  $A2 = mergeSort((B[mid + 1], \ldots, B[n]))$
  $A = merge(A1, A2)$
  return A

Mergesort is an order $n \log n$ method (Exercise 9.9). Its worst-case efficiency is also $O(n \log n)$ as compared to n^2 for quicksort. The disadvantage of mergesort is the need for temporary storage when sorting arrays. Linked lists are another story. In that case the use of pointers makes it possible to dodge the extra storage requirement as we will eventually illustrate.

It will now be useful to look at a specific example of linked list encoding. For this purpose we will consider a simple linked list struct in C++. For specificity, the objects in the list will also be structs of the form

```
struct Node{
  int dataValue;
  int keyValue;
  Node* Next;

  Node (){}
  Node (int data, int key, Node* next = 0)
  : dataValue(data), keyValue(key), Next(next) {}
};
```

The data member (`dataValue`) and key (`keyValue`) of a `Node` object are both integers and the `Next` member is a pointer to another `Node` object. There are two constructors: a default constructor and another that uses an initializer list to specify all three class members with `Next` being given a default value of the null pointer or 0.

The declaration of our linked list `struct` for `Node` objects is provided by the next listing.

Listing 9.2 *linkedList.h*

```
struct linkedList{
  Node* Head;
  int Size;
  void mergeSortList(Node** pHead);
```

```
Node* merge(Node* a, Node* b);
void split(Node* head, Node** midNode);

linkedList(){Head = 0;}
linkedList(const linkedList& L);
~linkedList();
linkedList& operator=(const linkedList& L);

void push(int newData, int newKey);
Node* lookUp(int key) const;
void insert(int newData, int newKey, int key);
void sort(){mergeSortList(&Head);}
};
```

The struct has two data members: a pointer to the head Node object and the integer Size that represents the number of nodes in the list. A constructor and copy constructor have been given; the constructor simply creates an "empty" linkedList object with a null pointer for its head node. There are three methods that correspond to the *push, lookUp* and *insert* Algorithms 9.3–9.5. There are also four methods that, in combination, will sort the nodes in a list via mergesort. The sort method is a wrapper function for mergesort with mergeSortList and merge corresponding to Algorithms 9.7 and 9.6, respectively.

The copy constructor was encoded as

```
linkedList::linkedList(const linkedList& L){

  if(L.Size == 0){
    Head = 0;
    Size = 0;
  }
  else{
    Size = L.Size;
    Node* temp = 0;
    for(Node* iter = L.Head; iter !=0; iter = iter->Next){
      if(temp == 0){//empty list
        Head = new Node(iter->dataValue, iter->keyValue);
        temp = Head;
      }
      else{
        temp->Next = new Node(iter->dataValue, iter->keyValue);
        temp = temp->Next;
      }
    }
    temp->Next = 0;//null pointer as Next for tail node
  }
}
```

The case of copying an empty list is dealt with initially. To copy a nonempty list, the value of size is initially set. Then, one uses a pointer to Node to move through the existing list while appending nodes to the new linkedList object. The appending is done using a pointer to Node, temp, that sequentially ties the new nodes to each other. After the head node is initialized using the head node of the list that is being copied, temp is used to create the next node in the list by defining the Next member of the Node object at its current location. It next moves to the node it just created and repeats the process.

The for loop in the linkedList copy constructor is an explicit example of the basic recursion for navigating through a list that was seen, e.g., in Algorithm 9.4. The role of the index variable is played by the pointer `iter` rather than an integer in this case; nonetheless, it works perfectly well because there is a well-defined place to begin (i.e., the head node), end (i.e., when the null pointer is encountered) and a natural choice for the increment operation (i.e., the `iter = iter.Next` step). An abstraction of this idea leads to the concept of iterators discussed in Section 9.5.

The `Node` objects in our `linkedList` class will be stored on the free-store. Thus, an explicit destructor is needed such as

```
linkedList::~linkedList(){
  Node* temp;
  while(Head != 0){
    temp = Head->Next;
    delete Head;
    Head = temp;
  }
}
```

An overloaded assignment operator must also be supplied (Exercise 9.10).

The definition of the `push` method is

```
void linkedList::push(int newData, int newKey){
  Node* newNode = new Node(newData, newKey, Head);
  Head = newNode;
  Size++;
}
```

This uses the nondefault `Node` constructor to initialize a new `Node` object with its `Next` member given the address of the current head node. The new node that is pushed onto the list becomes the head node, its `Next` member points to the former head node and the `Size` member is incremented.

The `lookUp` method for the `linkedList` struct looks like

```
Node* linkedList::lookUp(int key) const {
  Node* iter = Head;
  while(iter != 0){
    if(iter->keyValue == key){
      return iter;
    }
    iter = iter->Next;
  }
  return 0;
}
```

This follows verbatim from Algorithm 9.4 and provides an example of the general way to iterate through a list using a `while`, rather than a `for`, loop.

The `insert` method takes the form

```
void linkedList::insert(int newData, int newKey, int key){
  if(lookUp(newKey) == 0){//key does not exits
    Node* iter = lookUp(key);
    if(iter != 0){//target does exist
      Node* newNode = new Node(newData, newKey);
      newNode->Next = iter->Next;
      iter->Next = newNode;
      Size++;
    }
    else
      std::cout << "Key not found!" << std::endl;
```

```
  }
  else
    std::cout << "Duplicate key!" << std::endl;
}
```

The new aspect in this implementation beyond that in Algorithm 9.5 is allowance for the possibility that there may be no node in the list with the key value specified in the search. In that instance an error message is returned.

In line with Algorithm 9.7, our code for the application of mergesort to our simple linkedList class is

```
void linkedList::mergeSortList(Node** pHead){
  Node* head = *pHead;
  Node* midNode;

  if (head == 0 || head->Next == 0)//nothing to do
    return;

   split(head, &midNode);

  //recursion
  mergeSortList(&head);
  mergeSortList(&midNode);

  //now merge
  *pHead = merge(head, midNode);
}
```

An oddity of sorts that arises in this listing is the Node** method argument. The reason for this is that mergeSort will actually rearrange the pointers that connect the nodes in the list. In particular, this requires it to be able to change the value of the Head member of a linkedList object. If we simply passed Head to the function, only a local copy of the pointer would be available and changing it would not actually alter the value of Head. Unlike the array setting, we do not know how to partition the list into sublists of roughly equal length. The method split accomplishes this; it takes the head node pointer as its first argument and returns the "middle" node pointer in its pointer to Node* second argument. The mergeSort method is then applied to references to both head and "middle" node pointers thereby setting off the recursion. The final step is an application of the merge method that produces the new head node pointer.

Our code for the split function given below employs a standard approach for finding the "middle" node in a linked list.

```
void linkedList::split(Node* head, Node** midNode){
  struct Node* fast;
  struct Node* slow;
  if (head == 0 || head->Next == 0)
    *midNode = 0;

  else{
    slow = head;
    fast = head->Next;

    while (fast != 0){
      fast = fast->Next;
      if (fast != 0){
          slow = slow->Next;
```

```
            fast = fast->Next;
        }
    }
    *midNode = slow->Next;//new list has slow->Next as head node
    slow->Next = 0;//terminate the other list at slow
    }
}
```

A pair of pointers to Node are defined that move down the list (i.e., toward the tail node) with the fast pointer progressing twice as fast as its slow counterpart. When the fast pointer reaches the end of the list, signaled by its Next member being the null pointer, the slow->Next pointer will be pointing to what can be viewed as a "middle" node. There is more here than it may seem. The "middle" node that has been located with the algorithm is returned in the function's pointer argument and will then serve as the head node of the resulting new sublist. The original list that begins at the head argument to split still runs to its current tail node and this must be rectified before there will actually be two separate sublists. Making slow->Next a null pointer effectively terminates the original list at a new tail node that immediately precedes the "middle" node located by the algorithm.

With arrays the merging process requires the use of temporary storage. However, with linked lists, all that is required is a rearrangement of the next pointers which can be done "in place". For our linkedList class this rearrangement can be accomplish with

```
Node* linkedList::merge(Node* a, Node* b){
  Node* newHead;

  if(a == 0)
    return(b);
  else if(b == 0)
    return(a);

  if (a->keyValue > b->keyValue){//a is new head node
      newHead = a;
      newHead->Next = merge(a->Next, b);
  }
  else{//b is new head node
      newHead = b;
      newHead->Next = merge(a, b->Next);
  }
  return newHead;
}
```

The merge method is in the same spirit as Algorithm 9.6 and the motivating mergesort example. The Head pointer for the head node with the larger key becomes the new Head pointer for a combined list. The algorithm is then applied again except now one of the two lists will be shorter by an element.

A portion of a program that was written to test the linkedList struct is given below.

```
void f(linkedList L){
  std::cout << L.Head->Next->dataValue << std::endl;
}

int main(){
  linkedList L;
  L.push(2, 15); L.push(3, 18); L.push(4, 28); L.push(5, 27);
  L.push(6, 19); L.push(7, 49); L.push(2, 38); L.push(4, 6);
  L.push(100, 2); L.push(-1, 40);
```

```
L.insert(-4, 1, 18);
f(L);//test of copy constructor
std::cout << L.lookUp(1)->dataValue << std::endl;
Node* iter;
for(iter = L.Head; iter != 0; iter = iter->Next)
   std::cout << iter->keyValue << " ";
std::cout << std::endl;

L.sort();
for(iter = L.Head; iter != 0; iter = iter->Next)
   std::cout << iter->keyValue << " ";
std::cout << std::endl;

return 0;
}
```

This listing begins with the definition of a function **f** that will print out the data module for the second node in a **linkedList** object. This function will be called from the **main** function which will force a call to and, hence, provide a check for the copy constructor. Upon entering **main**, a **linkedList** struct with 10 node objects is created via the **push** method. Then, another **Node** object is incorporated into the list with **insert**, the function **f** is called and the **lookUp** method is tested. Next, a pointer to **Node** is used to iterate through the list and print out the keys for each of the **Node** objects it contains. Finally, the list is sorted and the keys are printed again. The output from the program is

```
100
-4
40 2 6 38 49 19 27 28 18 1 15
49 40 38 28 27 19 18 15 6 2 1
```

9.2.3 Stacks and queues

The *stack* and *queue* ADTs are essentially special types of linked lists that have restrictions on the way their nodes can be inserted or removed. Stacks operate on a last-in first-out basis while queues provide a first-in first-out option.

Objects may only be added and removed from the head of a stack. This restriction leads to a simple set of operation that can be performed by the data structure: viz,

- *makeStack()* which creates a new empty stack,
- *push()* that adds a new object to the front of the stack,
- *pop()* that removes the objects at the front of a nonempty stack,
- *preview()* a function that allows for examination, without removal, of the stack's head node and
- *isEmpty()* that returns true if the stack contains no objects and false otherwise.

For queues, objects can only enter the structure at the tail of the list while removal occurs only at the head. The operations that are supported by a queue ADT are

- *makeQueue()* that creates a new empty queue,
- *enqueue()* which adds a new object to the tail of the queue,
- *dequeue()* for removing an object at the head of a nonempty queue and
- *isEmpty()* that returns true if the queue contains no objects and false otherwise.

Both stack and queue ADTs are easily implemented using either dynamic arrays or linked lists. The creation of C++ code for this purpose is the subject of Exercises 9.38–9.40.

The R filehash package provides implementations of stack and queue ADTs. Stack and queue databases are created on disk using the functions createS and createQ. The databases are then linked to R with initS or initQ. For example, the listing below corresponds to an R session where both stack and queue databases were created.

```
> library(filehash)
filehash: Simple key-value database (2.2 2011-07-21)
> system("ls")
filehashEx.r
> createS("myStack")
<stack: myStack>
> mS <- initS("myStack")
> createQ("myQueue")
<queue: myQueue>
> mQ <- initQ("myQueue")
> system("ls")
filehashEx.r     myQueue          myStack
> class(mS)
[1] "stack"
attr(,"package")
[1] "filehash"
> class(mQ)
[1] "queue"
attr(,"package")
[1] "filehash"
```

The system function is used to check the contents of the current working directory with the Unix ls command. We see that two "empty" database files myQueue and myStack have been created on disk. After linking the databases to R (with initQ and initS) the result is that two new objects of type stack and queue are present in the R workspace that can be used to read and write from their respective databases in the current working directory.

Both the stack and queue objects obtained through filehash support the operations push, pop, top and isEmpty. The push and pop functions represent the *enqueue* and *dequeue* operations for queue objects while top is the analog of *preview*. The isEmpty function returns TRUE when applied to empty objects and FALSE otherwise.

Some experimentation with the stack object myStack produced

```
> isEmpty(mS)
[1] TRUE
> push(mS, 1)
> isEmpty(mS)
[1] FALSE
> push(mS, matrix(rnorm(16), 4, 4))
> push(mS, expression(1 + pi))
> push(mS, 1:5)
> print(pop(mS))
[1] 1 2 3 4 5
> top(mS)
expression(1 + pi)
```

Among other things, this illustrates the last-in first-out property of the stack object.

Similar operations performed on the queue object led to

```
> set.seed(123)
> isEmpty(mQ)
```

```
[1] TRUE
> push(mQ, 1)
> isEmpty(mQ)
[1] FALSE
> push(mQ, matrix(rnorm(16), 4, 4))
> push(mQ, expression(1 + pi))
> push(mQ, 1:5)
> print(pop(mQ))
[1] 1
> top(mQ)
             [,1]        [,2]        [,3]        [,4]
[1,] -0.56047565   0.1292877 -0.6868529   0.4007715
[2,] -0.23017749   1.7150650 -0.4456620   0.1106827
[3,]  1.55870831   0.4609162  1.2240818  -0.5558411
[4,]  0.07050839  -1.2650612  0.3598138   1.7869131
```

The output is consistent with the first-in first-out behavior that is expected from a queue object.

The databases created with `createS` and `createQ` continue to exist beyond the end of the R session where they were created. To access them in a future session one again uses the `initS` or `initQ` functions. To demonstrate this we ended our R session and then started a new one in the same working directory with the results

```
> library(filehash)
filehash: Simple key-value database (2.1 2010-02-04)
> mS <- initS("myStack")
> top(mS)
expression(1 + pi)
> mQ <- initQ("myQueue")
> invisible(pop(mQ))
> top(mQ)
expression(1 + pi)
```

The filehash library was loaded again and the existing databases `myStack` and `myQueue` were connected to `stack` and `queue` objects. A preview of the top element on the `stack` object reveals it is the same as before. Then the `pop` function was used to remove the `matrix` object from the `queue`. The command was "wrapped" in the `invisible` function to suppress the printed output this would have otherwise produced. The next component of the queue is found to be the expected `expression` object.

9.2.4 Hash tables

A hash table is a data structure that solves the problem of efficiently mapping a sparse set of keys, or more generally a set of keys that are not integers, into an array while still conserving storage. Knuth (1998b) is one of the best references for hashing. An overview of the topic from a C++ perspective is provided by Sedgewick (1998).

The two main ingredients of a hash table are the *hash function* and a mechanism for *collision resolution*. The hash function is a mapping $h : K \to \mathbb{N}_m$, where K is the key space (the set of all possible keys) and $\mathbb{N}_m = \{1, \ldots, m\}$ for some integer m. A given key k is then said to *hash* to $h(k)$. The data is stored in an array $A = (A[1], \ldots, A[m])$ indexed by $\mathbb{N}_m$ with the consequence that m is also the size of the table.

To insert an object a with key $a.key$ into the hash table, first compute the value $h(a.key)$ and then stores a in $A[h(a.key)]$. If there is already an object stored in $A[h(a.key)]$, either the object a has already been inserted into the table or there has been a *collision* where two different keys map to the same slot in the table. Collisions are generally impossible to avoid. But, by careful design their frequency can be kept to a minimum. *Chaining* and *open addressing* are two methods that can be used to deal with collisions.

In the chaining collision resolution system the entries of the table (i.e., $A[1], \ldots, A[m]$) store not individual objects, but rather linked lists that are sometimes called *buckets*. Initially empty, each bucket contains all the objects whose keys map to the same location. The contents of a bucket are searched linearly with the consequence that the average access time for an object is half the length of the list where it resides; that is, half the number of all the objects that have hashed into the bucket (Exercise 9.16). On the other hand, after the initial look-up, insertion takes constant time.

For a hash table consisting of linked lists $A[1], \ldots, A[m]$, an implementation of the look-up operation might look like

---

Algorithm 9.8 $lookUp(key)$: Look-up in a chaining hash table

---
$i := h(key)$
return $A[i].lookUp(key)$

Note that the $lookUp$ method called by $A[i]$ is the one in Algorithm 9.4. Thus, the look-up operation in a chaining hash table is primarily an application of the look-up operation for a linked list.

Insertion in a chaining hash table first requires an application of Algorithm 9.8. Then, a check is made to see whether the location $lookUp$ returns already contains an object with the given key. If so, an error is reported. The idea is summarized in Algorithm 9.9.

Algorithm 9.9 $insert(a)$: Insert in a chaining hash table

$b := lookUp(a.key)$
if $b.next = NULL$ **then**
    $b.next := a$
else if $b.key = a.key$ **then**
    return Error: duplicate key
end if

Assume that the hash function maps each key to a location chosen uniformly at random and independently of the locations to which other keys are mapped. If there are n elements already in a table of size m, the average length of a list is equal to the table's *load factor* $f = n/m$ giving $O(n/m)$ as the average number of queries in a look-up operation. Thus, the effort required for both look-up and insert operations, if implemented correctly, is decreased by a factor of m^{-1} relative to the linked list case.

Depending on the load factor, insertion/look-up operation in a chaining hash table can take (roughly) constant time on the average. In fact, Knuth (1998b) demonstrates that the average number of operations for look-up and insertion are approximately $1 + f/2$ and $1 + f^2/2$, respectively.

Of course, the uniform location condition is not possible to achieve with a well-defined deterministic function. Still, it turns out that there exist even very simple functions that for practical purposes satisfy this criterion.

In open addressing, no other storage is used beyond the array itself; i.e., only the m slots of the array A are available to store the data. The slots of the array will either be filled with objects that are accessible by the indexing operator [] or they will evaluate as $NULL$. In contrast to the chaining scenario, when a collision happens, the object is not inserted in the location indicated by the hash value. Instead, it is placed in an alternate location determined by another rule. The simplest rule is *linear probing*. After a collision occurs at

$h(k)$, the positions $h(k) + 1$, $h(k) + 2$, etc., are tried until an empty location is found. A slightly more complex alternative is to try the sequence $h(k) + s$, $h(k) + 2s$, etc., for some s relatively prime to m. A generalization of this approach, called *double hashing*, is the topic of Exercise 9.17.

Algorithm 9.10 implements the look-up operation using linear probing with shift s. The search continues until either the target key is found or an empty slot is located. Until one of these two conditions is satisfied, we continue shifting the location by s. The choice of s for best performance depends on the size m of the array in which the table is stored. For example, consider the insertion of many elements that all hash to the same value. Only if s and m are relatively prime will it be possible to actually fill the whole array (Exercise 9.29).

Algorithm 9.10 *lookUp(key)*: Hash table look-up with linear probing and shift s

---

$i := h(key)$
while $A[i] \neq NULL$ and $A[i].key \neq key$ **do**
    $i := i + s$
end while
return $A[i]$

With open addressing the load factor is at most 1. Under the assumption of uniform hashing Knuth (1998b) shows that the average number of operations is approximately $\frac{1}{2}(1 + (1 - f)^{-2})$ for insertion and $\frac{1}{2}(1 + (1 - f)^{-1})$ for look-up when $s = 1$. So, (near) constant time insertion and look-up can also be expected for linear probing.

9.2.4.1 Choosing a hash function

A good hash function should possess two important properties: it should be easy to compute and its values should in some sense distribute uniformly across the range of the table size. Normally, a hash function should be designed for the individual type of keys that are used for the objects that will be stored. If more detailed knowledge of the distribution of the keys is available, this can often be used to improve performance.

The two examples below provide specific instances of hash functions that are commonly used in practice.

Example 9.1. Suppose the keys are floating-point numbers in the range $(0, 1)$. In this instance, the function defined by $h(x) = \lfloor mx \rfloor$ is a natural candidate for the hash function to use with an m-element table.

Example 9.2. A hash function for integers that is very efficient and usually distributes the values well is given by the modulo operation; to hash the key k into a table of size m, just take $h(k) = k \bmod m$.

The table size may have an influence on the behavior of a hash function. Suppose the simple modular hash function of Example 9.2 is used on b-bit integers with a table of size 2^k. In that case, the k bits of lowest significance will be ignored, which is clearly undesirable. As noted in the previous section, this and other bad configurations for linear probing can be avoided by having a table size and shift that are relatively prime. Satisfying this condition lets us avoid cycling before the table is completely full. Most such problems can also be avoided by picking a prime number for the table size.

9.2.4.2 A simple C++hash table

In this section we will create an open addressing hash table that uses linear probing. The objects in the table will derive from a built-in C++ template struct called **pair** that allows two different data types to be packaged in one object. The key and data pairing that occurs in dictionaries in general and in hash tables in particular makes `pair` objects convenient data containers. As a result, `pair` objects will have a recurring role in this and the next chapter. An extension of `pair` that can hold multiple objects of multiple types is discussed in Appendix C.

First, the `pair` class is made available via the include directive

```
#include <utility>
```

The basic structure of the `pair` class can then be described as in the next listing.*

```
template<class dataType1, class dataType2>
struct pair {
  dataType1 first;
  dataType2 second;
  pair();
  pair(const dataType1& t1,  const dataType2& t2);
};
```

As the name suggests, a `pair` object is composed of two member objects; the `first` member is an object of type `dataType1` and the `second` member is an object of type `dataType2`. The `pair struct` has a default constructor that initializes `first` and `second` members as the defaults for their respective classes. A second constructor initializes `first` and `second` using specified values for the two data types. In keeping with our previous examples, both the key and data components for our hash table will be of type `int`. Thus, we will be storing `pair<int, int>` objects in the table although this particular `pair` struct is not entirely suited for our purpose. The problem surfaces when one attempts to apply a look-up method to `pair` objects. A key step in Algorithm 9.10 is the check to see that an array slot is empty in the sense of containing something that can be uniquely distinguished as a *NULL* entry. Being able to locate such empty slots is what makes the insertion operation feasible and efficient. If the hash table is implemented using an array structure as will be done here, the empty slots in the array will typically be filled using the default class constructor. This produces an "empty" `pair<int, int>` object that has no obvious way of making its "empty" status known to a look-up function.

To provide a true *NULL* object we will slightly extend the `pair` class via inheritance to obtain the `newPair` class described in the listing below.

Listing 9.3 newPair.h

```
struct newPair : public pair <int, int>
{
  bool isNull;
  newPair() : pair<int, int>() {
    isNull = true;
  }
  newPair(const int& key, const int& data) :
    pair<int, int>(key, data){
    isNull = false;
  }
  newPair(const newPair& pairObj) :
  pair<int, int>(pairObj.first, pairObj.second)
    {this->isNull = pairObj.isNull;}
};
```

---

* There are additional constructors that allow for implicit type conversion. See Chapter 21 of Vandervoorde and Josuttis (2003).

Since `pair` is a struct, `public` inheritance allows `newPair` objects to have the same access to the members `first` and `second` as would a `pair` object. Notice how the `newPair` constructors invoke the base class (i.e., `pair`) constructors via the mandatory initializer lists. This leaves them with the task of defining the Boolean variable `isNull` that represents the main point of departure from the base class. This new class member provides the "fix" for our problem by evaluating to `true` whenever the default constructor is used. An alternative approach that would work equally well here would be to use composition (i.e., imbed a `pair` object in the `newPair` class) rather than inheritance. This idea is explored in Exercise 9.20.

The class declaration for a class `hashTable` that will hold `newPair` objects is now given by

Listing 9.4 hashTable.h

```
class hashTable {
  int Size, tableSize;
  int Shift;
  newPair* pTable;

  bool empty(int i) const {
    cout << std::boolalpha << " isNull is "
         << pTable[i].isNull << endl;
    return pTable[i].isNull;
  }

public:
  hashTable() {};
  hashTable(int tablesize, int shift = 1);
  hashTable(const hashTable& hT);
  ~hashTable();

  hashTable& operator=(const hashTable& hT);
  bool operator==(const hashTable& hT) const;
  int insert(newPair a);
  int lookUp(int a) const;
  int hash(int a) const;
  int getData(int a) const;
};
```

The class has four private data members: the number of slots in the table `tableSize`, the current number of objects in the table `Size`, the shift parameter for linear probing `Shift` and a pointer to `newPair` that points to the beginning location where `newPair` objects will be stored in memory. The utility function `empty` that checks whether a designated array slot is empty is for internal use and, accordingly, has been made `private`. Note the use of the `isNull` member of the `newPair` class in the definition of the `empty` method. The `boolalpha` format flag from Exercise 3.1 is used here to convert the 0 and 1 values for a Boolean variable into a `true/false` format.

The `public` methods for class `hashTable` include the constructors, destructor, hash function and functions for filling in and looking up values in the table. A default "do nothing" constructor is given inline in Listing 9.4. The other constructor definition is

```
hashTable::hashTable(int tablesize, int shift):
  tableSize(tablesize), Shift(shift), Size(0) {
  pTable = new newPair[tableSize];
}
```

The input values for the shift and table size are used in an initializer list to set the values of the class members `Shift` and `tableSize`. The `Size` member is set to 0 and a pointer to `newPair` is initialized to hold `tableSize newPair` objects.

As dynamic memory allocation is involved, an explicit destructor is needed such as

```
hashTable::~hashTable(){
  if(pTable != 0)
    delete[] pTable;
}
```

For the same reason a copy constructor and an overloaded assignment operator are needed. The task of creating these functions as well as the overloaded comparison operator `==` is relegated to Exercise 9.18.

The hash function that will be used is the one from Example 9.2. The code that implements it is

```
int hashTable::hash(int a) const {
  cout << a << " hashes to " << a % tableSize << endl;
  return a % tableSize;
}
```

This uses the C++ modulus operator that requires us to include the cmath header file.

The look-up operation is carried out with

```
int hashTable::lookUp(int a) const {

  int ind = hash(a);
  while (!empty(ind) and !(pTable[ind].first == a)) {
      cout << "Collision at " << ind << endl;
      cout << "Next try " << ind + Shift;
      ind = (ind + Shift) % tableSize;
      cout << " that hashes to " << ind << endl;
  }
  return ind;
}
```

First, the supplied key value is hashed. Then, linear probing according to the value of the `Shift` class member is used to find either a nonempty array slot corresponding to the hashed key value or one whose index hashes to the given key. Some additional output statements have been included here for the purpose of a later example. With the `lookUp` method in place the actual retrieval of a data value is straightforward to accomplish using

```
int hashTable::getData(int a) const {
  return pTable[lookUp(a)].second;
}
```

A `hashTable` object is filled using the `insert` method below.

```
int hashTable::insert(newPair a) {

  int ind = lookUp(a.first);
  if (pTable[ind].first == a.first) {
      cout << "Already inserted this value" << endl;
      return -1;
  }
  cout << "Inserting at " << ind << endl;
  pTable[ind] = a; Size++;
  return ind;
}
```

The `lookUp` method is first applied to the key value for the input `newPair` object to obtain a target storage location. There are two things that can happen: either `lookUp` returns the index of an empty slot or the slot is occupied which means that the input key already exists. In the latter instance an "error" index of -1 is returned. Otherwise, the new object is inserted at the hashed key value and the `Size` member is augmented by 1. As with the `lookUp` method, some unnecessary output has been built into the function to allow us to track the hashing process in a test case.

A driver program with the `main` function below was used to illustrate the operation of our `hashTable` class.

```
int main(){
    hashTable ht(53, 13);
    ht.insert(newPair(213, 10));
    ht.insert(newPair(54, 11));
    cout << ht.getData(54) << endl;
    return 0;
}
```

A request is made for a table with 53 slots and the `Shift` value is set to 13. Then, two `newPair` objects are inserted into the table and the `Data` component of the second object is accessed. The output produced by the program is

```
213 hashes to 1
 isNull is true
Inserting at 1
54 hashes to 1
 isNull is false
Collision at 1
Next try 14 that hashes to 14
 isNull is true
Inserting at 14
54 hashes to 1
 isNull is false
Collision at 1
Next try 14 that hashes to 14
 isNull is false
11
```

Since $213 = 53 \times 4 + 1$, $213 \bmod 53 = 1$ and the first key value hashes to 1. The array slot with index 1 is empty (i.e., its `isNull` member evaluates to `true`) and a `newPair` object with a data component of 10 is inserted at that point in the table. The second object has a key of 54 that also initially hashes to 1. But, the slot is full (`isNull` is `false`) and the key value of 54 for the new object is not the same as that of the occupying object (i.e., 213). So, the search continues with a new shifted hash value $1 + 13 = 14$ that hashes to 14. This slot is empty and the `newPair` object with data component 11 is inserted into the table at that location. Up to this point the `lookUp` method has been used to find an empty slot with the `while` loop in `lookUp` being terminated when a slot's `isEmpty` member is found to be `true`. The request for the data value of the object with key 54 goes in the other direction; the search is for a specific, nonempty slot. The hashing process proceeds through exactly the same steps as when the targeted object was inserted: initially hashing to 1 before the probe shifts it to slot 14. In this case the `isNull` member still evaluates to `false`; but, the `while` loop ends when a comparison of the element's key value with the one being sought returns `true`. The requested data value is then retrieved successfully.

9.2.4.3 Hash tables in R

Hash tables are present in the internals of most programming languages, either inside the compiler in the case of compiled languages such as C++ or inside the interpreter in the case of interpreted languages such as R or Python. An interpreter must be able to retrieve information about any object previously defined and, from that perspective, it is easy to imagine that a hash table may be useful if only as a means to obtain information about an object given its name. Both R and Python use a general hash table mechanism to store the information about objects defined and manipulated by the user, as well as for internal use by the interpreter. In fact, a good portion of the time used in executing both R and Python code is spent in looking up objects in dictionaries. While *dictionaries* (or *associative arrays*, another synonym for a hash table) are a prominent language feature in Python, R's hash tables are much less obvious to a casual user. Nonetheless, its internal hash table mechanism is available and easy to use for storage purposes. There are at least two ways to make use of this R feature: directly, using `environments`, or through the hash package. Another option is provided by the filehash package that creates an external (from R) hash table database. All three approaches will be discussed in this section.

The key to creation of hash tables in R is the R `environment` concept. An `environment` consists of a set of symbol-value pairs (called a *frame*), and a pointer to another environment (called the *enclosure*). The details of the implementation or internal usage of `environments` are not of interest for us here; instead, we want to examine how to use them to create hash tables for storage of R objects.

The constructor for the `environment` class has prototype

```
new.env(hash = FALSE, parent, size)
```

The first argument in the constructor gives us the option of having the `environment` be a chaining hash table. The second argument `parent` is the `environment` that is to contain the object created by the constructor. As this plays no role in the current development, we will set `parent` to be `emptyenv()` which, according to the R environment help page, is the "the ultimate enclosure of any environment [...] to which nothing may be assigned". The `size` argument is the number of buckets to be used in the hash table and defaults to 29.

The insert and look-up operations for an `environment` are carried out using the `assign` function and either the `get` or `mget` functions; `get` allows access to one object while `mget` returns (as a list) the objects whose names/keys are specified in its list argument.

An abbreviated prototype for `assign` is

```
assign(x, value, pos)
```

The x argument is a character string that represents the name that will be given to the object specified by the `value` argument. The `pos` argument specifies the `environment` where the assignment is to take place. Similarly, a condensed prototype for `get` looks like

```
get(x, pos)
```

with x the name of the object to be retrieved from the `environment` specified by pos.

To determine if an object with name x is present in an `environment` one uses the `exists` function in the form

```
exists(x, pos)
```

with the `pos` argument set to the `environment` that is the target for the search. The `ls` function will list the contents of an `environment`, `environmentName` via `ls(environmentName)`. Finally, `rm(objectName, envir = environmentName)` can be used to delete an object `objectName` from `environmentName`. All elements are deleted by the command `rm(list = ls(), envir = environmentName)`.

Some experimentation with the R hash table facility produced the following results.

```
> set.seed(123)
> hT <- new.env(hash = TRUE, parent =    emptyenv(), size = 3)
> env.profile(hT)
$size
[1] 3

$nchains
[1] 0

$counts
[1] 0 0 0

> assign("x", 1, envir = hT)
> assign("mat", matrix(rnorm(16), 4, 4), envir = hT)
> assign("expr", expression(1 + pi), envir = hT)
> assign("intSeq", 1:5, envir = hT)
> env.profile(hT)
$size
[1] 3

$nchains
[1] 2

$counts
[1] 2 2 0

> eval(get("expr", envir = hT))
[1] 4.141593
> hT[["intSeq"]]
[1] 1 2 3 4 5
> hT$intSeq
[1] 1 2 3 4 5
> rm("intSeq", envir = hT)
> exists("intSeq", envir = hT)
[1] FALSE
```

First, an empty chaining hash table hT is created and examined using the env.profile
function. The output from env.profile tells us that hT has three buckets (i.e., size is
three) with no chains (or linked lists) and no elements in any of the buckets as indicated by
the values of counts. Next, four R objects are inserted into hT using assign and its profile
is examined again. The hash table now has two chains or linked lists with two elements in
each bucket and a third bucket that is empty. The indexing operator [[]] and extraction
operator $ provide alternatives to get for accessing elements in an environment as illustrated
by retrieving the expression and sequence objects from hT. Note that the "index" that is
used with [[]] must be the object's name or key. Finally, the sequence object is deleted
from hT using rm and exists is used to check that the deletion has been successful.

A more user-friendly front end for R hash tables is provided by the hash package where
hash tables have been implemented as an S4 class called hash. The class constructor can
take the form

```
hash(key, values)
```

with key a set of keys (in the form, e.g., of a list) whose corresponding data objects are
given in a list as the values argument. For example, we can recreate some aspects of the
previous example with

```
> library(hash)
hash-2.1.0 provided by Decision Patterns

> hT <- hash(list(1, 2, 3, 4), list(1, matrix(rnorm(16), 4, 4),
+ expression(1 + pi), 1:5))
> class(hT)
[1] "hash"
attr(,"package")
[1] "hash"
> eval(hT[["3"]])
[1] 4.141593
```

First the hash package is loaded into the current environment. Then, a hash table object is formed using a group of R objects. A check reveals that the new hash table is an object of the hash class. Evaluation of the expression object demonstrates that the entries in a hash object can be accessed with the indexing operator [[]] using the object's key as the index value.

The functions delete and has.key provide the same utility for hash objects as we obtained with rm and exists for environments. There is also a clear function that empties the table. The use of these functions in terms of our running example produces

```
> delete(c("3", "4"), hT)
> has.key("4", hT)
      4
FALSE
> clear(hT)
> hT
<hash> containing 0 key-value pair(s).
   NA : NULL
```

The third and fourth entry are removed from the table with delete and has.key is employed to check that removal actually occurred. The clear function is then applied to the hash object and an application of show indicates the table is now empty.

Hash tables created with environments (and, hence, those produced with the hash package) store both the keys and their corresponding data objects in memory. For large data sets this can become problematic. One solution to problems of this nature can be obtained by using the filehash package from Section 9.3 where the key values are retained in memory while the corresponding data is stored on disk and then accessed as needed.

The listing below demonstrates the steps in creation of a filehash object.

```
> library(filehash)
> dbCreate("myData")
[1] TRUE
> hT <- dbInit("myData")
[1] "hT"
> class(hT)
[1] "filehashDB1"
attr(,"package")
[1] "filehash"
```

First the filehash library is loaded and an empty database called myData is created on disk using the dbCreate function. The myData file is "linked" to a filehash object hT with the dbInit function using the database filename (myData in this case) as its argument. An application of the class function reveals that a filehash object has been created. The DB1 that appears in the class name corresponds to the default storage options where keys and data are stored in a single file.

The functions `dbInsert`, `dbFetch` (or the indexing operator `[[]]`) and `dbDelete` perform the hash table insertion, look-up and deletion operation while `dbExists` can be used to check for the existence of a specified key. For example, the listing below involves the same types of computations that were carried out previously using `environments`.

```
> set.seed(123)
> dbInsert(hT, "a", 1)
> dbInsert(hT, "b", matrix(rnorm(16), 4, 4))
> dbInsert(hT, "c", expression(1 + pi))
> dbInsert(hT, "d", 1:5)
> eval(hT[["c"]])
[1] 4.141593
> dbExists(hT, "d")
[1] TRUE
> dbDelete(hT, "d")
> dbExists(hT, "d")
[1] FALSE
```

As before, four objects are inserted into the hash table and the third `expression` object is evaluated using the index operator to perform the table look-up. The existence of the fourth object is tested using the `dbExists` function, it is removed from the table with `dbDelete` and the success of removal is checked using `dbExists`.

The databases created using the `filehash` package persist and can be accessed in future R sessions using the `dbInt` function. To illustrate this we created a "large" data file and ended the R session with

```
> dbCreate("bigfile")
[1] TRUE
> hT <- dbInit("bigFile")
> set.seed(123)
> invisible(sapply(1:4, FUN = function (i){
+ dbInsert(hT, paste("a", i, sep = ""),
+ rnorm(10^7, .05*i))}))
> q()
```

A new session was started in the same directory and a new `filehash` object was linked to the file with

```
> library(filehash)
filehash: Simple key-value database (2.1-1 2010-10-04)
> hT <- dbInit("bigFile")
> lapply(hT[c("a1", "a2", "a3", "a4")], FUN = mean)
$a1
[1] 0.05011525

$a2
[1] 0.1002785

$a3
[1] 0.1499748

$a4
[1] 0.1997805
```

The filehash library must be loaded again. But, the `dbInit` function makes the necessary connection with `bigFile` and `lapply` is used to compute the means of the four arrays that are held in the database. There is a bit of subtlety here in the use of `[]` rather than `[[]]`.

The former returns a list that contains the requested elements from the database thereby making it possible to use the `lapply` function.

9.2.4.4 Pairwise independent and universal hash families

In real life, input data may obey some regular pattern or its distribution may contain some unknown bias. Our uniform hashing assumption may not be satisfied in such a situation, leading to poor performance. In this section we explain an approach that avoids this problem.

Imagine you are the administrator of a site that uses an open-source program that contains a hash table implementation. The behavior of the hash table and its efficiency depends on the input to the program. A good hash table (and in particular, its hash function) is designed to work well for typical input sequences. For example, the results we discussed that support the claim that the expected number of steps for a hash table look-up is nearly constant require that the input come from a "typical" input sequence. Now imagine an adversary whose goal is to disrupt the performance of your program. Having inspected the code and knowing the hash function, your malicious foe may be able to choose a sequence of keys that all hash to the same slot, thus causing the hash table to perform much less efficiently or even causing errors that, with a "typical" input sequence, would be extremely unlikely to happen.

A simple solution to this problem is to avoid specifying the hash function in the code. Instead, choose the hash function randomly from a family of hash functions that is both large and diverse enough to guarantee a high probability of good behavior on any input. A family $\mathcal{H}$ of hash functions with range $\{0, 1, \ldots, m-1\}$ is said to be *universal* if, for every two keys k_1 and k_2, when an $h \in \mathcal{H}$ is randomly selected we have $P[h(k_1) = h(k_2)] \leq 1/m$. Universal hash families were introduced by Carter and Wegman (1979).

A stronger condition, useful in a large variety of settings (e.g., in de-randomization of randomized algorithms), is to require that for every key pair (k_1, k_2) and every pair (x, y) of hash values, when h is chosen uniformly at random from the family, we have $P[h(k_1) = x, h(k_2) = y] = 1/m^2$. In other words, every pair of values $(h(k_1), h(k_2))$ is equally likely. Such a family of hash functions is called *strongly 2-universal* or *pairwise independent*.

Universal hash families have other interesting and useful properties. For example, if n objects are hashed into a table of size $m \geq n$, the expected number of collisions per hashed key is no more than 1.

Theorem 9.1. *Suppose there are $n \leq m$ keys in a table of size m with the hash function chosen uniformly at random from a universal family. Then, for every key k,*

$$E[number\ of\ collisions\ on\ insertion\ of\ k] \leq 1.$$

Proof. For every key $\tilde{k} \neq k$, the probability that $h(\tilde{k}) = h(k)$ is at most $1/m$. Thus, the expected number of keys in the table with hash values equal to $h(k)$ is at most $n/m \leq 1$. $\square$

Efficient universal hash families are not difficult to find as demonstrated by the following two examples.

Example 9.3. A frequently used hash function for strings is defined by $h(S) = h(s_1) + h(s_2) + \cdots + h(s_n) \bmod m$, for the string $S = s_1 s_2 \cdots s_n$, where m is the size of the hash table and h is a hash function mapping the set of all characters into $\{0, \ldots, m-1\}$. An extension of this idea is to randomly and independently select N functions $h_1, \ldots, h_N$ from the set of all functions mapping the set of characters into $\{0, \ldots, m-1\}$ and define

$$h(S) = h_1(s_1) + h_2(s_2) + \cdots + h_n(s_n) \bmod m \tag{9.1}$$

for any string of length $n \leq N$. It is easy to show that this gives a universal hash family (Exercise 9.26).

Example 9.4. Consider hashing the set of integers $\{0, 1, \ldots, r - 1\}$. Let $p \geq r$ be a prime number and take $h(k) = ((ak + b) \bmod p) \bmod m$, where a and b are chosen randomly from $\{0, 1, \ldots, p - 1\}$ with $a \neq 0$. Carter and Wegman (1979) show that this gives a universal hash family.

9.2.4.5 Extensions

We have not discussed how to remove elements from a hash table. Some choices in designing a hash table lead to easier implementation of this operation than others. If the hash table uses chaining for collision resolution, it is easy to delete an element: just remove it from the linked list in which it is found (e.g., Exercise 9.11). Open addressing presents a more difficult challenge; after finding the table entry x that is to be deleted, it is necessary to move any elements hashing to the same value as x that may have been inserted after x. This becomes complicated due to the possibility that additional collisions may have happened after the probe at the location of x. We must at least find the end of the probe sequence and relocate the last inserted entry with hash value equal to that of x. This is not all; that entry may have caused other collisions which must now be tracked down. The simplest approach is to replace the deleted element by a *sentinel* value indicating a deletion has occurred. This lets the look-up operation skip over this location while also letting the insert procedure use the location (Exercise 9.19).

A hash table may need to be resized if it gets close to full. There is no cheap way to do this; every table entry must be rehashed into a larger table. On the other hand, this is not done frequently. A standard way to maintain a hash table of dynamically changing size is to double its size when it becomes half full and to halve its size when it becomes less than a quarter or eighth full. This prevents frequent resizing and in an amortized sense does not increase the number of operations by more than a small constant factor.

9.3 ADT priority queue

In many situations a set that needs to be represented or managed in a computer program has additional structure and is used according to this structure. One of the simplest examples is a *priority queue*. In this setting components are both inserted and removed during operation. Associated with each element is a numerical attribute that represents its *priority* relative to the other members of the queue and thereby dictates the sequencing of its removal.

The objects being stored will again be assumed to consist of key and data modules with an element's key value now being synonymous with its priority. Under this formulation, each removal step will result in the object with the smallest key (highest priority) being removed from the queue.

Situations where priority queues are useful are frequent in practice. For example, the way a business distributes products to dealerships may be prioritized in terms of past sales performance. The scheduling of jobs on a computer cluster is likely to be managed with a priority queue. The future event queues that arise in a discrete event simulation are often handled using priority queues as implemented through a heap (discussed in the next section) or one of its variants (e.g., Mansharamani 1997). The problem of finding the shortest travel route between two locations is an example of the *shortest path problem* from graph theory. In that setting, Dijkstra's algorithm for finding the optimal path benefits greatly from an efficient priority queue implementation (Exercise 9.32).

The operations that are supported by a priority queue are

- *insert(key)* that inserts an object with priority level *key* into the queue,
- *minimum()* which returns the object with the smallest key in the queue and
- *extractMin()* that both returns and extracts the object with the smallest key.

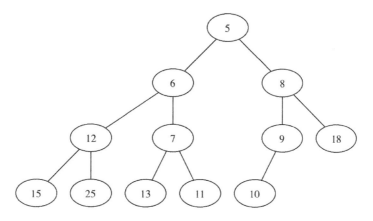

Figure 9.1 *A heap*

Here the assumption is that higher priority corresponds to a smaller key. A similar formulation is possible for the case where an object's priority increases with its key value.

Priority queues are easiest to visualize as binary trees. This perspective also makes it easier to develop algorithms and code for the data structure's operations. A binary tree is an abstract combinatorial object that is useful for representing a number of different data structures. Thus, it is worthwhile to invest a moment now on introducing the idea.

The definition of a binary tree is recursive. A binary tree is either 1) empty or 2) consists of a *node* called *root* and two binary trees that are called the *left* and *right* subtrees of the root, respectively. The roots of a node's subtrees are called its *children* and the root node is their *parent*. We visualize a tree as a graph whose vertices are nodes and whose edges link nodes to their children as in Figure 9.1. Note that, by definition, a node in a binary tree may have zero, one or two children and a node's child is either its *left child* or its *right child*. A node for which both subtrees are empty is called a *leaf*. A node that is not a leaf is an *internal node*. A *level* of a tree consists of all the nodes that are the same number of generations/steps removed from the root.

The binary tree in Figure 9.1 has four levels and all but the fourth level are full. The node containing the value 5 is the *root*. It has both a left and right child: i.e., the nodes with values 6 and 8, respectively. Both of these latter two nodes are internal. All the nodes on the last level of the tree are leaves as well as the node with value 18.

9.3.1 Heaps

The simplest efficient implementation of a priority queue uses the data structure called a *heap*. A heap is a binary tree that at each node stores a member of the set and satisfies the properties

- every node has two children, except for the nodes in the lowest two levels of the tree,

- the lowest level of the tree is filled left to right (see Figure 9.1) and

- the key value of every node is no greater than that of either of its children (known as the *heap property*).

Figure 9.1 shows an example of a heap with nodes having the integer keys 5, 6, 8, 12, 7, 9, 18, 15, 25, 13, 11 and 10.

It is easy to see how to find the node with the minimum key in a heap by the heap property; it is just the root of the heap. Less obvious is that a heap can be efficiently implemented using an array and this is what we describe next.

To store an n-element heap in an array one begins by allocating an array A of size at least n. This array in then filled in order, level by level from the root towards the leaves and left to right within every level. Thus, the root is stored in $A[1]$, its left and right children, respectively, are placed in $A[2]$ and $A[3]$, the nodes at the third level are placed in $A[4], A[5], A[6]$ and $A[7]$, etc. The idea is illustrated in Table 9.1 for the binary tree in Figure 9.1.

Table 9.1 *Array representation of the heap in Figure 9.1*

Array Index	1	2	3	4	5	6	7	8	9	10	11	12
Key Value	5	6	8	12	7	9	18	15	25	13	11	10

In general, for a node at $A[i]$, its children are placed at $A[2i]$ and $A[2i + 1]$ and its parent at $A[\lfloor i/2 \rfloor]$. This, of course, assumes that the array index starts at 1. For 0-offset indexing, the left and right child of $A[i]$ are $A[2i + 1]$ and $A[2i + 2]$ and its parent is $A[\lfloor (i - 1)/2 \rfloor]$.

If the value of a single node in the heap changes, it is easy to rearrange the structure back into a heap. If the kth node's key value is decreased, its subtree is still a heap. The only possible violation of the heap property is if its key is now smaller than that of its parent. In this case, swap the node with its parent and continue the process up the tree until the heap property is restored. If, instead, the node's key is increased, swap it with the one of its children that has the smallest value and continue the process down the tree until the heap property is again satisfied. The operation of moving nodes up and down the tree in order to correct violations of the heap property is called *heapifying*. The process of heapifying upward toward the root is described by Algorithm 9.11 and Figures 9.2–9.5.

Algorithm 9.11 *upHeapify(i)*: Float $A[i]$ up to its correct location

---
    while $i > 0$ and $A[i].key < A[\lfloor i/2 \rfloor].key$ **do**
        Swap the values of $A[i]$ and $A[\lfloor i/2 \rfloor]$
        $i := \lfloor i/2 \rfloor$
    end while

Figures 9.2–9.5 illustrate Algorithm 9.11 for the heap in Figure 9.1. A new node is added in the first open position on the fourth level of the tree (Figure 9.2) or, equivalently, the object is inserted in the thirteenth slot of the array in Table 9.1. The new node's key value of 4 is less than that of its parent (i.e., 9) and they are swapped (Figure 9.3). This corresponds to swapping entries 13 and $\lfloor 13/2 \rfloor = 6$ in the heap's array representation. Again, the new node is seen to have a key that is smaller than its parent; so, we swap the elements with indices 6 and $6/2 = 3$ in the array (Figure 9.4). Finally, 4 is smaller than the key of the root (or $\lfloor 3/2 \rfloor = 1$) node and once these two nodes are swapped (Figure 9.5) the heap property has been restored.

The up-heapifying procedure must terminate at or before it reaches the root and array indices used in making the swaps will always be valid. Some caution is required going down the heap as there is no assurance that an object occupies every node at the bottom level. If *size* is the number of objects that are currently in a heap, then a down-heapifying analog of Algorithm 9.11 is provided by Algorithm 9.12 below.

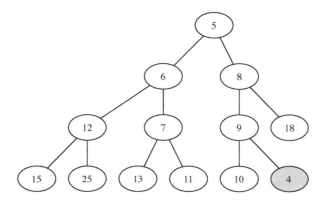

Figure 9.2 *New node "arrives": shaded gray with a key of 4*

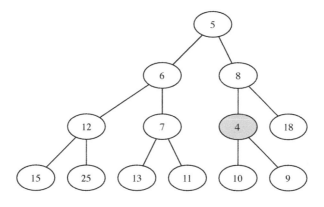

Figure 9.3 *Swap the new node with its parent that has a key of 9*

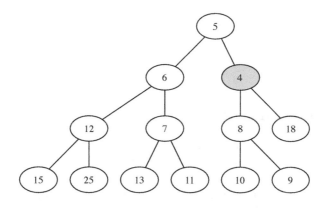

Figure 9.4 *Swap the nodes with keys of 4 and 8*

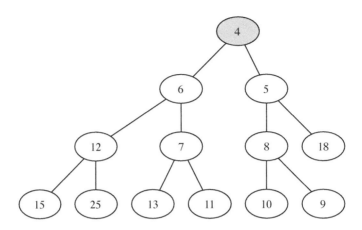

Figure 9.5 *Swap the root and the node with the key of 4 to restore the heap property*

Algorithm 9.12 *downHeapify(i)*: Float $A[i]$ down to its correct location

while $2i < size$ do
    if $A[2i].key > A[2i + 1].key$ then
        $j = 2i + 1$
    else
        $j = 2i$
    end if
    if $A[i].key > A[j].key$ then
        Swap $A[i]$ and $A[j]$
        $i := j$
    else
        Break
    end if
end while
if $2i = size$ and $A[i].key > A[2i].key$ then
    Swap $A[i]$ and $A[2i]$
end if

The number of operations involved in an application of either `downHeapify` or `upHeapify` is proportional to the number of steps from the kth node to some leaf below it, or to the root, respectively. The number of nodes is at most $n = 2^{r+1} - 1$ for a tree with r levels and there can be at most r nodes on any branch of the tree. Therefore, the number of operations is proportional to $\log_2(n)$.

To insert a node into a heap, the size of the heap is first enlarged by one and the new component is placed in this new position. Then, the new object is floated up the heap with the `upHeapify` method. Algorithm 9.13 describes the procedure. Insertion operations takes only linear time because, on average over the sequence of inserted elements, each call to `upHeapify` takes constant time.

Algorithm 9.13 *insert*(*a*): Insert a node into a heap stored in an array A

$size := size + 1$
$A[size] := a$
$upHeapify(size)$

Algorithm 9.14 shows how to extract a node from a heap. One swaps the highest-indexed element and the root, reduces the size by one, down-heapifys the root and then returns the former root as output. The idea is illustrated in Figures 9.6–9.8 for the heap in Figure 9.1. The first step is to swap the elements in the first and thirteenth (or last) positions producing Figure 9.6. Then, the first node is swapped with the second node in Figure 9.7 since this is its smallest child. Next, the new second node is swapped with its smallest child in Figure 9.8 and the heap property is restored.

Algorithm 9.14 *extractMin*(): Extract the minimum from a heap in an array A

Swap $A[1]$ and $A[size]$
$size := size - 1$
$downHeapify(1)$
return $A[size + 1]$

This same process used for removing the minimum can be used to sort objects that are stored in a heap. For a *size*-element heap, all that is necessary is to make *size* consecutive calls to `extractMin` as in Algorithm 9.15.

Algorithm 9.15 *heapSort*(): Sort objects in a heap by their key values

---

  for $i = 1$ to $size - 1$ **do**
    $extractMin()$
  end for

This procedure can actually be carried out in place as will be seen in the next section. Like any other comparison sorting method, heap sort requires on the order of $n \log n$ flops (Exercise 9.23).

9.3.2 A simple heap in C++

In this section we provide a simple C++ version of a heap for data of type `int`. In particular, this will allow us to use the `dArray` class of Section 9.2.1 to provide the array structure that will hold the objects in the heap.

The declaration for our `Heap` class is given in Listing 9.5 below.

Listing 9.5 heap.h

```
//simpleHeap.h
#ifndef HEAP_H
#define HEAP_H
#include "dArray.h"

class Heap {

  dArray A;

  void swap(int i, int j){
```

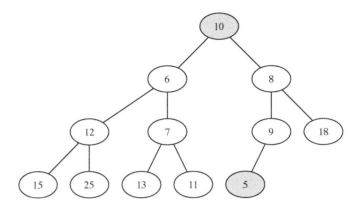

Figure 9.6 *Shaded nodes from the first and thirteenth positions have been swapped*

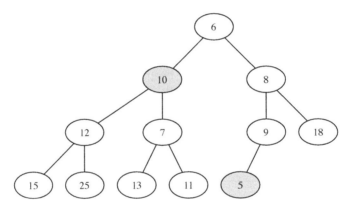

Figure 9.7 *First position and second position nodes are swapped*

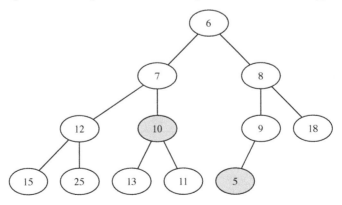

Figure 9.8 *The second position node is swapped with its smallest child*

```
    int temp = A.pA[i]; A.pA[i] = A.pA[j];
    A.pA[j] = temp;
  }
  void upHeapify(int i);
  void downHeapify(int i, int size);

public:

  Heap(int x) : A(dArray(1, &x)) {}
  Heap(const dArray& a);
  void insert(const int& x);
  int minimum() const {
    return A.pA[0];
  }
  int& extractMin();
  void heapSort();
  void printHeap() const;
};
#endif
```

The Heap class uses the array representation for a heap that was discussed in the previous section with the array storage being furnished by a dArray object. It is important for the Heap class methods to be able to manipulate the objects in A as they must be reordered to preserve the heap property when elements are either added or extracted from the queue. An expedient way to gain the access that is needed for such manipulations is to make Heap a friend of dArray. This is accomplished by adding

```
    friend class Heap;
```

to the dArray class declaration in Listing 9.1.

The Heap class constructor only initializes a Heap object. The actual process of building the heap is accomplished with the insert method. There are private methods swap, upHeapify and downHeapify that perform the re-arrangements of objects that are required for internal operations in a heap. These method are used in the public methods extractMin and heapSort that remove the smallest value from the heap and sort the heap elements, respectively. In contrast to extractMin, the minimum method simply returns the node at the top of the heap without removing it. As a result, no rearrangement of the heap elements is necessary. There is also a utility printHeap that will print the contents of the heap to standard output.

The upHeapify method proceeds along the path laid in Algorithm 9.11.

```
void Heap::upHeapify(int i){
  int temp = i;
  while(A.pA[(temp - 1)/2] > A.pA[temp]){
    swap(temp, (temp - 1)/2);
    temp = (temp - 1)/2;
  }
}
```

The main difference between the code here and Algorithm 9.11 is the use of 0-offset indexing when accessing the elements of the array. Also, note that integer division automatically performs the floor operation in Algorithm 9.11.

The downHeapify method is slightly more complicated. Also, rather than giving a rote implementation of Algorithm 9.12 we will allow downHeapify to only work on a portion of the tree. This feature will become useful in our heap sort code.

```
void Heap::downHeapify(int i, int size){
  int temp = i, newTemp;
  bool done = 0;
  while(2*temp + 1 <= size && !done){//still a remaining node
    if(2*temp + 1 == size){//no right child: bottom of the heap
      if(A.pA[temp] > A.pA[2*temp + 1])
        swap(temp, 2*temp + 1);
      done = 1;
    }
    else if(A.pA[2*temp + 1] < A.pA[2*temp + 2] && !done){
      if(A.pA[temp] > A.pA[2*temp + 1])
        swap(temp, 2*temp + 1);
      newTemp = 2*temp + 1;
    }
    else if(A.pA[2*temp + 1] >= A.pA[2*temp + 2] && !done){
      if(A.pA[temp] > A.pA[2*temp + 2])
        swap(temp, 2*temp + 2);
      newTemp = 2*temp + 2;
    }
    else
      done = 1;
    temp = newTemp;
  }
}
```

As indicated by Algorithm 9.12 there are several cases that must be considered. If we are at the bottom of the heap and there is no right child, simply swap the two elements if the parent is smaller. Otherwise, there will be two children and the parent is swapped with the child having the smallest key provided its key is the smaller of the two. The whole process is managed by a while loop that will terminate the downward movement of the i-indexed node at the level determined by the size argument. The other option is that none of the if or else if statements evaluate as true which means that the heap property has been restored. When that happens the last else statement activates a Boolean flag done that terminates the while loop when it evaluates to true. This is an alternative to using a break statement as in Algorithm 9.12. There is a subtle, but important, step where a temporary variable newTemp is used to store the current value of the index temp as the while loop progresses. This allows the value to be reset in the first else if block without altering the temp loop variable itself until the end of an iteration.

The downHeapify and upHeapify methods really do all the work in a heap and the other methods use them to achieve their particular goals. For example, the insert method takes the form

```
void Heap::insert(const int& x){
  A.append(x);
  upHeapify(A.Size - 1);
}
```

The append method from the dArray class is used to append the new object x as the last node in the heap. Then, upHeapify does the rest and moves the node up to its proper position. Similarly, extractMin uses downHeapify in the listing below to perform its function of removing the element with the smallest key from the heap.

```
int& Heap::extractMin(){
  if(A.Size == 1)//only 1 object in the heap
    return A.pA[0];
```

```
    swap(A.Size - 1, 0);
    A.Size--;

    downHeapify(0, A.Size);

    return A.pA[A.Size];
}
```

If the heap has only one element, then that is returned to the calling program. Otherwise, the top (or 0 index) node is swapped with the bottom (or A.Size - 1 index) node and the new top node is down-heapified. Note that the "minimum" node is not actually extracted but allowed to remain in the dArray object A. But, the Size parameter for A is decremented meaning that the slot that is now occupied by the "minimum" node is viewed as free space by A. Its content will therefore be overwritten by the next item that is inserted into the heap.

Finally, let us consider the heapSort method in the next listing.

```
void  Heap::heapSort(){
  for(int i = A.Size - 1; i > 0; i--){
    swap(i, 0);
    downHeapify(0, i - 1);
  }

  for(int i = 0; i <= (A.Size - 1)/2; i++)
    swap(i, A.Size - 1 - i);
}
```

This method sequentially removes objects from the top of the heap and inserts them on the bottom where they accumulate in decreasing order. The re-arranging is done using progressively smaller subtrees with a new subtree being formed on each step by excluding the current bottom node. So, for example, on the first iteration the bottom node that represents the object with the smallest key value is "excluded" to give a subtree of A.Size - 1 nodes that is converted back into a heap using downHeapify. The process is repeated and this time the object with the second smallest key will be placed in the A.Size - 2 slot of A and the new subtree will consist of all but the last two (non-null) elements of A, etc. At the end of the sorting process the objects will be arranged in a dArray object with keys that decrease from left to right. The last step in the heapSort method reverses the order so that the object with the minimum key is the first entry in the array.

As a simple application of our Heap class let us use it to sort the values in Figure 9.1. The main function for the program that was written to accomplish this is given below.

```
int main(){
  Heap H(5);
  H.insert(6); H.insert(8); H.insert(12); H.insert(7);
  H.insert(9); H.insert(18); H.insert(15); H.insert(25);
  H.insert(13); H.insert(11); H.insert(10);
  H.printHeap();
  H.heapSort();
  H.printHeap();
  return 0;
}
```

First, a Heap container is initialized. Then, the remaining elements from our array are inserted into the heap with the insert method. The resulting heap is printed to standard output, sorted and printed again. The output produced by the program is

5 6 8 12 7 9 18 15 25 13 11 10

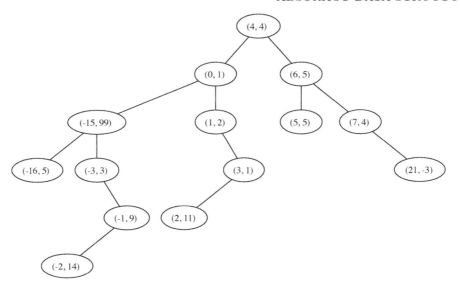

Figure 9.9 *A binary search tree*

5 6 7 8 9 10 11 12 13 15 18 25

Note that the order of entry for the key values has produced the heap shown in Figure 9.1. An application of `heapSort` rearranges its elements in numerically ascending order.

9.4 ADT ordered set

Heaps are used to store ordered sets when only a few operations are necessary. In some situations it may be useful to have more information about the elements stored in the data structure. For example, the operations `next` or `previous` that return an element's successor or predecessor, respectively, are not easy to implement efficiently on a heap. The ADT *ordered set* provides a structure that is more amenable to these and several other operations. Specifically, an ordered set requires the operations

- *makeOrderedSet*() that creates a new ordered set,
- *getMax*() and *getMin*(), methods that return the ordered set elements with the largest and smallest keys,
- *lookUp*(*key*) that performs a look-up of the specified key value in the data structure and
- *insert*(*key*, *object*) that inserts a new object with a specified key.

The basic structure that implements ADT ordered set is called a *binary search tree*. This can be visualized as a binary tree in which each node contains a member of the ordered set. The elements are arranged in the tree according to their keys so that they satisfy an order property: the key value of the node's left child is no larger than that of the node itself and the key value of a node's right child is no smaller than that of the node itself. In contrast to the previous ADTs we have considered, binary search trees are nonlinear in nature; that is, the data is not stored in a sequence as is the case for, e.g., dynamic arrays and lists.

Figure 9.9 shows an example of a binary search tree. The nodes are composed of (key, data) pairs with both the key and data objects being integers.

To find the node with the minimum key in a binary search tree, descend from the root to its left child and continue descending until reaching a node without a left child. This node's key value is the minimum of the set. The idea is summarized in Algorithm 9.16.

Algorithm 9.16 $getMin()$: Find the node with minimum key value in a binary search tree

---

$a = root$
while $a.left \neq NULL$ **do**
  $a := a.left$
end while
return a

The algorithm begins at the top or root node and descends along the left-hand edge of the tree. In the case of Figure 9.9 the progression is from the root node $(4, 4)$ to the $(0, 1)$ and $(-15, 99)$ nodes before stopping at the node with (key, data) value $(-16, 5)$. The same basic idea produces a function that returns the node with the largest key in the tree.

Searching in a binary search tree consists of moving into the left or the right subtree, depending on the outcome of the comparison between the key of the sought-for node and that of the node at the current location. Algorithm 9.17 gives the details.

Algorithm 9.17 $lookUp(key)$: Find a node with key value key in a binary search tree

---

$a := root$
while $a \neq NULL$ and $a.key \neq key$ **do**
  if $key < a.key$ **then**
    $a := a.left$
  else
    $a := a.right$
  end if
end while
return a

Suppose, for example, a search is to be conducted to find the node with key value 3 in Figure 9.9. The starting point is the root node that has the key value 4. As $3 < 4$, the algorithm moves to the left node with a key of 0. The target key is larger than the key of this node. So, the search must move to the right child with key value 1. This node has no left child. But, even if it had, the fact that 3 is larger than the current node's key would have moved the search to its right child where the target key is actually located.

Insertion of an object into a binary search tree is much like the look-up operation. One proceeds as in Algorithm 9.18 below.

Algorithm 9.18 $insert(a)$: Insert a node a into a binary search tree

---

$b := root$
while $b \neq NULL$ **do**
  if $a.key < b.key$ **then**
    $b := b.left$
  else
    $b := b.right$
  end if
end while
if $a.key < b.key$ **then**
  $b.left := a$
else
  $b.right := a$
end if
return b

Similar to the `lookUp` method, the algorithm moves down the tree by stepping to the left child of a node if it has a smaller key than that of the node to be inserted and to the right child otherwise. The difference is that the process continues until a node without a left or right child, depending on the value of its key, is located. The node to be inserted then becomes the appropriate child. The parent node must be updated to account for the changes in its children.

9.4.1 A simple C++ binary search tree

In this section we will develop a C++ binary search tree class `Tree`. The beginning point is creation of code for the node objects. A variation of the `newPair` struct from Listing 9.3 will be used for this purpose. It takes the form

```
struct Node : public pair <int, int>
{
  Node *Left, *Right;
  Node() : pair<int, int>() {
    Left = 0; Right = 0;
  }

  Node(const int& key, const int& data) : pair<int, int>(key, data){
    Left = 0; Right = 0;
  }
};
```

The members `Left` and `Right` are pointers to the `Node` objects that are the left and right children. Both constructors initialize these pointers to a null pointer; the tree will be built by insertion and the values of the pointers will be set at that juncture. There is a default constructor that takes no arguments and a constructor that creates a node with a specified value for its key and data component that will be held in the `first` and `second` member of the base class `pair` object.

We have also created an overloaded output insertion operator `<<` to work with a `Node` struct (cf. Exercise 3.19). A prototype for one version of the output insertion operator `<<` is

```
  ostream& operator<<(ostream& out, dataType& object)
```

Its first argument is a reference to an `ostream` object and its second is a reference to an object that has been targeted for output. The reason for the `ostream&` return type is that successive applications of `<<` will be used to chain together output; e.g.,

```
  cout << object1 << object2
```

translates to

```
 operator<<(operator<<(cout, object1),  object2)
```

The definition of `<<` that resulted from these considerations is

```
std::ostream& operator<<(std::ostream& out, const Node& node) {
  out << "(" <<  node.first << ", " << node.second <<")";
  return out;
}
```

Note that if `Node` was a class rather than a struct, it would be necessary to make `<<` a friend function as in Exercise 3.19.

The declaration for the `Tree` class is provided in Listing 9.6.

Listing 9.6 *tree.h*

```
class Tree{
  Node* Root;
  int Size;

 public:

  Tree() : Root(0), Size(0){}

  const Node* insert(const pair<int, int>& pairObj);
  const Node* lookUp(const int& key) const;
  const Node* getMin() const;
  const Node* getMax() const;
  int getSize() const {return Size;}
};
```

The class members are an integer variable `Size` that will contain the number of nodes in the tree and a pointer to a `Node` object `Root` that will be the first node to be inserted in the tree. The class constructor is of the default variety and initializes the class members to 0 (for `Size`) and the null pointer (for `Root`).

The methods for the `Tree` class include functions to locate the nodes in the tree with the largest (`getMax`) and smallest (`getMin`) keys. The tree is built using a method `insert` that takes a `pair` object as an argument while a tree entry with a specified key value is located with the `lookUp` method. All four of these methods return a pointer to a `const Node` that points to the memory location where the operation was performed or the requested key was found.

To find that largest or smallest key values in a `Tree`, one traverses the far right- or left-hand branch of the tree as in Algorithm 9.16. The next listing contains the code that implements these two procedures.

```
const Node* Tree::getMin() const {
  Node* temp = Root;
  if(temp != 0){
    while(temp->Left != 0)
      temp = temp->Left;
  }
  return temp;
}

const Node* Tree::getMax() const {
  Node* temp = Root;
  if(temp != 0){
    while(temp->Right != 0)
      temp = temp->Right;
  }
  return temp;
}
```

The insert and look-up methods below both move down the tree by descending from the root while comparing keys to determine whether a left or right child should be the next step in the descent. The difference is in what is accomplished once the sought-after location has been found.

```cpp
const Node* Tree::lookUp(const int& key) const {

  const Node* temp = Root;
  while(temp != 0 && temp->first != key){
    if(key < temp->first)
      temp = temp->Left;
    else
      temp = temp->Right;
  }
  return temp;
}

const Node* Tree::insert(const pair<int, int>& pairObj){
  Node *Previous, *Next = Root;
  Node* newNode = new Node(pairObj.first, pairObj.second);

  if(Next == 0){
    Root = newNode;
    Size++;
    return Root;
  }

  while(Next !=0){
    Previous = Next;
    if(pairObj.first < Next->first){
      Next = Next->Left;
    }
    else
      Next = Next->Right;
  }

  //an empty spot has been found
  if(newNode->first < Previous->first){
    Previous->Left = newNode;
  }
  else{
    Previous->Right = newNode;
  }

  Size ++;
  return newNode;
}
```

In the case of lookUp, a pointer that points to the location of the node with the specified key value is returned. For the insert method, a new Node object must first be created using the pair object that has been passed into the function. The search begins with the Root node pointer and inserts the new node there if the tree is empty. Otherwise, the same descent method as in lookUp is used to find the location where the node can be added. In doing this the pointer to Node, previous, is used to keep track of the previous node that was encountered at each step on the way down. This process necessarily ends with a node that has at most one child thereby allowing a new node to be appended. The only question is whether it should be appended as a left or right child. The trailing pointer previous contains this information. After insertion the Size member is incremented by one.

A program with the main function below was written to provide a test for our Tree class.

```
int main(){
  Tree T;
  T.insert(pair<int, int> (4, 4));
  T.insert(pair<int, int> (0, 1));
  T.insert(pair<int, int> (1, 2));
  T.insert(pair<int, int> (3, 1));
  T.insert(pair<int, int> (2, 11));
  T.insert(pair<int, int> (6, 5));
  T.insert(pair<int, int> (7, 4));
  T.insert(pair<int, int> (5, 5));
  T.insert(pair<int, int> (21, -3));
  T.insert(pair<int, int> (-15,99));
  T.insert(pair<int, int> (-16, 5));
  T.insert(pair<int, int> (-3, -3));
  T.insert(pair<int, int> (-1, 9));
  T.insert(pair<int, int> (-2, 14));

  cout << "The minimum is at " << *T.getMin() << endl;
  cout << "The maximum is at " << *T.getMax() << endl;

  cout << "The data value for key -1 is "
       <<  T.lookUp(-1)->second << endl;
  return 0;
}
```

An empty `Tree` object is initialized and the `insert` method is used to build the tree in Figure 9.9. The nodes with the smallest (`getMin`) and largest (`getMax`) keys are then accessed and `lookUp` is used to find the data component for a `node` object with a specified key value. The output from our test program is

```
The minimum is at (-16, 5)
The maximum is at (21, -3)
The data value for key -1 is 9
```

9.4.2 Balancing binary trees

Binary search trees can become unbalanced which makes them computationally inefficient. Figure 9.10 illustrates such a case. Here sequential insertion of the integers from 1 to 15 has led to an inefficient, unbalanced search tree that is actually a simple path. Instead of a maximum of fifteen comparisons, if the same set is represented by the perfectly balanced search tree in the figure, four comparisons suffice to find any element.

There are several ways to maintain the balance of a binary search tree under an arbitrary sequence of insertions and deletions. One of the more frequently used approaches is known as a *red-black tree* and requires each node in the tree to have an associated "color" attribute. The arrangement of these "colors" in the tree must satisfy a certain invariant. Specifically,

1. the root is "colored" black,

2. any child of a red node is black and

3. every path from the root to a leaf has the same number of black nodes.

These conditions imply a balance property of the red-black tree: namely, the lengths of any two paths from the root to a leaf of a red-black tree are within a factor of 2. Indeed, take any two root–leaf paths. By the second invariant, they contain the same number of black nodes, so their lengths may differ only in the number of red nodes. Between any two red

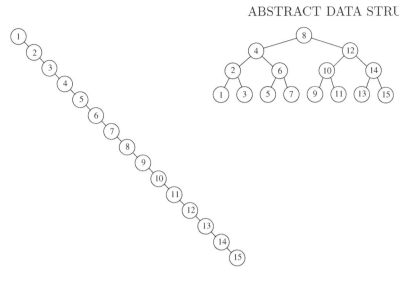

Figure 9.10 *An unbalanced and perfectly balanced search tree*

nodes in either of these two paths there is a black node. Thus, the longer of these paths has at most twice as many nodes as the shorter one.

To insert a node into a red-black tree, we begin by inserting it as if the tree were an ordinary binary search tree and coloring the node red. At this point, actions become contingent on restoring any of the invariant conditions that may have been violated.

If the node was inserted at the root, then the node must have been inserted into an empty tree. The first invariant is therefore violated and restored by simply recoloring the node as black.

If the new node is inserted as a leaf, there are two possibilities: its parent is black or red. In the first instance, all invariants are satisfied and there is nothing to do. If the parent node is red, matters become considerably more complicated as we now describe.

First, let us consider the scenario in Figure 9.11. The subtree there consists of four nodes: a new node to be inserted (indicated by c, for child, in the figure) that is initially given the color red (indicated by r with b for black), the parent node (indicated by p in the figure), its parent or, equivalently, the grandparent (or gp) of the new node and a sibling (i.e., aunt/uncle denoted by au) for the new node's parent. The parent and aunt/uncle are both colored red in this case while the grandparent is colored black. To restore the invariant property in this subtree, one recolors the parent and aunt/uncle as black and designates the grandparent as now being red. This provides a local fix and restores the third invariant. But, there will be a problem if the parent of the grandparent is red. This is resolved recursively by treating the grandparent as if it were a new node being added to the tree at its location. In that respect, the general scenario will involve the case where the "new" node as well as its parent and aunt/uncle have other children. Unless stated otherwise these relationships will not need to be changed and, as a result, do not explicitly appear in the figures or enter the discussion.

The next possibility is that the parent node is red while the aunt/uncle node is black. The procedure for restoring the invariant involves either a right or left rotation depending on whether the parent is a left or right child of the grandparent. We will deal with the case that the parent is a left child here and pose the treatment of the other scenario as an exercise (Exercise 9.34).

Suppose we are in the setting of Figure 9.12 where the new node is a left child of its parent. Then, the solution is to

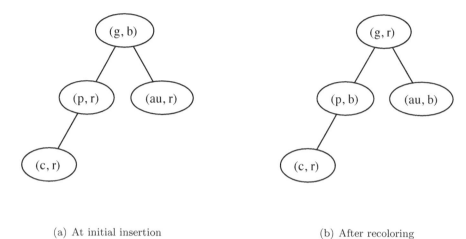

(a) At initial insertion (b) After recoloring

Figure 9.11 *Red node with red parent and aunt/uncle in a red-black tree: c, g, p, au stand for child, grandparent, parent, aunt/uncle and b and r indicate red and black*

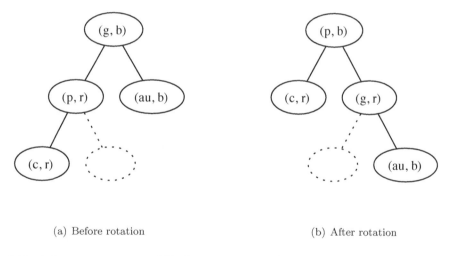

(a) Before rotation (b) After rotation

Figure 9.12 *Right rotation in a red-black tree: c, g, p, au stand for child, grandparent, parent, aunt/uncle, b and r indicate red and black and the dotted node is a possible child*

1. replace the grandparent with the parent while coloring it black,
2. replace the parent with the child,
3. replace the aunt/uncle with the grandparent with the aunt/uncle as its right child and recolor it red and
4. make the right child of the parent, should one exist (e.g., the dotted node in Figure 9.12), the left child of the grandparent.

These steps restore the third invariant while not changing the number of black nodes on either branch of the tree. The process of rearranging the nodes in Figure 9.12(a) to have the form in Figure 9.12(b) is called a right rotation.

All that remains is the case where the new node is a right child of its parent. In that instance, the steps are to

1. swap the child node with its parent,

2. make the former parent the left child of its former child,

3. attach the left child of the "new" node, should one exist (e.g., the dotted node 2 in Figure 9.13(a)), as the right child of the former parent node and

4. perform a right rotation with the former parent now occupying the role of the "new" node.

Figure 9.13(b) shows the result from applying Steps 1–3 to the tree in Figure 9.13(a). The outcome is analogous to the tree in Figure 9.12(a) except that the former parent has taken the place of the "new" node. Thus, the invariant can now be restored by a right rotation.

The removal of a node from a red-black tree is somewhat more complicated than insertion although it involves similar operations. Details are provided, for example, in Cormen, et al. (2003).

It turns out that the additional information used in red-black trees is not needed if the efficiency requirement for each operation is replaced with an *amortized bound*. One example of a data structure that achieves an amortized logarithmic time per operation is the *splay tree* discovered by Sleator and Tarjan (1985). Unlike the red-black tree case, splay tree operations are quite simple. The basic outline is that after every access to a node, a series of *rotations* is used to float the node up to the root of the tree. For example, if a small number of nodes is accessed quite frequently, over time they will tend to stay close to the root for quick access. Also, given a random sequence of accesses to the elements of a splay tree, the tree tends to balance itself and its depth is reduced appreciably. See Sedgewick (1998) for more details.

9.5 Pointer arithmetic, iterators and templates

In this final section we lay the stage for transition into the C++ Standard Template Library (STL) container classes in the next chapter. These classes have two features of particular importance: they are all implemented as templates and they all contain internal classes of pointer-like objects called *iterator* that provide the means to navigate inside a container.

To appreciate and understand iterators one must write code that creates and uses them. That is a task that is undertaken in this section and in the exercises. Our development mirrors the STL in that we will create an iterator class that is nested inside an ADT class. This requires a bit of new C++ syntax that will expand our language skills.

To this point all our C++ ADT classes have consisted of simple integer key and/or data components. The second goal of this section is to illustrate how easy it is to eliminate such restriction by using templates. For that purpose we will take the simple linked list class of Section 9.2.2 and alter it to allow the nodes in the container to have keys and data components of arbitrary types.

9.5.1 Iterators

On one hand pointers are probably the most difficult aspect of the C and C++ languages. On the other, they provide a powerful tool for creating compile time adaptive programs through the use of an (at least somewhat) intuitive array structure. We have used the array connection to access the information that resides in the memory location that represents the actual value of a pointer. Specifically, if p is a pointer, p[0] will return the value stored in the memory location that is pointed to by p. The transition from p to p[0] is called dereferencing. In many cases we have considered, p[0] corresponds to the beginning of

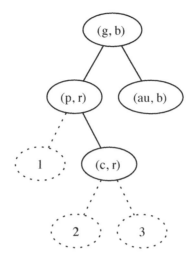

(a) Initial tree

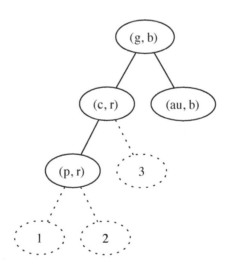

(b) After operations 1–3

Figure 9.13 *Initial rotation for dealing with a "new" right child: c, g, p, au stand for child, grandparent, parent, aunt/uncle, b and r indicate red and black and the dotted nodes are possible children*

consecutive blocks of memory. The `ith` block in this sequence can be accessed or dereferenced by the array type syntax `p[i]`.

In Section 3.5 we mentioned that a pointer can also be dereferenced with the $*$ operator; for a pointer `p` both `*p` and `p[0]` produce the same result. There is an alternative way to access the `ith` component in a sequence of consecutive blocks of memory, as well. If the sequence begins at the value of the pointer p, the syntax *(p + i) gives the same result as p[i]. This approach uses what is generally referred to as *pointer arithmetic*. The idea is that p + i is a pointer to the memory location i blocks removed from p and *(p + i) dereferences this pointer. In contrast to the array dereferencing approach, this makes a pointer look more like an index variable and we will use that perspective to generalize the pointer concept.

The simple program below illustrates the use of pointer arithmetic.

```
int main(){
  int* p = new int[5];
  for(int i = 0; i < 5; i++)
    *(p + i) = 3*i;

  for(int i = 0; i < 5; i++)
    cout << *(p++) << " ";
  cout << endl;

  return 0;
}
```

Both the addition and postfix increment operators are used here to traverse the memory that has been allocated to a pointer to int. The output from the program is

```
0  3  6  9  12
```

A somewhat different spin on the previous code uses the pointer directly in the indexing process.

```
int main(){
  int* begin = new int[5];
  int* end =   begin + 5;
  int i = 0;
  for(int* iter = begin; iter != end; iter++){
    *iter = 3*i;
    cout << *iter << " ";
    i++;
  }
  cout << endl;

  return 0;
}
```

Here the roles of the integer index and the pointer to int are reversed. The pointer iter now travels across the memory locations allocated to the pointer begin with its value restricted to lie between begin and another pointer end. The end pointer refers to the one-past-the-end memory location discussed in Section 3.5 that always exists and can be pointed to but not dereferenced. An equivalent formulation of the previous listing that uses a while loop is

```
int main(){
  int* begin = new int[5];
  int* end = begin + 5;
  int* iter = begin;
  int i = 0;
  while(iter != end){
    *iter = 3*i;
    cout << *iter << " ";
    iter++;
```

```
    i++;
  }
  cout << endl;
  return 0;
}
```

Other than providing information about a (possibly) interesting feature of C/C++ the development thus far would seem to have little to do with the ADT focus of this chapter. A connection can be made through a re-examination of the lookUp method for the linkedList class in Section 9.2.2. The basic recursion that was used there was

```
Node* iter = Head;
while(iter != 0){
  if(iter->Key == key)
    return iter;
  iter = iter->Next;
}
```

A comparison of this with the while loop version of our pointer arithmetic example reveals a certain similarity. They are, in fact, conceptually the same with Head playing the role of begin, the 0 or null pointer end node representing end and iter = iter->Next performing the function of the postfix increment operator for the pointer iter. Both our pointer arithmetic example and the lookUp function represent particular solutions to the more general problem of traversing and accessing a collection of objects; for the pointer arithmetic example the objects were data of type int while for lookUp the objects are of a more abstract nature.

The iterator (sometimes called a cursor) concept has developed as a standardized approach to navigation of ADTs. An iterator is an abstract, generic analog of a pointer. The generic part of this description manifests as a "guaranteed" set of operations provided by any iterator class: namely,

- ++, a postfix increment operator that advances the iterator to the next object in the ADT,

- *, a dereferencing operator that returns the current object being "pointed" to by the iterator and

- !=, a logical operator that allows for the comparison of two iterator objects.

The details behind the implementation of iterators for various ADTs will, of course, be (possibly very) different. Nonetheless, the user can expect these three basic operations to be provided and, in that sense, need not be familiar with the structure of the underlying ADT in order to explore and access its contents.

In many cases iterators possess other pointer-like operation such as -- (to move the iterator back to the previous object) or -> to access an object or some aspect of an object. Because iterators are ADT specific, the operations they will and can provide are determined by their underlying ADT. For example, in a singly linked list the decrement operator -- becomes problematic since the connection between nodes only moves from the head to the tail of the list. However, both ++ and -- are options if the list is doubly linked so that nodes have connections to both the next and preceding node in the list.

9.5.2 A linked list template class

Iterators require a close connection to the ADTs they traverse. In particular, an iterator must have access to the private members of its associated ADT. There are two ways that are typically used to create such a connection: a friend relationship between the iterator and ADT class or having the iterator class nested within the ADT class. We will take the latter approach here since it allows us to touch on a new topic (i.e., nested classes) and coincides with the way the C++ container classes treated in the next chapter have been implemented.

To illustrate the construction of a nested iterator class we will again work with a singly linked list class. In addition, to give the class some real utility, it will be implemented as a template.

The first step is to give the Node class a template structure. This is accomplished in the listing below.

```
template <class Key, class Data>
struct Node {
  Data dataValue;
  Key keyValue;
  Node* Next;

  Node(){}
  Node (Data data, Key key, Node* next = 0)
    : dataValue(data), keyValue(key), Next(next) {}
};
```

This is very similar to the definition of the Node struct from Section 9.2.2. The only substantive change is that the int data types for the key and data components have been replaced by the generic template parameters Key and Data.

The declaration for our linked list template container class LTC is given in Listing 9.7.

<hr>
Listing 9.7 *LTC.h*
<hr>

```
template <class Key, class Data>
struct LTC{

  Node<Key, Data>* Head;
  int Size;

  LTC(){Head = 0;}
  LTC(const LTC<Key, Data>& L);
  ~LTC();
  LTC<Key, Data>& operator=(const LTC<Key, Data>& L);

  //nested iterator class
  struct iterator{

    Node<Key, Data>* current;

    iterator(Node<Key, Data>* node = 0){current = node;}
    iterator(const LTC<Key, Data>& L) : current(L.Head) {}

    Node<Key, Data>& operator*();
    Node<Key, Data>* operator->();
    iterator& operator=(const iterator& iter);
    bool operator==(const iterator& iter) const;
    bool operator!=(const iterator& iter) const;
    iterator operator++();
    iterator operator++(int){return operator++();}
  };

    struct const_iterator : public iterator{
```

```
    const_iterator(Node<Key, Data>* node = 0) : iterator(node) {}
    const_iterator(const LTC<Key, Data>& L) : iterator(L){}

    const Node<Key, Data>& operator*();
    const Node<Key, Data>* operator->();
    const_iterator& operator=(const const_iterator& iter);
  };

  iterator begin();
  iterator end();
  const_iterator begin() const;
  const_iterator end() const;

  void push(const Data& newData, const Key& newKey);
  iterator lookUp(const Key& key);
  iterator insert(const Data& newData, const Key& newKey,
      const Key& key);
};
```

It looks like there are many new features here and that is certainly the case. But, a comparison between Listing 9.7 and Listing 9.2 reveals that this new class is not far removed conceptually from the linkedList class of Section 9.2.2. To be a bit more specific, if the two internal iterator classes and the begin and end methods are ignored, the two classes have the same basic elements and methods with the same number (if not type) of arguments. As with the Node struct, the changes required to adapt the LTC class from its linkedList predecessor mostly involve replacing the int key and data types with template parameters to allow for indexing and storage using objects of general types.

Let us begin the discussion of Listing 9.7 by examining the internal iterator struct for the LTC struct. As the phrase "nested" suggests, the entire declaration for class iterator is contained within the LTC class declaration.

The operative aspect of the iterator class in Listing 9.7 is its current member. This is a cursor-like object that moves between the different nodes in an LTC object. There are two class constructors: a default constructor that takes a pointer to Node argument with current initialized as the null pointer by default and a constructor that takes an LTC object as its argument and initializes current as the Head member of the object. Class methods are provided for dereferencing or object access (i.e., *), assignment, comparison of iterator objects (i.e., == and !=) and prefix and postfix increment operators that move current forward. In this latter respect, the postfix and prefix operators are the same with the latter being obtained by application of the former. Note the presence of a dummy int argument that allows the postfix operator to be overloaded.

The syntax for giving definitions of nested class methods outside of a template class can become somewhat involved. For example, the definitions for the prefix increment operators is

```
template <class Key, class Data> typename LTC<Key, Data>::iterator
      LTC<Key, Data>::iterator::operator++(){
  if(current != 0)
    current = current->Next;
  return *this;
}
```

The body of the method is straightforward in that the current member of the iterator object is set to the Next pointer thereby moving it to the next node in the list. It is the code leading up to that point that requires a bit of deciphering. First, there is the by now familiar template <class Key, class Data> code segment that informs the compiler that Key and `Data` are template parameters that will be defined elsewhere. The fact that `iterator` is nested in the `LTC` class means that the scope resolution operator must be used twice; the syntax `LTC<Key, Data>::iterator::operator++` states that `operator++` is a member of the `iterator` class which, in turn, is a member of the `LTC` class. The final issue is the return type. In this case `LTC<Key, Data>::iterator` indicates that the return type is an `iterator` object where the identity of `iterator` is contained in the class `LTC<Key, Data>`. In such cases the keyword `typename` tells the compiler that `LTC<Key, Data>::iterator` is, in fact, a class (e.g., Vandervoorde and Josuttis 2003). The same consideration applies to the assignment operator whose definition is

```
template <class Key, class Data> typename LTC<Key, Data>::iterator&
      LTC<Key, Data>::iterator::operator=(const iterator& iter){
  current = iter.current;
  return *this;
}
```

The definitions for the two comparison operators for the `iterator` class are given in the next listing.

```
template <class Key, class Data> bool
      LTC<Key, Data>::iterator::operator==(const iterator& iter)
      const {
  return (current == iter.current);
}

template <class Key, class Data> bool
      LTC<Key, Data>::iterator::operator!=(const iterator& iter)
      const {
  return !(*this == iter);
}
```

The `==` operator involves a straightforward comparison of the `current` member pointers of two `iterator` objects. Seeing that `!=` is the logical complement of `==`, the latter operator can be obtained directly from the former. Note that the `typename` keyword does not appear in either return type because `bool` is already a recognized data type.

Finally, the dereferencing operators are defined as

```
template <class Key, class Data> Node<Key, Data>&
      LTC<Key, Data>::iterator::operator*(){
  if(current != 0)
    return *current;
}

template <class Key, class Data> Node<Key, Data>*
      LTC<Key, Data>::iterator::operator->(){
  if(current != 0)
    return current;
  else return 0;
}
```

If the "cursor" is not at the end of the list, the object pointed to by the `current` pointer is returned as either a pointer or reference to `Node`.

Now that one of the two iterator classes is in place we can give the forms for some of the `LTC` class methods. Let us consider the `begin` and `end` methods of the `LTC` class. These return starting and stopping point iterator objects for movement inside a list. They look like

```
template <class Key, class Data> typename LTC<Key, Data>::iterator
        LTC<Key, Data>::begin(){
  return iterator(*this);
}
```

```
template <class Key, class Data> typename LTC<Key, Data>::iterator
        LTC<Key, Data>::end(){
  return iterator();
}
```

The begin method returns an iterator object whose current pointer points to the head node of the list while end produces an iterator object with a null pointer as its current member.

The push method for the LTC class is given in the next listing.

```
template <class Key, class Data> void
        LTC<Key, Data>::push(const Data& data, const Key& key){
  Node<Key, Data>* newNode = new Node<Key, Data>(data, key, Head);
  Head = newNode;
  Size++;
}
```

Apart from the template parameters, this is just the push method for the linkedList class. It makes no use of iterators as there is no need to perform actions anywhere inside the class. In contrast, the lookUp method is iterator driven as seen from

```
template <class Key, class Data> typename LTC<Key, Data>::iterator
        LTC<Key, Data>::lookUp(const Key& key){
  iterator iter(*this);
  while(iter != end()){
    if(iter->keyValue == key)
      return iter;
    iter++;
  }
  return 0;
}
```

This has the same basic form as the lookUp method of the linkedList class. Except now the details behind explicitly using a pointer to Node to iterate through the list have been encapsulated in an iterator object iter. The object is created with the constructor that takes an LTC object as its argument with the consequence that its current member points to the head node. It is moved through the list via the increment operator until either the requested key value is located or the end of the list is reached. The key values that are compared to the target are obtained by dereferencing the iter object.

The insert method for the LTC struct was encoded as

```
template <class Key, class Data> typename LTC<Key,Data>::iterator
        LTC<Key, Data>::insert(const Data& newData, const Key& newKey,
        const Key& key){
  if(lookUp(newKey) == end()){//key does not exits
    iterator iter = lookUp(key);
    if(iter != end()){//target does exist
      Node<Key, Data>* newNode =
        new Node<Key, Data>(newData, newKey);
      newNode->Next = iter->Next;
      iter->Next = newNode;
      Size++;
```

```
      return iter;
    }
    else{
      std::cout << "Key not found!" << std::endl;
      return end();
    }
  }
  else{
    std::cout << "Duplicate key!" << std::endl;
    return end();
  }
}
```

For the most part this repeats the theme of this section in that the code for the LTC class insert method mimics that for our previous linkedList class apart from some modifications to incorporate template parameters and the replacement of pointers to Node objects with iterators. A slight alteration has been made in that, upon completion of the insertion operation, the LTC insert method returns an iterator object that points to the spot of insertion.

Now let us return to the other iterator class: const_iterator. As the name suggests, this class provides iterators that can be used for read-only access to const LTC objects. To appreciate the need for this type of iterator consider the nonoperative implementation below of the LTC class copy constructor.

```
template <class Key, class Data> LTC<Key, Data>::LTC(const LTC& L){
  if(L.Size == 0){
    Head = 0;
    Size = 0;
  }
  else{

    Size = L.Size;
    iterator iter;
    Node<Key, Data>* temp = 0;
    for(iter = L.begin(); iter != L.end(); iter++){
      if(temp == 0){//empty list
        Head = new Node<Key, Data>(iter->dataValue,
        iter->keyValue);
        temp = Head;
      }
      else{
        temp->Next = new Node<Key, Data>(iter->dataValue,
        iter->keyValue);
        temp = temp->Next;
      }
    }
    temp->Next = 0;//null pointer as Next for tail node
  }
}
```

An attempt at compilation using this form of the constructor will produce a complicated error message including the standard admonishment that one receives when attempting to handle a const object with a method that can potentially alter its arguments: namely, that passing the this pointer to a non-const method (i.e., begin) "discards qualifiers".

The solution to the problem is to create another iterator class that can access but not alter nodes. Motivated by the analogous STL concept, the class has been named const_iterator.

The basic `iterator` class methods for comparing and incrementing will work perfectly well for our new class. Hence, `const_iterator` has been made a derived class from class `iterator`. The changes occur for the dereferencing operators which now take the form

```
template <class Key, class Data> const Node<Key, Data>&
        LTC<Key, Data>::const_iterator::operator*(){
  if(iterator::current != 0)
    return *iterator::current;
}

template <class Key, class Data> const Node<Key, Data>*
        LTC<Key, Data>::const_iterator::operator->(){
  if(iterator::current != 0)
    return iterator::current;
  else return 0;
}
```

Both methods have been specified to have `const` return types which forbids alteration of the data in the underlying `LTC` object. In addition to these two methods, overloaded versions of the `begin` and `end` methods are needed that will return `const_iterator`, rather than `iterator`, objects: namely,

```
template <class Key, class Data>
        typename LTC<Key, Data>::const_iterator
        LTC<Key, Data>::begin() const {
  return const_iterator(*this);
}

template <class Key, class Data>
        typename LTC<Key, Data>::const_iterator
        LTC<Key, Data>::end() const {
  return const_iterator();
}
```

To fix the LTC copy constructor it now suffices to replace the previous `iterator iter;` code segment with `const_iterator iter;`. The same basic idea works for the assignment operator shown below.

```
template <class Key, class Data> LTC<Key, Data>&
        LTC<Key, Data>::operator=(const LTC& L){
  if(this == &L)
    return *this;

  if(L.Head == 0){
    Head = 0; Size = 0;
    return *this;
  }

  Size = L.Size;
  Node<Key, Data>* temp = 0;
  for(const_iterator iter = L.begin(); iter !=L.end(); iter++){
    if(temp == 0){//empty list
        Head = new Node<Key, Data>(iter->Data, iter->Key);
        temp = Head;
    }
    else{
        temp->Next = new Node<Key, Data>(iter->Data, iter->Key);
```

```
        temp = temp->Next;
    }

    temp->Next = 0;//Next for tail node
    }
    return *this;
}
```

Code was written to test the LLC class parts of which are shown in the next listing.

```
void f(const LTC<int, int>  L){
    std::cout << L.Head->Next->dataValue << std::endl;
}

int main(){
    LTC<int, int> L;
    L.push(2, 15); L.push(3, 18); L.push(4, 28); L.push(5, 27);
    L.push(6, 19); L.push(7, 49); L.push(2, 38); L.push(4, 6);
    L.push(100, 2); L.push(-1, 40);
    L.insert(-4, 1, 18);
    f(L);//test of copy constructor
    std::cout << L.lookUp(1)->dataValue << std::endl;
    LTC<int, int>::iterator iter(L);
    for(iter = L.begin(); iter != L.end(); iter++)
        std::cout << iter->keyValue << " ";
    std::cout << std::endl;

    LTC<int, int> newL = L;//test of = operator
    std::cout << newL.Head->Next->keyValue << std::endl;

    return 0;
}
```

An LTC object is created using the push and insert methods. The function f is then used to force a call to the copy constructor after which the lookUp method is used. An iterator object is created next and used to explore the LTC object. Finally, the assignment operator is used to create another LTC object whose second node is examined. The output produced by this code is the expected

```
100
-4
40 2 6 38 49 19 27 28 18 1 15
2
```

9.6 Exercises

9.1. Write the definitions for the copy constructor, destructor and assignment operator in Listing 9.1 and demonstrate through an example that they work as expected.

9.2. Write a program that allows for adaptive selection of dArray class elements through user input. Then, rework the [] operator by using try and catch blocks to handle inappropriate index choices with prompts to the user for new values.

9.3. Write a method for the dArray class of Listing 9.1 that will delete the data at a specified index value and restructure the array so that none of its Size entries are empty.

9.4. Reformulate the dArray class to where it both grows and shrinks dynamically. To accomplish this the reSize method should be renamed grow and called whenever the array object becomes full. Then, add a new method `shrink` that reduces the memory allocated to a `dArray` object by a fixed proportion whenever the number of objects in the container drops sufficiently far below the current value of `containerSize`.

9.5. Create a dynamic array class in R with associated constructor and method functions for insertion, deletion and resizing.

9.6. Let $a_1, \ldots, a_n$ be the n elements of an array. Consider a randomized version of quicksort in which the pivot is chosen uniformly at random at every level of the recursion and in every subarray.

a) Show that the expected number of comparisons in an execution of the algorithm is $O(n \log n)$. For simplicity, assume that all the a_i are distinct. [Hint: let $i < j$ be an arbitrary pair of indices. Show that the probability that a_i and a_j are compared during a complete execution of this randomized version of quicksort is $2/(j - i)$.]

b) Use your result from part a) to show that if the elements of the array are input in a uniformly random order, the expected running time of the deterministic quicksort algorithm from Section 9.2.1 is $O(n \log n)$.

9.7. Write C++ code for the recursive mergesort Algorithm 9.7 for arrays from Section 9.2.1.

9.8. Write R code that uses the `Compare` group generic functions to implement both quicksort and mergesort for the `g5List` objects of Section 6.4 with ordering based on draw dates.

9.9. Show that when $n = 2^k$ the worst-case running time for mergesort is $O(n \log n)$.

9.10. Write the code for the overloaded = operator for the `linkedList` class of Section 9.2.2.

9.11. Write code for `remove` and `clear` methods in the `linkedList` class of Section 9.2.2 that will delete a specific node or remove all the nodes from a list object.

9.12. Develop a version of Algorithm 9.5 that will insert an object to the left of (i.e., immediately prior to) an object with a designated key in the singly linked list class of Section 9.2.2.

9.13. Add a `pop` method that removes and returns the head node from the `linkedList` class of Section 9.2.2.

9.14. Write an alternative version of the `push` method for the `linkedList` class of Section 9.2.2 that builds a linked list by appending new objects to the tail of the list.

9.15. In some instances the look-up of items in a list is driven by their "popularity": e.g., some nodes have data that needs to be accessed more often than others. In such instances the look-up operation can be made more efficient by moving the "popular" nodes to the head of the list. The result is referred to as a *self-organizing list*. Create a self-organizing version of the `linkedList` class in Listing 9.2. To achieve the self-organizing property, alter the `lookUp` method to where it moves the node with the requested key to the head of the list.

9.16. Show that the average access time (where average is taken over all permutations of the sequence of insert and look-up operations) for an element a in a chaining hash table is half the number of all elements that are hashed into the same location as a.

9.17. *Double hashing* uses two hash functions, h_1 and h_2, in an open addressing scheme where the ith attempt to find an empty slot for an item with key k is made at $(h_1(k) + ih_2(k))$ mod m where m is the size of the table. Discuss the pros and cons of this approach. In particular, can you envision a case where this approach can fail to produce a useful probe sequence?

9.18. Write the code for the copy constructor and overloaded = and == operators for the `hashTable` class of Section 9.2.4.2. Why is the == operator needed here?

9.19. Implement a `delete` method for the `hashTable` class in Section 9.2.4.2 that uses a sentinel to replace a deleted value.

9.20. Create an alternative version of the `newPair` class that uses composition rather than inheritance; i.e., `newPair` is not a derived class of `pair` but, instead, has a `pair` object as a class member. How does this affect the `hashTable` class in Section 9.2.4.2?

9.21. Create a chaining hash table for integer key and data components based on the `linkedList` class of Section 9.2.4.2.

9.22. Write a function for the `hashTable` class in Listing 9.4 of Section 9.2.4.2 that will resize an existing `hashTable` object. Create a similar function that will work with your chaining hash table from Exercise 9.21.

9.23. An algorithm is said to sort an n-element set *by comparison* if the only operations used to determine the ordering are direct element-to-element comparisons (for example, no arithmetic, manipulation of set members or counting is allowed). Prove that any algorithm that sorts by comparison requires at least order $n \log n$ operations. [Hint: Consider the change in the set of potential solutions when a single comparison is made and use Stirling's approximation formula.]

9.24. Suppose keys come from the set of floating-point numbers in the open interval (a, b). Generalize the simple hash function of Example 9.1 to this range.

9.25. Suppose keys are integers represented by b binary digits. Show how to reduce the problem of finding a hash function to the floating-point case of Exercise 9.24.

9.26. Show that (9.1) defines a universal hash family.

Knuth (1998b), Carter and Wegman (1979) and Wegman and Carter(1981)

9.27. A Bloom filter consists of an m-bit array (whose entries are all initialized to 0) and d hash functions $h_1, \ldots, h_d$ with range $\{0, \ldots, m - 1\}$. It is used to compactly represent a finite set $A = \{x_1, \ldots, x_n\}$. To insert x into A, set the d bits $h_1(x), \ldots, h_d(x)$ to all be 1. To check if $x \in A$ one then checks to see if all of these d bits are equal to 1. Clearly, the only possibility for error is a false positive; even if x does not actually belong to A, it may be that $h_1(x), \ldots, h_d(x)$ are all 1 because one or more elements of A map into this same bit pattern. Assume that the d hash functions are random (i.e., h_i maps each x uniformly, at random into $\{0, \ldots, m - 1\}$).

a) Estimate the probability of a false positive. [Hint: First show the expected number of bits that are still equal to 0 after the insertion of all n elements of A into the data structure is about $p = e^{-dn/m}$. Then, show that the probability of a false positive is $(1 - p)^d$ by conditioning on the fraction of zero bits.]

b) If n and m are fixed, show that to minimize the false positive probability we should have $p = 1/2$: i.e., $d = (m/n) \ln 2$.

Broder and Mitzenmacher (2005)

9.28. Surprisingly, it is possible to implement hashing so that look-up operations take *guaranteed* constant time. One of the simplest such schemes is called *cuckoo hashing* (e.g., Pagh and Rodler 2004). This implementation of a dictionary consists of two tables, T_1 and T_2, each of size m and two hash functions, h_1 and h_2, each with range $\{0, \ldots, m - 1\}$. Each key of the dictionary is stored in exactly one of the two tables. Thus, to check if a is in the dictionary, we only need check if a can be found at location $h_1(a.key)$ of T_1 or at location $h_2(a.key)$ of T_2. To insert a, first try $T_1[h_1(a.key)]$. If this entry is empty, insertion is done. Otherwise, a is still inserted and the current occupant is moved to T_2 using the same process with hash function h_2. Should that spot be occupied, the resident is replaced leaving a new homeless object at which point the roles of T_1 and T_2 are reversed.

If the process of kicking out elements to replace them by new ones continues for too long, it is possibly in a loop. So, we keep track of the number of iterations and if some prespecified upper bound is exceeded, we pick a new pair of hash functions and rehash everything. While drastic, this final step happens only rarely in practice. Give a pseudo-code implementation of cuckoo hashing.

9.29. Consider the problem of repeatedly inserting objects with different keys into a hash table with m array slots. Show that if collision resolution is by linear probing, the table will eventually be filled if and only if m and the shift parameter are relatively prime.

9.30. Add the following new methods to the `Heap` class in Listing 9.5.

a) A constructor that takes a reference to a `dArray` object as its argument.

b) A constructor that takes a reference to a linked list object as its argument.

c) A method `clear` that empties the heap.

d) A `delete` method that finds an object with a specified key and removes if from the heap.

9.31. Rework the `Heap` class of Section 9.3.2 to produce a *max-heap*: i.e., the key value of every node is no smaller than that of its left and right child.

9.32. The *shortest path problem* is usually defined on a graph $G = (V, E)$. Each edge in $e \in E$ has a nonnegative *length* $c(e)$ associated with it. Given two vertices in V, the *source* s and the *destination* t, the shortest path problem is to find a path from s to t of minimum total length.

In Dijkstra's algorithm we maintain a set S of "explored" vertices. Together with each vertex $v \in S$, we store the length $d(v)$ of the shortest path from s to v. In each step, we select an edge $e = uv$ such that $u \in S$, $v \notin S$ and $d(u) + c(uv)$ is minimized. The vertex v is then added to S and for each edge vw, we check if going from s to v and then to w produces a shorter path than the shortest one currently known to w. Every time such a shortcut is found, the relevant information is updated. This process repeats until t is added to S.

a) Prove correctness of this algorithm by establishing the following invariant: at any point during the execution of the algorithm, the value $d(v)$ for any $v \in S$ is the length of the shortest path from s to v.

b) Give a pseudocode description of Dijkstra's algorithm based on the outline just presented. Use a priority queue to efficiently implement the updates.

9.33. One application of the shortest path problem of Exercise 9.32 is to finding the most reliable paths in networks. Suppose instead of length, the value associated with each edge gives the probability that the edge is "active". Show how to use any algorithm for the shortest path to find the "most reliable" path between a pair of vertices, where the reliability of a path is expressed as the product of the reliabilities of its constituent edges.

9.34. Write pseudo-code for the insertion of a node into a red-black tree for the case where the parent of the new node is a right child.

9.35. Another simple data structure that is in a sense equivalent to a (randomized) binary search tree is a *skip list* (e.g., Pugh 1990). The idea here is to begin with an ordinary linked list containing elements in a sorted order and correct its inefficiencies. The only real problem is the search operation for which a simple trick is to add another list that works at a *coarser resolution*. This second list would contain only a subset of the elements, and its entries would be linked directly to the corresponding ones in the first (exhaustive) list. The second list could then be used to roughly locate an element by walking through it until two entries are found that "bracket" the target of the search. At this point a descent can be made into the exhaustive list with the search now restricted to the smaller range indicated by the first list.

The natural generalization of this idea with a view towards efficiency quickly leads to representing an n-element set by using $\log_2 n$ lists ("levels"), each of which contains roughly half of the elements of the previous one. To implement this in practice, we use randomization. To insert a node a, start in the top list and walk from the smallest element until two elements are found with keys that surround the key for a. Then, descend to the next lower level and continue the process. Once a location for a is found in the lowest (exhaustive) list, it may be inserted there. Insertion needs to consider the higher levels as well. To maintain useful properties, always insert the new entry into the lowest list, and then, for each level above it, flip an unbiased coin to decide whether to add it to the list in the higher level. This ensures that every list is roughly half the size of the list immediately below it and that with high probability there are about $\log_2 n$ lists. Prove that the search time for a skip list is $O(\log_2 n)$. Then, implement a C++ skip list version of the `Tree` class of Section 9.4 and compare the running times of look-up operations for these two classes.

9.36. Alter the `LTC` class in Listing 9.7 to obtain a doubly linked list where every node has two pointers: one that points to the next list object and one that points to the previous element. Add further functionality to this class by creating the new methods/operators

a) `sort` that uses mergesort to arrange the nodes in the list with either increasing or decreasing values for their keys,

b) `--` for the nested iterator classes that allows for reverse iteration or backward movement of an iterator,

c) `rbegin` and `rend` that provide the reverse iterator analogs of `begin` and `end` and

d) `pushFront`, `pushBack`, `popFront` and `popBack` that allow nodes to be added or removed from either end of a list object.

9.37. Create a C++ dynamic matrix container that will hold objects from a generic class `T` in a row and column format with both the number of rows and columns being allowed to expand as needed to hold data. Include an appropriate nested iterator class.

9.38. Write C++ code for a template class stack data structure.

9.39. Use the stack class from Exercise 9.38 to implement Algorithm 2.4 from Chapter 2.

9.40. Create an internal iterator class/struct for the `Tree` class of Section 9.4.1 that will traverse the nodes of a `Tree` object by moving in sequence from elements with smaller to larger key values. [Hint: Use your stack class from Exercise 9.38.]

9.41. Use a template approach to generalize your chaining hash table from Exercise 9.21 to hold an arbitrary (i.e., not necessarily integers) data type. Create an internal `iterator` class/struct that will work with this hash table.

Chapter 10

Data structures in C++

10.1 Introduction

The previous chapter presented a variety of abstract data structures (ADTs) that were implemented in C++ and R. This introduction served three purposes:

- it provided an overview of the ADTs that are available in R,

- it gave a general treatment of several important ADTs and their properties and

- it provided illustrations of how one could originate their own code for an ADT of interest.

The present chapter takes a perspective that is similar to our use of the Template Numerical Toolkit for numerical linear algebra in Section 7.6. That is, rather than create code for an ADT from scratch, in most instances it is more advisable to use an existing, third-party ADT implementation. The C++ Standard Template Library (STL) provides a readily available source with support for many of the common ADTs.

The STL includes three basic components: container classes, iterators and algorithms. The phrase *container class* refers to a class that can hold objects from some general class. The container classes in the STL are all ADTs.

Iterators are pointer-like objects that provide the means to explore and access the objects in an ADT. We discussed the concept of iterators in Section 9.5. All the STL containers implement iterators that have been tailored to work effectively with the different ADTs.

The STL algorithm component is a collection of functions that can be used with containers for purposes that include inserting, finding or removing objects and sorting. Its applications are not restricted to just the STL containers. The algorithm library is implemented in a generic fashion that allows it to work with, e.g., user-defined containers as well as many other applications. In fact, we have already used the `max` and `min` functions from the algorithm library in previous chapters.

In subsequent sections we will examine some of what the STL has to offer. The objective is to demonstrate how to use the essential features of the different containers and algorithms while also applying them in example settings that illustrate their utility. More detailed discussions can be found in a myriad of books and on-line tutorials. In particular, Josuttis (1999) provides a thorough and very readable overview of all three of the STL's components.

10.2 Container basics

The standard STL containers are

- the `vector` and `deque` dynamic array classes,

- the linked list container `list` and

- `map`, `multimap`, `set` and `multiset`.

The `vector` class is an implementation of a dynamic array ADT. The name `deque` stands for "double-ended queue". Thus, as might be expected, objects can be inserted or removed from either end of this container. The random access that `deque` provides to its contents as well as other features make it a dynamic array rather than a queue in the usual first-in first-out sense of the term. The `list` container is a doubly-linked list class. The `map` container is a dictionary where the objects in the container have both a key and a data component. For map the keys must be unique while `multimap` allows for repeated keys. The `set` containers are simplified versions of `map` containers where objects have a single member that serves as both the key and data component.

There are certain operations that are provided by all the standard container classes in the STL. In this section we will discuss several of these "global" methods. Later sections will illustrate their use for specific container types as well as other methods that are container-type specific.

To use a particular container, `ADTName`, from the STL one needs the `include` directive

```
#include <ADTName>
```

All the container classes are equipped with a default constructor, a constructor that takes arguments appropriate for the specific ADT, a copy constructor, an assignment operator and a destructor. Thus, for example, the default constructor for `ADTName` would be invoked with a command such as

```
ADTName<T> ADTObject;
```

Of course, a specific value (i.e., a data type name) must be supplied for `T` to define an actual `ADTName` object. That choice is at the user's discretion. This serves the same purpose as in our developments in Section 9.5: namely, it makes the ADT a container in the sense that it can hold objects from an arbitrary class.

The methods `empty` and `size` can be used with any of the STL containers. If `ADTObject` is an `ADTName` object, then

```
ADTObject.empty();
```

will return `true` if `ADTObject` contains no elements and `false`, otherwise. Similarly,

```
ADTObject.size();
```

returns the number of elements in `ADTObject`.

Movement through STL containers (and in some cases other indexing-type operations) is carried out via iterators. In this regard, every STL container has the two methods `begin()` and `end()` that return iterators pointing to the first element of the container and to one position past its last element, respectively. The reverse iterators `rbegin()` and `rend()` are also available. The iterator returned by `rbegin()` points to the last element in the container and the one returned by `rend()` points to one location before the first element in the container. As might be expected, reverse iterators work the same as regular or forward iterators except that `++` and `--` have the opposite effect of moving the iterator backward and forward, respectively. An example of using a reverse iterator is given in Section 10.4.1.

The comparison operators `==` and `!=` return values of a Boolean variable. For two `ADTName` objects `ADTObject1` and `ADTObject2`, the expression

```
ADTObject1 == ADTObject2
```

will evaluate to `true` if the two objects contain the same elements arranged in the same order. The `<`, `>`, `<=`, `>=` operators will also work in the sense of providing lexicographical comparisons between two containers. For example, the `<` operator compares two containers element-by-element until

- two elements are found that are not equal; the result of applying `<` to the two containers is then taken to be the result of applying `<` to these two elements,

- one of the containers has fewer elements than the other; the one with fewer elements is viewed as being "smaller",

- the two containers are found to be equal in which case `<` returns `false`.

The common methods that actually modify containers are `insert`, `erase` and `clear`. For an ADT object, `ADTObject`,

```
ADTObject.insert(iter, object);
```

will insert `object` into the container at the position specified by the iterator `iter`. The command

```
ADTObject.erase(iter1, iter2);
```

will remove all the elements in `ADTObject` in the positions starting with `iter1` and ending one position prior to `iter2`. The use of

```
ADTObject.clear();
```

is equivalent in its effect to `ADTObject.erase(begin(), end())` (apart from return values for some containers) in the sense that all elements in the container will be removed and the container will be empty; i.e., `ADTObject.empty()` will evaluate to `true`.

In addition to the standard containers, the STL includes the *container adapters* `stack`, `queue` and `priority_queue`. These are adaptations of the standard containers that function like stacks, queues and priority queues. That is, the `stack` container operates under a last-in first-out paradigm while `queue` containers remove elements on a first-in first-out basis. The `priority_queue` container implements a queue where objects are withdrawn according to their "priority" (see Section 9.3). At any given point, the only member of a stack or queue that can be accessed is the one on top and the same is true for the `stack`, `queue` and `priority_queue` containers. Thus, the internal elements of these containers cannot be accessed and, accordingly, iterators are not defined for these classes.

The starting point for our discussion of specific container types is the `vector` container that provided the model for our `dArray` class in Section 9.2.1. This is the subject of the next section.

10.3 Vector and deque

The C+ `vector` container class is a very flexible data structure that, as its name suggests, possesses similar features to a one-dimensional array. However, as a type of dynamic array it represents a major upgrade of the array concept. In particular, it provides storage flexibility in the sense that the size of a `vector` object can be expanded adaptively at run-time. This is precisely what is typically needed when trying to access data files of unknown size and/or structure.

Although the `deque` title is an acronym for "double ended queue", the implementation of this ADT is through a dynamic array structure that allows for random access to its elements. In that sense it is more closely related to the `vector` container than to the other STL queue containers. Accordingly, we will also discuss the `deque` class briefly at the end of this section.

Access to the `vector` container class is obtained via the directive

```
#include <vector>
```

As noted in the previous section, to create a `vector` container object v holding objects of type `className` one would employ the syntax

```
vector<className> v;
```

So, for example,

```
vector<double> v;
```

would produce something that would work just like an ordinary double precision array in that the ith component of v would be accessed using the familiar syntax v[i]. This dereferencing works quite generally in that regardless of the type of object that is stored in the vector container, the use of v[i] will always return the object that is stored in its ith slot.

Of particular interest for data analysis purposes is the case

```
vector<vector<double> > v;
```

that produces a container where v[i] now corresponds to an "ordinary" array of doubles in its own right. So, for all practical purposes, v is a matrix with ith row v[i] and its (i, j)th entry is accessed with v[i][j]. Note that the extra space after the first > in the declaration for v is necessary.

The advantage of using something like vector<vector<double> > v in lieu of double v[][] or double** v, is in the infrastructure provided by the STL. For example, when using pointers or arrays, a programmer must, at some point, make a commitment in terms of the size of the structure. Subsequent resizing is, at best, tedious and error-prone. In contrast, class vector provides built-in methods that allow the user to adaptively add elements to the structure: either one-by-one or by inserting multiple elements at a time.

The specific methods for the vector class that will be of most immediate utility for us are capacity, size, pop_back, push_back, insert and assign. The function size returns an integer value that represents the number of objects that are currently being held in a vector object. The capacity method gives the number of slots in a vector object. The difference between the value returned by capacity and size is the number of objects that can be added to a vector object before more memory must be allocated.

The push_back function adds an additional element onto the end of the vector. So, the code snippet

```
vector<int> v;
int a = 1;
v.push_back(a);
```

creates a new empty vector object to hold variables of type int and creates a new slot at the end of v that is filled with the value in the variable a (or 1, in this case). To remove the last component from a vector we proceed similarly using the function pop_back. So,

```
v.pop_back();
```

would undo what we had previously accomplished with push_back.

Insertion of an object obj into a vector object requires us to use the iterator class that is an internal (i.e., nested) feature of class vector. In general, such iterators are obtained via the syntax

```
vector<className>::iterator iter;
```

Then, as in Section 10.2, the insert method for class vector takes the form

```
v.insert(iter, obj);
```

where obj is the object to be inserted into the container. The effect of this command will be to insert the object in the slot immediately prior to that specified by iter and return an iterator that points to the inserted object. Thus, for example,

```
vector<className>::iterator iter = v.insert(begin() + 2, obj);
(*iter).methodName();
```

will insert obj in the third slot of v while shifting all the other objects one slot to the right. The iterator returned from insert is used to call a method methodName for the class of which obj is a member. The last line of code is therefore equivalent to `obj.methodName();`. The fact that `begin() + 2` will work in the previous code snippet is not immediately obvious. It resembles what can be done with pointer arithmetic (cf. Section 9.5.1). In general this type of syntax would not be meaningful. Random access iterators (as provided by the `vector` and `deque` classes) are the special case where it is allowed. Similarly, expressions such as `iter[i]`, `iter - k`, `iter -= k` and `iter += k` for k an integer are valid in the random access case.

The code below illustrates using the methods that have been described so far.

Listing 10.1 *vecEx1.cpp*

```
//vecEx1.cpp
#include <iostream>
#include <vector>

using std::cout; using std::endl; using std::vector;

void printSize(int size, int cap){
  cout << "The size is " << size <<
    " and the capacity is " << cap << endl;
}

int main(){
  vector<int> v(1);
  v[0] = 12;
  printSize(v.size(), v.capacity());
  v.push_back(1);
  printSize(v.size(), v.capacity());
  v.push_back(2);
  printSize(v.size(), v.capacity());
  for(int i = 0; i < v.size(); i++)
    cout << "v[" << i << "] = " << v[i] << " ";
  cout << endl;

  v.pop_back();
  vector<int>::iterator iter = v.insert(v.begin() + 1, 17);
  cout << "Dereferenced iterator = " << *iter << endl;
  for(iter = v.begin(); iter != v.end(); iter++)
    cout << *iter << " ";
  cout << endl;
  v.clear();
  printSize(v.size(), v.capacity());

  return 0;
}
```

Listing 10.1 begins with a print utility program. Then, the `main` function initializes a `vector` container for `int` variables, places the integer 12 in its first slot and uses the `push_back` method to append two integers to the array. After each addition, the values returned from `size` and `capacity` are printed to standard output before finally examining the contents of the `vector` object. Next, the last inserted element is removed with `pop_back` and the `insert` method is employed to insert the integer 17 between the first and second `int` in the container (i.e., in the slot with index 1). The position for insertion is designated by incrementing the `begin()` iterator for the `vector` object. A new iterator is defined to hold the one returned from `insert` and this new iterator is dereferenced to view the `int` value

that was inserted. The contents of the `vector` object are checked again at this point with a
`for` loop. In contrast to the previous traversal of the container that used the [] operator and
an integer index, this time it is done using an iterator that i) is initialized with `begin()`, ii) is
moved to the next object in the container via the postfix ++ operator and iii) terminates the
loop when it reaches the object at the slot immediately before the `end()` iterator. Finally,
the `clear` method from Section 10.2 is called and the values of `size` and `capacity` are
checked again to see its effect.

Upon execution, our program produced the output

```
The size is 1 and the capacity is 1
The size is 2 and the capacity is 2
The size is 3 and the capacity is 4
v[0] = 12 v[1] = 1 v[2] = 2
Dereferenced iterator = 17
12 17 1
The size is 0 and the capacity is 4
```

As with our `dArray` class of Section 9.2.1, the capacity of the `vector` object is doubled each
time a resizing operation is required. The last component of the array (i.e., 2) is successfully
removed with `pop_back` and then the integer 17 is inserted between the first and second
elements that have values of 12 and 1, respectively. After calling `clear` we see that the
capacity of the array has not changed; but, its content has been removed so that the value
of `size()` is 0.

An illustration of the creation of a `vector<vector<> >` construct is provided in the next
listing.

Listing 10.2 vecEx2.cpp

```cpp
//vecEx2.cpp
#include <iostream>
#include <vector>

using std::cout; using std::endl; using std::vector;

int main(){
  int nrows = 3, ncols = 2;
  vector<vector<int> > v;
  v.resize(nrows);
  for(int i = 0; i < nrows; i++){
    v[i].resize(ncols);
    for(int j = 0; j < ncols; j++)
      v[i][j] = (i + 1)*(j + 1);
  }

  for(int i = 0; i < v.size(); i++){
    for(int j = 0; j < v[i].size(); j++)
      cout << "v[" << i << "][" << j << "] = "
           << v[i][j] << " ";
    cout << endl;
  }

  return 0;
}
```

In this example a 3×2 array v of integers is created. Initially, the default `vector` constructor is used to obtain an "empty" array v with no slots. Then, slots are added to v using the `resize` method.

In general, for a `vector` object v the statement

```
v.resize(k);
```

will change its number of elements from `v.size()` to k. If the container has more than k objects, the first k elements are retained and the others are deleted. If, as is the case in Listing 10.2, k (i.e., 3) is larger than the current value of `v.size()` (i.e., 0), the contents of v are expanded by an additional `k - v.size()` objects created with their default class constructor while the existing content of v remains unchanged. Thus, for our case where v contains `vector` objects, the command

```
v.resize(nrows);
```

will expand the `vector` object v to have `nrows` slots holding `vector` objects that all have the default length 0. A similar outcome is obtained from

```
v[i].resize(ncols);
```

in Listing 10.2 in that the `vector` object v[i] is dimensioned to have `ncols` slots or, equivalently, the array corresponding to v will be of dimension `nrows` × `ncols`. The `ncols` elements of v[i] are then filled using a `for` loop. The output from running Listing 10.2 is

```
v[0][0] = 1 v[0][1] = 2
v[1][0] = 2 v[1][1] = 4
v[2][0] = 3 v[2][1] = 6
```

Finally, let us discuss copying of one `vector` object into another. If v is a `vector<T>` object for some given class T, the simplest way to create a new `vector<T>` object that contains the content of v is with the class copy constructor using syntax like

```
vector<T> v1(v);
```

This produces an independent copy of v in the `vector<T>` object v1. Alternatively, to copy one `vector` object or, more generally, some portion of the object, into another, existing `vector` object the `assign` method for the vector class can be used. For example, the same result that was obtained with the copy constructor is produced by

```
vector<T> v1;
v1.assign(v.begin(), v.end());
```

Our choice of the range `v.begin()` to `v.end()` here is arbitrary and a subset of the elements of v can be copied by replacing these iterators with ones that point to other location in the `vector` that is being copied.

Listing 10.3 below illustrates copying using the `vector` class copy constructor and `assign` method.

Listing 10.3 *vecEx3.cpp*

```
//vecEx3.cpp
#include <iostream>
#include <vector>
#include <string>

using std::cout; using std::endl; using std::vector;
using std::string;

void printVec(vector<vector <int > > v, string s);
```

```
int main(){
  vector<vector<int> > v, v1;
  int nrows = 3, ncols = 2;
  v.resize(nrows);
  for(int i = 0; i < nrows; i++){
    v[i].resize(ncols);
    for(int j = 0; j < ncols; j++)
      v[i][j] = (i + 1)*(j + 1);
  }

  v1.assign(v.begin() + 2, v.end());
  printVec(v1, "v1");
  vector<vector<int> > v2(v);
  printVec(v2, "v2");

  return 0;
}

void printVec(vector< vector<int> > v, string s){
  cout << "The number of rows in " << s <<  " is "
       << v.size() << endl;
  for(int i = 0; i < v.size(); i++){
    cout<< "The length of row " << i << " is " << v[i].size()
        << endl;
    for(int j = 0; j < v[i].size(); j++)
      cout << s + "[" << i << "][" << j << "] = "
           << v[i][j] << " ";
    cout << endl;
  }
}
```

The main function creates a vector<vector<int> > object that is then populated with six integers arranged in three rows and two columns. The begin() iterator is incremented by two in an application of assign which should have the consequence that only the last row of the vector object v will be copied into the object v1. Next, the copy constructor is used to copy all of v into a new vector object v2. The results of this copying are printed to standard output using a utility function printVec created for that purpose. The program produces

```
The number of rows in v1 is 1
The length of row 0 is 2
v1[0][0] = 3 v1[0][1] = 6
The number of rows in v2 is 3
The length of row 0 is 2
v2[0][0] = 1 v2[0][1] = 2
The length of row 1 is 2
v2[1][0] = 2 v2[1][1] = 4
The length of row 2 is 2
v2[2][0] = 3 v2[2][1] = 6
```

The output demonstrate that both copying procedures were successful.

The STL container class deque is similar to the vector class. The difference is that new elements can be prepended as well as appended in constant time. The deque containers become available through the include statement

```
#include <deque>
```

Then, a **deque** object, **Q**, is obtained via

```
deque <className> Q;
```

with `className` some specific choice for the template data-type parameter. The class methods include all of those given above for the `vector` class as well as `push_front` and `pop_front` that add and remove elements from the front of the container. Exercise 10.3 and Exercise 10.16 explore these and other properties of the `deque` class.

10.3.1 Streaming data

In applications that require the analysis of huge data sets it may be impractical to store the data in memory or even examine any portion of it multiple times. The *streaming* model of computation assumes that data arrive in a stream of observations. Depending on the particular application, each observation may consist of several attributes. The program must perform any computation on each individual observation immediately. The standard assumptions are that

- the size of the data set is too large to store and

- the diversity of the attributes or keys precludes the use of counters to tabulate the frequencies of their individual values.

Muthukrishnan (2005) gives an introduction to data streaming models, applications and results from an algorithmic standpoint. Aggarwal (2007) contains a collection of surveys that are oriented toward data mining applications.

Cormode and Muthukrishnan (2005) propose a simple but effective ADT called *count-min sketch* that can be used to answer queries about the distribution of variables in a streaming data set. This structure contains a $d \times w$ array B of nonnegative integers $B[i,j], 1 \leq i \leq d, 1 \leq j \leq w$ that will hold estimates of the frequency of occurrence of the values of a variable. The row and column dimensions of B correspond to d hash functions $h_1, \ldots h_d$ that map $\{1, \ldots, m\}$ into $\{1, \ldots, w\}$. The integers $\{1, \ldots, m\}$ represents the set of, typically, discretized or grouped values that the stream elements may take. The discretization occurs, for example, when a key/attribute is continuous and rounded to fall into a finite number of groups with some desired precision. In practice, the discretization need not produce integer values; but, there is no loss here in assuming that to be the case. The d hash functions are chosen uniformly at random from a pairwise independent hash family.

The elements of B are all initialized to zero and then updated as the data values arrive. Specifically, if a key value k is found to occur in the stream, $B[j, h_j(k)], j = 1, \ldots, d$ will all be incremented by one. At any point in time, an estimate of the frequency of that key value is

$$\bar{n}_k = \min_j B[j, h_j(k)]. \tag{10.1}$$

Let n_k denote the actual number of observations with the key k that have been observed in the first N elements from a data stream. An indication of the performance of the count-min sketch data structure is then provided by Theorem 10.1.

Theorem 10.1. *For $\epsilon, \delta > 0$ let $w = \lceil e/\epsilon \rceil$ and $d = \lceil \ln(1/\delta) \rceil$. Then, $n_k \leq \bar{n}_k$ and* $\text{Prob}(\bar{n}_k \leq n_k + \epsilon N) \geq 1 - \delta$.

Thus, the sketch estimator always overestimates its target and, with probability at least $1 - \delta$, exceeds it by no more than ϵN. Although the parameters δ and ϵ in Theorem 10.1 can be set arbitrarily, their choices will have fundamental consequences for the size of the resulting data structure.

Proof. Let the random variable $I_{k,i,j}$ be the indicator for the event that $(k \neq i)$ and $(h_j(k) = h_j(i))$ both occur: i.e., that the jth hash function hashes the integer i to the same value as

k. Since the hash functions are pairwise independent $E[I_{k,i,j}] = P[h_j(k) = h_j(i), k \neq i] \leq 1/w \leq \epsilon/e$.

Now define

$$X_{k,j} = \sum_{\substack{i=1 \\ i \neq k}}^{m} I_{k,i,j} n_i$$

which gives the number of times that other keys have hashed to the same value as k for the jth hash function. By definition of the data structure, $B[j, h_j(k)] = n_k + X_{k,j}$ which implies that $B[j, h_j(k)] \geq n_k$.

To establish the form of the upper bound, observe that $E[X_{k,j}] \leq N/w \leq \epsilon N/e$. Then, by Markov's inequality and the random choice of the d hash functions

$$
\begin{aligned}
\text{Prob}\left(\overline{n}_k - n_k \leq \epsilon N\right) &= \text{Prob}\left(\min_j B[j, h_j(k)] - n_k \leq \epsilon N\right) \\
&= \text{Prob}\left(B[j, h_j(k)] - n_k \leq \epsilon N \text{ for some } j = 1, \ldots, m\right) \\
&= 1 - \Pi_{j=1}^{d}\text{Prob}\left(B[j, h_j(k)] - n_k > \epsilon N\right) \\
&= 1 - \Pi_{j=1}^{d}\text{Prob}\left(X_{k,j} > \epsilon N\right) \\
&\leq 1 - e^{-d} = 1 - \delta.
\end{aligned}
$$

$\square$

In addition to estimating the number of times a certain key value has been seen, count-min sketch can be used to approximate other statistics of interest for an input sequence. These include *range queries* (approximately counting all key values in a given range), quantiles of the key distribution and *heavy hitters* (most frequent key values). See, e.g., Cormode and Muthukrishnan (2005) for further discussion of how this can be accomplished.

To illustrate the basic premise, we will create a C++ version of the count-min sketch structure to work with **string** data. Rather than deal with actual streaming data, the test for our data structure will be carried out by opening and reading from a file. The data that will be used for this purpose derive from a list of about 7700 English words and their frequencies in the British National Corpus: a 100,000,000 word database taken from a variety of sources of both spoken and written English (see, e.g., Leech, et al. 2001). This combination of words and frequencies was used to produce a "stream" of about 900,000 words consisting of these 7700 words appearing in different frequencies via the R commands

```
> newWords <- rep(words, freq)
> set.seed(123)
> newWords <- sample(newWords)
> length(newWords)
[1] 878717
```

The 7700 words and their frequencies were read into R and stored in the arrays **words** and **freq**, respectively. The (word, frequency) pairs were then converted back to "raw" data using the R replication function **rep**. In terms of the way it is used here, **rep** takes two vector arguments and replicates the first according to the frequencies specified in the second. For example, the first component of the vector **words** is the word "the" that occurred 61,847 times as specified by the first element of the vector **freq**. The **newWord** array that is returned from **rep** will therefore contain 61,847 instances of "the". After all the replication has taken place, the **newWord** vector contains 878,717 words. These were shuffled randomly using the R **sample** function. The result was output to a file called newWords.txt and this is what will be used to mimic reading from stream input.

The next listing gives the declaration for our count-min sketch class **CMS**.

```
class CMS
{
  int W;
  int D;
  unsigned long Seed;
  vector<vector<int> > B;
  vector<int> a;
  vector<int> b;

  void insertKey(string key);
  int fnvHash(int i, string key);

public:

  CMS(int w, int d, unsigned long seed);
  ~CMS() {};
  void readData(ifstream& inFile);

  unsigned long getFreqEst(string key);
};
```

The class has six private members including the number of hash functions D, the number of hashed values W and the `vector<vector<int> >` array structure B that will hold the frequencies of the hashed values. The universal hash family that produces the values in B will be created along the lines of Example 9.4 of Section 9.2.4.4. This requires a random number generator provided by the `ranGen` class from Appendix E whose seed corresponds to the Seed data member. The coefficients produced by the generator will be stored in the `vector<int>` class members a and b. There are three public methods: a class constructor, a function to read data from a stream and the function `getFreqEst` that computes the estimated frequencies from the B array. The private methods `fnvHash` and `insertKey` evaluate the hash functions at a given key value and update the B array, respectively.

The CMS class constructor takes the form

```
CMS::CMS(int w, int d, unsigned long seed): W(w), D(d), Seed(seed){
  a = vector<int> (D);
  b = vector<int> (D);
  B = vector<vector<int> >(W, vector<int>(D, 0));
  double* aAndb = new double[2*D];
  ranGen rng;
  rng.setSeed(Seed); rng.ranUnif(2*D, aAndb);
  for(int i = 0; i < D; i++) {
    a[i] = floor(W*aAndb[2*i]);
    b[i] = floor(W*aAndb[2*i + 1]);
  }
  delete [] aAndb;
}
```

The values for the class members D, W and Seed are specified as arguments to the constructor and then set in an initializer list. The two `vector<int>` objects that will hold the random coefficients for the hash functions are initialized next followed by initialization of the frequency array B. In this latter case an alternative, two argument, `vector` class constructor is called; the first argument is an integer that determines how many times the second argument is replicated in the container. Thus, `vector<int>(D, 0)` creates a D-element array of all zero entries and the second call to this constructor produces a W component `vector` object with this array in every slot. The end result is just a W × D "matrix" of zeros. The values for the elements of a and b are computed using uniform random deviates generated by an object from the `ranGen` class of Appendix E with seed set to the value of the class member `Seed`. The `ranUnif` method for the `ranGen` class produces numbers in the interval [0, 1] that are multiplied by `W` and then converted to integers between 0 and `W - 1`.

Flaming (1995) describes a good hash function for strings as

> "a magic formula that chops, mixes, slices and dices the input character string, using just a few statements".

There are numerous hash functions that can be categorized in this way some of which have been found to perform quite well in practice. One of these that has received good reviews is the public domain Fowler/Noll/Vo or FNV hash function that will be used as the base for our hash function family. The resulting hashing method for the `CMS` class is defined in the next listing.

```
int CMS::fnvHash(int i, string key){
  unsigned long int h = 2166136261UL;
  for (unsigned int j = 0; j < key.length(); j++ )
    h = (h*16777619)^key[j];

  return (a[i]*h + b[i]) % W;
}
```

The method takes two arguments: the `string` variable `key` that is the key to be hashed and an integer that specifies which one of the `D` hash functions to use for hashing the key. The body of the method is an implementation of the FNV hash algorithm. It produces a hash value that is combined with the appropriate coefficients from the `a` and `b` arrays (as determined by the `int` argument) to create the actual hash value that is returned by the method. The actions the FNV method performs on a key are straightforward: an initial hash value (2166136261) is sequentially multiplied by a constant (16777619) with a bit-wise exclusive OR operation (e.g., Exercise 2.4) being used to combine the result with each element of the key value. The process of breaking the key into parts (i.e., `key[0]`, ..., `key[key.length() - 1]` in this case) and then combining the values via exclusive OR operations (or addition, multiplication, etc.) is called *folding* and is one of the standard methods that is used in creating hash functions for strings (see, e.g., Chapter 13 of Budd 1994 or Chapter 10 of Drozdek 2005).

When the compiler initially encounters a long (or any other) integer literal like 2166136261 (prior to assigning it to `h`) in the `fnvHash` function, it must decide on a storage type. There are some differences between how compilers (e.g., from different versions of a standard) will handle this and warning messages may be generated as a result. One way to avoid this is to give the compiler a hand by stating up front what the storage size should be. This is accomplished by adding a suffix to the numeric value. For integers the suffixes `U` and `L` stand for `unsigned` and `long`, respectively. In combination (as used in the `fnvHash` function) they signify that the constant is of type `unsigned long`.

To insert a data value from the stream into the "hash table", its hash values are computed under all the `d` hash functions while the appropriate elements of `B` are incremented. This is accomplished with

```
void CMS::insertKey(string key){
  for(int i = 0; i < D; i++)
    B[i][fnvHash(i, key)]++;
}
```

Then, at any point in time the estimated frequency of occurrence of a key value can be computed using

```
unsigned long CMS::getFreqEst(string key){
  unsigned long minSoFar = numeric_limits<unsigned long>::max();
  unsigned long thisCount = 0;
  for(int i = 0; i < D; i++) {
    thisCount = B[i][fnvHash(i, key)];
    if(thisCount < minSoFar) minSoFar = thisCount;
  }
  return minSoFar;
}
```

The for loop computes the value of (10.1). The only problem is obtaining a value for the estimator that can initialize the search for the minimizer. This is provided by the numeric_limits template class function max with template parameter unsigned long that returns the largest unsigned long integer that can be represented on the machine.

We are now ready to take the CMS class for a "trial run". The function that will facilitate the test is

```
void CMS::readData(ifstream& inFile){
  string key; int theCount= 0, totalCount = 0;
  while(!inFile.eof()) {
    getline(inFile, key);
    if(key == "the") theCount++;
    insertKey(key);
    totalCount++;
    if(!(totalCount%100000))
      cout << theCount << " " << getFreqEst("the") << endl;
  }
}
```

This method takes an ifstream object as its argument that corresponds to an input file; for our specific application the file of interest will be newWords.txt that was created previously in R from the British National Corpus word data. Data from the file will be read until the end of the file is encountered: i.e., inFile.eof() evaluates to true. Each word is read in using the getline function from Section 3.8.5. The resulting key is then "inserted" into the B array with the insertKey method. To emulate what might transpire while reading from an actual stream, code has also been included to monitor the frequency of the word "the" and report on its estimated and actual frequency after every 100,000 words are read from the "stream".

The main function for a driver program that was created for our "data-streaming" problem is shown in the next listing.

```
int main(){
  unsigned long seed = 123;
  int w = 307;
  int d = 29;
  CMS A(w, d, seed);
  ifstream words("newWords.txt");
  A.readData(words);
  words.close();
  string key;
  while(1) {
    cout << "The word of interest? ";
    getline(cin, key);
    if(key.length() == 0) break;
    cout << "Estimated frequency is " << A.getFreqEst(key)
      << endl;
```

```
    }
    return 0;
}
```

A `CMS` object is created with a seed value of 123 for the random number generator, 29 hash functions and 307 bins/slots for the hashed key values. An `ifstream` object is then created for the newWords.txt file and the `readData` method is called with that object as its argument. After closing the input stream, the frequency array B will persist and we are given the option of asking about the frequencies of specific words that may be of interest.

An example of results from running the program are

```
6866 6996
13972 14240
20955 21384
28023 28587
35004 35720
41981 42846
49053 50075
56189 57365
The word of interest? the
Estimated frequency is 63135
The word of interest? of
Estimated frequency is 30063
```

The actual frequency of "the" over all 878,717 words in newWords.txt is 61,847 while the sketch estimate is 63,135. Similarly, the estimated frequency of "of" over all 878,717 words is found to be 30,063 while its actual frequency is 29,391. Table 10.1 reports the results from similar queries for the eight most frequent words in newWords.txt. Both the true frequency and estimated frequencies are listed in the table.

Table 10.1 *Actual versus estimated word frequency*

Word	Actual frequency	Estimated frequency
the	61847	63135
of	29391	30063
and	26817	27829
to	25627	28481
a	21626	25916
in	18214	20074
is	9982	10450
was	9236	10098

10.3.2 Flexible data input

Section 3.10 discussed using a template approach to create classes for input of general types of numeric data. While this is satisfactory in many cases, there are limitations to its utility. Specifically, it will not work in cases where the structure of the input file is unknown. Thus, one must know *a priori* the number of rows and columns in the file and, if there are rows of uneven length, the code must be altered and tailored to work with whatever pattern may exist.

With the `vector` class in hand, we are now in a position to create a general file input scheme that allows us to read in a file of numeric (or more general types of) data with essentially unknown formatting. A class will be designed specifically to acquire data from such a text file and store it in a suitable array format.

The assumption is that data will be read from a file where the numbers are delimited or separated in some known fashion. This could be by commas, semicolons or whatever. For specificity it will be assumed for now that the delimiter is just a blank (or white) space. The number of rows and columns in the file are not known and it may be that rows are of unequal length. From this perspective it becomes necessary to treat the content of the input file as being strings separated by white spaces. The strings can then be manipulated using **string** class methods (see Section 3.8) and converted to appropriate types of numeric (or other) values using functions (e.g., **atoi** and **atof**) that are available in C++ for that purpose.

The process of taking a line of input that contains a group of strings separated by some delimiter and breaking it into its individual substring components (or tokens) is sometimes called *tokenizing*. We will develop a specific tokenizer for use in our setting. The details involved in producing this function are somewhat irrelevant to our immediate purpose. Thus, the **tokenizer** method will be treated as a "black box" that takes a line of input and returns it to us in a **vector** container of **string** variables. The details of how the tokenizer works will be revealed at the end of the section.

If the form of the **tokenizer** method is ignored, the essential aspects of the plan for class **fileIn** can be seen from the class header file in the listing below.

Listing 10.4 fileIn.h

```
// fileIn.h
#ifndef FILEIN_H
#define FILEIN_H

#include <vector>
#include <string>

class fileIn{
  std::vector<std::vector<double> > v;
  void tokenizer(const std::string& str,
       std::vector<std::string>& tokens);

public:

  fileIn(const char* fname);
  ~fileIn(){};
  int getVNRows(){return v.size();}
  int getVLength(int i){return v[i].size();}
  double getV(int i, int j){return v[i][j];}
};
#endif
```

Here the data to be accessed are assumed to be in the form of rows in a file that will then be stored in a vector container of vector containers, v, of **double** variables very similar to the construction in Listing 10.2. This will allow us to process the data without knowing *a priori* the number of rows or columns in the file. The use of **double** as the data type, although likely the most typical choice, is still rather arbitrary. Minor tweaks of the basic idea can be used to address more general types as will be illustrated in the next section.

The constructor function in Listing 10.4 will take the input file name as an argument. In addition, three accessor functions are given inline: **getVNRows** that returns the number of rows in the data array v, `getVLength` that gives the number of columns for a given row of v and `getV` that returns the value of the specified element of v.

The definitions of the remaining two class methods are in the next listing.

_____Listing 10.5 _fileIn.cpp_____

```
//fileIn.cpp
#include "fileIn.h"
#include <fstream>
#include <iostream>

using std::string; using std::vector; using std::cout;
using std::endl; using std::ifstream;

fileIn::fileIn(const char* fname){
  vector<string> words;
  string line;

  ifstream in(fname);

  if(in.is_open()){
    while(!in.eof()){
      getline(in, line);
      if(line.length() != 0){
        tokenizer(line, words);
        if(words.size() != 0){
          v.push_back(vector<double>());
          for(int i = 0;i < words.size(); i++)
            v[v.size() - 1].push_back(atof(words[i].c_str()));
        }
        words.clear();
      }
    }
  }
  else {
    cout << "Unable to open file" << endl;
    exit(1);
  }
}

void fileIn::tokenizer(const string& str, vector<string>& tokens){
  string delimiters=" ";
  //Skip delimiters at beginning
  string::size_type lastPos = str.find_first_not_of(delimiters, 0);
  //Find the end of the first token
  string::size_type pos = str.find_first_of(delimiters, lastPos);

  //Loop while there are tokens to be found: i.e., one of the
  //searches for a delimiter (pos) or nondelimiter (lastPos)
  //does not fail.

  while(string::npos != pos || string::npos != lastPos){
    //Found a token, add it to the vector
    tokens.push_back(str.substr(lastPos, pos - lastPos));
    //Skip next set of delimiters and go to next token
    lastPos = str.find_first_not_of(delimiters, pos);
```

```
      //Find end of token or start of next delimiter          51
      pos = str.find_first_of(delimiters, lastPos);            52
   }                                                           53
}                                                              54
```

After first creating the `ifstream` object `in` on line 13 of the listing, a check is made to see that the targeted file was successfully opened (on line 15) using the `is.open` method for the `ifstream` class. Assuming that `is.open` returns `true`, the function `getline` for `ifstream` objects is then used on line 17 to extract the contents of the file one line at a time. This is done adaptively through use of a `while` loop that continues to execute until the end of the file is reached as indicated by the `ifstream` `eof` method that is called by `in`. Each line of input brought in by `getline` is passed to the tokenizer. The tokenizer writes the individual substrings (or numbers in this case) that make up the line into the string vector container `words`. An extra row is appended to the vector of data `v` with the `push_back` method on line 21 to hold the contents of `words`. This new row is then filled with the `words.size()` elements of `words` also using `push_back`. In the process of doing this the `atof` function is employed to translate a null-delimited character array to its numerical representation which first requires that we translate each string to a character array (or C-string) using the `c.str` function for the string class. Note that `atof` will work for both floating point and integers in that it will translate strings of integers without introducing a decimal point.

To conclude our discussion of data input, let us briefly discuss the "mysterious" `tokenizer` function in Listing 10.5. As previously discussed the `tokenizer` function breaks a string into the components that differ from the `delimiters`. In our particular case the value of `delimiters` has been chosen to be a white space on line 36 of Listing 10.5. But, this could be any character value and could even be included as an argument to the function to allow more flexibility in the file reading process. The most common delimiters are commas, white spaces and the tab delimiter `\t`.

Our tokenizer uses the string class iterators `find_first_of` and `find_first_not_of` that return the position of the character that satisfies their respective search conditions. The return type from each of these iterators is something called `size_type` which is basically an unsigned integer except for the addition of the special value `string::npos` that is returned to indicate failure. Note that the container vector of string `tokens` is being passed by reference. When `tokenizer` is called from the `fileIn` constructor the role of `tokens` is occupied by the local variable `words`. Thus, since `words` is passed by reference, `tokenizer` is able to write directly onto the area in memory that holds the content of `words` and effectively returns a new value of `words` to `fileIn` as its "output".

10.3.3 Guess5 revisited

In this section we will illustrate the use of the `fileIn` class of the previous section as well as using `vector` containers to hold user-defined data types. The discussion here carries forward into our treatment of the algorithm library in Section 10.7.

It will be useful to first recall some of the details concerning the fictional Guess5 lottery game that was introduced in Section 5.6. The game consisted of five balls being selected from a set of 40 balls without replacement. The drawings occur on Monday and Thursday using one of the two drawing machines, A or B, and one of the 10 ball sets, numbered 1 to 10. A series of nine test draws are conducted prior to each draw thereby producing a total of 10 selections of five balls every Monday and Thursday.

A set of data corresponding to 100 Guess5 draws was created in Section 5.6 and stored in the file guess5.txt. From the discussion in that section we know that the first row of data for each draw will contain the date of the draw (a `string` variable), the letter designation of the machine that was used to conduct the draw (a `string` variable), the number of the ball set that was used in the draw (a variable of type int) and the day of the draw indicated by M for Monday and T for Thursday (a string variable). The next nine rows consist of integers that represent the five ball numbers that were drawn for the nine test draws and the eleventh row contains the outcome of the actual Guess5 drawing for that night. This pattern continues throughout the file.

Before we can actually analyze the data in guess5.txt, it must be imported into our program. One approach to this can be based on the developments in Section 3.10. Our path will be somewhat more general and instead use a slight modification of the class fileIn from the previous section.

The guess5.txt file contains a mix of values for string and int variables with the consequence that the fileIn class that transforms all the data to type double is not directly useful. This is easy enough to fix by

- replacing

    ```
    std::vector<std::vector<double> > v;
    ```

    and

    ```
    double getV(int i, int j){return v[i][j];}
    ```

    in the fileIn class declaration in Listing 10.4 with

    ```
    std::vector<std::vector<std::string> > v;
    ```

    and

    ```
    std::string getV(int i, int j){return v[i][j];}
    ```

- replacing

    ```
    v.push_back(vector<double>());
    ```

    and

    ```
    v[v.size() - 1].push_back(atof(words[i].c_str()));
    ```

    on lines 21 and 23 of of Listing 10.5 with

    ```
    v.push_back(vector<std::string>());
    ```

    and

    ```
    v[v.size() - 1].push_back(words[i]);
    ```

This entails that all the data will be imported as string objects. These objects will then be converted to numeric values as necessary using the pattern that is known to exist in the recorded data for each drawing.

Proceeding along the lines of our work in Section 6.4, the information for individual drawings will be stored in objects from a class called g5. The objects will then be placed in an appropriate container that will facilitate the analysis of the combined data as well as subsets corresponding to each of the two machines, each of the 10 ball sets and the 20 machine/ball set combinations. This latter container class will be called g5Array.

The declaration for the g5 class is

```
class g5{
  std::string Date, Machine;
  int SetNo;
  std::string Day;
  short** pData;
```

```
  int* pFreq;
  void getFreq();

public:
  g5() : pData(0), pFreq(0) {}
  g5(std::string Date, std::string machine, int setNo,
       std::string day, short** pdata);
  g5(const g5& g5Obj);
  ~g5();

  g5& operator=(const g5& g5Obj);
  bool operator==(const g5& g5Obj) const;

  int getFreq(int i) const {
    return pFreq[i];
  }
  std::string getDate() const {
    return Date;
  }
  void printG5() const;
};
```

The class has six members: `string` variables `Date`, `Machine` and `Day` for the drawing date, name of the machine that was used and day of the week for the drawing, an `int` variable that will hold the set number used in the drawing, a pointer-to-pointer `short` integer "array" that will hold the actual drawing information and a pointer to `int` that will hold the frequencies of occurrence for the 40 balls for a particular g5 object. The numbers being drawn are all between 1 and 40 and the full four bytes of `int` storage would be wasted here; the two byte `short int` type will work just as well. There is a default constructor that sets the two pointer variables to null pointers. This is needed, for example, to create an empty `vector` container of a specified `size` for holding g5 objects. There is also a constructor that takes input values for five of the member variables. Accessor methods are given for two of the class members and a utility print method has been provided. Dynamic memory allocation is involved and, accordingly, a copy constructor, overloaded assignment operator and explicit destructor are provided. An overloaded == comparison operator has been included for work that will arise in Section 10.7. Creation of the code for these methods is addressed in Exercise 10.9.

The nondefault constructor for class g5 looks like

```
g5::g5(std::string date, std::string machine, int setNo,
        std::string day, short** pdata) : Date(date),
        Machine(machine), SetNo(setNo), Day(day) {
  pFreq = new int[40];
  for(int i = 0; i < 40; i++)
    pFreq[i] = 0;

  pData = new short*[10];
  for(int i = 0; i < 10; i++){
    pData[i] = new short[5];
    for(int j = 0; j < 5; j++)
      pData[i][j] = pdata[i][j];
  }
  getFreq();
}
```

The values of all the member elements, except the two pointers, are set via an initializer list. The `pFreq` pointer is initialized and the current contents of the corresponding memory locations are overwritten with zeros. For the actual drawing data there will be 10 draws with five balls selected on each draw. Memory is acquired to hold this information using the pointer `pData` and the memory locations are filled with the drawing data.

The role of setting the values corresponding to the `pFreq` pointer falls to the private member function `getFreq` that is called by the constructor. It takes the form

```
void g5::getFreq(){

  for(int i = 0; i < 10; i++)
    for(int j = 0; j < 5; j++)
      pFreq[pData[i][j] - 1]++;
}
```

This method fills the necessary memory locations by moving through the `pData` array while augmenting the frequencies of the numbers that are encountered.

We are now ready to create the container for storing objects from the g5 class. The basic scheme is to use a `vector<vector<g5> >` object for this purpose. The array will be chosen to have 20 rows: one for each of the 20 machine/ball set combinations for the game. This configuration will make analysis of the various data subsets more efficient than performing a search through the entire collection of data as in Section 6.4.

The declaration for the `g5Array` class is given in the next listing.

```
class g5Array{

  std::vector<std::vector<g5> > v;
  int nDraws;

public:

  struct const_iterator : public
        std::iterator<std::forward_iterator_tag, g5> {
    int Index;
    const g5Array* const_pA;
    std::vector<g5>::const_iterator const_current;
    const_iterator(){}
    const_iterator(const g5Array* pa) : const_pA(pa) {}
    const g5& operator*() const {return *const_current;}
    const g5* operator->() const {return &(*const_current);}
    const_iterator& operator=(const const_iterator& iter);
    bool operator==(const const_iterator& iter) const;
    bool operator!=(const const_iterator& iter) const {
      return !(*this == iter);
    }
    const_iterator operator++();
    const_iterator operator++(int){return operator++();}
    int nextIndex();
  };

  g5Array(const char* fileName);
  bool operator==(const g5Array& gA) const {
    if(v == gA.v) return true;
    return false;
  }

  const_iterator begin() const;
```

```
  const_iterator end() const;
  double chiSquare() const;
  std::vector<g5> g5Sort() const;
  const_iterator lookUp(const std::string s) const;
};
```

There are two class members: the vector<vector<g5> > array that will hold the drawing data and an integer variable nDraws that gives the number of g5 objects being held in the container. An internal iterator class has been provided for exploration of the data structure. As in Section 9.5.1 iterators from this class are returned by the begin and end methods that point to the first and one past the last element of the container. The three methods chiSquare, g5Sort and lookUp will be discussed in Section 10.7.

The definition of the constructor for class g5Array is

```
g5Array::g5Array(const char* fileName){

  int s, drawIndex = 0;
  //initialize v
  v = std::vector<std::vector<g5> >::vector(20);
  //determine the number of draws
  fileIn fIn(fileName);
  nDraws = fIn.getVNRows()/11;
  //temporary storage for draw data
  short** pTemp = new short*[10];
  for(int i = 0; i < 10; i++)
    pTemp[i] = new short[5];

  for(int k = 0; k < nDraws; k++){
    //first copy in the draw data
    for(int i = 0; i < 10; i++){
      for(int j = 0; j < 5; j++){
        pTemp[i][j] = atoi(fIn.getV(drawIndex + 1 + i, j).c_str());
      }
    }

    s = atoi(fIn.getV(drawIndex, 2).c_str());
    if(fIn.getV(drawIndex, 1) == "A")
      v[s - 1].push_back(g5(fIn.getV(drawIndex, 0),
        fIn.getV(drawIndex, 1),
        atoi(fIn.getV(drawIndex, 2).c_str()),
        fIn.getV(drawIndex, 3), pTemp));
    else
      v[9 + s].push_back(g5(fIn.getV(drawIndex, 0),
        fIn.getV(drawIndex, 1),
        atoi(fIn.getV(drawIndex, 2).c_str()),
        fIn.getV(drawIndex, 3), pTemp));

    drawIndex += 11;
  }

  //memory clean-up
  for(int i = 0; i < 10; i++)
    delete[] pTemp[i];
  delete pTemp;
}
```

The only argument for the constructor is the name of an input file that holds the drawing data. A (modified as indicated above) fileIn object that will adaptively determine the length of the input file is used to import the data. There are 11 rows of data for each drawing. So, the number of draws is just the size of the vector container v in the fileIn class (that is returned by its accessor function getVNRows) divided by 11.

After initializing the fileIn object, the g5Array class member v is initialized along with a pointer-to-pointer pTemp for data of type short that will be used for temporary storage of the drawing results. Although the number of rows in the array v is fixed at twenty, the vector<g5> objects that comprise each of these rows can grow as needed. They are initially of size 0; but, they will grow by (at least) one slot each time a new object is pushed onto the vector via the push_back method.

The process of filling the container involves working our way through the string data from the fileIn object one drawing at a time. The contents of the first row of data for a draw are copied into variables of type string or int (after transformation via the atoi function), as appropriate, and the values from the next 10 rows are transformed from string to int and copied into the memory allocated to pTemp. All this information is then used to create a new g5 object that is pushed onto the row of v that corresponds to its machine and ball set; the first 10 rows of v have been used to store draws produced by a combination of the 10 ball sets with machine A while the remaining 10 serve the same purpose for drawings that used machine B.

The first new thing to notice concerning the g5Array iterator class is the inheritance relation indicated in the class declaration by

```
public std::iterator<std::forward_iterator_tag, g5>
```

The STL allows for several types of iterators that all derive from a basic template class. To work effectively in the STL, user-defined iterators need to conform to this class and the easiest way to accomplish that is by inheritance from the appropriate STL iterator class. This is accomplished through a statement such as*

```
class newIterator : public std::iterator<iteratorType, dataType>
```

Here dataType is the class for objects that will be pointed to by the iterators. The parameter iteratorType can take several values: namely,

- output_iterator_tag,
- input_iterator_tag,
- forward_iterator_tag that inherits from input_iterator_tag,
- bidirectional_iterator_tag that inherits from input_iterator_tag and
- random_access_iterator_tag that inherits from bidirectional_iterator_tag.

The forward, bidirectional and random access parts of the iterator type names are self-descriptive in terms of the iterator's properties. Input and output iterators are forward iterators with, respectively, read-only and write access to elements in the container. In view of this discussion, we now see that the iterator class for the g5Array class is a child/derived class of the basic STL forward_iterator_tag class whose iterators will point to g5 objects.

The g5Array const_iterator struct has three members. The pointer const_pA to const g5Array is used to tie an iterator to a specific container as is necessary for the g5Array methods begin and end. The const_current member is just a const_iterator (cf. Section 9.5.2) from the STL associated with a vector<g5> object. In particular, const_current will be able to work with the vector<g5> object that will be accessible through const_pA. The const_current iterator and the integer variable Index work in tandem to give the current

---

* The STL iterator class has five template parameters. The last three have default values that will suffice for our purposes.

iterator location. The object being traversed is of type `vector<vector<g5 > >` which means we can think of it as a double array with 20 rows of possibly unequal length; the `Index` value tells us the row number and `const_current` then points to the column inside the row. Some care has been taken to protect against alteration of the actual draw data by a `g5Array` object. In this setting, all we need to do is process the data and untoward changes in the input data could prove problematic (or worse). The use of a `const g5Array` data type for the iterator class pointer member and the `const_iterator` for the `vector` class iterator give us protection against such possibilities.

Creation of code for the nondefault constructor and the `*`, `->`, `=`, `==` and `!=` operators for the `g5Array const_iterator` class is the subject of Exercise 10.10. Our focus will instead be directed towards development of the prefix increment operator.

In a general sense, navigation of a `g5Array` object is straightforward; one proceeds across the elements in a given row of a container until the end and then moves to the next row and repeats the process. However, the rows can be empty and some care needs to be taken to avoid going past the last element in a `g5Array` object. The formulation of the `++` operator given below allows for such contingencies.

```
g5Array::const_iterator g5Array::const_iterator::operator++(){

  const_current++;
  if(const_current == const_pA->v[19].end()){
    *this = const_pA->end();
    return *this;
  }

  if(const_current == const_pA->v[Index].end()){
    Index = nextIndex();

    if(Index == -1){
      *this = const_pA->end();
      return *this;
    }

    const_current = const_pA->v[Index].begin();
  }
  return *this;
}
```

Our first action is to increment the `vector<g5>` iterator. A check is then made to see if we are already at the end of the container. The next step is to determine if we are at the end of a row. If that is not so, then no further action is needed. On the other hand, if the end of a row is encountered, a search is made to find the next nonempty row using

```
int g5Array::const_iterator::nextIndex(){

  int temp = Index + 1;
  while(temp < 19 && this->const_pA->v[temp].empty())
    temp++;

  if(const_pA->v[temp].empty())
    return -1;//all remaining slots are empty

  return temp;
}
```

This function moves from row to row until it finds one that is not empty. It returns the row index or −1 if no nonempty rows are found. Depending on the return value from nextIndex the ++ operator will return either an iterator that points to the begin() iterator for the vector<g5> object that represents the next row to be explored or the g5Array container's end() iterator. We define this latter iterator by

```
g5Array::const_iterator g5Array::end() const {
   const_iterator iter;
   iter.const_pA = this;
   if(v.empty()){
      iter.Index = 0;
      iter.const_current = v[0].end();
      return iter;
   }

   int i = 19;
   while(v[i].empty())
      i--;
   iter.Index = i; iter.const_current = v[i].end();

   return iter;
}
```

This method returns the natural choice of an iterator that points to one past the last element of the last nonempty row in a container.

The begin companion method for end is defined in the listing below.

```
g5Array::const_iterator g5Array::begin() const {

   const_iterator iter;
   iter.const_pA = this;
   if(v.empty()){
      iter.Index = 0;
      iter.const_current = v[0].begin();
      return iter;
   }
   int i = 0;
   while(v[i].empty())
      i++;
   iter.Index = i; iter.const_current = v[i].begin();

   return iter;
}
```

The iterator returned by this method points to the first object in the first nonempty row of the container.

A program that tests the g5Array class has the main function

```
int main(){

   g5Array g5A("guess5.txt");
   g5Array::const_iterator iter = g5A.begin();
   iter->printG5(); iter++; iter->printG5();
   return 0;
}
```

The "test" is simple; a g5Array container is created using the guess5.txt data file from Chapter 5 and an iterator is defined to point to its first element. The iterator is used to

print the first g5 object in the container after which it is incremented and used to print the
next object it encounters. The output this produces is

```
Guess5 Drawings for 1952-06-12 used Machine A with Ball Set 1
The drawing results are:
20 35 7 23 27
8 23 26 35 12
22 1 15 24 40
33 5 26 8 22
28 21 15 26 18
10 18 4 40 20
1 9 25 27 28
3 4 36 35 34
26 7 35 38 39
3 23 14 18 17
Guess5 Drawings for 1952-06-30 used Machine A with Ball Set 1
The drawing results are:
24 7 20 16 30
23 18 8 39 9
13 34 17 38 18
34 1 12 4 30
19 21 16 17 5
2 1 3 14 28
30 4 37 6 18
23 37 32 20 14
22 32 24 4 7
34 19 30 31 11
```

The first row in the vector<vector<g5> > component of a g5Array object will, by construction, contain only information about drawings that used machine A and ball set number 1.
So, the output is consistent with what would be expected from that perspective. A manual
check of the guess5.txt file reveals that June 12 and 30 of 1952 were the first, in order, days
where that machine and ball set combination was used in a drawing.

A more thorough test of the g5Array class will be given in Section 10.7. There we will, for
example, add a method that will compute a chi-square statistics for a set of Guess5 drawing
data.

10.4 The C++ list container

The STL list container provides an implementation of a doubly linked list along the lines
of Exercise 9.36. Double linkage can be viewed as meaning that every node has both a
next and previous member that point to the next and previous object in the list. Thus,
movements are possible in both forward and backward directions inside the container.

Use of the list container requires the include directive

```
#include <list>
```

Then, an empty list container L holding objects of type className can be created with

```
list<className> L;
```

A list with a specified number, size, of (empty) nodes is produced by

```
list<className> L(size);
```

The empty nodes are obtained from the default constructor for the className class.

New objects can be appended and existing objects removed from the end of a list object
using the push_back and pop_back methods that work the same as for the vector container.

It is also possible to prepended or remove objects from the front of a `list` object. This is accomplished with the push_front and pop_front methods. As an example, consider the code segment below.

```
list<int> L(5);
int a = 1;
L.push_front(a);
L.pop_front();
```

The first action is creation of a `list` container that can hold five `int` objects. The first node in the container is filled with an `int` variable having the value 1 that is then removed thereby emptying the `list`.

Both pop methods simply remove the node from the list and do not return a value. The first and last components of a `list` object L are accessed (but not removed) with L.front() and L.back().

Like the `vector` container, the `list` containers have an `assign` method that can be used for copying subsets of one list into another. For example,

```
L.assign(begin, end);
```

will destroy the current content of the `list` L and replace them with elements from a "range" specified by the iterators begin and end. In particular,

```
L1.assign(L.begin(), L.end());
```

will copy the contents of a list object L into another `list` object L1.

The assign operation will even work with iterators from other classes as illustrated in the listing below.

```
//listEx.cpp
#include <iostream>
#include <vector>
#include <list>

using std::cout; using std::endl; using std::vector;
using std::list;

int main(){
  vector<int> v(3);
  v[0] = 12; v[1] = 1; v[2] = 2;

  list<int> L;
  L.assign(v.begin(), v.end());
  for(list<int>::iterator iter = L.begin(); iter != L.end(); iter++){
    cout << L.front() << " ";
    L.pop_front();
  }
  cout << endl;
  cout << "Empty? " << std::boolalpha << L.empty() << endl;

  L = list<int>(3, 12);
  for(list<int>::iterator iter = L.begin(); iter != L.end(); iter++)
    cout << L.front() << " ";
  cout << endl;
  L.clear();
  cout << "Empty? " << std::boolalpha << L.empty() << endl;
  return 0;
}
```

Here a `vector` container `v` is created and filled with three integers. Then, the `begin()` and `end()` iterators for the `vector` object are used in the `assign` function to fill a new `list` object L. The contents of L are examined and removed via the `front` and `pop_front` methods using a `for` loop with the `list` class iterators `begin()` and `end()`. A check is made to see if the `list` object is empty (see Exercise 3.1 for discussion of the `boolalpha` format flag) after which it is refilled using another constructor. This latter constructor takes two arguments: the container size (3 in this example) and an object (12 in this instance) that will be copied, repeatedly, to fill all the nodes. The elements of the `list` object are examined again and the `clear` and `empty` methods are called.

The output from our `list` example program is

```
12 1 2
Empty? true
12 12 12
Empty? true
```

The list was filled as expected first using the contents of the `vector` object and then with three copies of the integer 12. It was then successfully emptied by both the `for` loop and `pop_front` combination and the `clear` method.

10.4.1 An example

This section provides an illustration of how one might make use of a `list` container in a data analysis context. For specificity, suppose that data is being observed on a daily basis: e.g., values of a stock index or a portfolio's worth. Information is acquired each weekday that needs to be stored in a way that makes at least the most recent results easily available. Also, in some cases, the information arrives out of sequence and needs to be conveniently placed into its proper order in the time series. Linked lists are, not surprisingly, a viable storage option for data of this nature.

To focus the discussion, a "stock index" data set will be created here whose form will dictate the storage scheme. The data were created in R with

```
> day <- as.Date("1952-01-17")
> day <- day + 0:139
> day <- day[!(weekdays(day) == "Saturday"
+ | weekdays(day) == "Sunday")]
> length(day)
[1] 100
> set.seed(123)
> indexData <- paste(day, 2500 + arima.sim(n = 100,
+ model = list(ar = .6), sd = 5))
> print(noquote(indexData[95:100]))
[1] 1952-05-28 2487.8268629974   1952-05-29 2490.794985197
[3] 1952-05-30 2499.0719741635   1952-06-02 2496.56644968506
[5] 1952-06-03 2500.97969142216 1952-06-04 2492.49840131185
> write.table(indexData[-96], file = "index.txt", quote = FALSE,
+ row.names = FALSE, col.names = FALSE)
```

The starting date for the data is 1952-01-17 that is stored in an R `Date` object. Vectorized addition of this object with the array of integers from 0 to 139 produces an array `day` of `Date` objects that correspond to 140 successive days beginning at 1952-01-17. As the "market" is not open on weekends, the elements of `day` that correspond to a Saturday or Sunday need to be removed. This is accomplished by applying the R subsetting/indexing operator to `day` with a logical array as its argument. The elements of this latter array evaluate as `FALSE` for weekend `Date` objects. The `weekdays` function from Section 5.6 is used to determine the weekday for each `Date` object. A check of the length of `day` after weekends have been removed reveals that the final date array will have 100 elements.

The (simulated) values for the "stock index" variable were generated with R's `arima.sim` function. This function is capable of producing data from autoregressive and moving average models. Here we have used it for a first order autoregressive model of the form

$$X_t = \phi X_{t-1} + \varepsilon_t, \ t = 1, \dots, 100,$$

with $\phi = .6$ and the ε_i zero mean, uncorrelated normal errors with standard deviation 5. The output from `arima.sim` was shifted by adding a factor of 2500 and then combined with the `day` vector using the `paste` function to create a 100 component character array named `indexData`. This data is shown in Figure 10.1 that was created with the R `ts.plot` function for plotting univariate and multivariate time series.

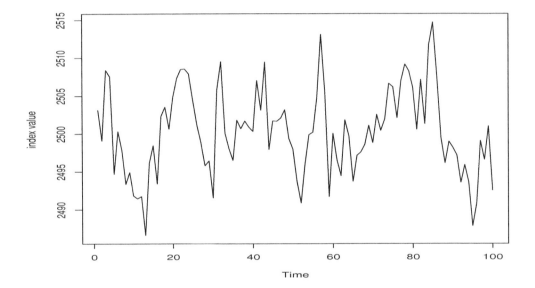

Figure 10.1 *Simulated "stock index" values*

We will use our "stock index" data to illustrate various features of the STL `list` class. In particular, we will use it to demonstrate insertion of a new node into a list. To make this (somewhat) more meaningful, a particular date 1952-05-29 was omitted when the data from R was written to a text file. This date corresponds to slot 96 of the `indexData` array and, hence, `indexData[-96]` is the same array sans the entry in row 96. For future reference the last six lines of the complete data set were printed during our R session before writing the data to its file.

The first step in analyzing the "index" data is to create a class that will hold the objects that make up the data. There are two aspects to consider: the observation time and the value of the response, or "stock index", variable. Seeing as the observation times are in a year/month/day format, we can store that information in a `Date` struct of the form

```
struct Date{
  int Year, Month, Day;

  Date(){}
  Date(int year, int month, int day)
```

```
    : Year(year), Month(month), Day(day) {}

  bool operator<(const Date& date) const;
  bool operator>=(const Date& date) const {
    return !(*this < date);
  }
  bool operator==(const Date& date) const;
  bool operator!=(const Date& date) const {
    return !(*this == date);
  }
};
```

The struct has three members that hold the day (Day), month (Month) and year (Year) for when the observation occurred. The nondefault constructor merely sets the member's values using an initializer list. To facilitate printing of Date objects, the output insertion operator has been overloaded as in Exercise 3.19: viz,

```
ostream& operator<<(ostream& out, const Date& date){
  cout << date.Month << "-" << date.Day << "-" << date.Year;
  return out;
}
```

Elements will be inserted into our list container for the "stock index" data in order of their temporal arrival. The Date objects will therefore provide the natural sorting and searching information that will be used for look-up and other purposes. Such operations depend on the comparison operators with the two key definitions being the ones for < and ==. Other operators can then be obtained by logical negation or composition of these two.

The < and == operators for Date objects are defined by

```
bool Date::operator<(const Date& date) const {
  if(Year < date.Year)
    return true;
  if(Year == date.Year && Month < date.Month)
    return true;
  if(Year == date.Year && Month == date.Month
    && Day < date.Day)
    return true;
  return false;
}

bool Date::operator==(const Date& date) const {
  if(Year == date.Year && Month == date.Month
    && Day == date.Day)
    return true;
  return false;
}
```

Observations with smaller Year members occur earlier while observations within the same year occur earlier if they have smaller Month members, etc. The definition of == is similarly straightforward.

The data that is recorded each day will be held in the members of the struct IndexData that has the declaration/definition

```
struct IndexData{
  Date T;
  double Value;
  IndexData(){}
```

```
   IndexData(Date t, double value) : T(t), Value(value) {}
};
```

This is just a container for an object of type `Date` and a `double` and the same utility could be realized using a `pair` struct from, e.g., Section 9.2.4.2. An overloaded output insertion operator was also created for this struct that took the form

```
std::ostream& operator<<(ostream& out, const IndexData& d){
   cout << d.T << " " << d.Value;
   return out;
}
```

We are now ready to formulate the `IndexList` class that will be used to manage an accumulated collection of "index" data such as the data created from R in the index.txt file. The perspective will be similar to that of the Guess5 data from Chapters 5–6 and Section 10.3.3. That is, the data resides on disk and must be read into our program to instantiate a list object. The reading will be carried out using the `fileIn` class from Sections 10.3.2– 10.3.3. Specifically, the version of `fileIn` from Section 10.3.3 that imported all data as `string` objects will be used here and, similar to the developments for the `g5Array` class in that section, methods from the `string` class will be used to reformat the data in a way that makes it suitable for analysis.

The declaration for the `IndexList` class is given in the next listing.

```
struct IndexList{
   list<IndexData> L;

   IndexList(){}
   IndexList(fileIn fIn);

   IndexData pop_front();
   list<IndexData>::iterator lookUp(const Date& date);
   list<IndexData>::iterator insert(const IndexData& d);
};
```

The struct has a single `list` member. There is both a default and nondefault constructor that takes a `fileIn` object as its only argument. There are also the standard `lookUp`, `insert` and `pop_front` methods that will require a bit of explanation. Before proceeding in that direction it should be noted that this struct is largely a wrapper for a `list` class object that adds on a few additional features that are useful for this particular application.

The nondefault constructor for the `IndexList` struct is

```
IndexList::IndexList(fileIn fIn){
   Date tempD;
   IndexData tempI;
   int idx1, idx2;
   for(int i = 0; i < fIn.getVNRows(); i++){
      idx1 = fIn.getV(i, 0).find_first_of("-");
      idx2 = fIn.getV(i, 0).find_last_of("-");
      tempD.Year = atoi(fIn.getV(i, 0).substr(0, idx1).c_str());
      tempD.Month = atoi(fIn.getV(i, 0).substr(idx1 + 1,
            idx2 - idx1 - 1).c_str());
      tempD.Day = atoi(fIn.getV(i, 0).substr(idx2 + 1).c_str());
      tempI.T = tempD;
      tempI.Value = atof(fIn.getV(i, 1).c_str());
      L.push_front(tempI);
   }
}
```

The `fileIn` methods `getVNRows` and `getV` return the number of observations and the elements of the `vector<vector<string > >` object that holds the data. The question that remains is how to effectively extract the information about the index from this object. To illustrate the problem, consider the first line of the index.txt file that looks like

```
1952-01-17 997.19762176724
```

The first `string` object returned by `getV(0, 0)` in this case will be 1952-01-17. We need to extract the values for `Day` (i.e., 17), `Month` (i.e., 1) and `Year` (1952) from this `string` and use them to create a `Date` object. There are various ways to deal with the individual elements in a `string` object. Array type indexing can be used and that will work here (along with the + operator for concatenation) if it can alway be assured that the month and day values will be recorded as two-digit integers, possibly with leading zeros. We have opted for a slightly more robust alternative adapted from the `tokenizer` method of Section 10.3.2.

An application of the `string` class `find_first_of` method to the `string` 1952-01-17 with – as its argument will return the location immediately to the right of the `Year` value 1952. A similar use of `find_last_of` will give us the location in the `string` that immediately precedes the `Day` value 17. All this is simple enough; but, it does not address the issue of how to actually obtain the value for `Day`, `Month` and `Year`. The `substr` method gives us an option that will work regardless of how many digits are used when recording the dates. As its mnemonic moniker suggests, this function creates a copy of a subset of the elements in a `string` object. It takes two arguments; the first is the location where the copying process should begin and the second is the number of consecutive elements that are to be copied starting at that location. If the second argument is not supplied, the end of the `string` object is used by default.

By using `string` class methods, the problem of extracting the `Date` data in the `IndexList` class constructor reduces to some simple bookkeeping. First, we pinpoint the locations of the two dashes and store them in the `int` variables `idx1` and `idx2`. Then, these are used to define the number of elements for `substr` to copy. The `substr` method returns a `string` object which requires an application of the `c_str` method to convert its output into a C-style string that can be passed to `atoi`. After obtaining the date information, it can be used directly to construct a `Date` class object.

Once the date information has been processed, the observed value for the response or "index" variable will be the next object in the `vector<vector<string > >` container member of the `fileIn` object. This can be converted to type `double` by combining the `c_str` method and `atof`. Then, the result can be used with the corresponding `Date` object to create an `IndexData` object for insertion into the `list` member of the `IndexList` class via the `push_front` method. This process continues until all `getVNRows()` observation in the data set have been exhausted.

The `list` class has both an `insert` and a `push_back` method that perform interior insertions and add elements to the tail of a `list` object, respectively. For this particular setting it is somewhat more convenient to combine these into a single method of the form

```
list<IndexData>::iterator IndexList::insert(const IndexData& d){
  list<IndexData>::iterator iter = L.begin();
  while(iter->T >= d.T && iter != L.end())
    iter++;
  if(iter == L.end())
    L.push_back(d);
  else
    L.insert(iter, d);
  return iter;
}
```

The `Date` member of the object being inserted is used to determine its place in the list. It is added to the front of the list if its date is later than that of the object returned by the `list.begin()` iterator. Otherwise, it is placed immediately in front of the object with the latest date possible prior to that of the inserted object. If the object occurred earlier than all the other list elements it is appended to the struct's `list` object using the `push_back` method. In any event, a `list` class iterator is returned that points to the insertion location in the `list` member of the `IndexList` object.

There is no `lookUp` method for the `list` class and, accordingly, one must provide their own.[†] A method that serves this purpose is given in the next listing.

```
list<IndexData>::iterator IndexList::lookUp(const Date& date){
  list<IndexData>::iterator iter = L.begin();
  while(iter->T != date && iter != L.end())
    iter++;
  return iter;
}
```

A `list` iterator is initialized with the iterator returned from the `begin` method of the `IndexList` member `list` object. The list is then traversed using this iterator until the target date is found or the one past the end `list` iterator is reached.

To both access and remove an object from the front of a `list` object the `front` and `pop_front` methods must be used in tandem. The alternative take on the `pop_front` method below combines these methods to achieve simultaneous access and removal.

```
IndexData IndexList::pop_front(){
  IndexData d;
  if(!L.empty()){
    d = L.front();
    L.pop_front();
    return d;
  }
  return IndexData();
}
```

The `main` function below is from a program that uses the `IndexList` class.

```
int main(){
  IndexList C(fileIn("index.txt"));
  Date tempD;
  IndexData tempI;
  tempD.Year = 1952; tempD.Month = 05; tempD.Day = 29;
  tempI.T = tempD; tempI.Value =   2490.794985197;
  list<IndexData>::reverse_iterator riter(C.insert(tempI));
  riter--;
  while(riter != C.L.rend()){
    cout << *riter << endl;
    riter++;
  }

  return 0;
}
```

First an `IndexList` object is created from the data we stored in the index.txt file. The goal is to insert the information for the intentionally omitted date 1952-05-29 into the `IndexData` object. The `insert` method will return a forward `list` iterator that points to the insertion

[†] The `algorithms` library does provide search methods that can be used with `list` or other STL objects.

location. To illustrate the success of the insertion operation it is more expedient to to move backward to the front of the container. This can be accomplished using a reverse iterator as mentioned in Section 10.2. A reverse iterator `riter` for the `list` container is obtained by using the forward iterator from `insert` (i.e., `C.Insert(tempI)`) as the argument to a `reverse_iterator` class constructor. The action produced by `++` will now be movement of the iterator toward the front of the container. This path is traversed using a `while` loop that terminates when the `rbegin` function returns the iterator that points to one before the beginning of the `list` object. Notice that the information about the first element in the list will be printed using this approach as part of the loop. An alternative route would have been to simply use the forward iterator `C.Insert(tempI)` and move back to the front of the list via the `--` operator. The problem is that termination of the loop would occur when the iterator `C.L.begin()` was encountered. Thus, the loop would stop one node short of the front of the list. Although this behavior is easy enough to work around in this case, reverse iterators provide the solution for instances where a backward iteration needs to terminate on, rather than before, the front component of a `list` object.

The output from running our program was

```
5-28-1952  2507.13
5-29-1952  2490.79
5-30-1952  2511.7
6-2-1952   2514.68
6-3-1952   2507.63
6-4-1952   2499.45
```

Referring back to the beginning of the section, reveals that the missing node has been inserted successfully.

10.4.2 A chaining hash table

Currently the official C++ STL does not include a hash table implementation.[‡] Hash tables are a part of the draft standard described in the C++ TR1 document that is already at least partly implemented in several compilers, most notably gcc, and is largely safe and portable. The next standard will include these features and many others, a few of which are discussed in Appendix C.

Despite the absence of a specific hash table container, the STL still provides all that is necessary to create effective hash tables as will be demonstrated in this section. The development here is motivated by Section 12.4 of Ford and Topp (2002).

Conceptually, we know from Section 9.2.4 that a chaining hash table consists of three components:

- an array structure that represents the buckets for the table,

- linked lists that hold the elements in each bucket and

- elements that are a composed of key and data modules.

Thus, the appropriate containers from the STL can simply be used in combination to create a composite structure that has these requisite features. In that regard, the `pair` container from Section 9.2.4.2 has already been used as the third component and there is no reason to deviate from that scheme. The `vector` container can provide the buckets; this comes with the added benefit of providing a built-in resizing mechanism for the table. Objects from the

---

[‡] Several hash table implementations are available through external libraries and STL extensions, including the SGI STL, and some versions of Microsoft's Visual Studio. In most cases, the template type is called `<hash_map>`. However, not only the implementations but also the interfaces differ among these nonstandard versions with the consequence that code using `<hash_map>` may be more difficult to port.

`list` container class will occupy the slots in the `vector` object with the overall container having the type

```
vector<list<pair<Key, Data> > >
```

Again, note that the spaces between the > symbols are a syntactic necessity. The declaration for the `chainHash` class provides the details on how we have implemented our hash table construction plan.

```
template<class Key, class Data>
class chainHash {
  int tableSize, Size;
  vector<list<pair<Key, Data> > > Table;
  int hash(Key a) const;

public:

  chainHash(int tablesize) : tableSize(tablesize), Size(0) {
    Table.resize(tableSize);
  }
  struct iterator{
    chainHash* pCH;
    int Index;
    typename list<pair<Key, Data> >::iterator current;
    iterator(){}
    iterator (chainHash<Key, Data>* pch, int index,
        typename list<pair<Key, Data> >::iterator iter)
        : pCH(pch), Index(index), current(iter){}
    pair<Key, Data>& operator*() const;
    pair<Key, Data>* operator->() const;
    bool operator==(const iterator& iter) const;
    bool operator!=(const iterator& iter) const;
    iterator operator++();
    iterator operator++(int){return operator++();}
    void nextBucket();
  };

  iterator insert(pair<Key, Data> a);
  iterator lookUp(Key a);
  iterator begin();
  iterator end();
  int getSize() const {return Size;}
  int getTableSize() const {return tableSize;}
};
```

The `chainHash` class itself is relatively simple. There are three private members: the number of buckets, `tableSize`, the number of objects in the table, `Size`, and the `vector` object that holds the `list` objects containing the `pair` objects that will populate the table. There are the basic `insert` and `lookUp` methods to build and find objects in the table as well as accessor functions to retrieve the values of the `Size` and `tableSize` class members. None of these involve new ideas at this point. Accordingly, most of the subsequent discussion will be directed toward the more novel aspect of fleshing out the internal iterator class that will provide the means to navigate inside a `chainHash` object.

The `chainHash` iterator class has three members. The `current` member plays a role similar to its pointer namesake in the LCC class of Section 9.5. However, `current` is now an iterator in its own right that is obtained from the nested iterator class for the `list` container. The formulation is actually quite similar to the one for the `g5Array` class of Section 10.3.3. As in that setting, there is an `Index` class member that gives the bucket or slot of the `vector` object `Table` where an iterator is currently located. In combination, `Index` and `current` uniquely determine where an object is located in the hash table. The pointer pCH points to the `chainHash` object on which the iterator is defined.

The operators `*` and `->` for the nested iterator class in the `chainhash` class take the expected forms

```
template <class Key, class Data> pair<Key, Data>&
        chainHash<Key, Data>::iterator::operator*() const {
  return *current;
}


template <class Key, class Data> pair<Key, Data>*
        chainHash<Key, Data>::iterator::operator->() const {
  return &(*current);
}
```

They simply return a reference or a pointer to the pair object at the current location. There is perhaps a bit more here than meets the eye in that, e.g., `*current` is an application of `*` to a `list` iterator which, in turn, will return the `pair` object at its current location.

The relational equality operator is formulated as

```
template <class Key, class Data> bool
        chainHash<Key, Data>::iterator::operator==(
        const iterator& iter) const {
  return (current == iter.current && Index == iter.Index
        && pCH == iter.pCH);
}
```

Thus, two iterator objects are considered the same if they both point to the same object in a list in the same bucket in the same `chainHash` object. The `!=` operator is defined as the negation of `==`.

The most difficult task is defining the increment operator for the `iterator` class. As in Section 9.5, the postfix increment operator is defined in terms of the prefix operator using the dummy `int` variable argument that distinguishes between the two function signatures. Thus, it suffices to concentrate on the prefix operator.

Our plan is to navigate through a `chainHash` object sequentially by

- traversing the indices of the `vector` object `Table` in numerically ascending order and

- by always moving through a list from head to tail (that is, from `begin()` to `end()`).

Two main issues arise here: movement through a `list` object for a given bucket and movement to a new bucket when the end of a `list` object is reached. Let us deal with the latter problem first.

The function `nextBucket` is an analog of the `nextIndex` method for the `g5Array` class of Section 10.3.3. It provides a way to move an iterator to a new bucket or slot of the `chainHash` vector object `Table`.

```
template <class Key, class Data> void
        chainHash<Key, Data>::iterator::nextBucket(){
  for(int i = Index + 1; i < pCH->tableSize; i++)
    if(!pCH->Table[i].empty()){
      Index = i;
      current = pCH->Table[i].begin();
      return;
    }
```

```
//no empty buckets were found
Index = -1; current = pCH->Table[0].end();
}
```

The basic premise here is that the **nextBucket** method will be called when the end of a **list** object has been reached making it necessary to find the next nonempty bucket in the table. The value of **Index** is the current slot for the **iterator** object which means that the next logical place to look for more table elements is in the next bucket to the right or **Table[Index + 1]**. Since this is a **list** object, it has the associated method **empty** that can be used to check if the list contains any elements. The search for a nonempty list continues until either one is found or there are no further slots left in **Table**. In this latter instance, the end of the table has been reached and we need some way to indicate this. The approach employed here is to return nonoperative values for **Index** and **current**. We will say a bit more about this choice shortly.

An alternative to our implementation of **nextBucket** is offered in Exercise 10.6. In that version, the hash table is augmented by a list that links the elements inserted in the table. This list directly provides iterators for moving between buckets in the order in which they were populated. The only overhead is space (instead of time) due to the additional data that must be stored besides the hash table's own array. This approach has a precedent in, for example, the Python data structure called *ordered dictionary*.

The **chainHash** class provides the standard STL type methods **begin** and **end** that return iterators to the first and one past the end elements in a container. We are now in a position that allows us to define these two methods. Specifically, the definition for the **begin** method appears in the next listing.

```
template<class Key, class Data>
      typename chainHash<Key, Data>::iterator
      chainHash<Key, Data>::begin(){
  iterator iter;
  iter.pCH = this;
  //start the search in a bucket with index 0
  iter.Index = -1;
  //now set Index and current
  iter.nextBucket();
  return iter;
}
```

The "beginning" of the hash table should be the first element in the first nonempty bucket and **begin** returns an iterator that points to that object if it exists. For this purpose an **iterator** object is created with its **chainHash** object pointer set to point at the **chainHash** object that will be the one to call the **begin** method. Its bucket index is then set to −1 before calling the **nextBucket** method. The first step in **nextBucket** is to advance the index by one, which sets it at the first slot in the **Table** vector container. The remainder of the work is carried out by **nextBucket** which sets a final value for both the **Index** and **current** class members.

The final entry in a **chainHash** object is the last object in the last (i.e., farthest right-hand) nonempty bucket. One can enact a scheme similar to the **g5Array** approach to physically locate this quantity. An alternative strategy employed by, e.g., Ford and Topp (2002) is to simply define an object that cannot actually be realized and return this whenever an attempt is made to go past the end of the container. Our choice here needs to agree with the object returned from **nextBucket** when no further empty buckets are found. This last consideration leads to the choice of the **end** method given in the next listing.

```
template<class Key, class Data>
       typename chainHash<Key, Data>::iterator
       chainHash<Key, Data>::end(){
  iterator iter;
  iter.pCH = this;
  iter.Index = -1;
  iter.current = Table[0].end();
  return iter;
}
```

All the pieces are now in place that allow us to give the definition for a prefix increment operator.

```
template <class Key, class Data>
       typename chainHash<Key, Data>::iterator
       chainHash<Key, Data>::iterator::operator++(){
  if(*this == pCH->end())
    return *this;

  current++;
  if(current == pCH->Table[Index].end())
    nextBucket();
  return *this;
}
```

The first step is to check if the end of the table has already been reached. In that case, an application of ++ will have no effect. Otherwise, the location of the iterator is advanced inside its current bucket. A check is made to see if the one past the end node of the current list object has been reached. In that event, the next nonempty bucket or pCH->end() is found via nextBucket.

The insert and lookUp methods for class chainHash given below involve similar operations to the ones that were used to define our iterator increment operator.

```
template<class Key, class Data>
       typename chainHash<Key, Data>::iterator
       chainHash<Key, Data>::insert(pair<Key, Data> a){
  int index = hash(a.first);
  Table[index].push_front(a);
  iterator iter(this, index, Table[index].begin());
  Size++;
  return iter;
}
```

```
template<class Key, class Data>
       typename chainHash<Key, Data>::iterator
       chainHash<Key, Data>::lookUp(Key a) {
  int index = hash(a);
  typename chainHash<Key, Data>::iterator
    iter(this, index, Table[index].begin());
  while(iter->first != a && iter.current != Table[index].end())
    iter++;
  return iter;
}
```

The methods return iterators to the locations of pair objects that have been either inserted or found. These iterator objects are created using the nondefault constructor that takes a pointer to chainHash, a bucket index and a list container iterator as its arguments. The bucket index is obtained by hashing a key provided in the method's argument while the list iterator is the one returned by the `begin` method of the chosen bucket's `list` object. For `insert` the new `pair` object is added to the `list` object using its push_front method and the number of elements in the table is also incremented. In contrast, `lookUp` does not alter the table but merely searches until the `pair` object with the designated key has been found. This key is presumed to exist and no mechanism has been provided to handle a misspecified key value.

The `main` function shown below is from a program that was written to try out our `chainHash` class.

```
int main(){
  chainHash<double, int> cH(3);
  cout << "The table size is " << cH.getTableSize() << endl;

  for(int i = 0; i < 10; i++)
    cH.insert(pair<double, int>(sqrt(i), i*i));
  cout << "The number of objects in the container is " <<
    cH.getSize() << endl;

  chainHash<double, int>::iterator iter = cH.lookUp(sqrt(5));
  cout << iter.Index << " " << iter->first << " "
       << iter->second << endl;
  iter++;
  cout << iter.Index << " " << iter->first << " "
       << iter->second << endl;

  for(iter = cH.begin(); iter != cH.end(); iter++)
    cout << iter.Index << " " << iter->first << endl;

  return 0;
}
```

A table with three buckets is created and then filled with 10 `pair` objects that have Key components of type `double` and `int` Data members. An `iterator` object is defined to point to the `pair` object with key value $\sqrt{5}$. The increment operator is applied to this `iterator` object and it is then used in a `for` loop to explore the `chainHash` object.

To make this all work a specific choice is needed for a hash function that can be used with non-integer keys. The one selected for this purpose is

```
template<class Key, class Data> int
        chainHash<Key, Data>::hash(Key a) const {
  return ((int) floor(a)) % tableSize;
}
```

The C++ `floor` function (along with a cast) is used here to transform double precision keys into integers. Although effective, this is not the best solution to the problem; a generic class should not really need adjustments to allow for different types of data. A more satisfactory solution is to overload the modulus operator for the Key class in a way that gives the desired result in the form of a generic hash function in the `chainHash` class. This approach will be examined in Exercise 10.21.

The output from our test program is

```
The table size is 3
The number of objects in the container is 10
2 2.23607 25
2 2 16
```

```
0 3
0 0
1 1.73205
1 1.41421
1 1
2 2.82843
2 2.64575
2 2.44949
2 2.23607
2 2
```

Both the table size and number of elements are as expected. The `lookUp` operation successfully returns the `pair` object whose key value of $\sqrt{5}$ hashed to the second bucket in the hash table. The increment operator then moves the `iterator` object one step further in the corresponding `list` object to the `pair` container that had the key $2 = \sqrt{4}$. The results of the `for` loop container traversal show that the keys have all been hashed correctly and that the `begin` and `end` methods have produced the desired results.

10.5 Queues

The queue-type containers in C++ come in two basic varieties: class `queue` and class `priority_queue`. Their respective include directives are

```
#include <queue>
```

and

```
#include <priority_queue>
```

Objects from the two classes are then specified using syntax such as

```
queue<className> Q;
```

```
priority_queue<className> Q;
```

with `className` a specified data type.

From the discussions in Section 9.2.3 and Section 9.3, we anticipate that objects in a `queue` container will be organized on a first-in first-out basis while those in a `priority_queue` container will be arranged (in a heap) according to some type of priority. This is indeed the case for the C++ queue containers with priority being determined by the `<` operator. Thus, for user-defined data types, this operator will have to be defined explicitly.

Both queue classes provide the same basic methods: namely,

- `push(object)` to push an element `object` onto the end of the queue,
- `pop()` to remove the top (i.e., first-in or highest priority) object from the queue,
- `front()` (respectively, `top`) to examine the top (respectively, highest priority) node of a `queue` (respectively, `priority_queue`) object without actually removing it from the queue,
- `empty()` that returns a Boolean variable evaluating to `true` when the queue is empty and
- `size()` that returns the number of objects in the container.

The `pop` method simply removes the queue's top entry and has no return value. To access the top element one must instead use either `front` or `top`. The behavior of `pop` is not defined when the `queue` object is empty. Thus, it is best to check for that possibility with `empty` before using the method.

The remainder of this section will be focused on illustrating the use of the `priority_queue` container in the context of a discrete event simulation wherein the activity in some physical system is modeled through a sequence of events that occur, sequentially, over a specified period of time. The setting we will use is that of an $M/M/n$ queue: i.e., a queue with n servers where there are independent Poisson customer arrivals having rate parameter λ_C and independent exponential service times with mean λ_T. The servers and customers are also assumed to operate independently. A similar development using the `queue` class will be explored in Exercise 10.15.

The basic elements of the model are the customers and servers and, accordingly, each will need to be developed in the form of a class. The customer class will be treated first. Its declaration is

```
struct Customer{
  int Arrive;
  bool Gold;
  Customer(int t = 0, bool gold = 0) : Arrive(t), Gold(gold) {}

  bool operator<(const Customer& x) const;
};
```

There are two members: the customer's arrival time `Arrival` and a Boolean variable `Gold`. The `Arrival` class member is the time that the `Customer` object enters the simulation. To simplify matters, time will evolve in discrete (e.g., one minute) intervals and, hence, `Arrival` will take on only integer values. The `Gold` class member represents the presence or absence of "gold" status in the sense associated with car rentals, airlines, etc., where frequent patronage can lead to a customer receiving a preferential "gold" level of service. The way that `Gold` is used here is in the definition of the < operator which takes the form

```
bool Customer::operator<(const Customer& x) const {
  if(Gold == 0 && x.Gold == 0 ||
     Gold == 1 && x.Gold == 1){
    if(Arrive < x.Arrive)//give earlier arrival higher priority
      return false;
    else
      return true;
  }

  if(Gold == 0 && x.Gold == 1)
    return true;
  if(Gold == 1 && x.Gold == 0)
    return false;
}
```

The STL priority queue is a max queue: i.e., elements are partially sorted in descending order as determined by the < operator. All things being equal, our overloaded version of this operator has the effect that earlier arrivals get higher priority and, conversely, later arrival leads to lower priority. On the other hand, a `Customer` object with a `Gold` value of `true` is always preferred (i.e., is >=) to a customer with `Gold` equal to `false`.

Output that allows us to track the progress of the simulation is created with an overloaded output insertion operator amenable for used with `Customer` objects. It takes the form

```
std::ostream& operator<<(std::ostream& out, const Customer& x){
  out << "Arrival time = " << x.Arrive << " Gold = " <<
      std::boolalpha << x.Gold << std::endl;
  return out;
}
```

Our definition for the `Server` class is

```
struct Server{
  bool Busy;
  int serviceTime;
  Server(){
    Busy = 0; serviceTime = 0;
  }
};
```

A particular `Server` object may be "busy" with a `Customer` object in which case its Boolean `Busy` member will evaluate to `true`. Otherwise, the `Server` object will be available for assignment to the next object from the queue. The other member of the `Server` struct is `serviceTime`. As with arrival time, this will be measured in discrete intervals.

With the `Customer` and `Server` classes at our disposal, the simulation class can now be defined as in Listing 10.6.

Listing 10.6 *MMNQueue.h*

```
class MMNQueue{

  std::priority_queue<Customer> Q;
  Server* pS;//array of servers
  ranGen Rng;//random number generator
  void upDate(int t);
  int rPoisson();
  double rExp(double lambda);
  bool isGold();
  void upDateServers();
  int freeServer();

  //simulation parameters
  int runTime, nServers;
  double lambdaC, lambdaT, goldProb;
  unsigned long Seed;

  //statistics
  int totalTime, numberCustomers;

public:

  MMNQueue(int runtime, int nservers, double lambdac,
        double lambdat, double goldprob, unsigned long seed) :
        runTime(runtime), nServers(nservers), lambdaC(lambdac),
        lambdaT(lambdat), goldProb(goldprob), Seed(seed){

    totalTime = 0; numberCustomers = 0;
    pS = new Server[nServers];
    Rng = ranGen();
    Rng.setSeed(Seed);
  }
  ~MMNQueue(){
    delete[] pS;
  }
  double runSimulation();
};
```

The parameters for the simulation are

- `runTime`: the length of time the simulation will be allowed to run.

- `nServers`: the number of servers for the system.

- `lambdaC` and `lambdaT`: the means for the Poisson customer arrival distribution and the exponential service time distribution.

- `goldProb`: the presence or absence of `Gold` "status" for a `Customer` object is modeled as a Bernoulli random variable with `goldProb` being the probability of `Gold` evaluating as `true`.

- `Seed`: generation of random numbers from the Poisson, exponential and Bernoulli distributions requires a random number generator of some sort. The one that will be used is the `ranUnif` method from the `ranGen` class in Appendix E. The value of `Seed` provides the seed that will initialize the generator.

The class constructor defines these parameters through an initializer list. The constructor also creates an `nServer` array of `Server` objects and initializes the `ranGen` object for generating uniform random deviates.

There are various measures that could be used to assess the performance of our customer-server system for a particular set of values for its parameters. Our attention will be directed toward customer waiting time prior to service. For this purpose the waiting time for all customers during the simulation `totalTime` is accumulated as well as the total number of customers that had to wait `numberCustomers`. These variables are initialized to 0 in the constructor. At the end of the simulation they will allow us to determine the average waiting time. The use of additional performance measures is considered in Exercise 10.14.

The operative, `private` members of the `MMNQueue` class are

- the `priority_queue` object `Q` with template parameter `Customer` that will hold the `Customer` objects that are waiting for service,

- a pointer to `Server`, `pS`, that will be used to hold and manage the system's `Server` objects and

- the `ranGen` object `Rng` for producing uniform random numbers.

There are also several utility functions that have been designated as `private`. The `rPoisson`, `rExp` and `isGold` methods are for generating Poisson, exponential and Bernoulli random deviates, respectively. The `upDate` method is used to update the system at each new time unit. It, in turn, uses `upDateServers` and `freeServer` to update the elements in the server "array" `pS` and to check for the availability of a server to serve the next customer object in the queue. These functions will be discussed in more detail shortly.

As the `MMNQueue` class uses dynamic memory allocation for its `Server` component, an explicit destructor as well as a copy constructor and assignment operator should be defined for uses that would involve manipulation of `MMNQueue` object. This will not be a feature of our example and we have chosen to forgo the specification of these methods here.

The method that starts the simulation is `runSimulation` whose definition is

```
double MMNQueue::runSimulation(){
  for(int t = 0; t < runTime; t++){
    cout << "For time = " << t << endl;
    upDate(t);
    cout << "queue size " << Q.size() << endl;
    cout << endl;
  }

  while(!Q.empty()){//may be some people still in the queue
    totalTime += runTime - Q.top().Arrive;
```

```
    Q.pop();
  }

  return ((double)totalTime/(double)numberCustomers);
}
```

This function serves as the simulation's "clock" in that it moves the process forward at one unit time intervals from the beginning time at 0 to its end at time runTime - 1. During this period its only function is to call the upDate method that handles initialization of new Customer objects, manages the service times for the Server objects, etc. When the simulation ends, the runSimulation function will empty the priority_queue container while adding the waiting time for its Customer elements (that waited and were not served) onto the process waiting time variable totalTime. Its last action is to return the average waiting time over the entire simulation.

Most of the work for a MMNQueue object will be done by the upDate function given in the listing below.

```
void MMNQueue::upDate(int t){
  upDateServers();

  int serverIndex = 0;
  //number of new customers that arrived in last minute
  int numberNew = rPoisson();
  cout << "number of new customers " << numberNew << endl;
  //serve as many as possible right away
  while(freeServer() != -1 && Q.empty() && numberNew > 0){
    serverIndex = freeServer();
    pS[serverIndex].Busy = 1;
    pS[serverIndex].serviceTime = ceil(rExp(lambdaT));
    numberNew--;
  }

  Customer temp;
  if(numberNew > 0){//add to the queue
    for(int i = 0; i < numberNew; i++){
      temp = Customer(t, isGold());
      cout << "New waiting customer has " << temp;
      Q.push(temp);
    }
    numberCustomers += numberNew;
  }

  while(freeServer() != -1 && !Q.empty()){
    serverIndex = freeServer();
    pS[serverIndex].Busy = 1;
    pS[serverIndex].serviceTime = ceil(rExp(lambdaT));
    totalTime += t - Q.top().Arrive;
    cout << "Customer being served has " << Q.top();
    Q.pop();//remove customer
  }
}
```

The first step in this method is to update the Server objects using

```
void MMNQueue::upDateServers(){
  for(int i = 0; i < nServers; i++){
    if(pS[i].serviceTime > 0)
```

```
        pS[i].serviceTime--;//reduce remaining service time
      if(pS[i].serviceTime == 0)
        pS[i].Busy = 0;//free server for new customer
    }
  }
```

which goes through the elements in the `Server` array decrementing the time remaining to serve their current `Customer` object and/or changing their `Busy` member to `false` if they have finished their current assignment.

The next task for `upDate` is to generate the number of new `Customer` arrivals. This is accomplished with the method

```
int MMNQueue::rPoisson(){
  double sum = rExp(1./lambdaC);
  int numberNew = 0;
  while(sum < 1){
    numberNew++;
    sum += rExp(1./lambdaC);
  }

  return numberNew;
}
```

that, in turn, calls

```
double MMNQueue::rExp(double lambda){
  double u = 0.;
  Rng.ranUnif(1, &u);
  return (-lambda*log(1. - u));
}
```

The latter method uses the `ranGen` object to generate a stream of random numbers from an exponential distribution with a given value for its mean `lambda`. As such it is used both to generate the service times for the `Server` objects as well as the number of arrivals from the Poisson distribution. The latter application stems from Algorithm 4.5 that derives from the fact that the waiting times for arrivals in a Poisson process with mean λ are independent and exponentially distributed with mean $1/\lambda$. As a result, by generating values from this exponential distribution until their sum passes some fixed time point t, we will have realized a set of arrivals from a Poisson process over the interval $[0, t]$. The corresponding value for the Poisson random variable is the penultimate number of terms in the sum.

Once the number of new `Customer` objects is set, there are two possibilities: either the queue is empty or it is not. If it is empty (as determined by the `empty` method for the `priority_queue` class), the new customer objects will be assigned to `Server` objects with `Busy` values of `false` as long as they are available. The availability or lack of availability is determined by

```
int MMNQueue::freeServer(){

  int serverIndex = 0;
  while(pS[serverIndex].Busy && serverIndex < nServers - 1)
    serverIndex++;

  if(pS[serverIndex].Busy)
    return -1;
  else
    return serverIndex;
}
```

This function goes through the Server objects and returns the index of the first available object. It returns −1 to indicate that all the objects are in use. When an available Server object is found, its index in the pS pointer is returned. At that point, upDate generates a value for serviceTime for that Server object and changes its Busy member to true. Service is immediate with the consequence that there is no reason to generate a specific Customer object for the chosen Server object.

After all the available Server objects (should any exist) have been assigned a Customer object, it becomes necessary to push any remaining new Customer objects onto the queue using the priority_queue push method. These customers count in the sense that they will have to wait and, hence, the numberCustomer member variable must be increased by the number of objects that have been added to the queue. The actual Customer objects are therefore initialized with the current time and a value for their Gold member determined by

```
bool MMNQueue::isGold(){
  double u = 0.;
  Rng.ranUnif(1, &u);
  if(u < goldProb)
    return true;
  return false;
}
```

This function uses the MMNQueue object's ranGen member to simulate a Bernoulli random variable with success probability goldProb.

The final step in upDate is to see if any Server objects have become available that can be assigned to the elements in the queue with the highest priority. The freeServer method is used to search for an available Server object and, if one is found, its Busy value is set to true and a value for its serviceTime member is generated from an exponential distribution with parameter lambdaS using rExp. For each available Server object, a Customer object is removed from the queue in two steps. First, the priority_queue top method is used to find the arrival time for the member at the top of the queue. The difference between this value and the current time is the length of time the Customer object waited and this value is added to the total waiting time variable totalTime. Next, the same highest priority element is removed from the queue with the pop method.

A program that tries out our MMNQueue class is given below.

```
int main(){
  MMNQueue bSim(10, 4, 3, 3, .1, 123);
  cout << bSim.runSimulation() << endl;
  return 0;
}
```

An MMNQueue object is created and used to start the simulation with its runSimulation method. The particular choices that have been made for the simulation parameters entail that it will run for 10 time units with four Server objects. The average number of arrivals for the Poisson process is three per time unit and the average service time is three units. The proportion of Customer objects with a true value for their Gold member is chosen to be .1 and the Seed for the ranGen object is set at 123.

Some enhanced output that was produced by our simulation program is given below.

```
For time = 0
number of new customers 3
queue size 0

For time = 1
number of new customers 2
New waiting customer has Arrival time = 1 Gold = false
```

```
queue size 1

For time = 2
number of new customers 3
New waiting customer has Arrival time = 2 Gold = false
New waiting customer has Arrival time = 2 Gold = false
New waiting customer has Arrival time = 2 Gold = false
queue size 4

For time = 3
number of new customers 4
New waiting customer has Arrival time = 3 Gold = false
New waiting customer has Arrival time = 3 Gold = false
New waiting customer has Arrival time = 3 Gold = false
New waiting customer has Arrival time = 3 Gold = false
Customer being served has Arrival time = 1 Gold = false
Customer being served has Arrival time = 2 Gold = false
queue size 6

For time = 4
number of new customers 2
New waiting customer has Arrival time = 4 Gold = false
New waiting customer has Arrival time = 4 Gold = false
Customer being served has Arrival time = 2 Gold = false
queue size 7

For time = 5
number of new customers 3
New waiting customer has Arrival time = 5 Gold = false
New waiting customer has Arrival time = 5 Gold = false
New waiting customer has Arrival time = 5 Gold = false
Customer being served has Arrival time = 2 Gold = false
Customer being served has Arrival time = 3 Gold = false
Customer being served has Arrival time = 3 Gold = false
queue size 7

For time = 6
number of new customers 2
New waiting customer has Arrival time = 6 Gold = false
New waiting customer has Arrival time = 6 Gold = false
queue size 9

For time = 7
number of new customers 4
New waiting customer has Arrival time = 7 Gold = true
New waiting customer has Arrival time = 7 Gold = false
New waiting customer has Arrival time = 7 Gold = false
New waiting customer has Arrival time = 7 Gold = false
Customer being served has Arrival time = 7 Gold = true
queue size 12

For time = 8
number of new customers 2
New waiting customer has Arrival time = 8 Gold = false
New waiting customer has Arrival time = 8 Gold = false
queue size 14

For time = 9
number of new customers 4
New waiting customer has Arrival time = 9 Gold = false
New waiting customer has Arrival time = 9 Gold = true
New waiting customer has Arrival time = 9 Gold = false
New waiting customer has Arrival time = 9 Gold = false
Customer being served has Arrival time = 9 Gold = true
Customer being served has Arrival time = 3 Gold = false
queue size 16
```

Four of the five `Customer` objects that arrive during the first two time periods of the simulation are assigned to `Server` objects. Apparently, no servers are free to handle the fifth arrival and it must be pushed onto the queue. The first `Customer` object with a `Gold` status of `true` arrives at time period 7 and is served immediately bypassing other objects that have arrived earlier and been waiting in the queue. At the end of the simulation, the queue contains sixteen `Customer` objects and the objects that have had to wait have done so for an average of 3.04 time units.

10.6 The map and set containers

The STL provides four different varieties of the ordered set data structure from Section 9.4: map, `multimap`, `set` and `multiset`. The `map` container is the most direct analog of the Section 9.4 discussion. The `multimap` container is the same as `map` apart from allowing for repeated keys. The `set` containers and `multiset` containers are similarly related and provide simplifications of maps that contain only key values. All four container types are implemented using red-black binary search trees.

The four container types provide the same basic set of operations so that it suffices to pick one for exposition purposes. Accordingly, our discussion will focus only on the `map` container class.

Access to the `map` container class is achieved through the include directive

```
#include <map>
```

Then, an empty `map` object is created with the syntax

```
map<Key, Data> m;
```

with `Key` and `Data` the names of the data types that will provide the key and data components of the nodes stored in the container. Objects of type `Key` are assumed to possess an inherent ordering that is quantified via a definition of the < operator for the class. It is also possible to provide a user-supplied function to be used in making comparisons through an argument to one of the alternative class constructors. That option will not be explored here.

As was true of the `Tree` class in Section 9.4, `map` containers can be filled using their `insert` method. The basic objects that are used in insertion are from the `pair` class that was described in Section 9.2.4.2. If `m` is a `map<Key, Data>` object,

```
m.insert(pair<Key, Data>(key, data));
```

will add a node to the tree with `Key` value `key` and `data` as its `Data` module and return an iterator that points to the position of insertion.

The look-up operation for a `map` container is carried out by the `find` method. This function takes an object of type `Key` as its argument and returns an iterator to either the position of the located object or `end()` if no object with that key value is found. An indexing operator `[]` is also provided that can be used as an alternative to `find`. For a `map` object `m`, `m[key]` will return a reference to the `Data` member of the `pair` object in the tree that has `Key` value `key` if it exists. Unlike `find`, if the specified `Key` value does not exist, a new node is inserted into the `map` object with the given `Key` value and its `Data` component determined from the default constructor for the `Data` class. Because it returns a reference to the `Data` member, `[]` provides a means to both modify an existing node or insert a completely specified new node into the tree. Specifically, the command

```
m[key] = data;
```

will produce one of two outcomes. If the `map` object `m` contains a node with `Key` value `key`, its `Data` member will be overwritten with the value `data`. Otherwise, a new node will be inserted with the Key value `key` and `Data` component `data`.

Removal of elements from a `map` is accomplished by the **erase** method. The method takes two forms. One version of **erase** takes an iterator that points to the position of the node that is to be erased. In this instance the node is removed and the function returns nothing. The other option is to use a key value for the **erase** function argument. In that latter instance, **erase** removes the object with the specified key and returns the integer 1 for a `map` object and, more generally, the number of erased objects for a `multimap` that allows duplicate keys.

The listing below illustrates the use of a `map` container.

```
//mapEx.cpp
#include <iostream>
#include <utility>
#include <map>

using std::cout; using std::endl; using std::map; using std::pair;

int main(){

   map<int, int> m;
   m.insert(pair<int, int> (4, 4)); m.insert(pair<int, int> (0, 2));
   cout << m[0] << endl;
   m[0] = 1;
   cout << m[0] << endl;
   m[1] = 2;
   cout << m[1] << endl;
   m[3] = 1; m[2] = 11; m[6] = 5; m[7] = 4; m[5] = 5; m[21] = -3;
   m[-15] = 99; m[-16] = 5; m[-3] = 3; m[-1] = 9; m[-2] = 14;

   map<int, int>::iterator iter;
   for(iter = m.begin(); iter != m.end(); iter++)
      cout << iter->first << " ";
   cout << endl;
   iter = m.find(-2);
   cout << iter->second << endl;
   m.erase(iter); m.erase(-16); m.erase(-15); m.erase(-1);

   map<int, int>::reverse_iterator riter;
   for(riter = m.rbegin(); riter != m.rend(); riter++)
      cout << riter->first << " ";
   cout << endl;
   return 0;
}
```

Here we revisit the binary search tree from Figure 9.9. The tree is constructed using both the insert method and the [] operator. After inserting a node with key 0 and data member 2, the value of the data member is both accessed and changed using the index method. Output is produced to see the effect of these two procedures as well as that of an insertion using []. An iterator is created next and used to explore the container while printing out the key values for each node in the process. The **find** method is then employed to locate the node with a key of −2. The iterator returned from **find** is used in the **erase** method to remove that particular node from the tree. Finally, several nodes are removed with **erase** using their key values and the container is explored again using an iterator to assess the changes that have been made. In this latter instance we have carried out the exploration with a reverse iterator. The starting point for a reverse iteration through a container is obtained from rbegin with rend providing the corresponding stopping point. The ++ operator will move the iterator backward from the current node. Of course, there is the question of how forward or backward movement will actually take place. This question is resolved by running the program.

The output from our program is

```
2
1
2
-16  -15  -3  -2  -1  0  1  2  3  4  5  6  7  21
14
21  7  6  5  4  3  2  1  0  -3
```

The insertion and indexing operations have produced the expected output. The "mystery" concerning the behavior induced by ++ is also solved; the internal map forward iterator class has been implemented so that the movement is from smaller to larger key values. Reverse iterators move from the larger to the smaller keys. The last line of output produced with the reverse iterator based for loop verifies that both forms of the erase method (i.e., the one with the iterator argument and the one that takes a key arguments) have worked correctly.

10.7 Algorithm basics

As noted in the introduction, the STL consists of three basic components: data structures, iterators and algorithms. In this final section, we describe various aspects of the latter of these three resources. A complete exposition is beyond the scope of the present text. An overview of the entire algorithm collection is provided, for example, in Malik (2007) with Josuttis (1999) also giving a detailed and thorough treatment. Our goal here is to merely demonstrate some of the features and power of the algorithm package that can pave the way for a more in depth study.

The STL algorithm library is composed of a group of data processing methods that perform various basic operation on data structures. These include searching, sorting and transforming as well as a few numeric calculations. Rather than being tied to specific container classes, the algorithm functions have an independent identity. This is achieved by having them work with iterators rather than container class objects. In this sense, their utility is not restricted to the STL and they can also be utilized with user-created ADTs. We will illustrate this last feature subsequently.

The algorithm library becomes available with the include directive

```
#include <algorithm>
```

This provides access to groups of functions whose purposes can be categorized as (e.g., Josuttis 1999) nonmodifying or modifying, removing and mutating. Nonmodifying algorithms do not alter the elements in the container. Searching methods fall into this group. Modifying algorithm include methods that transform and replace container elements while mutating algorithms change the element order but not their values. A small set of numeric methods are available that include summation, inner products and partial sums. These latter methods require the inclusion of a different header file via

```
#include <numeric>
```

In general, the invocation of an algorithm, algorithmName, from the algorithm library can take one of several forms. The simplest is

```
algorithmName(begin, end)
```

where **begin** and **end** are STL "compliant" iterators that provide the beginning and end of a range of elements that are to be processed. In some instances two ranges may be involved with the second range providing a destination for the results of the algorithm's application. In such a case, the syntax will look like

```
algorithmName(beginSource, endSource, beginDestination)
```

with **beginSource** and **endSource** iterators for the data that will be processed by the algorithm and **beginDestination** an output iterator that points to the beginning of the range where the processed data is to be placed. A second "end" iterator is not necessary as that position is determined by the length of the range from **beginSource** to **endSource**. A third type of argument structure for an algorithm is

```
algorithmName(begin, end, functionObject)
```

or

```
algorithmName(beginSource, endSource, beginDestination,
              functionObject)
```

These latter forms differ from the previous two only in terms of the addition of a function object or functor (see, e.g., Section 8.2) that provides the tool to be used in processing the elements of the source container. Not any function object will do and, as was the case for iterators, function objects need to inherit from an STL class. The two possibilities in this case are **unary_function** and **binary_function**.

To illustrate some of the basic ideas, let us work with the **g5Array** class of Section 10.3.3. Our goal will be to create three new methods for the class that are all implemented using methodology from the algorithm library. The methods that will be created are ones that will look up drawing results for a specified date, order the **g5** objects in the container by the date of their drawings and compute a chi-square statistic.

The most straightforward method to implement is that for sorting. It takes the form

```
std::vector<g5> g5Array::g5Sort() const {
  std::vector<g5> vSort(nDraws);
  std::copy(this->begin(), this->end(), vSort.begin());
  g5Less f;
  std::sort(vSort.begin(), vSort.end(), f);

  return vSort;
}
```

This method uses two algorithms: copy and sort. There are three main sorting methods in the algorithm library: **sort**, **partial_sort** and **stable_sort**. The **sort** method is akin to quicksort while **partial_sort** and **stable_sort** are relatives of heapsort and mergesort. All three of these algorithms require random access iterators which means that we must first transfer the elements from the **g5Array** object to, e.g., a **vector** object. This object will need to have **nDraw** slots. Once it is created, the **begin()** and **end()** iterators from the **g5Array** object provide the source of the data to be copied and the **begin()** iterator for the **vector** object gives the beginning of the range where the data is to be replicated.

Once all the g5 objects from the **g5Array** object are in the **vector** format, they are ready for sorting. But, sorting requires specifying a method for comparison. This can be done by overloading the comparison operator < (as in Section 10.5) or by providing an explicit function object to be used in making comparisons. We will take the latter route.

The functor that will be used for comparison of g5 objects is

```
struct g5Less : public std::binary_function<g5, g5, bool> {

  g5Less(){}
  bool operator ()(const g5& g1, const g5& g2){

    std::string temp1, temp2;
    temp1 = g1.getDate(); temp2 = g2.getDate();
    if(atoi(temp1.substr(0, 3).c_str()) < atoi(temp2.substr(0, 3).c_str()))
      return true;
    if(atoi(temp1.substr(0, 3).c_str()) == atoi(temp2.substr(0, 3).c_str())
       && atoi(temp1.substr(5, 6).c_str()) < atoi(temp2.substr(5, 6).c_str()))
      return true;
    if(atoi(temp1.substr(0, 3).c_str()) == atoi(temp2.substr(0, 3).c_str())
       && atoi(temp1.substr(5, 6).c_str()) == atoi(temp2.substr(5, 6).c_str())
       && atoi(temp1.substr(8, 9).c_str()) < atoi(temp2.substr(8, 9).c_str()))
      return true;
    return false;
  }
};
```

The sorting is to be done by draw date and, accordingly, we need to supply a function object that provides an analog of <. Thus, it should be a binary function that returns true when the draw date of one g5 object is earlier than that of another. The dates that are used to make the comparisons are extracted from the Date members of the two g5 class arguments using string class methods in a similar manner to the developments in Section 10.4.1. The novel aspect is the use of inheritance from the STL binary_function template class. There are three template parameters that represent the type of the first and second function argument and the return type, in that order. Thus, the parameters are specified as g5 for both function arguments and bool. It now suffices to apply the algorithm sort with the vector object's begin() and end() iterators.

The main function of a program that will apply our sort method to the data in guess5.txt is shown in the next listing.

```
int main(){
  g5Array g5A("guess5.txt");
  std::vector<g5> vSort = g5A.g5Sort();
  std::cout << "Sorting results" << std::endl;
  for(int i = 0; i < 3; i++)
    vSort[i].printG5();
  std::cout << std::endl;

  return 0;
}
```

It produces the output

```
Sorting results
Guess5 Drawings for 1952-01-17 used Machine A with Ball Set 6
The drawing results are:
10 38 23 20 15
36 15 11 7 37
20 10 9 25 2
29 14 16 31 34
12 38 28 26 2
16 19 22 26 33
25 17 21 3 10
16 8 32 6 29
22 26 7 24 12
29 16 37 36 27
Guess5 Drawings for 1952-01-21 used Machine B with Ball Set 4
```

```
The drawing results are:
11  9  23  10  20
32  7  16  18  40
38  35  26  36  19
24  14  39  1  19
35  1  3  7  28
30  38  18  3  24
31  6  16  9  3
16  3  9  39  25
12  4  3  33  28
33  39  4  38  29
Guess5 Drawings for 1952-01-24 used Machine A with Ball Set 5
The drawing results are:
32  1  30  27  23
20  7  1  17  18
16  19  28  3  13
33  40  10  14  31
35  12  6  27  4
2  39  40  13  33
25  12  29  31  39
20  28  25  24  36
17  5  20  9  18
15  39  40  9  23
```

Only the first three g5 objects from the sorted array are printed and their dates are seen to be correct. Note that sorting here is not immediate. Although the data in guess5.txt is in chronological order, the **g5Array** structure stores it according to the machine and ball set that were used in the drawing and this is where the sorting algorithm was applied.

Next up is the method for looking up a g5 object that corresponds to a specified draw date. The code created for this purpose looks like

```cpp
g5Array::const_iterator g5Array::lookUp(const std::string s) const {

    g5Equal f;

    const_iterator iter = std::find_if(begin(), end(),
        std::bind2nd(f, s));
    return iter;
}
```

It returns an iterator that points to the location of the target object in the g5Array container.

The find_if function from the algorithm library is used to carry out the search in lookUp. The functor object supplied to find_if needs to be a unary operator that plays the role of the == operator. In this case we want comparisons to be carried out in terms of draw dates which suggests using a functor such as

```cpp
struct g5Equal : public std::binary_function<g5, std::string, bool>{

    g5Equal(){}

    bool operator() (const g5& g, const std::string s) const {
        if(g.getDate() == s)
            return true;
        return false;
    }
};
```

Again, the function will inherit from the STL `binary_function` class with its first template parameter being type g5, a second parameter of type `string` and a `bool` return type. The problem is that the function is a binary function and a unary function is what `find_if` requires.

The STL provides the *function adapter* `bind2nd` that sets the second argument for a binary function at some specified value thereby converting a binary function into a unary function. There is also a `bind1st` adapter that serves the same purpose for the first argument of a binary function. In terms of the way it is used in the `lookUp` method, `bind2nd` will call the overloaded () operator for the g5Equal class with its second argument set at the value of the `string` argument supplied to `lookUp`.

Finally, the g5Array `chiSquare` method is given by

```
double g5Array::chiSquare() const {

  int* freq = new int[40];
  for(int i = 0; i < 40; i++)
    freq[i] = 0;

  sumFreq f;
  freq = std::accumulate(begin(), end(), freq, f);

  double e = 10*nDraws*5/40;
  double chiStat = 0;
  double* temp = new double[40];
  std::transform(&freq[0], &freq[40], &temp[0],
         std::bind2nd(std::minus<double>(), e));
  std::transform(&temp[0], &temp[40], &temp[0],
         std::bind2nd(std::divides<double>(), sqrt(e)));
  chiStat = std::inner_product(&temp[0], &temp[40], &temp[0], 0.);

  return (39./35.)*chiStat;
}
```

All that is being computed here is the "global" chi-square statistic for the entire data set. The creation of a more selective method that allows targeting of particular machine/ball set combinations is left as an exercise (Exercise 10.11).

The first step in the `chiSquare` method is to calculate the frequency of occurrence of the different balls. This is accomplished by first creating a pointer to `int` that will be used to collect the data from the `pFreq` pointers that are members of each of the g5 objects in the container. The actual accumulation of frequencies will employ the `accumulate` algorithm from the STL numeric library.

The basic idea for `accumulate` derives from the now (hopefully) familiar summing recursion

```
double sum = b;
for(int i = 0; i < n; i++)
  sum += a[i];
```

that will sum the entries of an n-element array a given a starting sum of b. The `accumulate` algorithm provides a way to create an abstract analog of this recursion.

A prototype for `accumulate` is

```
accumulate(iterator begin, iterator end, dataType initialValue,
      binaryFunction f)
```

Here `begin` and `end` are iterators for a container object that point to the begin and end of a region where the "accumulation" is to be performed. The `initialValue` object is an instance of the class `dataType` whose objects are to be accumulated. It plays a role similar to b in our simple sum recursion example. In that sense, it can be viewed as a starting value for the recursive process that will be directed by `accumulate`. The function f is a binary function object that determines the operations that will be used to perform the "accumulation": e.g., abstract versions of +, -, * or any other procedure that is amenable to an accumulation process.

For the g5 setting, our choice for the binary functor is

```
struct sumFreq   : public std::binary_function<int*, g5, int*> {

  sumFreq(){}

  int* operator()(int* freq, const g5& g){
    for(int i = 0; i < 40; i++)
      freq[i] += g.getFreq(i);
    return freq;
  }
};
```

For any given g5 object argument, the `sumFreq` functor will access the frequency of occurrence for each of the 40 balls using the g5 class accessor method `getFreq` and then add them to the appropriate elements of the array represented by its first pointer to `int` argument. The recursion is begun by giving `accumulate` the initial value of a pointer whose memory locations have all been set to 0. The range is specified by the `begin()` and `end()` iterators from the `g5Array` class which results in the collection of frequencies for every object in the container.

For a bit of variety, the actual chi-square statistic was evaluated using the `transform` and `inner_product` algorithms with the latter coming from the numeric library. The requisite calculations can be broken into three steps: i) subtract the expected frequency from each observed frequency, ii) divide this difference by the square root of the expected frequency and then iii) square and sum the results from steps i) and ii). The `transform` algorithm is used to carry out the first two steps. This function simply transforms every element in a specified range (given by its first two arguments) using a user-supplied functor object (given as its third argument). In this instance the code writing labor can be reduced by using some of the predefined functors that are available from the STL. Most of the standard arithmetic (e.g., addition, subtraction, division and multiplication as `plus`, `minus`, `divides` and `multiplies`, respectively) and logical operations (e.g., logical equal, logical AND, logical OR and logical NOT as `equal_to`, `logical_and`, `logical_or` and `logical_not`, respectively) are available in prepackaged forms.

The `transform` algorithm is used in the `chiSquare` method first to subtract off the expected value from all the frequencies. The subtraction is carried out using the `minus` functor with template parameter set to `double`. Since `minus` encapsulates a bivariate function, its second argument is set as the expected value using the `bind2nd` function adapter. This basic process is repeated on the output from `transform` except with `minus` replaced by `divides` and the expected value replaced by its square root. Note that the use of `double` as the template parameter for `minus` casts the integer frequencies to double precision.

The output of the second application of transform is used in the `inner_product` algorithm to get the nonadjusted chi-square statistic. A prototype for `inner_product` is

```
inner_product(iterator begin, iterator end, iterator beginA,
        initialValue)
```

The elements in the range designated by `begin` to `end` are multiplied by those in a range starting at `beginA` and the results are summed with `initialValue` being the starting value for the sum. For our case, we want to square the elements which is accomplished by choosing the `begin` and `beginA` iterators to be the same. The initial value that is given for the sum is 0.

A program that tests the new method is given below.

```
int main(){
  g5Array g5A("guess5.txt");
  std::cout << "Chi-square results" << std::endl;
  std::cout << g5A.chiSquare() << std::endl;
  std::cout << std::endl;
  std::cout << "Lookup  results" << std::endl;
  g5Array::const_iterator iter
    = g5A.lookUp(std::string("1952-01-17"));
  iter->printG5();

  return 0;
}
```

The output from the program is

```
Chi-square results
27.7769

Lookup  results
Guess5 Drawings for 1952-01-17 used Machine A with Ball Set 6
The drawing results are:
10 38 23 20 15
36 15 11 7 37
20 10 9 25 2
29 14 16 31 34
12 38 28 26 2
16 19 22 26 33
25 17 21 3 10
16 8 32 6 29
22 26 7 24 12
29 16 37 36 27
```

Among other things, comparison with results from Section 5.6 shows that the chi-square statistic has been evaluated correctly.

10.8 Exercises

10.1. Build a large `vector` object by repeatedly adding elements of type `pair<double, int>` to it. First, allocate only a single entry. Then, insert many elements into this vector (say, a million). Do this in a loop so that in each iteration you insert one more element. Let each pair consist of a `double` returned by `(double)clock()` and the integer computed as the difference in the capacity of the vector now and in the previous iteration (this may be found by invoking the `capacity` method). After the loop is over, print only the iterations in which the capacity of the vector changed. When does this happen? How much time does an automatic resize take? Try the same idea but with a vector whose entries are objects with a representation more complex than simple integers or pairs.

10.2. Compare the performance of the `insert` method for `vector` containers in two cases: when the elements are inserted at the beginning and when they are inserted at the end.

10.3. If elements must be inserted at both the beginning and end of a dynamic array, the deque container is more efficient than the vector class. This container is optimized for amortized constant-time cost of both push_back and push_front, as well as for use of the subscripting operator []. Compare the running time of a long sequence of front and back insertions for a vector and deque object.

10.4. Generate a large file of random numbers from some probability distribution: e.g., 10,000,000 pseudo-random numbers from the standard normal distribution. Now use a version of the count-min sketch ADT to process the data in the file as if it were a stream of numbers similar to the approach taken in Section 10.3.1 for the newWords.txt file. Experiment with both the number and form of the hash functions as well as the size parameter w for the hash tables. Use grouped means and variances to obtain estimators for the true (sample) mean and variance of the entire data set as well as during several points while reading from the input file.

10.5. Use the vector class to develop a template analog of the hashTable class of Section 9.2.4.2 that used linear probing.

10.6. Modify the chainHash class in Section 10.4.2 to provide a guaranteed constant-time iteration by augmenting it with a linked list that connects the buckets of the hash table in the order in which they were established. [Hint: Each time a new bucket is opened in the table, add it to the head or the tail of the list; then re-implement nextBucket to use the list iterators instead of manually traversing the table.]

10.7. Provide a C++ implementation of the cuckoo hashing concept from Exercise 9.28.

10.8. Write a program that reads a sequence of integers from an input stream and stores them in a map. As each integer is read from the stream, the map is checked and if the integer has not been seen before, it is inserted with a count of 1. If the integer is already an element of the map, only its count should be incremented.

After the map-based code is working correctly, rewrite it but use a hash table instead (e.g., the one from Section 10.4.2 or the TR1 hash table described in Appendix C). Compare the running times when they are used to process a large file. The idea is that hash table look-up and insertion operations are expected to take (near) constant time, while a map is implemented using a binary search tree for which the same operations require up to $O(\log n)$ computations. This should be seen in your run-time comparisons.

10.9. Write the code for the copy constructor, destructor, overloaded assignment operator =, comparison operator == and print utility for the g5 class of Section 10.3.3.

10.10. Write the code for the nondefault constructor and overloaded *, ->, =, ==, != operators for the const_iterator class of Section 10.3.3.

10.11. Modify the chiSquare method from the g5Array class of Section 10.7 to where it will calculate the chi-square statistic for any specified range of ball set numbers and either or both of the two machines.

10.12. Modify the g5 class so that information is available from a class member on the actual game drawings (i.e., sans test draws). Modify your chiSquare method from Exercise 10.11 to produce results for the game drawings alone.

10.13. Consider the data on the G5 bonus ball game from Exercise 5.17.

a) Create a g5 class that will hold the information from a drawing for this game.

b) Use the fileIn class from Section 10.3.3 to read the data in from a file and place it in a g5 objects.

c) Store the g5 objects as they are created in a vector<vector<g5> > dynamic array object.

d) Modify your work from Exercise 10.11–10.12 to obtain similar results for your g5 dynamic array class.

10.14. Modify the simulation code from Section 10.4 to also monitor the variables service time, waiting time and total time in the system (i.e., the waiting time plus service time) for the subsets of Customer objects that have Gold values of true or of false.

10.15. Rework the simulation example from Section 10.4 using the queue container adapter with service rendered on a first-in first-out basis according to arrival time.

10.16. Use the deque container class to modify your code from Exercise 10.15 to allow the last person in the queue to leave at random.

10.17. A company has a total of n machines that it uses to produce its product; a machines are required for daily production while $n - a$ machines are held in reserve. The failure time for a machine has an exponential distribution with mean λ_F and the times to failure of the n machines are independent. When a machine fails, it is replaced by a reserve machine, assuming that one is available, and sent out for repair. Repair times are also exponentially distributed with mean λ_R. Use the discrete event simulation technique of Section 10.5 to approximate the expected time before production comes to a halt in that an active machine fails with no machines left in reserve for various values of the system parameters n, a, λ_F and λ_R.

10.18. Add a remove method to the chainHash class of Section 10.4.2.

10.19. Implement the -- operator for the iterator class in the chainHash class.

10.20. Add a method for resizing an object from the chainHash class.

10.21. Rework the example in Section 10.4.2. First, use

```
template<class Key, class Data> int
        chainHash<Key, Data>::hash(Key a) {
    return a % tableSize;
}
```

as the definition for the hash function in class chainHash. Then, create a class/struct Key that has a single member of type double. Equip your class/struct with the overloaded modulus operator

```
int operator%(double a, int m) {
    return ((int) floor(a)) % m;
}
```

A number of other operators will need to be defined before this new class/struct will work effectively with the chainHash class.

10.22. Take the stock index example of Section 10.4.1 and create a database that stores the list in order by value of the index rather than time. Write a C++ program that manages the database by accessing/adjusting it for new entries.

10.23. There are cases where it is necessary to alter a heap entry in a way that will change its priority. For example, if the heap is storing personnel records arranged by salary, there will (hopefully) be raises that would change the order of individuals in the heap. To change the priority of an item, we need to be able to access it somehow by name or value and the basic heap does not provide this functionality. One way to resolve this problem is to use a hash table to keep track of the indices of the elements in the heap's array; that is, every object in the heap has two components: a key and a data member by which it is ordered. An object's key value is then paired with its array index to create a table that tells us how to locate and, e.g., alter a specific element in the heap. Use the chaining hash table from Section 10.4.2 in conjunction with the vector class to create a priority queue that allows for changes in the key values. Is this development necessary for the STL priority_queue class?

10.24. Create a custom `matrix` container. The data is to be stored in a `vector<vector<>`
`>` construct and a nested iterator class should be provided for exploring the container.

10.25. Use a `priority_queue` container to manage minimization of the function $f(\theta) = \theta \sin\left(4\pi\theta^2\right)$, $\theta \in [0,1]$ using a blind random search over the interval $(0,1)$. To accomplish this use a `pair` data type with the `first` and `second` member elements being the value for the variable and the function, respectively. The comparison operators then need to be defined to work only with the `second` member of the struct.

10.26. Create a class that will store sparse matrices in a list structure. The storage format consists of a linked list with a "head" node for each row that has nonzero entries. Associated with each "head" node is another linked lists of nodes that contain the column indices and values for any nonzero element in the row. Provide your class with

a) a constructor that will take an object from class `Matrix` of Section 3.9 and Appendix D and converts it to compressed storage mode,

b) a copy constructor and assignment operator,

c) matrix addition methods (both `+` and `+=`),

d) matrix multiplication methods (both `*` and `*=`) and

e) a method that will insert a new element into a specified row and column of the matrix.

10.27. Create a class that will store polynomials in a list structure. The nodes in the list represent the terms in the polynomials with member elements being the power and coefficient. The power provides the key for each node. Your class should include

a) a copy constructor and assignment operator,

b) methods for polynomial addition (both `+` and `+=`),

c) methods for polynomial multiplication (both `*` and `*=`) and

d) a method that will insert a new term into a polynomial.

10.28. The STL includes a `vector<bool>` specialization of the `vector` class that is optimized to use only one bit of memory for each element. In addition to other `vector` class methods, it includes a `flip` method that will toggle a specified bit to the complement of its current value. Thus, if `v` is a `vector<bool>` object, `v.flip()` will replace all bit values by their complements while `v[i].flip()` performs the same operation on the bit at the `i`th slot in the container.

Now, consider the problem where a random number generating device is supposed to generate pairs of integers over a grid $\{1,\dots,n\} \times \{1,\dots,m\}$. Your task is to verify that the coding has been done correctly and, in particular, that every grid point can be selected. While you may generate arbitrarily many pairs with the device, you are not allowed to actually look at the underlying code. Use a `vector<vector<bool>` `>` container to create an algorithm with average $O(mn)$ running time for checking that every number can appear.

Chapter 11

Parallel computing in C++ and R

11.1 Introduction

The developments thus far have focused on running serial code: i.e., programs where there is only one task being executed at any point in time. For some situations the computation time associated with this approach may be prohibitive. One obvious way to resolve difficulties of this nature is by splitting the problem into smaller subtasks and then solving each of these tasks on different computers or processors. Parallel computing techniques provide one avenue for realizing the benefits of this divide-and-conquer strategy.

For our purposes we will focus on parallel computing in environments where there are multiple processors that will work simultaneously or concurrently on their assigned subtask. A measure of performance improvement is then provided by the level of *speedup* as defined by

$$\text{speedup}(k) = \frac{\text{serial execution time}}{\text{parallel execution time for } k \text{ processors}},$$

where the serial execution time is the amount of time required to perform the computations using one processor and parallel execution time is the time required when multiple processors are used. The best possible result is linear speedup where speedup(k) is a linear function of the number of processors k with a slope of 1. In practice, linear speedup is not realized due to time that is required for, e.g., inter-processor communication.

Parallel processing "languages" provide ways of managing the work performed by different processors in a multi-processor environment. Their function is primarily to oversee and facilitate communication between the different CPUs. In this respect they are better referred to as APIs rather than languages. In particular, the MPI and OpenMP parallel programming APIs considered in this chapter can be viewed as a collection of procedures or directives that are specifically designed for their respective parallel computing environments.

Parallel computing APIs can be broken into a dichotomy corresponding to whether they deal with memory (i.e., RAM) that is shared or distributed. In both situations there will be a number of separate processors; but, for shared memory there is a common block of RAM that can be accessed by a group of processors. This is now the typical scenario for desktop and laptop machines with the advent of multiple-core processors. Communication between the different processors is relatively simple in this case because "messages" that need to be passed from one processor to another can simply be left in the shared memory region. The standard API for use in shared memory parallel computing is OpenMP.

Distributed memory is now the dominant architecture for clusters and super-computers. Here processors are grouped into nodes containing a few CPUs (that share memory). There are many nodes and jobs are farmed out across the nodes to solve the overall computing problem. The nodes exist in different physical locations and do not share memory. As a result, special equipment and software is required to allow them to communicate in a way that will facilitate whatever data sharing is needed to complete their appointed tasks. The resulting hardware involves high-speed serial links between nodes. The API that is generally employed for inter-node communication is MPI.

We will introduce some of the basic features of OpenMP in the next section. Then, in Section 11.3 a few essential MPI commands will be examined. Section 11.4 illustrates how parallel processing can be accomplished in R through the Rmpi and multicore packages. Finally, in Section 11.5 the problem of random number generation in multiple-processor settings is addressed. There are some issues that arise in this context concerning the creation of independent random number streams for the different processors.

There are numerous books that have been devoted to parallel computing in general and OpenMP and MPI in particular. Thus, it is unrealistic to attempt a full treatment of these topics here. Instead, the goal of this chapter is to provide an introductory treatment that will leave the reader with the ability to write simple, yet functional, programs that employ a minimal number of the essential functions for the different parallel APIs. In this sense the intent is to provide an entry point for additional study using the many excellent learning resources that exist for parallel programming such as the texts of Quinn (2003) and Chandra, et al. (2001).

11.2 OpenMP

The OpenMP API provides a collection of compiler directives that produce a type of add-on paralellization to serial code. These directive, called pragmas (for "pragmatic information"), furnish a means for encapsulating a region that is designated for parallel treatment.

The different parts of a program that run in parallel are called *threads*. This term is indicative of how parallelization is actually implemented. At the beginning and end as well as possibly at many other points throughout an OpenMP program, execution will occur in serial mode: i.e., a single master thread of execution will be all that is being handled by the processing units. However, at certain stages of the program a pragma will be encountered at which juncture the single thread branches or *forks* into multiple threads or processes that execute simultaneously. At the end of the parallel region the threads rejoin or collapse back into the master thread and the program executes in serial mode until another pragma is encountered. The number of threads and the number of processors need not coincide although that will be assumed to be the case.

The general pragma syntax is

```
#pragma omp directiveName   [clauses ]
   {
          statements and expressions
   }
```

where `directiveName` designates one of the OpenMP directives and `clauses` are optional components that further specify the behavior/actions associated with the directive. Note that a new line must be entered after the clauses. So, the curly braces after the clauses must appear on a new line in the program's file. Our discussion will be restricted to two OpenMP directives: namely,

- `parallel` that informs the compiler to treat the subsequent code block using multiple threads and

- `parallel for` which produces parallel treatment of a standard `for` loop.

Since OpenMP is for shared memory computing one of the main issues that arises is protection of variables and objects that need to be the property of only a single thread. The default behavior is that memory is shared. Memory protection is achieved through the use of appropriate clauses. In this regard our interest will focus on the clauses

- `private` that forces allocation of separate, nonshared, memory for the objects and variables that are named in the list accompanying the clause. These `private` regions cannot be accessed by other processors which protects the data being stored in these areas from outside corruption.

- `firstprivate` which has a similar function to `private` except that the objects/variables are initialized in the parallel section using initial values from a serial code segment.

The code listing below provides an example of using a pragma to parallelize a block of code.

Listing 11.1 *ompEx.cpp*

```
//ompEx.cpp
#include <omp.h>
#include <iostream>

int main () {
  int nP = omp_get_num_procs();
  omp_set_num_threads(nP);
  int myNumber;
#pragma omp parallel private(myNumber)
  {
    myNumber = omp_get_thread_num();
    std::cout << "Hi ya'll from thread " <<
      myNumber << "!" << std::endl;
  }
  return 0;
}
```

The first thing to observe in Listing 11.1 is the inclusion of the OpenMP header file omp.h. Then, two OpenMP functions

```
omp_get_num_procs
```

and

```
omp_set_num_threads
```

are used that return the number of processors and set the number of threads, respectively. In this case the number of threads to be used in each parallel section has been specified as being the same as the number of processors.

The parallel block of code in Listing 11.1 is created using a `parallel` directive with the variable `myNumber` being given a unique memory location for each of the nP threads through use of a `private` clause. When the master thread forks at the beginning of a pragma every thread is given a unique ID number starting with 0. The value of a particular thread's identifier is obtained using omp_get_thread_num which is written into the variable `myNumber` in the program. Had `myNumber` not been designated as `private` it would have been treated as `shared` and its value would be stored in a single memory location that was accessible by all threads. The resulting behavior of the program would be unpredictable with one thread being able to write its value for `myNumber` into memory in the space of time prior to another accessing the value and using it to write its message to standard output.

Compilation of OpenMP programs requires the fopenmp compiler option. The command line input for a program programName.cpp would look like

```
g++ -fopenmp programName.cpp -o executableName
```

This has the effect of creating executableName as the executable version of the program. For example, Listing 11.1 was compiled and executed with

```
$ g++ -fopenmp ompEx.cpp -o ompEx
$ ./ompEx
Hi ya'll from thread Hi ya'll from thread Hi ya'll from thread
   Hi ya'll from thread 1!302!!!
```

This is certainly not the output we expected. But, it is not difficult to see what has transpired. The object `cout` is shared by all four threads that have been created to execute the parallel block of code. The result is a so-called *race condition* where all four threads attempt to access the shared object simultaneously. This problem can be resolved through use of a critical directive which has the form

```
#pragma omp critical
```

This pragma forces the execution of the corresponding block of code to take place one processor at a time. Thus, if the previous output statement is replaced with

```
#pragma omp critical
    {
    std::cout << "Hi ya'll from thread " <<
        myNumber << "!" << std::endl;
    }
```

the result is

```
Hi ya'll from thread 1!
Hi ya'll from thread 2!
Hi ya'll from thread 0!
Hi ya'll from thread 3!
```

Notice from this that the threads do not necessarily execute the program in sequence with their thread number.

Although using `critical` solved our output problem, it had the side effect of making the program execute in a serial, rather than parallel, mode when it reached the point of writing to the shell. From that perspective there was nothing gained by using the `parallel` construct. This observation holds more generally with the implication that the use of `critical` directives will have adverse performance consequences.

An alternative parallel `for` implementation of our greeting program might include code such as

```
#pragma omp parallel for private(myNumber)
    for(int i = 0; i < n; i++){
        myNumber = omp_get_thread_num();
#pragma omp critical
        std::cout << "Hi ya'll from thread " <<
            myNumber << "!" << std::endl;
    }
```

with n a value obtained from user input. In particular, the choice of n = 7 produced

```
Hi ya'll from thread 0!
Hi ya'll from thread 1!
Hi ya'll from thread 2!
Hi ya'll from thread 3!
Hi ya'll from thread 0!
Hi ya'll from thread 1!
Hi ya'll from thread 2!
```

on a particular four processor machine.

To conclude this section we will consider a simple optimization problem involving the function in Figure 8.1 from Chapter 8. As before the aim is to find the global minimum of this function and the golden section algorithm from a non-template version of the `Optim` class in Chapter 8 will be employed to accomplish that goal. What will make the present approach different from our previous development is that it will be carried out in parallel. The idea is quite simple in that the domain of the function, or [0, 1] in this case, will be partitioned into nP contiguous intervals of equal size with nP being the number of processors/threads. The threads are then assigned different intervals from the partition and find their respective local minima simultaneously using the golden section method. The code that was used to accomplish this is collected in Listing 11.2.

Listing 11.2 *ompOptDriver.cpp*

```
//ompOptDriver.cpp
#include <iostream>
#include <cmath>
#include <omp.h>
#include "functor.h"
#include "optim.h"

using std::cin; using std::cout; using std::endl;
double pi = 2.*acos(0.);

double f(double theta){
  return theta*sin(4*pi*theta*theta);
}

int main(int argc, char** argv){
  Functor func(&f);

  int nEvals;
  cin >> nEvals;

  int nP = omp_get_num_procs();
  omp_set_num_threads(nP);

  double myMin;
  double* pMin = new double[nP];
  double myLow, myUp;
  int myRank;
  Optim Opt(nEvals);

#pragma omp parallel private(myRank, myMin, myLow, myUp)
#pragma omp firstprivate(Opt, func)
  {
    myRank = omp_get_thread_num();
    myLow = (double)myRank/(double)(nP);
    myUp = (double)(myRank + 1)/(double)(nP);
    myMin = Opt.golden(func, myLow, myUp);
    pMin[myRank] = myMin;
  }
  double min = pMin[0];
  cout << min << endl;
  for(int i = 1; i < nP; i++){
    if(func(pMin[i]) < func(min)) min = pMin[i];
    cout << pMin[i] << endl;
  }
  cout << "The minimizer is " << min << endl;
  return 0;
}
```

There are a number of subtle aspects about this program that merit comment. First note that for an object or variable to be referenced in the `private` clause it must first have been declared or initialized. That is the reason that `Opt`, `myRank`, `myMin`, `myLow` and `myUp` were all declared in the serial portion of Listing 11.2 before the `parallel` construct. Once a parallel block is encountered each thread receives its own versions of all the private variables and objects. In the case of objects, these thread-specific versions are created using the default class constructor. As a result, an object must come from a class having a default constructor for it to be used in a `private` clause. On the other hand, the default destructor is called at the end of a parallel block. This means that objects for individual threads have scope restricted to a parallel block and do not persist throughout the program or carry over to other parallel regions.

The `pMin` pointer in Listing 11.2 provides access to as many memory locations in shared memory as there are threads. The individual threads then perform their respective optimization tasks and write the outcomes into distinct locations dictated by their thread number. Because the scope of `pMin` is all of `main` it continues to exist after the end of the parallel block thereby allowing its minimum element to be determined as a global minimizer.

For minimization purposes each thread should have its own `Optim` and `Functor` class object and all objects should be initialized with the same values. However, problems arise if we try to assign a pre-initialized object as `private` rather than one that has merely been declared. The objects for threads other than the master (or thread number 0) are created using the default constructor which means they will not be initialized correctly. On the other hand, if an object of type `Optim` and `Functor` were simply declared prior to the parallel region there would be no way to initialize it inside the region. The `firstprivate` clause that was used here furnishes the solution to such difficulties. In general, a statement of the form

```
#pragma omp firstprivate(variableName)
```

will cause each thread to receive their own copy of the variable `varName` with the value that it had prior to the beginning of the parallel block. For Listing 11.2 the use of the `firstprivate` clause with the `Optim` class object ensures that every thread receives an independent object with the number of iterations set at `nEvals` for each thread (cf. Listing 8.7). Similarly, each thread will have its own `Functor` object initialized with the address of `f`.

To compile our optimization code we use

```
$ g++ -c -fopenmp ompOptDriver.cpp
$ g++ -c -fopenmp optim.cpp
$ g++ -c -fopenmp ranGen.cpp
$ g++ -fopenmp ompOptDriver.o optim.o ranGen.o -o ompOpt
```

This could, of course, have been placed in a make file to provide more automation in the compilation process. Execution of our program on a four processor machine then produces

```
$ ./ompOpt
100
1.95069e-22
0.25
0.619058
0.937337
The minimizer is 0.937337
```

11.3 Basic MPI commands for C++

The dominant API for distributed memory computing environments such as clusters is MPI. The MPI acronym stands for Message Passing Interface and that is precisely what MPI functions are created to accomplish. Messages are passed between processors through the creation of *communicators* which represent groupings of processors that are capable of communicating with each other. In C++ the communicator concept is implemented through the COMM class. In particular, the COMM_WORLD communicator object that connects all the processors being used by a program will always exist. This is the only communicator object that will be discussed here.

To use MPI functions in C++ programs we need to include the header file mpi.h that provides all the MPI-specific prototypes. The MPI classes and functions then reside in the MPI namespace and can be accessed via the scope resolution operator. The MPI part of our code must be placed between the two commands

- MPI::Init()

- MPI::Finalize()

The Init here stands for "initialize" while Finalize terminates inter-processor communication. Accordingly, MPI calls are illegal before Init or after Finalize.

Associated with the COMM_WORLD object are a number of methods for communication and related purposes. Of particular interest are

- MPI::COMM_WORLD.Get_rank()

- MPI::COMM_WORLD.Get_size()

that give the MPI versions of the omp_get_thread_num and omp_get_num_procs functions that are used with OpenMP. The two functions return integers representing the unique integer rank that has been assigned to each processor and the number of processors being managed by the COMM_WORLD communicator object.

The code listing below provides a more formal version of our standard "Hello world!" application using just the four functions we have learned thus far.

```
//mpiEx1.cpp
#include <iostream>
#include <mpi.h>

int main(){

  MPI::Init();

  int myRank = MPI::COMM_WORLD.Get_rank();
  int nP = MPI::COMM_WORLD.Get_size();

  std::cout << "Hello ya'll from processor " << myRank;
  std::cout << " of " << nP << std::endl;

  MPI::Finalize();
  return 0;
}
```

To execute this code a command is issued to the operating system that will result in the placement of an executable version of the program in RAM on each processor. At that point the individual processors work autonomously to carry out their tasks of printing out the greeting. The processor's rank can be used to specialize the computation on each processor as we see here and in later examples.

The specifics of compilation for our program are system-dependent. An illustration of how this might be accomplished in an interactive mode is

```
$ mpicxx mpiEx1.cpp -o mpiEx1
$ mpiexec -np 5 mpiEx1
Hello ya'll from processor 4 of 5
Hello ya'll from processor 2 of 5
Hello ya'll from processor 0 of 5
Hello ya'll from processor 1 of 5
Hello ya'll from processor 3 of 5
```

The syntax mpicxx (on the cluster where this code was compiled) invokes a C++ compiler for MPI programs. This produces the executable mpiEx1 that is run using the mpiexec command with the –np 5 option requesting that five processors be used to perform the computations. Notice that the processors do not finish in order which is not surprising as they are working independently.

In practice the set of commands that were used here to compile and run mpiEx1.cpp may or may not be appropriate for a given system using MPI. In some cases a single node can be obtained on a cluster that will allow for an interactive session and it is not terribly difficult to install one of the MPI implementations such as Open MPI on a desktop machine for interactive use. In such cases something similar to our approach can be expected to work. But, both of these options entail using MPI in shared memory environments which is not really its purpose. The power and utility of MPI is best realized in distributed-memory, high-performance computing systems. In such settings job submission will generally take place in batch mode using job scripts that can take various forms depending on the software that is being used to manage job scheduling. The details involved can usually be obtained from the computing center that administers the cluster of interest.

Our mpiEx1.cpp program is not particularly interesting for several reasons. From a parallel perspective it falls short in that there is no interaction between processors. Inter-processor communication is useful for carrying out the "boss/worker" or "master/slave" paradigm frequently employed for parallel programming. The simplest way to communicate between processors is through the Send and Recv (an abbreviation for receive) methods. The forms of these functions are

```
void MPI::COMM_WORLD.Send(void* buf, int count, MPI_Datatype datatype,
          int dest, int tag, MPI_Comm comm)
```

```
void MPI::COMM_WORLD.Recv(void* buf, int count, MPI_Datatype datatype,
          int source, int tag, MPI_Comm comm)
```

The first argument for both of these functions is a pointer to void. This is just a way of passing generic data into and out of the functions. The quantity buf appears in both the Send and Recv functions. This is the data that is being sent by Send and received by Recv. The (third) datatype argument is used to describe the contents of buf as some specific data type that is recognized by MPI. Such data types include all the common ones in C++ as well as those that are user defined. The latter option will not be explored here. For our purposes it is enough to know that, for example, MPI::CHAR, MPI::INT, MPI::LONG, MPI::UNSIGNED_LONG, MPI::FLOAT, MPI::DOUBLE, MPI::LONG_DOUBLE and MPI::BOOL correspond to the C++ primitive data types char, int, long, unsigned long, float, double, long double and bool, respectively.

The (second) count argument to the Send and Recv functions specifies how many of the datatype values are being sent or received. The count arguments do not need to match for a send and receive. But, the receive count must be at least as large as what has been sent in order to avoid an error. The comm argument is the communicator object that is managing the inter-node communication and will always be COMM_WORLD for our applications.

The dest and source arguments in Send and Recv represent the rank of the node for which the message is intended and the rank of the node from which a message is to be received, respectively. The remaining tag argument is a parameter that provides some sorting facility for incoming and outgoing messages. It comes in handy if one is sending, for example, two different doubles to another processor. Using a different tag for each one will make certain they are identified correctly and then stored in a suitable variable by the receiving processor.

With the above as preliminaries we are now ready to write our second MPI program. It uses inter-node communication to create the "Hello world!" application.

```
//mpiEx2.cpp
#include <iostream>
#include "mpi.h"

int main(){

  MPI::Init();
  int myRank = MPI::COMM_WORLD.Get_rank();
  int nP = MPI::COMM_WORLD.Get_size();

  char msg[20];
  if(myRank == 0){
    strcpy(msg, "Hello ya'll");
    for(int i = 1; i < nP; i++)
      MPI::COMM_WORLD.Send(msg, 20, MPI::CHAR, i, 0);
  }
  else{
    MPI::COMM_WORLD.Recv(msg, 20, MPI::CHAR, 0, 0);
    std::cout << msg << " from processor " << myRank << std::endl;
  }

  MPI::Finalize();
  return 0;
}
```

The idea behind mpiEx2.cpp is that a portion of our greeting message will now be relayed from the "master" or rank 0 processor to the other processors using the Send and Recv functions. This illustrates how branching statements can be used to determine the action of each of the processors; in this case, the 0 processor is told to send while all the others are told to receive via use of an if/else block.

The message segment "Hello ya'll" will be stored in the char array msg which has been chosen to have length 20. The choice of 20 is arbitrary. It could be any value large enough to provide storage for the portion of the greeting it must hold. To actually place "Hello ya'll" in msg the 0 rank processor uses the strcopy function whose purpose is copying C strings into character vectors. An array of (at most) 20 elements of type MPI::CHAR is then sent and received. Processor 0 sends the array to each of the other processors using the Send function in a for loop with the dest argument being the receiving processor's rank and the value of tag set arbitrarily to 0. The other processors receive the message by using 0 as the source argument for the Recv function. In this latter instance the value of tag must agree with the value used in the Send function. The results obtained from running the program with seven processors is

```
$ mpicxx mpiEx2.cpp -o mpiEx2
$ mpiexec -np 7 mpiEx2
Hello ya'll from processor 4
```

```
Hello ya'll from processor 6
Hello ya'll from processor 1
Hello ya'll from processor 2
Hello ya'll from processor 3
Hello ya'll from processor 5
```

As a somewhat more complicated application let us develop code for performing a balanced simple one-way analysis of variance (AOV) in parallel. The premise is that (very large) samples have been taken from each of nP populations and the necessary AOV statistics for each sample will be computed using separate processors. Listing 11.3 provides one possible implementation of this idea.

Listing 11.3 *anovaMPI.cpp*

```cpp
//anovaMPI.cpp
#include <iostream>
#include <fstream>
#include <cstdlib>
#include "mpi.h"
#include "summary.h"

using std::cout; using std::endl;
using std::ifstream;

int main(int argc, char** argv){
  double** pData;
  double* pMyData;

  MPI::Init(argc, argv);

  int myRank = MPI::COMM_WORLD.Get_rank();
  int nP = MPI::COMM_WORLD.Get_size();
  int n = atoi(argv[1]);

  if(myRank == 0){
    pData = new double*[nP];

    ifstream fIn;
    fIn.open(argv[2]);
    for(int j = 0; j < nP; j++){
      pData[j] = new double[n];
      for(int i = 0; i < n; i++){
        fIn >> pData[j][i];
      }
    }

    fIn.close();

    for(int i = 1; i < nP; i++)
      MPI::COMM_WORLD.Send(pData[i], n, MPI::DOUBLE, i, 0);
    Summary mySummary(n, pData[0]);
    double* stat = new double[2];
    mySummary.west(stat);
    cout << "For processor " << myRank
         << " the mean and variance are "
         << stat[0] << " and " << stat[1] << endl;
```

```
   double num = 0;
   double xBar = stat[0];
   double den = ((double)(n - 1))*stat[1];

   for(int i = 1; i < nP; i++){
     MPI::COMM_WORLD.Recv(stat, 2, MPI::DOUBLE, i, 1);
     num +=  ((double)(i)/(double)(i + 1))*(stat[0] - xBar)*
       (stat[0] - xBar);
     xBar = ((double)(i)/(double)(i + 1))*xBar +
       stat[0]/((double)(i + 1));
     den += ((double)(n - 1))*stat[1];
   }
   num *=(double)n/((double)(nP-1));
   den /= (double)(nP*n - nP);
   cout << "The grand average is " << xBar << endl;
   cout << "The numerator mean square is " << num << " on "
       << (nP - 1) << " degrees-of-freedom" << endl;
   cout << "The denominator mean square is " << den << " on "
       << (n*nP - nP) << " degrees-of-freedom" << endl;
   if(den != 0)
     cout << "The F-ratio is " << (num/den) << endl;
 }

 else{
   double* mypData = new double[n];
   MPI::COMM_WORLD.Recv(mypData, n, MPI::DOUBLE, 0, 0);
   Summary mySummary(n, mypData);
   double* stat = new double[2];
   mySummary.west(stat);
   cout << "For processor " << myRank
       << " the mean and variance are "
       << stat[0] << " and " << stat[1] << endl;
   MPI::COMM_WORLD.Send(stat, 2, MPI::DOUBLE, 0, 1);
   }

 MPI::Finalize();

 return 0;
}
```

The first new feature encountered in Listing 11.3 is the presence of `argc` and `argv` in the `Init` function. This alternative form allows the command line input obtained at execution to become available to all the processors being managed by the `COMM_WORLD` communicator. This is used explicitly here by having `argv[1]` contain the sample size of the data sets that the processors will be given to analyze.

The input/output of data becomes complicated when more than one processor is involved and the ability of different processors to establish file connections is system-dependent. For this program we have made the minimal assumption that the master or 0 rank processor is capable of reading from a file. Thus, the first step in the analysis is for the master processor to read the data from a file whose name is contained in `argv[2]`. This data is stored in the double-precision pointer-to-pointer format with row indices corresponding to populations/treatments and column indices representing the observation number. The result is an `nP` × `n` "array" `pData`. The storage layout for `pData` runs counter to the usual statistical format that would have columns representing the treatments. This is necessary (or at least convenient) due to the row-major storage order employed by C/C++. As the goal is to ultimately pass the data sampled from each population to the different processors, the nP elements of pData need to point to the initial locations of the n storage blocks that are allocated for each of the nP samples.

Once the master processor has read in the data it uses a for loop to pass out the samples by sending each processor an element of pData that is determined by the receiving processor's rank. All the processors, including the master, then analyze their respective samples using an object from a class Summary. This class is structured along the lines of Exercise 3.34. In particular, it contains an implementation (in the method called west) of Algorithm 2.3 for computing a sample mean and variance. These two statistics are stored in locations accessed through pointers that are sent back to the master processor once a processor's work is completed. Finally, the master processor combines the results (also via Algorithm 2.3) from the other processors to obtain the F-statistic for testing the hypothesis that all nP populations have the same mean.

To try out our parallel ANOVA code data were generated in R with

```
> set.seed(123)
> A <- replicate(4, rnorm(1000000))
> apply(A, 2, mean)
[1] -5.214370e-04 -1.555513e-03  1.541233e-06  1.775802e-04
> apply(A, 2, var)
[1] 0.9998541 0.9995122 1.0001920 1.0012544
> write.table(t(A), file = "aov.txt", quote = FALSE,
+ row.names = FALSE, col.names = FALSE)
```

Here we have used the replicate function in a similar manner to Section 5.6 for the purpose of random number generation. Specifically, 1,000,000 samples were produced from each of four "treatments". Some summary statistics were also generated that will provide a check for certain aspects of our C++ results. The parallel ANOVA program was then compiled using the commands

```
$ mpicxx -c summary.cpp
$ mpicxx -c anovaMPI.cpp
$ mpicxx anovaMPI.o summary.o -o anovaMPI
```

Upon execution it returns the output

```
$ mpiexec -np 4 anovaMPI 1000000 aov.txt
For processor 1 the mean and variance are -0.00155551 and 0.999512
For processor 2 the mean and variance are 1.54123e-06 and 1.00019
For processor 0 the mean and variance are -0.000521437 and 0.999854
For processor 3 the mean and variance are 0.00017758 and 1.00125
The grand average is -0.000474457
The numerator mean square is 0.607539 on 3 degrees-of-freedom
The denominator mean square is 1.0002 on 3999996 degrees-of-freedom
The F-ratio is 0.607415
```

Comparison with the results from R suggests that the output is correct. As expected, the F-statistic is not significant with a p-value of about .6.

11.4 Parallel processing in R

It remains to consider how parallel processing can be carried out in R. There are several packages that endow R with multithreading capabilities. These include R/Parallel, nws, snow and Rmpi. The CRAN website has a task view for high-performance computing that provides an indication of future developments and the current state of parallel computing using R. The review article by Schmidberger, et al. (2009) contains a thorough treatment of parallel computing in R covering sixteen different packages that can be used for code parallelization. A companion article by Euster, et al. (2011) provides a tutorial on using four of these packages.

Our focus in this section will initially be on the Rmpi package that provides an interface to MPI. This can be used to produce an interactive parallel processing environment via the "master/slave" form of command structure.

To begin Rmpi we load the Rmpi package using

```
> library(Rmpi)
```

To create n slave processors the function `mpi.spawn.Rslaves` can be used with argument `nslaves = n`. For example, the choice of n = 3 produces

```
> library(Rmpi)
> mpi.spawn.Rslaves(nslaves = 3)
        3 slaves are spawned successfully. 0 failed.
master (rank 0, comm 1) of size 4 is running on: eubank-2
slave1 (rank 1, comm 1) of size 4 is running on: eubank-2
slave2 (rank 2, comm 1) of size 4 is running on: eubank-2
slave3 (rank 3, comm 1) of size 4 is running on: eubank-2
```

All slaves are closed via the command `mpi.close.Rslaves()`. To quit R one should now use `mpi.quit()` rather than `quit()` since the former command will make certain that `mpi.finalize()` is called.

As was true for threads with OpenMP, it is possible to have more slaves than the physical number of processors on a machine. This will not generally improve program performance in terms of running time. Thus, to simplify discussion we will treat each slave as corresponding to a different processor.

The Rmpi versions of the MPI functions `Get_rank` and `Get_size` are

```
    mpi.comm.rank()
```

and

```
    mpi.comm.size()
```

These functions return the rank of the calling slave processors and the total number of processors: i.e., the number of slaves plus one for the master processor.

Objects are sent from the master to a slave using an analog of the MPI broadcast function `Bcast` discussed in Exercise 11.4. This function sends an R object to all the slaves. The syntax is

```
    mpi.bcast.Robj2slave(object)
```

with `object` any type of R object including a function.

The slaves are instructed to run a specific set of R code, `rCode`, through the command

```
    mpi.remote.exec(rCode)
```

Upon executing this command on the master processor the slaves will carry out the tasks specified in `rCode` and return control to the master processor once they have completed their assignments.

Our first example of using Rmpi involves the obligatory greeting message with a minor twist. The R function `hello` is created first in the form

```
hello <- function(index){
  if(is.element(mpi.comm.rank(), index))
    print(paste("Hello ya'll from processor ", mpi.comm.rank(),
              "!", sep=""))
  else
    print(paste("Processor", mpi.comm.rank(), "has nothing to say!"))
}
```

The `hello` function is distributed and executed on each of our three slave processors with

```
> mpi.bcast.Robj2slave(hello)
> mpi.remote.exec(hello(1:2))
$slave1
[1] "Hello ya'll from processor 1!"

$slave2
[1] "Hello ya'll from processor 2!"

$slave3
[1] "Processor 3 has nothing to say!"
```

This illustrates that `mpi.remote.exec` returns a list consisting of the objects returned from each slave.

As a more complicated example let us consider a regression variable selection problem. The artificial data set that will be used for this purpose is generated with

```
> set.seed(123)
> n <- 1000
> p <- 10
> nDelete <- sample(1:p, size = 1)
> indicesToDelete <- sample(1:p, nDelete)
> beta <- rnorm(p, sd = 2)
> beta[indicesToDelete] <- 0
> beta
 [1]  3.1174166  0.1410168  0.2585755  0.0000000  0.9218324 -2.5301225
 [7] -1.3737057  0.0000000  2.4481636  0.0000000
> X <- matrix(rnorm(p*n), n, p)
> y <- X%*%beta + rnorm(n)
```

The data consist of 1000 observations on a response (stored in the vector y) and 10 "predictor variables" (stored in the `matrix` object X). The response and predictor variables are related through a linear model with standard normal random errors. The coefficient vector for the predictors is constructed in two step. First, an integer between 1 and 10 is chosen at random to determine the number of variables that will have 0 coefficients and then the indices for these variables are randomly selected. Next, the coefficient vector is generated from a zero mean normal distribution with standard deviation two and the selected coefficients are set to zero. From the output we saw that in this particular instance the fourth, eighth and tenth "predictors" will be independent of the response variable.

We will now treat our simulated data as if its properties are unknown to us and attempt to determine which variables should be used in a fit of a linear model involving the 10 predictors. To accomplish this all $2^{10} = 1024$ possible variable subsets (including just a constant term) will be fit to the data. The criterion that will be used to rank the fitted models is the Bayesian Information Criterion (see, e.g., Chapter 2 of Eubank 1999)

$$\mathrm{BIC}(i) = n \log(\mathrm{RSS}(b_i)) + \log(n)\#(i),$$

where $\mathrm{RSS}(b_i)$ is the residual sum-of-squares for fitting the ith model and $\#(i)$ is the number of variables in that model. A "best" fitting model is one that provides a minimum value for the criterion.

The R function that can be used to evaluate BIC is `extractAIC` with (simplified) prototype

```
extractAIC(fit, k)
```

The `fit` argument can be an `lm` object returned from the R `lm` function (discussed, for example, in Section 3 of Appendix B). The `extractAIC` function returns a two-component numeric vector consisting of the number of variables in the fitted model and (apart from an additive constant) the value of

$$n \log(\mathrm{RSS}(b_i)) + k\#(i),$$

for a constant k that defaults to 2. For our purpose we will use `extractAIC` in the form

```
extractAIC(fit, k = log(n))
```

to produce values that are equivalent to BIC.

Our plan is a simple divide-and-conquer scheme; a group of the possible model subsets will be assigned to each processor. The individual processors will then evaluate the BIC criterion over their model subsets and report the best-fitting model back to the master processor. For this purpose we will use the Rmpi `send` and `receive` commands. They have the basic form

```
mpi.send.Robj(object, destination, tag)
```

and

```
mpi.recv.Robj(mpi.any.source(), mpi.any.tag())
```

In the `mpi.send.Robj` function `object`, `destination` and `tag` represent the R object to be sent, the rank of the slave that should receive the object and, analogous to the MPI case, an integer `tag` that can be used to differentiate between messages when more than one object is being sent. A simplified version of the receive function is employed here where the "wild cards" `mpi.any.source()` and `mpi.any.tag()` are used to allow a processor to receive an object from any `source` with any tag. This will suffice because only the master processor will send and it will send only one object.

The variable subsets are constructed and sent to the processors using

```
> nSubSets <- 2^{p - 2}
> subSets <- expand.grid(rep(list(0:1), p))
> indexList <- lapply(1:4, FUN = function(i)
+ subSets[((i - 1)*nSubSets + 1):(i *nSubSets), ])
> for(i in 1:3) mpi.send.Robj(indexList[[i + 1]], i, 0)
> mpi.bcast.cmd(mySubSets <- mpi.recv.Robj(mpi.any.source(),
+ mpi.any.tag()))
```

Four processors will be used in the analysis. Thus, each processor will receive $2^p/4 = 2^{10}/4 = 256$ variable subsets to analyze. The entire collection of subsets are enumerated in the data frame `subSets` using the `rep` and `expand.grid` function. The latter function creates a data frame using all combinations of the supplied vectors/factors: i.e., a data frame with 2^{10} rows having 10 elements that are all zeros and ones. The code excerpt below illustrates the idea for $p = 3$.

```
> expand.grid(rep(list(0:1), 3))
  Var1 Var2 Var3
1    0    0    0
2    1    0    0
```

3	0	1	0
4	1	1	0
5	0	0	1
6	1	0	1
7	0	1	1
8	1	1	1

The zeros and ones in the array indicate which variable subsets should be in the fitted model: 0 signifies that the variable should be excluded and 1 means it should be included in the fit. The first row consists of all zeros. It has the effect of producing a fit with only the constant term: i.e., the fit is just the average of the response values.

The `lapply` function is used to create a list whose components are comprised of consecutive blocks of 256 rows from the `subSets` data frame. These are then distributed to the processors via a for loop. The processors are told to call `mpi.recv.Robj` to receive their respective data using the `mpi.bcast.cmd` function. This latter function is a version of `mpi.remote.exec` that produces no output.

The function that will be used to fit the variable subsets is

```
myRegTask <- function(mySubSets){
  n <- length(y)
  aicVec <- apply(mySubSets, MARGIN = 1, function(a){
      ind <- as.vector(a, mode = "logical")
      if(sum(a) == 0)
        extractAIC(lm(y ~ 1), k = log(n))
      else
        extractAIC(lm(y ~ X[, ind]), k = log(n))})
  location <- which.min(aicVec[2, ])
  list(mySubSets[location, ], aicVec[, location])
}
```

The input argument `mySubSets` will contain a subset of the `subSets` data frame held by the master processor. There is a minor problem of determining how to translate the rows of this object into something that can be used with the indexing operator when it is applied to the `X` matrix to select the variable subset that will be used in the regression. The solution is to translate each row into a logical vector using the vector class constructor with `mode` argument set to `logical`. An if statement is also needed to make certain that the case of fitting the constant is handled correctly; the sum of the elements of the `ind` vector will evaluate to 0 only when all the predictor variables are excluded from the fit and this is the condition that is checked to determine if that is the case. The second component of the vector returned from `extractAIC` is the value of the criterion function. Thus, the best fit for a given processor will be the one whose row index corresponds to the column of `aicVec` that contains the smallest BIC value. This index is found by applying the `which.min` function[*] to the second row of `aicVec`. The processor then returns its optimal subset along with the corresponding value of BIC to the master processor.

Finally, the data and the `myRegTask` function are sent to the processors which are ordered to carry out their calculations by

```
> mpi.bcast.Robj2slave(y)
> mpi.bcast.Robj2slave(X)
> mpi.bcast.Robj2slave(myRegTask)
> mpi.remote.exec(myRegTask(mySubSets))
```

The resulting output from the slaves and the master processor is

---

[*] As might be expected, there is also a `which.max` function for finding the location of the largest value.

```
$slave1
$slave1[[1]]
    Var1 Var2 Var3 Var4 Var5 Var6 Var7 Var8 Var9 Var10
376    1    1    1    0    1    1    1    0    1     0

$slave1[[2]]
[1]   8.00000 54.56251


$slave2
$slave2[[1]]
    Var1 Var2 Var3 Var4 Var5 Var6 Var7 Var8 Var9 Var10
630    1    0    1    0    1    1    1    0    0     1

$slave2[[2]]
[1]    7.000 2088.777


$slave3
$slave3[[1]]
    Var1 Var2 Var3 Var4 Var5 Var6 Var7 Var8 Var9 Var10
888    1    1    1    0    1    1    1    0    1     1

$slave3[[2]]
[1]   9.00000 58.72511

> myRegTask(indexList[[1]])
[[1]]
    Var1 Var2 Var3 Var4 Var5 Var6 Var7 Var8 Var9 Var10
120    1    1    1    0    1    1    1    0    0     0

[[2]]
[1]    7.000 2082.128
```

The best-fitting model is therefore the one that generated the data; i.e., the one with the fourth, eighth and tenth variables deleted.

The R leaps package contains a function **regsubsets** that will also carry out all-subsets (as well as other options) variable selection. An application of it to our simulated data set produced

```
> library(leaps)
> summary(regsubsets(x = X, y = y, nbest = 1, nvmax = 10))[c(1, 6)]
$which
   (Intercept)    a     b     c     d     e     f     g     h     i     j
1         TRUE TRUE FALSE FALSE FALSE FALSE FALSE FALSE FALSE FALSE FALSE
2         TRUE TRUE FALSE FALSE FALSE FALSE FALSE FALSE FALSE  TRUE FALSE
3         TRUE TRUE FALSE FALSE FALSE FALSE  TRUE FALSE FALSE  TRUE FALSE
4         TRUE TRUE FALSE FALSE FALSE FALSE  TRUE  TRUE FALSE  TRUE FALSE
5         TRUE TRUE FALSE FALSE FALSE  TRUE  TRUE  TRUE FALSE  TRUE FALSE
6         TRUE TRUE FALSE  TRUE FALSE  TRUE  TRUE  TRUE FALSE  TRUE FALSE
7         TRUE TRUE  TRUE  TRUE FALSE  TRUE  TRUE  TRUE FALSE  TRUE FALSE
8         TRUE TRUE  TRUE  TRUE  TRUE  TRUE  TRUE  TRUE FALSE  TRUE FALSE
9         TRUE TRUE  TRUE  TRUE  TRUE  TRUE  TRUE  TRUE FALSE  TRUE  TRUE
10        TRUE TRUE  TRUE  TRUE  TRUE  TRUE  TRUE  TRUE  TRUE  TRUE  TRUE

$bic
 [1]  -457.0253  -953.7679 -1879.9000 -2558.5660 -3136.6823 -3210.0709
 [7] -3236.5377 -3232.6831 -3228.3021 -3221.8135
```

First the leaps library is loaded into the workspace. The **regsubsets** function is then applied to the data with the maximum number of variables to include (nvmax) set to 10 and with only the single best subset of each size (nbest = 1) being returned. The output from regsubsets is not directly usable but must instead be wrapped inside a call to its corresponding summary method function. A variety of information can be obtained in this fashion including the variables in the (best) subsets and the values of the BIC criterion as have been extracted here. The BIC values differ due to the use of different but equivalent formulas. The ordering remains the same; the best model contains seven variables and excludes the fourth, eighth and tenth variable.

As is the case for MPI, Rmpi is designed for use in distributed memory settings. At present, there is no direct extension of OpenMP for use in R. An effective way to use R in a shared memory context is available through the multicore package that provides a parallel version of the basic R lapply function called mclapply. It allows different elements of an input list to be handled by different processors. The basic syntax for mclapply is

```
mclapply(X, FUN)
```

with X the list to which the function FUN is to be applied. The default behavior is to portion the entire job sequentially across the processors/cores that are available in the system. If there are nP processors, the first one will apply FUN to the first component of the list X while the second processor applies FUN to the second, etc. If X has more elements than there are cores, multiple elements will be assigned to some (or all) of the processors.

The mclapply function can be used, for example, to carry out the variable selection calculations that were featured in the Rmpi example. The first step is to load the library and determine how many processors are available on the system.

```
> library(multicore)
> multicore:::detectCores()
[1] 4
```

The ::: syntax that appears here is another form of namespace/scope resolution operator similar to the one discussed in Section 5.3. This one allows access to the internal variables in a namespace while :: accesses only those variables in a package that have been exported or made public. Here we are using ::: to obtain access to the function detectCores in the multicore package. The output illustrates that the machine used for this example has four cores. Similar to our Rmpi formulation, each of these cores will be used to analyze different collections of variable subsets from the indexList list object. The myRegTask function can again be used to compute the BIC values with the result

```
> mclapply(indexList, myRegTask)
[[1]]
[[1]][[1]]
    Var1 Var2 Var3 Var4 Var5 Var6 Var7 Var8 Var9 Var10
120    1    1    1    0    1    1    1    0    0     0

[[1]][[2]]
[1]    7.000 2082.128


[[2]]
[[2]][[1]]
    Var1 Var2 Var3 Var4 Var5 Var6 Var7 Var8 Var9 Var10
376    1    1    1    0    1    1    1    0    1     0

[[2]][[2]]
[1]  8.00000 54.56251
[[3]]
```

```
[[3]][[1]]
    Var1 Var2 Var3 Var4 Var5 Var6 Var7 Var8 Var9 Var10
630    1    0    1    0    1    1    1    0    0    1

[[3]][[2]]
[1]    7.000 2088.777


[[4]]
[[4]][[1]]
    Var1 Var2 Var3 Var4 Var5 Var6 Var7 Var8 Var9 Var10
888    1    1    1    0    1    1    1    0    1    1

[[4]][[2]]
[1]  9.00000 58.72511
```

This agrees with what we previously obtained from Rmpi.

Of course the real question is whether or not the multicore approach reduces the computation time. The `system.time` function discussed in Section 5.4 provides a means to address this issue. In the case of our variable selection example this leads to

```
> timeVec <- vector(length = 2, mode = "numeric")
> for(i in 1:100){
+ timeVec[1] <- timeVec[1] +
+ system.time(mclapply(indexList, myRegTask))[[3]]
+ timeVec[2] <- timeVec[2] +
+ system.time(myRegTask(subSets))[[3]]
+ }
> timeVec[2]/timeVec[1]
[1] 2.43891
```

The `system.time` function returns a five-element list with the third component being elapsed time. This function is used to evaluate the elapsed computation times for `mclapply` and serial execution of `myRegTask` across all 2^{10} subsets. The calculations are repeated 100 times with the total elapsed time being accumulated for each approach. The outcome suggests that the speedup produced by `mclapply` exceeds 2 in this instance.

11.5 Parallel random number generation

Simulation studies are in many respects the ideal application for parallel processing. They are "embarrassingly parallel" in the sense that computations may generally be conducted without the need for inter-processor communication. Thus, the potential exists for near linear speedup in cases that can occur in practice. However, there are fundamental problems that arise from the fact that the random number streams produced and used by each processor must behave as if they are independent in a probabilistic sense for the division of labor between processors to be productive. An overview of issues that arise in parallel random number generation and methods for producing effective parallel generators is provided by Coddington (1996) who concludes that good parallel random number generators should a) produce intra-processor streams of high quality in the usual statistical sense while exhibiting minimal inter-processor dependence, b) be scalable (i.e., adaptable for use in systems that have many processors) and c) require no data movement between processors.

Two noteworthy sources for generator packages that meet Coddington's criteria are the SPRNG package of Mascagni and Srinivasan (2000) and the RngStreams package discussed in L'Ecuyer, et al. (2001, 2002). The SPRNG package produces different random number sequences for the processors via parameterization while RngStreams employs a sequence splitting approach wherein a long random number stream is partitioned across the different processors. While there has been some debate as to the relative merits of the two approaches, in practice it appears that both packages can be effective and provide viable solutions to the problem of obtaining something that can be viewed as independent streams. Practical difficulties that arise in using SPRNG stem from its ties to MPI and a rather complex implementation that necessitates the creation and linking of a compiled library. In contrast, the source code for the RngStreams package is compact and quite easy to incorporate into C/C++ programs using either OpenMP or MPI through simple include directives. This property extends its applications from clusters down to desktop environments and, accordingly, it is the parallel pseudo-random number generator option that will be pursued here.

The "backbone" generator for RngStreams is the combined multiple recursive generator Mrg32k3a defined in (4.11)–(4.12). The period of the generator exceeds 2^{191}. For use in a parallel context this long cycle is divided into 2^{64} nonoverlapping streams each of length 2^{127}.

The key to accessing different streams produced by (4.11)–(4.12) is the technique described in L'Ecuyer (1990) that allows movement between streams through linear transformations of the initial seeds using known matrices. Specifically, if $\tilde{x}_{1,0}, \tilde{x}_{2,0}$ are the initial seeds/states for the two generators, at the nth step of the recursion the state for generator i is $\tilde{x}_{i,n} = (A_i^n \bmod m_i)x_{i,0} \bmod m_i, i = 1, 2$, for known 3×3 matrices A_1 and A_2 with $m_1 = 4294967087$ and $m_2 = 4294944443$. The powers of the matrices A_1 and A_2 can be computed explicitly and, in particular the matrices A_1^{127} and A_2^{127} that are needed for movement between the different streams of the generator are found to be

$$A_1^{127} = \begin{bmatrix} 2427906178 & 3580155704 & 949770784 \\ 226153695 & 1230515664 & 3580155704 \\ 1988835001 & 986791581 & 123051564 \end{bmatrix} \tag{11.1}$$

and

$$A_2^{127} = \begin{bmatrix} 1464411153 & 277697599 & 1610723613 \\ 32183930 & 1464411153 & 1022607788 \\ 2824425944 & 32183930 & 2093834863 \end{bmatrix}. \tag{11.2}$$

Section 4.5 demonstrated how to use the `SetPackageSeed` function to set the initial state for `RngStream` objects. This process requires the user to supply six initial integers to serve as the beginning seed for the generator. The first `RngStream` object that is created will use this initial seed. Subsequent objects will then be started on the next stream; i.e., they will have initial "seeds" that are 2^{127} states removed from the initial state of the previous object's generator.

Our first objective is to perform some simple experiments that will illustrate how the seeds are changed (and how to guarantee that they are changed) for different processors. For this purpose the `GetState` function will be needed. It has prototype

```
void GetState (unsigned long seed[6]) const
```

This function provides a means to return the current state of an `RngStream` object at any point during random number production. In particular, it allows us to determine an object's initial seed and that is how it will be used in our examination of the behavior of the `RngStream` class constructor.

The listing below illustrates the use of RngStreams in an OpenMP program.

Listing 11.4 ranOpenMP.cpp

```
//ranOpenMP.cpp
#include <omp.h>
#include "RngStream.h"
#include <iostream>

using std::cout; using std::endl; using std::cin;

int main(){

  int nP = omp_get_num_procs();
  omp_set_num_threads(nP);//set number of threads

  unsigned long seed[6] ={1,2,3,4,5,6};
  RngStream::SetPackageSeed (seed);
  RngStream RngArray[nP];//array of RngStream objects

  int myRank;
#pragma omp parallel private(myRank)
  {
    myRank = omp_get_thread_num();
#pragma omp critical
    {
    cout << "For thread " << myRank << endl;
    RngArray[myRank].WriteState();
    cout << "The random number is "
        << RngArray[myRank].RandU01() << endl;
    }
  }

  return 0;
}
```

The first step is to include the **RngStream** class header file RngStream.h via an include directive. The **SetPackageSeed** function is then used to set the six integer seeds, of type **unsigned long**, as the integers from 1 to 6. An array of **nP RngStream** objects is created next with **nP** being the number of processors. The default behavior is for the seed of each array element to be 2^{121} states removed from the initial state of its predecessor and their states are written to standard output to allow for comparison with our other parallel treatments. Independent streams will now be produced if each thread uses a different entry from **RngArray** when generating random numbers. To accomplish this we simply use the number assigned to each thread by the OpenMP API as the index it employs for accessing the array of **RngStream** objects. To avoid garbled output the output segment from the parallel region is wrapped in a **critical** region.

Listing 11.4 was compiled and executed on a four processor machine via

```
$ g++ -fopenmp ranOpenMP1.cpp RngStream.cpp -o ranOpenMP
$ ./ranOpenMP
For thread 1
The current state of the Rngstream:
 Cg = {3847595764, 542750874, 3358998068, 4025640956, 701604884, 2546910389}

The random number is 0.701702
For thread 0
The current state of the Rngstream:
 Cg = {1, 2, 3, 4, 5, 6}
```

```
The random number is 0.0010095
For thread 2
The current state of the Rngstream:
 Cg = {311773008, 2901318700, 433058656, 3749492613, 2059732357, 994549473}

The random number is 0.476142
For thread 3
The current state of the Rngstream:
 Cg = {3522494900, 2524210175, 3812848698, 4095818817, 2057726304, 1219287084}

The random number is 0.0469012
```

These numbers will recur in our other examples.

The basic premise of the previous example is readily extended to produce something useful. For example, suppose that the goal is to fill an n-element array held in a pointer pU to double with independent pseudo-random uniforms. A code segment that accomplishes this is

```
#pragma omp parallel private(myRank)
    {
        myRank = omp_get_thread_num();
#pragma omp for
        for(int i = 0; i < n; i++)
            pU[i] = RngArray[myRank].RandU01();
    }
```

The computing effort is distributed across nP threads with each thread producing values from a different RngStream object.

The basic idea behind the previous listing was used to do some performance comparisons that are shown in Table 11.1. The values in the table are the (machine dependent) times that were needed to generate approximately a billion random numbers using one, two, four, eight and sixteen processors in OpenMP.

Table 11.1 *Time required to generate 1 billion random uniforms*

Number of Processors	Time (in seconds)	speedup
1	130.74750	1
2	111.25400	1.175216
4	62.95395	2.076875
8	37.61315	3.476111
16	22.05340	5.928678

The use of four processors cuts the computation time in half while sixteen processors carry out the work in a sixth of the serial time.

Let us now consider how to construct an analog of our OpenMP program that can be used with MPI. There are various ways to accomplish this. Perhaps the simplest is to have all processors create an array of RngStream objects and then use the object that corresponds to their rank. This approach is left as an exercise (Exercise 11.10). Our particular solution is somewhat more involved and requires the use of some of the features that are implicitly provided in the RngStreams package.

As noted at the beginning of this section, the seeds/states of RngStream objects are advanced by multiplication involving the matrices (11.1)–(11.2). A function that can be used to carry out these multiplications $\mod(2^{32} - 209)$ and $\mod(2^{32} - 22853)$ is contained in an anonymous namespace in the RngStream.cpp file that is downloaded with the RngStreams package. An anonymous namespace is created using syntax such as

```
namespace
{
  functions, constants, etc.
}
```

This creates a namespace that is available only to functions that inhabit the same file. In this respect, it provides an alternative to the creation of global static functions and variables. Appendix E provides another illustration of when an anonymous namespace can prove useful.

The anonymous namespace in the RngStream.cpp file holds the matrices in (11.1)–(11.2) in the 3×3 arrays A1p127 and A2p127 and the moduli $2^{32} - 209$ and $2^{32} - 22853$ are stored in the unsigned long integers m1 and m2. It also includes the function matVecModM with prototype

```
void MatVecModM (const double A[3][3], const double s[3], double v[3],
        double m)
```

This function will perform multiplication of a 3×3 array A times a three-element array s modulo m and return the result in the three-element array v. Thus, by taking A to be A1p127 or A2p127 and m to be m1 or m2, a given seed array can be advanced 2^{127} states to obtain a seed for the next random number stream. With this in mind we added

```
static void AdvanceSeed (unsigned long seedIn[6],
        unsigned long seedOut[6]);
```

to the header file RngStream.h and then appended

```
void RngStream::AdvanceSeed (unsigned long seedIn[6],
        unsigned long seedOut[6]){
  double tempIn[6]; double tempOut[6];
  for(int i = 0; i < 6; i++)
    tempIn[i] = seedIn[i];
  MatVecModM (A1p127, tempIn, tempOut, m1);
  MatVecModM (A2p127, &tempIn[3], &tempOut[3], m2);
  for(int i = 0; i < 6; i++)
    seedOut[i] = tempOut[i];
}
```

to the RngStream.cpp file. By declaring AdvanceSeed as static it can be used directly without an RngStream object which is precisely what is needed for the application that we have in mind.

Our MPI implementation of the RngStreams methodology now takes the form

Listing 11.5 *ranMPI.cpp*

```
//ranMPI.cpp
#include <iostream>
#include <fstream>
#include "mpi.h"
#include "RngStream.h"

using std::cout; using std::endl;

int main(){
  unsigned long seed[6] = {1,2,3,4,5,6};
```

```cpp
MPI::Init();
int myRank = MPI::COMM_WORLD.Get_rank();
int nP = MPI::COMM_WORLD.Get_size();

if(myRank == 0){
  unsigned long tempSeed[6];
  double ranValue = 0;
  //start with the 0 processor
  RngStream::SetPackageSeed(seed);
  RngStream Rng;

  double myNumber = Rng.RandU01();
  cout << "For processor 0 the state is" << endl;
  for(int j = 0; j < 6; j++)
    cout << seed[j] << endl;
  cout << endl;

  //create and send the seeds to the other processors
  for(int i = 1; i < nP; i++){
    RngStream::AdvanceSeed(seed, tempSeed);
    MPI::COMM_WORLD.Send(tempSeed, 6, MPI::UNSIGNED_LONG, i, 0);
    cout << "For processor " << i << " the state is" << endl;
    for(int j = 0; j < 6; j++){
      cout << tempSeed[j] << endl;
    }
    cout << endl;
    for(int j = 0; j < 6; j++)
      seed[j] = tempSeed[j];
  }

//now collect the results
  for(int i = 1; i < nP; i++){
    MPI::COMM_WORLD.Recv(&ranValue, 1, MPI::DOUBLE, i, 0);
    cout << "The random number for processor " << i
         << " is "<< ranValue << endl;
  }
  cout << "The random number for process 0 is " << myNumber
       << endl;
}

else{
  unsigned long mySeed[6];
  //get my seed
  MPI::COMM_WORLD.Recv(mySeed, 6, MPI::UNSIGNED_LONG, 0, 0);

  //create the rngstream object
  RngStream::SetPackageSeed(mySeed);
  RngStream Rng;

  //generate my number and send it back to the 0 processor
  double myNumber = Rng.RandU01();
  MPI::COMM_WORLD.Send(&myNumber, 1, MPI::DOUBLE, 0, 0);
}

MPI::Finalize();
```

```
    return 0;
}
```

The idea behind Listing 11.5 is straightforward. The master (or rank 0) processor uses our AdvanceSeed function along with an initial seed to determine seeds that will produce "independent" random number streams for the other processors. These seeds are communicated to the processors using the Send and Recv functions. The processors then use these seeds to initialize their individual RngStream objects. After finishing their tasks, the processors report the random uniform value they generated to the master processor for output.

To check that our scheme for using RngStream objects produces the desired results both the seeds that are sent to each processor and the random numbers they generate are printed out by the master processor. With four processors the results are

```
$ mpicxx -c RngStream.cpp
$ mpicxx -c ranMPI.cpp
$ mpicxx RngStream.o ranMPI.o -o ranMPI
$ mpiexec -np 4 ranMPI
For processor 0 the state is
1
2
3
4
5
6

For processor 1 the state is
3847595764
542750874
3358998068
4025640956
701604884
2546910389

For processor 2 the state is
311773008
2901318700
433058656
3749492613
2059732357
994549473

For processor 3 the state is
3522494900
2524210175
3812848698
4095818817
2057726304
1219287084

The random number for processor 1 is 0.701702
The random number for processor 2 is 0.476142
The random number for processor 3 is 0.0469012
The random number for process 0 is 0.0010095
```

This output agrees with our prior results from our OpenMP program as it should.

It remains to describe how "independent" random number streams can be obtained for simulations conducted using Rmpi or multicore. In keeping with previous developments this will be accomplished using an R interface to RngStreams.

There are two packages that provide an RngStreams interface: rlecuyer and rstream. The former is actually a built in feature for Rmpi and readily accessed as in

```
> library(Rmpi)
> mpi.spawn.Rslaves(nslaves = 4)
        4 slaves are spawned successfully. 0 failed.
master (rank 0, comm 1) of size 5 is running on: cos
slave1 (rank 1, comm 1) of size 5 is running on: cos
slave2 (rank 2, comm 1) of size 5 is running on: cos
slave3 (rank 3, comm 1) of size 5 is running on: cos
slave4 (rank 4, comm 1) of size 5 is running on: cos
> mpi.setup.rngstream(seed = 1:6)
Loading required package: rlecuyer
> mpi.remote.exec(runif(1))
          X1         X2        X3         X4
1 0.001009498 0.7017015 0.476142 0.04690119
```

Note that the rlecuyer package needs to be installed before this approach will work. If that is the case, the effect of `mpi.setup.rngstream` is to replace R's default Mersenne twister generator with the backbone generator from RngStreams. It appears that the rlecuyer implementation of RngStreams is slower than that for the rstream package. As our result, our discussions will focus on rstream from this point forward. Further discussion of both packages and their use with nws and snow can be found in Karl, et al. (2011).

The first step is to load the rstream package with

```
> library(rstream)
```

This has the effect of introducing the S4 virtual class `rstream` into the current R workspace. Instances of the `rstream` class may be created from the derived class `rstream.mrg32k3a` that provides access to the now familiar RngStreams backbone generator. A new `rstream` object, `newRstreamObj`, is created using syntax of the form

```
newRstreamObj <- new("rstream.mrg32k3a", seed = Seed)
```

with Seed a six-component vector of integers. Subsequent `rstream.mrg32k3a` objects created with `new` will automatically have initial seeds that are 2^{127} states removed from the most recently created object. The `rstream.sample` function returns a sample of n uniform random numbers from the `rstream` object `newRstreamObj` via the command

```
rstream.sample(newRstreamObj, n)
```

Because `rstream` objects are stored as pointers in the underlying foreign language code, it is necessary to "pack" them before they can be sent to another node in Rmpi. The same process is also needed to save an `rstream` object so that it will be available in a future R session. Packing is accomplished with

```
rstream.packed(newRstreamObj) <- TRUE
```

Before they can be used packed objects must be unpacked by

```
rstream.packed(newRstreamObj) <- FALSE
```

Beyond this, the process of actually using rstream with Rmpi is relatively straightforward. For the case of p processors, one carries out the following sequence of steps:

1. create p instances of the `rstream.mrg32k3a` class,

2. pack each object using `rstream.packed`,

3. load the rstream package on each processor,

4. send the `rstream` objects to the `p` processors using `mpi.send.Robj`,

5. receive the objects on each processor with `mpi.recv.Robj` and

6. unpack the `rstream` objects on each processor with the `rstream.packed` function.

At this point the random numbers on each node will correspond to "independent" streams and can be used in whatever other tasks the processors may be assigned.

An illustration of the scheme we have laid out for using rstream with Rmpi is given in the excerpt from an R session shown below.

```
> library(rstream)
> library(Rmpi)
> mpi.spawn.Rslaves(nslaves = 3)
        3 slaves are spawned successfully. 0 failed.
master (rank 0, comm 1) of size 4 is running on: localhost
slave1 (rank 1, comm 1) of size 4 is running on: localhost
slave2 (rank 2, comm 1) of size 4 is running on: localhost
slave3 (rank 3, comm 1) of size 4 is running on: localhost
> #create a list of rstream.mrg32k3a objects
> rngList <- vector(length = 4, mode = "list")
> rngList <- c(new("rstream.mrg32k3a", seed = c(1, 2, 3, 4, 5, 6)),
+ replicate(3, new("rstream.mrg32k3a")))
> #pack the objects that will be sent to the processors
> for(i in 2:4) rstream.packed(rngList[[i]]) <- TRUE
> #load the package and send the packed rstream objects
> mpi.bcast.cmd(library(rstream))
> for(i in 1:3) mpi.send.Robj(rngList[[i + 1]], i, 0)
> mpi.bcast.cmd(myRng
+ <- mpi.recv.Robj(mpi.any.source(), mpi.any.tag()))
> #unpack the objects on the processors
> mpi.bcast.cmd(rstream.packed(myRng) <- FALSE)
> #now generate the random numbers
> mpi.remote.exec(rstream.sample(myRng, 1))
        X1          X2          X3
1 0.7017015 0.476142 0.04690119
> rstream.sample(rngList[[1]], 1)
[1] 0.001009498
```

The initial steps are to import the rstream and rmpi packages into the workspace. Three slaves are then spawned and a vector of lists `rngList` is created to hold the `rstream` objects that will be sent to the processors. The first `rstream` object is created using the `new` function with the seed vector $(1, 2, 3, 4, 5, 6)$ and placed in the first component of `rngList`. The remaining three `rstream` objects are created similarly via `replicate` and used to fill the remaining `rngList` entries. This packaging allows us to automate (via a `for` loop) the packing of the list elements. Note that `rngList[[1]]` belongs to the master processor and does not have to be packed. The next step is to load the `rstream` package on each of the slaves. This is accomplished with the `mpi.bcast.cmd` function. A `for` loop is used in conjunction with the `mpi.send.Robj` function to send the packed `rstream` objects to the slaves after which the processors receive them with `mpi.recv.Robj`. The `mpi.bcast.cmd` function is employed to tell each slave to unpack its `rstream` object followed by an application of `mpi.remote.exec` with the `rstream.sample` functions to produce the first random uniform of each random number stream. A comparison of these values with those obtained in our MPI and OpenMP examples shows that we have reproduced the streams that were previously obtained in those settings.

The use of rstream is even simpler with the multicore package. The same random numbers that were produced with Rmpi can be obtained by using `mclapply` as in the code sequence below.

```
> library(rstream)
> library(multicore)
> rngList <- vector(length = 4, mode = "list")
> rngList <- c(new("rstream.mrg32k3a", seed = c(1, 2, 3, 4, 5, 6)),
+ replicate(3, new("rstream.mrg32k3a")))
> mclapply(rngList, function(a) rstream.sample(a, 1))
[[1]]
[1] 0.001009498

[[2]]
[1] 0.7017015

[[3]]
[1] 0.476142

[[4]]
[1] 0.04690119
```

Thus, we have now produced the same random number streams with OpenMP, MPI, Rmpi and multicore.

11.6 Exercises

11.1. Create an alternative version of Listing 11.2 that uses Newton's method.

11.2. Develop an analog of the parallel ANOVA program for MPI in Listing 11.3 that can be used in OpenMP.

11.3. In Exercise 11.2 suppose that the data is generated in R with

```
> set.seed(123)
> a <- rnorm(4000000)
> write(a, "aovVector.txt", ncolumns = 1)
```

The data in aovVector.txt is then read into the OpenMP program and reshaped to conduct the analysis. Code that was written to carry out the reshaping is shown below

```
    int j;
#pragma omp parallel for
    for(int i = 0; i < 4; i++)
      for(j = 0; j < 1000000; j++)
        pA[i][j] = pa[i + 4*j];
```

with pa and pA, respectively, the pointers that hold the initial array created from R and a 4×1000000 array for the reshaped data with rows representing the different "treatments". Demonstrate that this approach fails, explain why it fails and fix the problem.

11.4. The MPI "broadcast" function has prototype

```
void MPI::COMM_WORLD.Bcast(void* buf, int count,
          MPI_Datatype datatype, int root)
```

This function provides more immediate communication between processors that allows us to avoid the loops that were used with `Send` and `Recv`. The arguments are the same as those for `Send` until the `root` argument is reached in the prototype. This latter parameter represents the rank of the processor whose data in `buf` is to be distributed to all the other processors. All processors must call `Bcast` with the same arguments.

Write an MPI program that issues a greeting message from each processor. The main text of the message should be sent from the processor with rank 0 using the `Bcast` function. Then each processor should provide its own customization by, for example, including its particular rank in the message it prints to standard output.

11.5. The opposite effect of `Bcast` can be obtained from the MPI function `Gather`. It collects information from all of the processors onto a single processor that is usually taken to be processor 0. The syntax for this latter case is

```
void MPI::COMM_WORLD.Gather(void* sendbuf, int count,
        MPI_Datatype datatype, void* recvbuf, int count,
        MPI_Datatype datatype, 0)
```

The quantity `sendbuf` is a pointer to `count` memory locations that is sent from each processor to processor 0. This is then gathered into the pointer `recvbuf` which needs to be defined only on processor 0 and have size equal to the product of `count` and the number of processors (including processor 0). As was the case with `Bcast`, the function `Gather` must be called by all processors. Use `Bcast` and `Gather` to provide an alternative version of Listing 11.3 that does not require send/receive loops.

11.6. Refer to problem 3.10. Write programs that split the computations across four (or, more generally, some power of 2) processors using OpenMP and/or MPI. Compare the running time of your parallel implementation to that of your serial code from Exercise 3.10.

11.7. Write an R function that will carry out ANOVA in parallel using Rmpi and return summary statistics, including an F-statistic, as output.

11.8. Extend the parallel analysis of variance programs from Listing 11.3 and Exercises 11.2 and 11.7 to allow for unbalanced designs.

11.9. Use RngStreams and the rstream package to create programs to perform random searches in parallel to find the minimizer of the function in Figure 8.1. Develop the programs in OpenMP, Rmpi and the multicore package.

11.10. Create an MPI program that uses `RngStream` objects to generate independent random number streams and requires no communication between the processors.

11.11. Rework the all-subsets regression example from Section 11.4 to allow for an odd number of processors.

11.12. Develop a vectorized function for packing and unpacking `rstream` objects.

An introduction to Unix

In this appendix we will give a brief overview of some of the basic Unix commands. More detailed treatments are provided in a myriad of on-line tutorials and texts such as Kochan and Wood (2003).

The focal setting in this appendix is the command line of a Unix shell; although, much of the work in this text can be accomplished using various graphical user interfaces, the command line is always available even in remote logins. So, it is worthwhile to know how to work in that environment.

The interface between a user and the Linux operating system is provided by a command line interpreter generally referred to as the *shell*. There are a number of shell options that include Bourne (sh, bash and zsh) type, C (csh and tsch) type and Korn (krn) shells. They each represent programming languages whose purpose is the interpretation of Unix commands that are entered on the shell's command line located at the shell prompt. The prompt often takes the form of a $ or %. We will employ the bash shell exclusively in our work. It may be started by entering

```
$ /bin/bash
```

on the command line.

Unix commands make it possible to transfer and copy files, compile code, start applications, etc. In general these commands have the form

```
commandName -option arguments
```

with `commandName` being a single, case-sensitive, "word" consisting of a few letters that often abbreviate the command's purpose. A command may or may not involve additional options. The details behind such additional specifications can be obtained by referring to the command's associated manual page obtain by entering

```
man commandName
```

on the command line. A related, but sometimes different, set of information can be accessed by `info commandName`. If you can remember a command's name but do not recall what it does, that information is returned by entering `whatis commandName`. If you only remember part of a command name `commandNamePart`, then `apropos commandNamePart` will search the `whatis` database for occurrences of the string `commandNamePart`.

Disk memory is organized in Unix as a tree structure. It begins with the root directory or / that represents the top of the tree and then subdirectories branch from the root. These subdirectories have further subdirectories, etc. When you first log in, the shell will open at your personal *home* directory which has a name something like

```
/home/yourUserName
```

or

```
/Users/yourUserName
```

with yourUserName the user name you have been given on the system. Your current location in the directory tree can always be obtained by entering `pwd` (for present working directory) on the command line.

Starting from our original login location we would like to go various places and accomplish various things. Commands for such purposes will be discussed in subsequent sections. The presentation is organized by groupings of commands with similar or related purposes.

A.1 Getting around and finding things

The main tool for moving from point A to point B in Unix is the *change directory* command cd. For example, by entering

```
cd pathToDirectory/directoryName
```

you will move from your current location along the path specified by pathToDirectory and end up at directoryName on the Unix directory tree. There are various shortcut ways to specify a path. But, what always works is a list, separated by /'s that starts at the root directory and details every stop along the way from the root to the location of interest. In general, this is what we will refer to as a *path* throughout the text. For example, the command cd /usr/local/share will take you from your current location in the directory tree to the share subdirectory by going from the root / to the subdirectory usr and then to its subdirectory local.

A very useful shortcut is provided by the $\sim$ symbol which can always be used in place of the name for your home directory. Some related shortcut commands that come in handy are:

cd .. which moves you one level closer to the root and

cd without an argument, which always returns you to your home directory.

A.2 Seeing what's there

So, now you have arrived at a particular location (or know of some location of interest) and would like to find out what is there. The list command, ls, can be used for this purpose. Specifically,

```
ls pathToDirectory/directoryName
```

will provide a list of all the files in the directory directoryName reached by pathToDirectory. More generally,

```
ls -l pathToDirectory/directoryName
```

gives a complete listing that details, from left to right, the following information for each file in the directory:

- read, write and execute permissions (that will be discussed in more detail below),
- the number of hard links (i.e., the number of directory entries that reference the file or directory),
- the owning user,
- the owning group,
- the size of the file or directory in bytes,
- creation date and
- the file or directory name.

It turns out that neither ls nor ls -l tell the whole story. Many times there are files in your directory that begin with a . such as .bashrc and .emacs. To see these you have to use ls -a. Using ls without a path specification returns the result of applying it to the current directory.

The `ls` command gives us what we need to determine the contents of a directory. For big directories (as well as for other reasons), it is sometimes preferable to automate the search for things of interest inside a directory. This can be accomplished with the `find` command. For example,

```
find pathToDirectory/directoryName -name filename
```

will locate any file with the name filename in directoryName. Often, instead of seeking a particular file, what is really needed is a list of all the files with certain name similarities such as those having the same file extension. This can be accomplished by combining `find` with *wildcards*. These are special symbols that provide placeholders or file name expansions thereby allowing find (as well as other commands) to just match some portion of a file's name. Perhaps the most common wildcards are * that matches all characters in any quantity and ? that will match any single character. Using wildcards one can create commands like

```
find pathToDirectory/directoryName -name *.cpp
```

which will produce a listing of every file in directoryName with a .cpp file extension. Similarly,

```
find pathToDirectory/directoryName -name ex?.cpp
```

would give a listing of files with names such as ex1.cpp, ex2.cpp, exE.cpp, etc., but not, e.g., ex.cpp or ex12.cpp. The latter two would both be returned if ex*.cpp was used as the search argument.

To see the actual contents of a text file you can edit it as will be discussed below. But, if you just want a quick peek at a few lines of a file fileName, either the `head` or `tail` commands can be used as in

```
head -n fileName
```

to list the first n lines of the file and

```
tail -n fileName
```

to see its last n lines. The entire contents of fileName will be written to the screen with

```
cat fileName
```

The commands `more` and `less` will allow you to scroll through a file. The space bar will advance the file one "page" while q returns you to the shell prompt. You can obtain information about a file without looking at it using the `wc` command. Entering

```
wc fileName
```

will return the number of lines, words and bytes in fileName.

A.3 Creating and destroying things

So far our interaction with Unix has been relatively static. But, things become much more interesting once you can create and populate directories of your own. There are a multitude of ways to go about doing this and we will consider just a few of the standard ones here.

The simplest way to create a file is through the `touch` command which creates an empty text file fileName upon entering

```
touch fileName
```

To delete fileName, employ the *remove* command, `rm`, like

```
rm fileName
```

One of the most important things to create is a directory and the *make directory* command, `mkdir`, can be used to accomplish this. For example,

```
mkdir directoryName
```

adds a new directory, directoryName, to the current directory. This directory may be removed using the *remove directory* command `rmdir` with

```
rmdir directoryName
```

provided directoryName is empty. If this is not the case, it must be emptied using a recursive application (i.e., the `-r` option) of rm. Specifically,

```
rm -r directoryName
```

removes directoryName as well as all of its contents.

Life would be quite complicated if all our files and directories had to be created from scratch. The *copy* command `cp` keeps us from having to do this by allowing us to copy files from one directory to another. To copy a file fileName in pathToSourceDirectory to a new file newName in pathToTargetDirectory use

```
cp pathToSourceDirectory/fileName   pathToTargetDirectory/newName
```

To move, rather than copy, the file one uses the move command `mv` in the form

```
mv pathToSourceDirectory/fileName   pathToTargetDirectory/newName
```

Output redirection is useful for a variety of purposes. One of these is creation of files with the output from a program. Output redirection is achieved with the $>$ symbol using syntax such as

```
command options arguments > fileName
```

For instance,

```
man pwd > temp.txt
```

will write the contents of the `man` file for the `pwd` command into the text file temp.txt.

The real purpose of the `cat` command is concatenation of files. When used in conjunction with output redirection a command sequence such as

```
cat temp1.txt temp2txt > temp.txt
```

would create a new file temp.txt consisting of the file temp1.txt with temp2.txt appended to its end.

In order to copy and move files around you must have permission to use them in such operations. Information about the permissions associated with a particular file or files in a directory can be obtained through the `ls -l` version of the `ls` command. As noted above, this produces a variety of output which includes the file permissions as the first, 10 character, component of the line associated with any particular file.

Permission information provides the details concerning who can and cannot access a file or directory in question. The first character in the permission string indicates the type of file it is with d indicating a directory, l meaning a symbolic link and – representing a common file. Characters 2–4 determine what the owner can do with the file. In particular, if characters 2–4 are rwx this means the owner can read (i.e., r), write (i.e., w) and execute (i.e., x) the particular file. A – in any of the slots means that permission is denied for that particular activity. The next three slots, characters 5–7, convey exactly the same information concerning permissions for the owning group with slots 8–10 giving the permissions allotted to all other users. Thus, a 10-character string corresponding to a file fileName of the form

```
-rwxr-xr--
```

would say that the owner of fileName can read, write and execute it, members of the group that own the file can read and execute it while it is read-only for all other users.

If you own a file you can change any or all of the permissions using the chmod command. Here the letters u, g and o indicate the user (owner), group owner and other users, respectively. To change permissions you then enter one of the letters u, g or o followed by a + or – before one of r, w or x to either add (with a +) or remove (with a –) the permission in question. A command such as

```
chmod g-x fileName
```

would convert the previous permissions for fileName to

```
-rwxr--r--
```

thereby removing the permission to execute fileName from the owning group members.

The creation of new files from scratch will often be accomplished via some type of text editor. There are several editors available for this purpose in Unix that include pico, vi and emacs. The pico editor is simple to use but primitive in terms of its functionality. The texts Robbins, et al. (2008) and Cameron, et al. (2004) provide extensive information about using the more advanced editors vi and emacs, respectively.

A.4 Things that are running and how to stop them

At any time on your system there may be several jobs or tasks that are being managed by your shell. A list of the processes that are running is returned by the ps command. Results from using ps might appear like

```
$ ps
  PID TTY           TIME CMD
29837 ttys000    0:00.08 -bash
29891 ttys000    0:00.07 emacs
29895 ttys000    0:06.69 ./ack
```

This says there are currently three processes that are active: the shell (bash), emacs and an executable with the name ack. The first column of the output is the process ID (a unique number that the shell has assigned to each of the processes), the next column is the "name" of the controlling terminal, the third column is the CPU time that has been used by each process and the final column is the name of the command. More information about activity can be obtained from jobs. For this same example, jobs produced

```
$ jobs
[1]-  Stopped                 emacs
[2]+  Stopped                 ./ack
```

The emacs and ack programs have been given job numbers 1 and 2, respectively, and both have been stopped or suspended. This suspension was accomplished by the Ctrl z combination. A suspended job may be restarted by entering fg (for foreground) prior to its job number.

To kill a job that is running in the foreground use Ctrl c. This is, unfortunately, quite useful as a way of stopping runaway output from a program that produced unexpected results due to coding or other errors. Such programs can also be suspended with Ctrl z and then killed using kill processID or kill %jobNumber with processID and jobNumber representing the process ID and job number, respectively, for the job to be terminated. In our example, kill 29895 or kill %2 will terminate the ack job after which ps produces

```
$ ps
  PID TTY           TIME CMD
```

```
29837 ttys000     0:00.10 -bash
29891 ttys000     0:00.20 emacs
```

verifying that ack is no longer being managed by the shell. To terminate the shell itself either **exit** or Ctrl d can be used.

APPENDIX B

An introduction to R

In this appendix we give a brief introduction to the use of R in a Unix environment. More detailed discussions can be found in the manual *An Introduction to R* available from the CRAN website (`http://cran.r-project.org`), Gentleman (2009) and various introductory texts such as Verzani (2005). The R package comes with an extensive collection of manual pages that can be accessed using the `help` or ? functions with the command or object of interest supplied as the argument. For instance, the manual page for the R function t that transposes a matrix is obtained via either

```
help(t)
```

or

```
?t
```

There are also examples corresponding to the help pages that can be accessed using the `example` function with, e.g., `example(t)` producing illustrations of using the transpose function. Some functions such as those that involve symbols must be quoted when using `help`, ? and `example`; e.g.,

```
> example("%%")

%%> x <- -1:12

%%> x + 1
 [1]  0  1  2  3  4  5  6  7  8  9 10 11 12 13

%%> 2 * x + 3
 [1]  1  3  5  7  9 11 13 15 17 19 21 23 25 27

%%> x %% 2 #-- is periodic
 [1] 1 0 1 0 1 0 1 0 1 0 1 0 1 0

%%> x %/% 5
 [1] -1  0  0  0  0  0  1  1  1  1  1  2  2  2
```

is obtained by applying `example` to the R modulus operator %%. The output includes results from applying the modulus operator and the integer division operator %/% to the integers from -1 to 12.

To begin using R, the first step is to obtain the most recent version of R from CRAN. Here one can obtain a precompiled binary distribution of the basic R system. The location of the R installation can be determined by issuing the command `R.home()` from within an R session. It is also possible to build R directly from source code. We discuss this option briefly in Section 4.9. Graphical user interfaces that can be used with R include ESS and R Commander. The simplest interface is through a Unix shell and that is the approach that will be used here.

To start an R session, enter R on the shell command line followed by a carriage return. After some copyright information and other messages, a new prompt > appears at which point you may begin entering commands. To suppress the copyright information, you can use a silent option as in

```
$ R --silent
>
```

Like Unix, R is case-sensitive.

To exit R use q() followed by a carriage return. After entering the q() command, R will ask if you want to save the workspace image. This is because R accumulates all the functions and data sets that have been created in a given session and will erase them unless the save option is chosen. By saving a workspace you can return to it at a future time and all the functions, etc., that were created during the corresponding session will still be available for use. In this respect, it is recommended that one create and save workspaces that store each new R project of substance. The command save.image("workspaceName") will save the current image of the active workspace to the file workspaceName. If no name is specified (i.e., the command is save.image()), the workspace is saved under the name .RData.* To reload this image in an R session the command is load("workspaceName"). Note that R uses quotes to designate an object as a character string.

In order to access a previously saved workspace workspaceName, one must either supply a full path to load or initialize R from the directory where the saved image resides. If R has been started in a different directory, by entering setwd("pathToDirectory") the current working directory can be changed to the one specified by pathToDirectory. The current working directory can be determined via the command getwd(). Somewhat more generally, commands to the underlying Unix system can be executed while in R using the system command. For instance, entering system("pwd") at the R command prompt will produce the same information as getwd.

Once an R session has begun it may be of interest to load various R packages or libraries. This is accomplished through the library function with, e.g., library(lattice) having the effect of making the objects in the R lattice graphics package accessible from the current R session. The command search() will return a list of all the packages that are currently loaded.

To see a list of all the help topics that are available for the installed packages use help.start(). This produces an html file for which the packages link will allow access to information about a specific package of interest. Some packages also have "vignettes" which are pdf files containing discussions (including code) of topics related to the package. The command vignette(package = "packageName") will return a list of the available vignettes for the package packageName.

There are many R packages that are not part of the standard installation but can be downloaded separately. To install a package one uses install.packages("packageName") that will install an R package packageName (if it exists). The user will be asked to select a package repository (or CRAN mirror) from a list after which installation should proceed automatically.

R has analogs of the basic Unix commands for listing the contents and removing objects from a workspace. The command ls() will provide a list of the objects in the current workspace. To remove objects from the current session use rm(o1, o2, ...) with o1, o2, ... a comma separated list of the names of the objects that are to be deleted. In particular, rm(list = ls()) will remove all objects from the workspace.

The three most common data structures in R are data frame, matrix and list. The matrix and list structures possess the expected properties; e.g., matrix(pi, 2, 3) creates a 2×3 matrix with every entry equal to pi the R approximation to the value of π. The command

---

* As noted in Appendix A, the . prepended to the file name will render it invisible to the ls command unless its –a option is used.

list(o1, o2, ...) returns a list object that contains the R objects o1, o2, ... that were supplied as its arguments. Data frames are hybrid data structures that possess aspects of both matrices and lists. They are constructed using the data.frame function.

There are a number of data sets that come as part of the standard R package. A list of all the ones that are available can be obtained by entering the command data(). To obtain a particular data set dataSetName from the list one enters data(dataSetName) at the R command prompt. This returns a data frame object that holds the contents of the specified data set.

R provides two assignment operators: = and <-. For our purposes they can (and will) be used interchangeably. We will tend to use <- in most programming applications as it clearly conveys the direction of assignment. For example, the sequence of commands

```
> data(mtcars)
> carData <- mtcars
> mtcars -> carData
```

loads an R data set named mtcars into the workspace and assigns it (twice) to the data frame carData.

There are certain system names that should be avoided when naming variables or constants. These include c, q, t, T and F. The q function terminates an R session, t transposes a matrix and T and F are abbreviated versions of the R logical constants TRUE and FALSE that provide the values for a Boolean variable. The c function is of particular importance and occupies a workhorse role in the R language. The c stands for "concatenate" which is what the function does. It takes a comma separated list of arguments and combines them into a vector. For example,

```
> a <- c(1, 3, pi)
> a
[1] 1.000000 3.000000 3.141593
> class(a)
[1] "numeric"
```

produces a vector object of the R class or data type "numeric". Note that all that was necessary for R to print out the object was simply to enter its name on the command line. This approach works quite generally for matrices and other R objects and produces the same results as would be obtained using the R show function: i.e., show(a) is equivalent to just entering a on the command line. The unix-like functions head and tail are available for matrices and have the effect of showing the first or last few rows of the matrix, respectively.

There is a point that is worth mentioning concerning the R logical constants. The T and F options corresponds to global variables whose initial values are set to TRUE and FALSE. However, T and F can be assigned other values (either intentionally or otherwise) that can cause errors and confusion. In contrast, TRUE and FALSE are reserved words whose values cannot be changed. Thus, it is safest to use the latter two choices when writing R code.

Another basic function that recurs frequently is the colon operator : that is a special case of the function seq() which has several possible arguments. For example, the commands 1:n and seq(1,n) will have the same effect of producing the vector of integers from 1 to n. More generally, seq allows for expressions such as seq(1, n, by = r) which selects every rth integer between 1 and n or seq(1, n, length = r) which would list the r partition points that are needed to break the interval 1 to n into $r - 1$ subintervals of equal length. There is nothing special about the use of integers here and, for example, a command such as seq(pi/2, -10*pi, length = 7) will have the intended effect of creating a seven component partition for the interval $[-10\pi, \pi/2]$ that breaks it into six subintervals.

B.1 R as a calculator

R is typically used in an interactive mode through command line input. When employed in this manner it is possible realize an effect much like an advanced scientific calculator through creation and execution of expressions involving the operators and functions that are available in the language. In this section we will examine R from this "calculator" perspective.

First, R can process algebraic command line input involving the operators

- `-` for binary minus,
- `+` for binary plus,
- `*` for binary multiplication,
- `/` for binary division,
- `^` for exponentiation,
- `%%` for modulus,
- `%/%` for binary integer division,
- `abs` for absolute value,
- `floor` for the next smallest integer and
- `ceiling` for the next largest integer.

Expressions can also employ standard mathematical function such as

- `sqrt` for the square root function,
- `cos` for the cosine function,
- `sin` for the sine function,
- `tan` for the tangent function,
- `exp` for the exponential function,
- `log`, `log10` and `log2` for the logarithm for base e, base 10 and base 2 and
- `gamma` for the gamma function.

This makes it easy to perform insightful calculations such as

```
> a <- 2*pi
> (log(exp(cos(a))) + sin(log10(10^{pi})))/cos(0)
[1] 1
```

Most of the basic mathematical operations and functions have been vectorized in the sense of working with matrix and vector arguments. In such cases, the function is applied to an array on an element-wise basis. The code snippet below demonstrates this by creating a vector of four integers and computing their values *mod* 7.

```
> x <- vector(mode = "integer", length = 4)
> x[1] <- 17
> x[2] <- -43
> x[3] <- 11
> x[4] <- 107
> x%%7
[1] 3 6 4 2
```

The example illustrates one of the most powerful features of R: its facility for subscripting arrays and lists. Subscripting is accomplished with [] for arrays while both [] and [[]] work with lists. The difference between the two in this latter instance is that [] will return a sublist while [[]] returns the actual object being held in the list. As seen from the example, when used with a one-dimensional array an integer argument to [] returns the array's element whose index agrees with the argument.

A vector in R can be converted to a matrix (or, more generally, to a multi-dimensional array) by specifying its dim attribute. In the case of our previous integer vector this produces

```
> dim(x) <- c(2, 2)
> x*x
       [,1]   [,2]
[1,]   289    121
[2,] 1849 11449
> (x*x)[1, 1]%%7
[1] 2
> (x*x)[, 2]%%7
[1] 2 4
```

First, the vector is converted to a 2×2 matrix by setting both the row and column arguments for dim(x) to two. The vectorization feature of the multiplication operator * is then illustrated with this matrix. Next, subsetting operations are used on the matrix in conjunction with the modulus and multiplication operators. In general, [] takes two comma separated arguments for working with a matrix that correspond to a row and column index, respectively. The (x*x)[1, 1] syntax, for example, gave us the top left-hand element of the x*x matrix. Either the column or row arguments for [] may be left unspecified in which case all the row or column entries will be returned for the subscript that is specified. The last expression using (x*x)[, 2] illustrated this feature.

Subsetting operations are not restricted to positive integers for arguments. Negative integers can be used to delete rows or columns while logical values can be used to pick out array elements that meet some specified criteria. To demonstrate this feature we again employ the matrix

```
> x
      [,1]  [,2]
[1,]    17    11
[2,]   -43   107
```

to produce the results

```
> x[-1,]
[1] -43 107
> (x > 0)
        [,1]  [,2]
[1,]    TRUE  TRUE
[2,]   FALSE  TRUE
> x[(x > 0)]
[1]    17    11   107
> x[(x[, 1] > 0), 2]
[1] 11
```

The use of -1 indicates that the row with index 1 is to be deleted. The result is a "matrix" consisting of only the second row of x. The next command reveals that a logical expression applied to a matrix returns a matrix of the same dimension with entries that are either TRUE or FALSE depending on whether or not they satisfy the specified condition. In this particular case, the condition is that an element be positive and the resulting matrix of logical values has a FALSE for the $(2, 1)$ element with the value -43. The utility of this result is realized when we use the logical array as an index argument for the original matrix to produce a subset of the matrix elements corresponding to those entries that evaluate to TRUE. Similar operations can be performed with specific rows or columns of a matrix. In the example, (x[,

1] > 0) evaluates as TRUE for the positive entry in the first column with the consequence that the first row is selected. The specification of 2 for the column index therefore returns the (1, 2) matrix element. A thorough discussion of R subscripting is provided by Spector (2008).

The + and − operators applied to matrices give matrix addition and subtraction assuming the matrices involved have the same row and column dimensions. Matrix transposition is accomplished with the t function while matrix multiplication is obtained through the operator %*%. To illustrate the idea we first create two matrix objects using

```
> set.seed(123)
> A <- matrix(runif(15), 3, 5)
> B <- matrix(runif(15), 3, 5)
> A
          [,1]      [,2]      [,3]      [,4]      [,5]
[1,] 0.2875775 0.8830174 0.5281055 0.4566147 0.6775706
[2,] 0.7883051 0.9404673 0.8924190 0.9568333 0.5726334
[3,] 0.4089769 0.0455565 0.5514350 0.4533342 0.1029247
> B
           [,1]      [,2]      [,3]      [,4]      [,5]
[1,] 0.89982497 0.3279207 0.6928034 0.6557058 0.5941420
[2,] 0.24608773 0.9545036 0.6405068 0.7085305 0.2891597
[3,] 0.04205953 0.8895393 0.9942698 0.5440660 0.1471136
```

Here the R uniform random number generator runif was used to create two 3×5 matrices A and B. At the outset the seed for the random number generator has been set to 123 using the set.seed function so that our calculations can be repeated from any incarnation of R provided the same seed and default random number generator (i.e., Mersenne twister) are used. Now R can be used to evaluate a matrix expression such as

```
> (cos(A) + B)%*%t(3*(B^2) - 2*A)
           [,1]       [,2]      [,3]
[1,]  2.83191033 -4.496181 3.364332
[2,]  0.17064156 -2.247529 4.930415
[3,] -0.03323455 -2.361368 6.782212
```

When applied to an array the max and min functions will return the largest and smallest array element. In the case of our two matrices this produces

```
> max(A); max(B)
[1] 0.9568333
[1] 0.9942698
> min(A); min(B)
[1] 0.0455565
[1] 0.04205953
```

Note the use of semicolon separation here to allow more than one command to be entered from a single prompt.

Both the seq and c functions can be used to specify array row or column indices. As demonstrated above with the > operator, logical operators can also be used in this capacity. The R logical operators include

   !   for unary not,

   <, <= and >=   for logical binary less than, etc.,

   ==   for logical binary equal to,

   &   for logical binary AND and

   | for logical binary OR.

The `rbind` and `cbind` functions are particularly useful for working with arrays. These functions combine the rows or columns, respectively, of the arrays that are provided as their arguments. This allows us to accomplish manipulations such as

```
> cbind(A[,1], B[,1], A[,c(2, 4)])
          [,1]       [,2]       [,3]       [,4]
[1,] 0.2875775 0.89982497 0.8830174 0.4566147
[2,] 0.7883051 0.24608773 0.9404673 0.9568333
[3,] 0.4089769 0.04205953 0.0455565 0.4533342
> C <- rbind(A[1:2,], B[3,])
> C
           [,1]      [,2]      [,3]      [,4]      [,5]
[1,] 0.28757752 0.8830174 0.5281055 0.4566147 0.6775706
[2,] 0.78830514 0.9404673 0.8924190 0.9568333 0.5726334
[3,] 0.04205953 0.8895393 0.9942698 0.5440660 0.1471136
```

Like any good scientific calculator, R has a `sum` function for adding up a group of numbers. But, it is somewhat more powerful than its hand-held relative in that it can carry out calculations like

```
> sum(seq(-pi, pi, by = .5))
[1] -1.840704
> sum(A)
[1] 8.545771
```

When used in conjunction with the `apply` function `sum` can produce marginal row or column sums for an array. In general the `apply` function has the form

```
    apply(X, MARGIN, FUN)
```

where X is an array, FUN is the function to be used and MARGIN is 1 or 2 depending on whether the function should be applied to the rows or the columns of the array. Using this idea with our previous matrices produces results like

```
> apply(A, 1, sum)
[1] 2.832886 4.150658 1.562227
> apply(B, 2, sum)
[1] 1.187972 2.171964 2.327580 1.908302 1.030415
```

There is nothing special about `sum` in this example in that other choices for the FUN argument to `apply` will work equally well. For example,

```
> apply(A, 1, min)
[1] 0.2875775 0.5726334 0.0455565
> apply(B, 2, max)
[1] 0.8998250 0.9545036 0.9942698 0.7085305 0.5941420
```

Other calculator-like features of R are the `choose` and `factorial` functions that calculate the binomial coefficients and factorials. The `choose` function has two arguments for the number of trials and number of successes in a binomial experiment while `factorial` takes one integer argument. An illustration of a typical calculation is

```
> choose(10,3); factorial(5)
[1] 120
[1] 120
```

Finally, R includes functions that compute the values of probability density/mass functions, cumulative probabilities, quantiles and random numbers for a rich collection of probability distributions. These functions all have names of the form dDist, pDist, qDist and rDist with Dist the name for the distribution and d, p, q and r indicating the density/mass function, cumulative distribution function, quantile function and random number generation. The choices for Dist include

beta   for the beta distribution,

binom   for the binomial distribution,

chisq   for the chi-square distribution,

exp   for the exponential distribution,

f   for the F distribution,

hypergeometric   for the hypergeometric distribution,

nbinom   for the negative binomial distribution,

norm   for the normal distribution,

pois   for the Poisson distribution,

t   for the Student's t distribution and

unif   for the uniform distribution.

The random number generator for the uniform distribution was used earlier to create the matrices A and B that were used in the examples. The commands below illustrate the use of functions for some of the other distributions.

```
> qchisq(.95, df = 9)
[1] 16.91898
> pchisq(qchisq(.95, df = 9), df = 9)
[1] 0.95
> set.seed(123)
> rpois(3, lambda = 5)
[1] 4 7 4
> dbinom(4, size = 4, p = .25)
[1] 0.00390625
> qnorm(.9, mean = 0., sd = 1.); qnorm(.9); qnorm(.95); qnorm(.975)
[1] 1.281552
[1] 1.281552
[1] 1.644854
[1] 1.959964
```

Notice from this that if a choice is not specified for the mean and standard deviation of the normal distribution the dnorm, pnorm, qnorm and rnorm functions will all provide values for the standard normal distribution.

B.2 R as a graphics engine

Undoubtedly one of the most useful features of the R package is the collection of functions it provides for generating graphics. The plot function is a key component of the available graphic utilities. It has the prototype

```
    plot(x, y, ... )
```

with x and y the variables for the x and y axes and the ellipsis ... indicating further possible arguments that can be used to add enhancements to a figure. When used in its simplest form commands like

```
set.seed(123)
x <- 1:100/101
y <- x + runif(100, -.25, .25)
plot(x, y)
```

produce a plot such as Figure B.1. Here random numbers have been generated from a uniform distribution on $[-.25, .25]$ then added to the value of a line with unit slope that was evaluated at uniformly spaced points over the unit interval. The plot is satisfactory but lacks standard features such as a title.

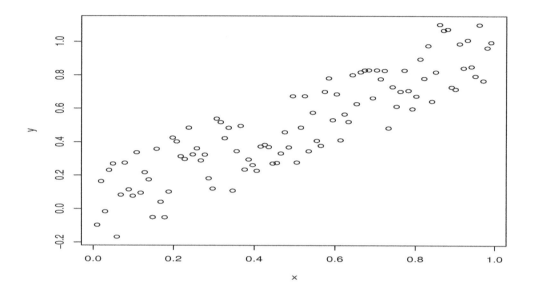

Figure B.1 *Uniform random numbers added to a line with unit slope*

The ellipsis argument for `plot` is an R standard that indicates there are other arguments that may be supplied depending on the classes for the objects x and y. For our purposes x and y can be assumed to be numeric arrays in which case the additional arguments of most interest are

> `type` for the type of plot with `l` and `p` producing lines and points (with points being the default option),

> `lty`, `lwd` for values that specify the line type and width,

> `pch` for an integer or specified symbol for plots using points,

> `main` for a quoted phrase to be used for the plot title and

> `xlab`, `ylab` for text, in quotes, to be placed on the horizontal and vertical axes.

There are six line types specified by the integers from 1 to 6. A solid line corresponding to `lty = 1` is the default. It is also possible to create custom line types as described in Murrell (2006). There are twenty-six symbol options with the selection of a specific symbol being accomplished by setting `pch` to one of the integers between 0 and 25. Open circles, for which `pch = 1`, are the default symbol. Any character can also be set as the symbol; for example `pch = "A"` will produces a plot with points that are represented by A.

Once a plot has been drawn, additional points or lines can be added using the `points` and `lines` functions. A legend can also be added to the plot using the `legend` function.

Figure B.2 was produced by Listing B.1 below. It illustrates a number of R's graphical capabilities that will be the focus of our discussion in the remainder of this section.

Listing B.1 *figEx.r*

```
parOld <- par()
#pdf("graphEx.pdf")
set.seed(123)
x <- (-100):100/101
y <- x + rnorm(201, sd = .25)    .
parOld <- par()
par(mfrow = c(2, 2))
hist(y, main = "Histogram of y data", xlab = "y",
        ylab = "frequency")
boxplot(y, main = "Boxplot of y data", xlab = "", ylab = "y")
plot(x, y, main = "Scatter plot", xlab = "x", ylab = "y", pch = 4)
lines(x, x, lty = 6, lwd = 2)
legend(.15, -.5, legend=c("y data", "mean"),
        pch = c(4, -1), lty = c(-1, 6), lwd = 2)
qqnorm(rstandard(lm(y ~ x)), xlab = expression(Q[N](u)),
        ylab = expression(tilde(Q)(u)))
lines((x = seq(-3, 3, by = .1)), x)
#dev.off()
par(parOld)
```

First, Listing B.1 creates a multiple figure plot using the graphics parameter `mfrow`. Specifically, the command `par(mfrow = c(2, 2))` states that we want a 2 × 2 array of figures and that the rows of the array are to be filled in first as new graphs are inserted. A similar parameter `mfcol` fills in the array by columns. A list of the R graphical parameters can be obtained by entering `par()` on the command line. By using `par` in the way it was employed in Listing B.1, the `mfrow` parameter is set globally and will remain that way unless it is explicitly changed again. In particular, `par(mfrow = c(1, 1))` has the effect of changing `mfrow` back to its original value.

In general several global parameter changes may be made in creating a figure. A simple way to return to the original settings is to save the current parameter values (e.g., as `parOld`) and then reset them using the `par` function (e.g., with `par(parOld)`) after a plotting task has been completed. That is what was done in Listing B.1.

The data that is used in Figure B.2 is created by generating a response vector y from a regression model with unit slope and zero intercept. The random error part of the data is generated from a normal distribution with mean zero and standard deviation .25. The independent variable (or x) values are uniformly spaced over the unit interval. The upper left-hand plot in Figure B.2 is a histogram of the y vector values that was created using the R function `hist`. The second plot on the first row is a boxplot for the same data created using the function `boxplot`. Appropriate titles and axis labels have been added to both figures using the `main`, `xlab` and `ylab` arguments.

The first figure on the second row is a scatter plot of the x and y vectors where the selected plotting symbol (via `pch = 4`) is x rather than the default open circle. The `lines` function has been used to overlay a dashed line with twice the default width on the scatter plot to indicate the true mean for the responses. A legend has been placed on the plot using the `legend` function with its first two arguments specifying the location of the upper left-hand corner of the legend box. The `legend` argument for the `legend` function gives the text to be used to provide information about the symbols and lines that appear in the graph. There is a bit of a problem here since the line ordinates were not plotted with symbols and the responses were not connected by lines. Nonetheless, R expects both a plot symbol and line type to be specified for each component of the value that is assigned to the `legend` argument. The problem is easily resolved by providing values (e.g., −1) that do not

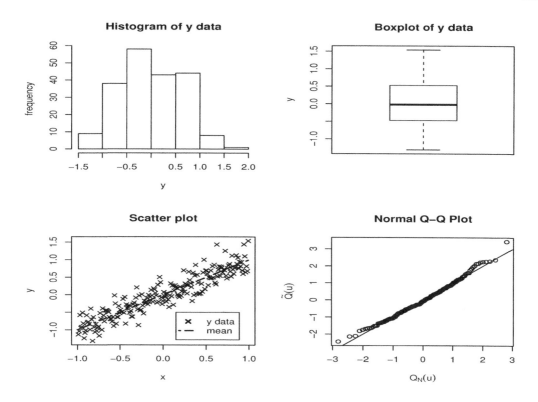

Figure B.2 *A figure array*

correspond to valid symbols or line types. Note that the line width must also be given if that is to be a feature of the legend.

The final figure is a normal distribution Q-Q plot of the standardized residuals obtained from fitting a linear model to our simulated data using the R `lm` function that will be described more fully in the next section. A line with unit slope is overlaid on the Q-Q plot using the `lines` function. The labels for both axes (Q_N for normal quantiles and $\tilde{Q}$ for the sample quantiles) are created using a LaTeX type feature of R that allows mathematical symbols and nomenclature to be included in axis labels, legends and titles. The R help page for `plotmath` describes this feature.

Listing B.1 contains two statements that have been rendered inactive by prefacing them by the R comment symbol `#`. If the comment markers are removed the result would be that all the graphics output would be written to a pdf file (called graphEx.pdf in this case). In this instance, the connection to this graphics "device" will remain open and all graphics output will be funneled to the pdf file until the device is closed using `dev.off()`.

It is possible to have multiple graphics devices available for use. The list of active devices is returned from `dev.list()` and switching between devices is accomplished with `dev.set`. To illustrate the idea consider the following commands from an R session.

```
> system("ls *.pdf")
ls: *.pdf: No such file or directory
> x <- 1:100
> plot(x, x)
> dev.list()
quartz
     2
```

```
> pdf("some.pdf")
> dev.list()
quartz     pdf
   2        3
> plot(x, x^2)
> dev.set(which = 2)
quartz
   2
> plot(x, x^2)
> dev.off(3)
quartz
   2
> system("ls *.pdf")
some.pdf
```

An initial check of the working directory shows that no pdf files are present. A plot is then created which initializes the X11 terminal. The output from `dev.list()` indicates that only this device (called quartz in this instance and assigned the number 2) is available for plotting. A new device is opened for creating a pdf file at which point there are two devices and the output from `plot` is sent to the new graphics device numbered 3 that writes it to the file some.pdf. A switch is made back to the X11 device and the same graphics output is sent to the X11 terminal where it now becomes visible. The pdf device is then closed and we are informed that the X11 device is the only one that remains active. A check reveals that the file some.pdf file has been created in the current working directory.

B.3 R for statistical analysis

As the previous two sections have hopefully demonstrated, R is much more than just a statistics software package. Of course, it does come equipped with a number of functions for carrying out many of the standard statistical calculations that arise in the analysis of data. Many, many more are available through packages that can be downloaded from the `CRAN` website.

There is a generic R function `summary` that produces some of the standard summary statistics when it is applied to an array or a data frame. The ensuing output was produced using one of the data sets that comes with R.

```
> data(mtcars)
> A <- as.matrix(mtcars[, c(1, 3, 6)])
> summary(A)
      mpg              disp             wt
 Min.   :10.40    Min.   : 71.1    Min.   :1.513
 1st Qu.:15.43    1st Qu.:120.8    1st Qu.:2.581
 Median :19.20    Median :196.3    Median :3.325
 Mean   :20.09    Mean   :230.7    Mean   :3.217
 3rd Qu.:22.80    3rd Qu.:326.0    3rd Qu.:3.610
 Max.   :33.90    Max.   :472.0    Max.   :5.424
> apply(A, 2, sd)
      mpg         disp          wt
 6.0269481  123.9386938   0.9784574
> apply(A, 2, IQR)
      mpg        disp         wt
 7.37500   205.17500    1.02875
```

The data set `mtcars` in this example contains information on several variables related to 32 cars that was reported in a 1974 issue of *Motor Trend* magazine. We have elected to work with three of the variables: miles per gallon (mpg), displacement (disp) and weight (wt) and have coerced them into the form of a matrix with the `as.matrix` function. An application of the `summary` function to the resulting array produces typical summary statistics corresponding to each column in the array. The standard deviations and inter-quartile ranges were also computed for each column using the `sd` and `IQR` functions in combination with the `apply` function. The sample variances could have been obtained similarly by using `var` instead of `sd`.

The values for the mean, median and quartiles can be obtained directly using the `mean`, `median` and `quantile` functions as illustrated by

```
> mean(A[, 1])
[1] 20.09062
> median(A[, 1])
[1] 19.2
> quantile(A[, 1], prob = c(.25, .75))
  25%    75%
15.425 22.800
> min(A[, 1]); max(A[, 1])
[1] 10.4
[1] 33.9
```

The `probs` argument for `quantile` designates the percentile or percentiles that are to be computed. The choice of `probs = c(.25, .75)` indicates that the 25th and 75th percentiles or, equivalently, the two quartiles are to be calculated. The `max` and `min` functions were also used to obtain the smallest and largest values of the mpg variable.

It is not uncommon for data to have missing values which appear in R as the logical constant `NA`. The default behavior of most R data analysis functions when they encounter an `NA` is to either return `NA` as output, report an error or omit the missing value and perform the requested analysis. The default behavior can often be overridden by specifying a value for a function argument with a name such as `na.rm` or `na.action`. In particular, the `na.rm` (for "NA remove") argument is what appears in the `mean` function as illustrated by the code snippet

```
> dat <- c(pi, NA, exp(pi), 1:5)
> mean(dat)
[1] NA
> mean(dat, na.rm = TRUE)
[1] 5.897469
```

By setting `na.rm` to `TRUE` a mean value for the nonmissing values in the `dat` array is returned by the `mean` function.

The `order()` function can be used for sorting elements in an array. This function takes a vector as input and returns a vector of integers representing the ranks of the corresponding input vector elements. More precisely, the ith component of the vector r returned from `r <- order(x)` is the rank of `x[i]` among the values `x[1]`,..., `x[length(x)]` with `length` the function that returns the number of elements in an array. Consequently, `x[order(x)]` is the vector x with elements rearranged in numerically ascending order. In this case the same effect could be accomplished with R's sorting function `sort`: i.e., `x[order(x)]` is the same as `sort(x)`. The `order` function works more generally in the sense that for a matrix A the expression `A[order(A[,i]),]` produces a matrix with rows rearranged so that the elements in its ith column are ordered.

R contains functions for performing most of the basic types of analyses that arise in statistics. For example, the `lm` function that was used in the previous section is available for fitting linear models. The primary argument for `lm` is a `formula` object that defines a relationship between dependent and independent variables. The variables are separated

by the ~ operator with the dependent variables on the left-hand side and the independent
variables on the right-hand side separated by +s. Suppose there are three arrays y, x1,
x2 of equal length and the formula log(y) ~ x1 + x1^2 + x1*x2 - 1 is used in the lm
function. This would specify that a linear model was to be fitted with data in a vector
y providing the independent variable values in the form of the natural logarithm of its
elements. The model is then to be fitted with terms involving x1 as well as its square and
product with the values in the x2 array. The presence of a −1 in the formula results in a fit
with no intercept term. Otherwise, an intercept would be included by default.

The lm function returns an object of class lm that has member elements that include
coefficients, residuals and fitted. These can be accessed using the $ operator. In the
case of the mtcar example this produces

```
> mtLm <- lm(A[, 1] ~ A[, 2] + A[, "wt"])
> mtLm$coefficients
(Intercept)        A[, 2]     A[, "wt"]
34.96055404   -0.01772474   -3.35082533
```

Note that both a column index and a column name have been used to access a specific
column of the array A that contains the data. An application of the summary function to
our lm object returns an analysis of variance table as well as other relevant information.

```
> summary(mtLm)

Call:
lm(formula = A[, 1] ~ A[, 2] + A[, "wt"])

Residuals:
    Min      1Q  Median      3Q     Max
-3.4087 -2.3243 -0.7683  1.7721  6.3484

Coefficients:
             Estimate Std. Error t value Pr(>|t|)
(Intercept) 34.96055    2.16454  16.151 4.91e-16 ***
A[, 2]      -0.01773    0.00919  -1.929  0.06362 .
A[, "wt"]   -3.35082    1.16413  -2.878  0.00743 **
---
Signif. codes:  0 '***' 0.001 '**' 0.01 '*' 0.05 '.' 0.1 ' ' 1

Residual standard error: 2.917 on 29 degrees of freedom
Multiple R-squared: 0.7809,     Adjusted R-squared: 0.7658
F-statistic: 51.69 on 2 and 29 DF,  p-value: 2.744e-10
```

There are functions such as coefficients, fitted and residuals that extract the information suggested by their names from lm objects. Along similar lines the function rstandard
(discussed on the R help page for influence.measures) was used to compute the standardized residuals that appeared in the Q-Q plot portion of Figure B.2. As another example the
information below was obtained about the residuals from our fit to the data set created
using the mtcars data frame.

```
> summary(residuals(mtLm))
     Min.    1st Qu.     Median       Mean    3rd Qu.       Max.
-3.409e+00 -2.324e+00 -7.683e-01  5.204e-17  1.772e+00  6.348e+00
```

Finally, it is good to know that R comes equipped with a function sample that can be
used for drawing random samples from an array that is specified as its first argument. The
prototype for sample appears as

```
 sample(x, size, replace = FALSE, prob = NULL)
```

The x argument is the array to be sampled while size is the number of elements to select. If replace is set to TRUE, sampling will be done with replacement and is carried out without replacement otherwise. The prob argument can be a vector of probabilities for selecting the elements of the x array. The default value for prob is the R reserved word NULL that represents the null or empty object. If no value is specified for prob the selection is made with uniform probability. To demonstrate the idea consider

```
> set.seed(123)
> sample(letters, 4)
[1] "h" "t" "j" "u"
> sample(LETTERS, 10, replace = TRUE)
 [1] "Y" "B" "N" "X" "O" "L" "Y" "L" "R" "O"
> sample(c(1, 2, 3), 10, replace = TRUE)
 [1] 1 3 1 1 1 3 3 3 2 3
> sample(c(1, 2, 3), 10, replace = TRUE, prob = c(.1, .3, .6))
 [1] 2 2 3 3 3 3 1 1 2 2
```

This example used the built-in R arrays letters and LETTERS that contain the lower- and upper-case letters of the alphabet.

APPENDIX C

C++ library extensions (TR1)

The C++ Technical Report 1 (Austern 2005), commonly referred to as TR1, is a draft document officially titled *ISO/IEC TR 19768, C++ Library Extensions*. While it has as yet to be adopted as the standard, it is likely that most of its features will be included in the next standard version of C++. Despite its unofficial status, several compilers, including gcc, currently implement most or all of the TR1 components. To distinguish the proposed components from the standard library, they have been placed in the namespace `std::tr1` that will appear in the programs in this appendix.

The text by Becker (2006) provides an accessible treatment of all the features of TR1. In this appendix we merely focus on those aspects that are most relevant to the material in Chapters 4, 9 and 10.

C.1 Pseudo-random numbers

One of the proposed extensions to the C++ standard library is the implementation of several template classes for the generation of pseudo-random numbers. At a high level, there are two kinds of classes: *engines* and *distributions*. An engine basically describes a random number generator (see Chapter 4) that produces pseudo-random integers that are intended to be uniformly distributed over the length of its period. They come in two varieties: a basic engine provides an implementation of a specific generator, while a compound engine either modifies a basic engine, or combines two of them into a single generator. A distribution transforms a sequence (or stream) of (uniformly distributed) integers into floating-point values that obey a particular distribution. It is also possible to explicitly combine an engine and a distribution in an object of type `variate_generator`. An example of syntax for this latter purpose looks like

```
variate_generator<engineName, distributionName<> >
        rngName(engineName(seed), distributionName<>(parameterValues))
```

Here `engineName` specifies the choice for the engine/generator while `distributionName` is the distribution to be simulated. The simplest approach is to use one of the predefined engines that include the "minimal standard" congruential generator of Park and Miller (1988) (`minstd_rand0`), the Mersenne twister (`mt19937`) and several subtract-with-carry generators (e.g., `ranlux3`). More generally, there are engine template classes that provide ways to specify custom congruential and subtract-with-carry generators as well as variants of the Mersenne twister. The distributions are also template classes as indicated by the `<>` after the class name. The default values for the template parameters are generally satisfactory which allows us to bypass their explicit specification.

Some of the options that are available for the distributions are described below.

- An object `dist` from the class `uniform_real` of uniform distributions on the interval a to b (with a and b both of type `double`) is constructed with

```
uniform_real<> dist(a, b);
```

The values of a and b default to 0 and 1, respectively.

- An object `dist` from the `normal_distribution` class with mean `mu` and standard deviation `sigma` (both parameters being of type `double`) is created with

  ```
  normal_distribution<> dist normal(mu, sigma)
  ```

  The default is the standard normal distribution.

- The `binomial_distribution` class has parameters `n` and `p` with `n` of type `int` and `p` of type `double`. The syntax

  ```
  binomial_distribution<> dist(n, p);
  ```

  will create an instance `dist` of the class. The parameters `n` and `p` default to 1 and .5, respectively.

- The `gamma_distribution` class (e.g., Section 4.6) is parameterized with a single shape parameter `alpha` of type `double` that defaults to 1. An instance of the class is obtained via

  ```
  gamma_distribution<> dist(alpha);
  ```

  To produce values for a gamma distribution with mean `alpha*beta`, multiply values from this generator by `beta`.

- An instance of the `poisson_distribution` class with double precision mean `lambda` is produced with

  ```
  poisson_distribution dist(lambda);
  ```

The following program illustrates the use of `variate_generator` in combination with several of the distributions mentioned above.

```cpp
//tr1RNG.cpp
#include <tr1/random>
#include <iostream>

using namespace std::tr1;

int main() {
  variate_generator<mt19937, uniform_real<> >
    rngU(mt19937(123), uniform_real<>());
  variate_generator<mt19937, normal_distribution<> >
    rngN(mt19937(123), normal_distribution<>(2, .5));
  variate_generator<mt19937, binomial_distribution<> >
    rngB(mt19937(123), binomial_distribution<>(3, .7));
  variate_generator<mt19937, poisson_distribution<> >
    rngP(mt19937(123), poisson_distribution<>(20));
  variate_generator<mt19937, gamma_distribution<> >
    rngG(mt19937(123), gamma_distribution<>(3));

  for(int i = 0; i < 5; ++i)
    std::cout << rngU() << " " << rngN() << " " <<
      rngB() << " " << rngP() << " " << 2*rngG() << std::endl;

  return 0;
}
```

First, we need to include the header file `tr1/random` that allows us to access the distributions and generators. Objects are then created to produce random numbers from the uniform (on $[0, 1]$), normal (with mean two and standard deviation .5), binomial (with 3 trials and success probability .7), Poisson (with mean 20) and gamma (with shape parameter 3) distributions all using the seed 123 and the Mersenne twister engine. Each object is used to produce a sequence of five random deviates. The resulting output is

```
0.696469  2.5429  1 24  9.01262
0.712955  2.50087 2 22  5.27469
0.286139  1.71696 2 23  8.59719
0.428471  1.15374 1 27  9.13627
0.226851  2.36488 2 10  5.90542
```

Note that the multiplication of numbers produced by the gamma generator by 2 has the effect of producing random deviates from the chi-square distribution with six degrees-of-freedom.

C.2 Hash tables

TR1 adds four new container classes:

- unordered_set,
- unordered_multiset,
- unordered_map and
- unordered_multimap.

These containers are direct analogs of the set, multiset, map and multimap containers from Section 10.6 that are implemented using hash tables rather than binary search trees.

The listing below gives a simple example of using unordered_map.

```cpp
//tr1hashEx1.cpp
#include <utility>
#include<iostream>
#include <tr1/unordered_map>

using namespace std::tr1; using std::pair;

int main() {

  unordered_map<int, int> ht;

  ht.insert(pair<int,int> (254,10));
  ht.insert(pair<int,int> (54,11));
  ht.insert(pair<int,int> (54,22));
  for(unordered_map<int, int>::iterator iter = ht.begin();
        iter != ht.end(); iter++)
    std::cout << "(" << (*iter).first << "," << (*iter).second << ")"
              << " has count " << ht.count((*iter).first)
              << std::endl;

    std::cout << "Data value for key 54 = " << (ht.find(54))->second
              << std::endl;

  return 0;
}
```

The unordered_map container becomes available through the tr1/unordered_map header file. All four hash table variants require at least two template parameters and up to as many as five, unless default values are used. The two parameters that are always required are the types for the key and data components of each object. Additional parameters may be used to specify the hash function itself (if the default is not appropriate or does not work with the key type) and a comparison function for keys (to allow checking if a key is already in the table). Thus, in this case, we have created an `unordered_map` container `ht` that uses integers as both key and data components. The `insert` method is used to attempt to add three `pair` objects to the table. The last insert involves a key that has already been used in the table. Then, an iterator is employed to explore the container while printing out the key and data values for its stored elements. Finally, the `find` method is used to locate the table entry with the key of 54. This method returns an iterator pointing to the table entry which explains the use of `->` to access the `second` member of the associated `pair` object.

The output produced by our program is

```
(254,10) has count 1
(54,11) has count 1
Data value for key 54 = 11
```

In addition to the key and data components, we have also printed the number of times each of the keys appear in the table. For the simple `unordered_map`, the only possible return values are 0 (if the key does not exist in the table) and 1 (if the key does exist in the table). In the next example we will see that the `unordered_multimap` container allows multiple elements with the same key to be inserted so that the `count` method may return arbitrary nonnegative integers. In contrast, the output from this example demonstrates that attempts to insert an entry with a duplicate key in a `unordered_map` structure are ignored and the first object entered with a given key is the one that is retained.

The result from the previous example can be contrasted with the output from the listing below where the hash table is an `unordered_multimap` object.

```cpp
//tr1hashEx2.cpp
#include <utility>
#include<iostream>
#include <tr1/unordered_map>

using namespace std::tr1; using std::pair;

int main() {
  unordered_multimap<int, int> ht;

  ht.insert(pair<int,int> (254,10)); ht.insert(pair<int,int> (54,11));
  ht.insert(pair<int,int> (54,22));

  for(unordered_multimap<int, int>::iterator iter = ht.begin();
      iter != ht.end(); iter++)
     std::cout << "(" << (*iter).first << "," << (*iter).second
            << ")" <<  " has count " << ht.count((*iter).first)
            << std::endl;

  pair<unordered_multimap<int, int>::iterator,
      unordered_multimap<int, int>::iterator>
  this_range = ht.equal_range(54);

  for(unordered_multimap<int, int>::iterator iter = this_range.first;
      iter != this_range.second; iter++) {
    std::cout << iter->second << " ";
  }
  std::cout << std::endl;
  return 0;
}
```

The program's output is

```
(254,10) has count 1
(54,11) has count 2
(54,22) has count 2
11 22
```

Both elements with key values of 54 have been inserted into the table since each has a `count` value of 2. To access them, the method `equal_range` can be used to get a pair of iterators pointing to the beginning and the end, respectively, of the part of the container that stores all the elements with the given key value.

Let us now consider a slightly more complicated example that demonstrates the usage of a hash table keyed on strings.

```cpp
//tr1hashEx3.cpp
#include <utility>
#include <iostream>
#include <string>
#include <tr1/unordered_map>

using namespace std::tr1;

int main() {

  unordered_map<std::string, int> ht;

  std::cout << "Empty? " << std::boolalpha << ht.empty() << std::endl;
  ht.insert(std::pair<std::string, int> ("example",10));
  ht.insert(std::pair<std::string, int> ("missing",9));
  ht.insert(std::pair<std::string, int> ("guess",15));
  ht["value"]=16;

  std::cout << "Empty? " << ht.empty() << std::endl;
  std::cout << "Size: " << ht.size() << std::endl;

  for(unordered_map<std::string, int>::iterator iter = ht.begin();
      iter != ht.end(); iter++)
    std::cout << iter->first << " " << ht[iter->first] << " "
              << ht.count(iter->first) << std::endl;

  ht.erase("value");
  std::cout << "value: " << ht.count("value") << std::endl;

  return 0;
}
```

First there is an initial query to test the emptiness of the hash table after which three (`string`, `int`) `pair` objects are inserted into the table. This is followed by an illustration of a simpler way to insert an element via the magic of C++ operator overloading. In this case, the subscripting operator [] has been overloaded in the TR1 implementation to act as an application of the `insert` method and thereby provide us with an intuitive way of initializing a new entry. This feature is also used in the `for` loop in a slightly different context to access the data component for a given key value. Other features demonstrated in this example are the use of the `size` method that returns the number of entries in an associative array and the `erase` method that will delete an entry with a specified key.

The program produces the output

```
Empty? true
Empty? false
Size: 4
example 10 1
value 16 1
missing 9 1
guess 15 1
value: 0
```

Initially the container is empty and then evaluates as not empty with four elements after the four insertions. The iteration through the container using the key values as indices produces the expected results and the application of the **erase** method has removed the entry with the targeted key.

To this point, we have worked only with fairly simple cases and been satisfied with whatever the hash table containers give us in terms of the number of buckets. More customization is possible by using the nondefault class constructors and by setting the hash table load factor (see Section 9.2.4) that represents the average number of objects per bucket.

A simplified prototype for the **unordered_map** class constructor looks like

```
unordered_map(size_type nBuckets, const Hash& hashFunctor())
```

Here nBuckets is the desired initial number of buckets and Hash is a template functor struct. The size_type and size_t designation that appear in this listing and the one below correspond to unsigned integers. The Hash functor struct has the generic framework

```
namespace std {
  namespace tr1 {
      template<>
      struct hash<keyType> : public unary_function<keyType, size_t>
      {
        std::size_t operator ()(const keyType& key) const
        {
          //hash function specification
        }
      };
  }
}
```

with keyType the class for the keys. A specific hash function must be supplied in the body of the () method. This can be constructed directly or by using specializations of the template such as hash<bool>, hash<char>, hash<int>, hash<double> and hash<std::string>. The hash<int> and hash<std::string> functors were used automatically in the previous examples when we specified int and string as the key classes for our unordered_map and unordered_multimap objects.

Merely specifying the number of buckets in the constructor will not insure that only that many buckets are used. The hash table containers grow dynamically in a way that is determined by the specified (or default) values for the number of buckets and the table's load factor. A value for the load factor can be set by the max_load_factor method as will be demonstrated below.

To illustrate some of the additional features of the TR1 hash table classes, let us consider a case where objects are stored according to keys from the struct

```
struct Key : public pair<std::string, int> {
  Key(std::string str, int id) : pair<std::string, int>(str, id) {}

  bool operator == (const Key&a){
```

```
      return this->first == a.first && this->second == a.second;
  }
};

std::ostream& operator<<(std::ostream& out, const Key& key) {
  out << "(" <<  key.first << "," << key.second <<")";
  return out;
}
```

This Key class is a simple extension of the pair class that provides an overloaded comparison operator. The == operator is needed for comparisons of objects during look-up operations in a table. An overloaded output insertion operator has also been constructed that will work on Key objects.

A key class such as Key requires an explicit specification of the Hash functor. In this regard, we have opted to use

```
namespace std {
  namespace tr1 {
      template<>
      struct hash<Key> : public unary_function<Key, size_t>
      {
        std::size_t operator ()(const Key& key) const
        {
          //functors for the string and int hash functions
          hash<std::string> stringHash; hash<int> intHash;
          size_t hashValue = (stringHash(key.first)
                            + intHash(key.second));
          return hashValue;
        }
      };
  }
}
```

Apart from the specific code for the hash function, this struct is obtained by merely plugging in the proper value for the keyType in the Hash functor template given above. The hash function itself is a simple combination of two of the Hash class specializations: the one for string objects and the one for integers.

The main function from a program that makes use of the Key struct in an unordered_map container is given in the next listing.

```
int main() {

  //3 buckets and hash<key> as the hash function
  unordered_map<Key, double> ht(3, hash<Key>());
  ht.max_load_factor(5);//average of 5 objects per bucket

  Key key1("s1", 2); Key key2("s2", 3); Key key3("s3", 4);
  Key key4("s4", 5); Key key5("s5", 6); Key key6("s6", 7);
  Key key7("s7", 8); Key key8("s8", 9); Key key9("s9", 10);

  ht[key1] = 1; ht[key2] = 2*ht[key1]; ht[key3] = 3*ht[key2];
  ht[key4] = ht[key3]/4; ht[key5] = ht[key4]/5; ht[key6] = ht[key5]/6;
  ht[key7] = 7*ht[key6]; ht[key8] = 8*ht[key7]; ht[key9] = 9*ht[key8];

  std::cout << "Size: " << ht.size() << ", Number of Buckets: " <<
    ht.bucket_count() << std::endl;
  for(int i = 0; i < ht.bucket_count(); i++)
```

```
    std::cout << "Bucket " << i << " has " << ht.bucket_size(i) <<
       " elements" << std::endl;

  std::cout << "Hash values" << std::endl;
  for(unordered_map<Key, double>::const_iterator iter = ht.begin();
       iter != ht.end(); iter++)
    std::cout << iter->first << " was placed in bucket "
           << ht.bucket(iter->first) << std::endl;

  std::cout << "The elements in bucket 0 are" << std::endl;
  for(unordered_map<Key, double>::local_iterator iter = ht.begin(0);
       iter != ht.end(0); iter++)
    std::cout << iter->first << " ";
  std::cout << std::endl;

  return 0;
}
```

A container with three (initial) buckets that uses our hash<Key> functor is created with the nondefault constructor for the unordered_map<Key, double> class. The maximum load factor is then set to 5. A group of nine Key objects are created next and inserted into the table using the overloaded indexing operator. After insertion we check the number of elements and number of buckets in the table and then determine the number of (Key, double) pair objects that have been placed in each of the buckets.

An iterator is used in conjunction with a for loop to travel through the table. The Key component of each element in the table is accessed by the iterator and printed along with its bucket number to standard output. Finally, the first bucket is examined. This is accomplished using a quantity called a local_iterator (or, for const containers, a const_local_iterator is available) that can be defined for a specific bucket. In particular, each bucket has a local version of the global begin and end methods that return local_iterator objects: namely, begin(bucketIndex) and end(bucketIndex) give iterators that are specific to the bucket with index bucketIndex.

The output produced by our program is

```
Size: 9, Number of Buckets: 3
Bucket 0 has 4 elements
Bucket 1 has 2 elements
Bucket 2 has 3 elements
Hash values
(s6,7) was placed in bucket 0
(s4,5) was placed in bucket 0
(s3,4) was placed in bucket 0
(s1,2) was placed in bucket 0
(s9,10) was placed in bucket 1
(s2,3) was placed in bucket 1
(s8,9) was placed in bucket 2
(s7,8) was placed in bucket 2
(s5,6) was placed in bucket 2
The elements in bucket 0 are
(s6,7) (s4,5) (s3,4) (s1,2)
```

By setting the maximum load factor sufficiently high, we have made sure that the table would not be expanded beyond the three buckets requested from the constructor. The first (index 0) bucket is the one with the most elements and these are examined in both the global and local exploration of the container.

C.3 Tuples

In Section 9.2.4.2 we introduced the C++ `pair` struct. This is a useful data structure that is particularly well suited for ADT settings as indicated by our work in the previous section. The TR1 extension contains a `tuple` template class that serves a similar purpose except that it can hold more than two objects all of which can be from different classes. In this latter respect the `tuple` class provides a C++ analog of the `list` class from R.

The `tuple` class becomes available through the include statement

```
#include <tr1/tuple>
```

There are then several ways to construct `tuple` objects. For ease of presentation, suppose we have four classes T1, T2, T3 and T4. The same approach will, of course, work for any finite number of classes. If all our classes have default constructors,

```
tuple<T1, T2, T3, T4> tupleObj1;
```

creates a `tuple` object with four slots all of which are filled via the corresponding default constructors of their respective classes. If t1, t2, t3, t4 are instances of our four classes

```
tuple<T1, T2, T3, T4> tupleObj2(t1, t2, t3, t4);
```

will give us a `tuple` that contains these specific object. Finally,

```
tuple<T1, T2, T3, T4> tupleObj3(tupleObj2);
```

produces a copy `tupleObj3` of the object `tupleObj2`. Assignment operations are also supported and one can even assign the elements of a `tuple` object through assignment to a `pair` object. Of course, satisfactory assignment and copying will require the presence of assignment operators and copy constructors for the classes used in the `tuple` and the `tuple` objects that are involved must have the same number of slots.

Access to the slots of a `tuple` object is obtained with the `get` template function. The slot index is passed as the template parameter. Syntax such as

```
get<n>(tupleObj)
```

will return a reference to the (n − 1)st slot of the object `tupleObj` with 0-offset indexing.

The code below tests some of the features of the `tuple` class.

```
//tupleEx.cpp
#include <tr1/tuple>
#include <iostream>
#include <utility>
#include "matrix.h"
#include "vector.h"

using std::tr1::tuple; using std::pair; using std::cout;
using std::endl; using std::tr1::get;

int main() {

  double** pMat = new double*[2];
  double* pVec = new double[3];
  for(int i = 0; i < 2; i++){
    pMat[i] = new double[3];
    for(int j = 0; j < 3; j++){
      pMat[i][j] = (double)((i + 1)*(j + 1) + j);
      pVec[j] = (double)(j + 1);
    }
  }
```

```
Matrix matrixObj(2, 3, pMat);
Vector vectorObj(3, pVec);

pair<Matrix, Vector> pairObj(matrixObj, vectorObj);
tuple<Matrix, pair<Matrix, Vector>, Vector >
  tupleObj1(matrixObj, pairObj, vectorObj);

get<0>(tupleObj1).printMatrix();
get<1>(tupleObj1).second.printVec();
Matrix newMat(2, 2, 1.);
get<0>(tupleObj1) = newMat;
get<0>(tupleObj1).printMatrix();

tuple<Matrix, pair<Matrix, Vector>, Vector> tupleObj2(tupleObj1);
get<0>(tupleObj2).printMatrix();

return 0;
}
```

Objects from three classes are created at the beginning of the program: class `Matrix`, `Vector` and `pair<Matrix, Vector>`. A `tuple` object is then created with slots corresponding to the three classes using these objects. The `get` function is used next to print out the contents of the first `Matrix` slot and the `Vector` slot of the `pair<Matrix, Vector>` object. It is used again to overwrite the `Matrix` slot with an identity matrix. Finally, a copy of the first `tuple` object is made and its `Matrix` slot is printed to standard output. The results from running the program are shown below.

```
        1           3           5
        2           5           8
1
2
3
        1           0
        0           1
        1           0
        0           1
```

The Matrix and Vector classes

The "complete" versions of the classes Matrix and Vector that were used, e.g., in Chapters 3 and 7 are given in this appendix. We begin with the class Matrix header file.

Listing D.1 *matrix.h*

```
//matrix.h
#ifndef MATRIX_H
#define MATRIX_H
#include "vector.h"

class Matrix{
  //class members
  int nRows, nCols;
  void pointerCheck() const;
  void pointerCheck(int i) const;

protected:
  double** pA;

public:
  //constructors and destructor
  Matrix(int nrows = 0, int ncols = 0, double a = 0.);
  Matrix(int nrows, int ncols, const double* const* pa);
  Matrix(const Matrix& A);
  virtual ~Matrix();

  //overloaded operators
  Matrix& operator=(const Matrix& B);
  Matrix operator+(const Matrix& B) const;
  Matrix& operator+=(const Matrix& B);
  Matrix operator-(const Matrix& B) const;
  Matrix& operator-=(const Matrix& B);
  Matrix operator*(const Matrix& B) const;
  Matrix operator*(double b) const;
  Vector operator*(const Vector& v) const;
  const double* operator[](int i) const {return pA[i];}

  //matrix operations
  Matrix trans() const;

  //solution of a linear system
  virtual Vector backward(const Vector& RHS) const;
  virtual Vector forward(const Vector& RHS) const;
  Vector gauss(const Vector& RHS) const;
  virtual Vector cholesky(const Vector& RHS) const;
  //banded systems
  Vector bandBack(const Vector& RHS, int bWidth) const;
  Vector bandFor(const Vector& RHS, int bWidth) const;
  Vector bandChol(const Vector& RHS, Matrix& G, int bWidth) const;
  void QR(Matrix& Q, Matrix& R) const;

  //computation of spectra
  Vector eigen(int Nvals, Vector& v, Matrix& U, int itMax = 38,
        double delta = .00000001) const;
  Vector SVD(int Nvals, Vector& v, Matrix& U, Matrix& V,
        int itMax = 38, double delta = .00000001) const;
```

```
//utilities and accessors
void printMatrix() const;
Vector matToVec(int j) const;//jth column to a vector
void vecToMat(int j, const Vector& v);//vector into jth column
int getnRows() const {return nRows;}
int getnCols() const {return nCols;}

friend Matrix operator*(const Vector& v1, const Vector& v2);
};

class pdBand : public Matrix{
//derived class member
int bW;

public:

pdBand(int nrows = 0, int bw = -1, double a = 0.);
pdBand(int nrows, int bw, const double* const* pa);
pdBand(const pdBand& A);

virtual ~pdBand(){};

pdBand& operator=(const pdBand& B);
pdBand operator+(const pdBand& B) const;
Matrix operator+(const Matrix& B) const;


pdBand trans() const {return pdBand(*this);}

virtual Vector backward(const Vector& RHS) const;
virtual Vector forward(const Vector& RHS) const;
virtual Vector cholesky(const Vector& RHS) const;

int getBW() const {return bW;}
Matrix bandToFull() const;
};

Matrix operator*(const Vector& v1, const Vector& v2);

Matrix operator*(double, const Matrix& A);

#endif
```

The method definitions are then provided by

Listing D.2 *matrix.cpp*

```
//matrix.cpp
#include<algorithm>
#include <iostream>
#include <cmath>
#include <cstdlib>
#include <limits>
#include <new>
#include "matrix.h"

using std::cout; using std::endl;

Matrix:: Matrix(int nrows, int ncols, double a){
  nRows = nrows;
  nCols = ncols;
  //set pA to null pointer in default case
  if(nRows == 0 || nCols == 0){
    pA = 0;
    return;
  }
  pA = new(std::nothrow) double*[nRows];
  pointerCheck();
```

```
  for(int i = 0; i < nRows; i++){
    pA[i] = new(std::nothrow) double[nCols];
    pointerCheck(i);
    for(int j = 0; j < nCols; j++){
      if(i == j)
        pA[i][j] = a;
      else
        pA[i][j] = 0.;
    }
  }
}

Matrix:: Matrix(int nrows, int ncols, const double* const* pa){
  nRows = nrows;
  nCols = ncols;
  pA = new(std::nothrow) double*[nRows];

  pointerCheck();
  for(int i = 0; i < nRows; i++){
    pA[i] = new(std::nothrow) double[nCols];
    pointerCheck(i);
  }

  for(int i = 0; i < nRows; i++)
    for(int j = 0; j < nCols; j++)
      pA[i][j] = pa[i][j];
}

Matrix::Matrix(const Matrix& A){
    nRows = A.nRows;
    nCols = A.nCols;
    pA = new(std::nothrow) double*[nRows];
    pointerCheck();

    for(int i = 0; i < nRows; i++){
      pA[i] = new(std::nothrow) double[nCols];
      pointerCheck(i);
    }

    for(int i = 0; i < nRows; i++)
      for(int j = 0; j < nCols; j++)
        pA[i][j] = A.pA[i][j];
}

Matrix::~Matrix(){
  if(pA != 0){
    for(int i = 0; i < nRows; i++)
      delete[] pA[i];
    delete[] pA;
  }
}

//Overloaded operators:

Matrix& Matrix::operator=(const Matrix& A){

  if(this == &A) //avoid self assignment
    return *this;
  //if dimensions match we can just overwrite; otherwise....
  if(nRows != A.nRows||nCols != A.nCols){

    if(pA != 0)//check first before releasing memory
      this->~Matrix();

    //define/redefine object's members
    nRows = A.nRows;
    nCols = A.nCols;
    pA = new(std::nothrow) double*[nRows];
```

```
    pointerCheck();

    for(int i = 0; i < nRows; i++){
      pA[i] = new(std::nothrow) double[nCols];
      pointerCheck(i);
    }
  }
  for(int i = 0; i < nRows; i++)
    for(int j = 0; j < nCols; j++)
      pA[i][j] = A.pA[i][j];

  return *this;
}

Matrix Matrix::operator+(const Matrix& B) const {
  if(nRows != B.nRows || nCols != B.nCols){
    cout << "Bad row and/or column dimensions in +!" << endl;
    exit(1);
  }
  Matrix temp(nRows, nCols);
  for(int i = 0; i < nRows; i++)
    for(int j = 0; j < nCols; j++)
      temp.pA[i][j] = this->pA[i][j] + B.pA[i][j];

  return temp;
}

Matrix& Matrix::operator+=(const Matrix& B){
  if(nRows != B.nRows || nCols != B.nCols){
    cout << "Bad row and/or column dimensions in +=!" << endl;
    exit(1);
  }
  for(int i = 0; i < nRows; i++)
    for(int j = 0; j < nCols; j++)
      pA[i][j] += B.pA[i][j];

  return *this;
}

Matrix Matrix::operator-(const Matrix& B) const {
  if(nRows != B.nRows || nCols != B.nCols){
    cout << "Bad row and/or column dimensions in -!" << endl;
    exit(1);
  }
  Matrix temp(nRows, nCols);
  for(int i = 0; i < nRows; i++)
    for(int j = 0; j < nCols; j++)
      temp.pA[i][j] = this->pA[i][j] - B.pA[i][j];

  return temp;
}

Matrix& Matrix::operator-=(const Matrix& B){
  if(nRows != B.nRows || nCols != B.nCols){
    cout << "Bad row and/or column dimensions in -=!" << endl;
    exit(1);
  }
  for(int i = 0; i < nRows; i++)
    for(int j = 0; j < nCols; j++)
      pA[i][j] -= B.pA[i][j];

  return *this;
}

Matrix Matrix::operator*(const Matrix& B) const {
  if(nCols != B.nRows){
    cout << "Bad row and column dimensions in *!" << endl;
    exit(1);
```

```
  }
  Matrix C(nRows, B.nCols);//matrix of all 0s
  for(int i = 0; i < nRows; i++)
    for(int j = 0; j < B.nCols; j++)
      for(int k = 0; k < B.nRows; k++)
        C.pA[i][j] += pA[i][k]*B.pA[k][j];

  return C;
}

Matrix Matrix::operator*(double b) const {

  Matrix C(nRows, nCols);//matrix of all 0s
  for(int i = 0; i < nRows; i++)
    for(int j = 0; j < nCols; j++)
      C.pA[i][j] = b*pA[i][j];

  return C;
}

Vector Matrix::operator*(const Vector& v) const {
  double temp;
  Vector c(nRows);//matrix of all 0s
  for(int i = 0; i < nRows; i++){
    temp = 0;
    for(int j = 0; j < nCols; j++)
      temp += pA[i][j]*v[j];

    c.pA[i]=temp;
  }

  return c;
}

//matrix operations

Matrix Matrix::trans() const {
  Matrix B(nCols, nRows, 0.);
  for(int i = 0; i < nCols; i++)
    for(int j = 0; j < nRows; j++)
      B.pA[i][j] = pA[j][i];
  return B;
}

//solution of a linear system

Vector Matrix::backward(const Vector& RHS) const {
  double temp;//temporary storage

  Vector b(nRows, 0.);//solution vector

  //initialize the recursion
  if(pA[nRows - 1][nRows - 1] != 0)
    b.pA[nRows - 1] = RHS.pA[nRows - 1]/pA[nRows - 1][nRows - 1];
  else{
    cout << "Singular system!" << endl;
    exit(1);
  }

  //now work through the remaining rows
  for(int i = (nRows - 2); i >= 0; i--){
    if(pA[i][i] != 0){
      temp = RHS.pA[i];
      for(int k = (i + 1); k < nRows; k++)
        temp -= b.pA[k]*pA[i][k];
      b.pA[i] = temp/pA[i][i];
    }
    else{
```

```
          cout << "Singular system!" << endl;
          exit(1);
      }
  }
  return b;
}

Vector Matrix::forward(const Vector& RHS) const {
  double temp;//temporary storage

  Vector b(nRows, 0.);//solution matrix

  //initialize the recursion
  if(pA[0][0] != 0)
    b.pA[0] = RHS.pA[0]/pA[0][0];
  else{
    cout << "Singular system!" << endl;
    exit(1);
  }

  //now work through the remaining rows
  for(int i = 1; i < nRows; i++){
    if(pA[i][i] != 0){
      temp = RHS.pA[i];
      for(int k = 0; k < i; k++)
        temp -= b.pA[k]*pA[i][k];
      b.pA[i] = temp/pA[i][i];
    }
    else{
      cout << "Singular system!" << endl;
      exit(1);
    }
  }
  return b;
}

Vector Matrix::gauss(const Vector& RHS) const {
  double multiplier = 0;
  Matrix G(nRows, nRows, pA);
  Vector h(nRows, RHS.pA);

  for(int j = 0; j < nCols; j++){//column to be swept

    for(int jj = j + 1; jj < nCols; jj++){//current operation column

      if(G[j][j] == 0){
        cout << "Oops! Division by 0!" << endl;
        exit(1);
      }

      for(int i = j + 1; i < nRows; i++){//work down rows
        multiplier = G[i][j]/G[j][j];
        G.pA[i][jj] = G.pA[i][jj] - multiplier*G.pA[j][jj];
      }
    }

    for(int i = j + 1; i < nRows; i++){//do the same to the RHS
      multiplier = G[i][j]/G[j][j];
      h.pA[i] = h.pA[i] - multiplier*h.pA[j];
    }
  }
  //now backsolve
  Vector b = G.backward(h);
  return b;
}

Vector Matrix::cholesky(const Vector& RHS) const {
  double temp = 0;
```

```
  Matrix G(nRows, nCols, 0.);
  for(int j = 0; j < nCols; j++){//proceed by columns
    if(pA[j][j] == 0){
      cout << "Singular system!" << endl;
      exit(1);
    }
    temp = pA[j][j];//starting value for diagonal element recursion

    if(j > 0)
      for(int k = 0; k < j; k++)
        temp -= G[j][k]*G[j][k];

    G.pA[j][j] = sqrt(temp);

    for(int i = (j + 1); i < nRows; i++){//now do the rest
      temp = pA[j][i];
      for(int k = 0; k < j; k++)
        temp -= G[i][k]*G[j][k];
      G.pA[i][j] = temp/G[j][j];
    }
  }
  cout << "Cholesky factor:" << endl;
  G.printMatrix();
  Vector h = G.forward(RHS);
  Vector b = G.trans().backward(h);
  return b;
}

Vector Matrix::bandBack(const Vector& RHS, int bWidth) const {
  double temp;//temporary storage
  int up;

  Vector b(nRows, 0.);//solution vector

  //initialize the recursion
  if(pA[nRows - 1][nRows - 1] != 0)
      b.pA[nRows - 1] = RHS.pA[nRows - 1]/pA[nRows - 1][nRows - 1];
  else{
    cout << "Singular system!" << endl;
    exit(1);
  }

  //now work through the remaining rows
  for(int i = (nRows - 2); i >= 0; i--){
    if(pA[i][i] != 0){
      temp = RHS.pA[i];
      up = std::min(nCols, i + bWidth + 1);
      for(int k = (i + 1); k < up; k++)
        temp -= b.pA[k]*pA[i][k];
      b.pA[i] = temp/pA[i][i];
    }
    else{
      cout << "Singular system!" << endl;
      exit(1);
    }
  }
  return b;
}

Vector Matrix::bandFor(const Vector& RHS, int bWidth) const {
  double temp;//temporary storage
  int low;

  Vector b(nRows, 0.);//solution vector

  //initialize the recursion
  if(pA[0][0] != 0){
    b.pA[0] = RHS.pA[0]/pA[0][0];
```

```
    }
    else{
      cout << "Singular system!" << endl;
      exit(1);
    }

    //now work through the remaining rows
    for(int i = 1; i < nRows; i++){
      if(pA[i][i] != 0){
        temp = RHS.pA[i];
        low = std::max(0, i - bWidth);
        for(int k = low; k < i; k++)
          temp -= b.pA[k]*pA[i][k];
        b.pA[i] = temp/pA[i][i];
      }
      else{
        cout << "Singular system!" << endl;
        exit(1);
      }
    }
    return b;
}

Vector Matrix::bandChol(const Vector& RHS, Matrix& G,
          int bWidth) const {
    double temp = 0;
    int low, up;

    for(int j = 0; j < nCols; j++){
      if(pA[j][j] == 0){
        cout << "Singular system!" << endl;
        exit(1);
      }
      temp = pA[j][j];

      low = std::max(0, j - bWidth);

      for(int k = low; k < j; k++)
        temp -= G[j][k]*G[j][k];

      G.pA[j][j] = sqrt(temp);

      up = std::min(j + bWidth, nRows - 1);

      for(int i = (j + 1); i <= up; i++){
        temp = pA[j][i];
        for(int k = low; k < j; k++)
          temp -= G[j][k]*G[i][k];
        G.pA[i][j] = temp/G[j][j];
      }
    }
    Vector h = G.bandFor(RHS, bWidth);
    Vector b = G.trans().bandBack(h, bWidth);
    return b;
}

void Matrix::QR(Matrix& Q, Matrix&R) const {
    if(nRows < nCols){
      cout << "The number of rows is less than the number of columns!" << endl;
      exit(1);
    }

    //make work matrix copy of *this
    Matrix Work = Matrix(nRows, nCols, this->pA);
    Vector temp;
    Vector q;
    for(int i = 0; i < nCols; i++){
      q = Work.matToVec(i);
```

```
      R.pA[i][i] = sqrt(q.dotProd(q));
      q = q*(1/R.pA[i][i]);
      Q.vecToMat(i, q);
      temp = Work.trans()*q;
      for(int j = (i + 1); j < nCols; j++)
        R.pA[i][j] = temp[j];
      Work -= q*temp;//outer product update
  }
  cout << "The Q matrix" << endl;
  Q.printMatrix();
  cout << "The R matrix" << endl;
  R.printMatrix();
  cout << "The product QR" << endl;
  (Q*R).printMatrix();
}

//eigenvalues, eigenvectors and SVD

Vector Matrix::eigen(int Nvals, Vector& v, Matrix& U, int itMax,
        double delta) const {
  double change, temp;
  int niter;

  Matrix ACopy(nRows, nCols, pA);//work copy of A
  Vector vCopy;//work copy of v

  Vector lambda(Nvals, 0.);
  for(int i = 0; i < Nvals; i++){
    change = std::numeric_limits<double>::infinity();
    niter = 0;
    vCopy = v;
    lambda.pA[i] = 0;

    while(change > delta && niter < itMax){
      vCopy = ACopy*vCopy;
      vCopy = (1./sqrt(vCopy.dotProd(vCopy)))*vCopy;
      lambda.pA[i] = vCopy.dotProd(ACopy*vCopy);
      if(niter == 0)
        temp = lambda.pA[i];
      else if(temp != 0.){
        change = fabs(1. - lambda.pA[i]/temp);
        temp = lambda.pA[i];
      }
      niter++;
    }
    U.vecToMat(i, vCopy);
    v -= (vCopy.dotProd(vCopy))*vCopy;
    ACopy -= lambda.pA[i]*(vCopy*vCopy);
  }
  return lambda;
}

Vector Matrix::SVD(int Nvals, Vector& v, Matrix& U, Matrix& V,
        int itMax, double delta) const {
  Vector temp;
  Matrix ACopy(nRows, nCols, pA);

  if(nRows > nCols){//work with A^TA
    temp = (ACopy.trans()*ACopy).eigen(Nvals, v, V, itMax, delta);
    Matrix Temp(Nvals, Nvals, 0.);
    for(int i = 0; i < Nvals; i++)
      Temp.pA[i][i] = 1./sqrt(temp[i]);
    U = ACopy*V*Temp;
  }
  else{//work with AA^T
    temp = (ACopy*ACopy.trans()).eigen(Nvals, v, U, itMax, delta);
    Matrix Temp(Nvals, Nvals, 0.);
    for(int i = 0; i < Nvals; i++)
```

```
        Temp.pA[i][i] = 1./sqrt(temp[i]);
    V = ACopy.trans()*U*Temp;
  }
  Vector lambda(Nvals, 0.);
  for(int i = 0; i < Nvals; i++)
    lambda.pA[i] = sqrt(temp[i]);
  return lambda;
}

// Utilities:

void Matrix::printMatrix() const {
  for(int i = 0; i < nRows; i++){
    for(int j = 0; j < nCols; j++)
      cout << " " << pA[i][j] << "   ";
    cout << endl;
  }
}

Vector Matrix::matToVec(int j) const {
  Vector v(nRows, 0.);
  for(int i = 0; i < nRows; i++)
    v.pA[i] = pA[i][j];
  return v;
}

void Matrix::vecToMat(int j, const Vector& v){
  for(int i = 0; i < nRows; i++)
    pA[i][j] = v.pA[i];
}

void Matrix::pointerCheck() const {
  if(pA == 0){
    cout << "Memory allocation for pA failed" << endl;
    exit(1);
  }
}

void Matrix::pointerCheck(int i) const {
  if(pA[i] == 0){
    cout << "Memory allocation for pA[" << i << "] failed" << endl;
    exit(1);
  }
}

//Overloaded multiplication operators
Matrix operator*(const Vector& v1, const Vector& v2){
  Matrix B(v1.getnRows(), v2.getnRows());
  for(int i = 0; i < v1.getnRows(); i++)
    for(int j = 0; j < v2.getnRows(); j++)
      B.pA[i][j] = v1[i]*v2[j];
  return B;
}

Matrix operator*(double b, const Matrix& A){
  return A*b;
}

//pdBand methods

pdBand::pdBand(int nrows, int bw, double a)
  : Matrix(nrows, bw + 1, a)
{
  bW = bw;
}

pdBand::pdBand(int nrows, int bw, const double* const* pa)
  : Matrix(nrows, bw + 1, pa)
```

```
{
  bW = bw;
}

pdBand::pdBand(const pdBand& A)
  : Matrix(A)
{
  bW = A.bW;
}

pdBand& pdBand::operator=(const pdBand& A){

  if(this == &A)//avoid self assignment
    return *this;
  this->Matrix::operator=(A);
  bW = A.bW;

  return *this;
}

Matrix pdBand::operator+(const Matrix& B) const {
  Matrix A = this->bandToFull();
  return (A + B);
}

pdBand pdBand::operator+(const pdBand& B) const {
  pdBand temp(getnRows(), B.bW);
  temp.Matrix::operator=(this->Matrix::operator+(B));

  return temp;
}

Vector pdBand::backward(const Vector& RHS) const {
  double temp;//temporary storage
  int up;
  int nrows = this->getnRows();
  Vector b(nrows, 0.);//solution vector

  //initialize the recursion
  if(pA[nrows - 1][0] != 0)
    b.pA[nrows - 1] = RHS.pA[nrows - 1]/pA[nrows - 1][0];
  else{
    cout << "Singular system!" << endl;
    exit(1);
  }

  //now work through the remaining rows
  for(int i = (nrows - 2); i >= 0; i--){
    if(pA[i][0] != 0){
      temp = RHS.pA[i];
      up = std::min(nrows, bW + 1);
      for(int k = 1; k < up; k++)
        temp -= b.pA[i + k]*pA[i][k];
      b.pA[i] = temp/pA[i][0];
    }
    else{
      cout << "Singular system!" << endl;
      exit(1);
    }
  }
  return b;
}

Vector pdBand::forward(const Vector& RHS) const {
  double temp;//temporary storage
  int low;
  int nrows = this->getnRows();
  Vector b(nrows, 0.);//solution vector
```

```
  //initialize the recursion
  if(pA[0][0] != 0){
    b.pA[0] = RHS.pA[0]/pA[0][0];
  }
  else{
    cout << "Singular system!" << endl;
    exit(1);
  }

  //now work through the remaining rows
  for(int i = 1; i < nrows; i++){
    if(pA[i][0] != 0){
      temp = RHS.pA[i];
      low = std::max(0, i - bW);
      for(int k = low; k < i; k++)
        temp -= b.pA[k]*pA[k][i - k];
      b.pA[i] = temp/pA[i][0];
    }
    else{
      cout << "Singular system!" << endl;
      exit(1);
    }
  }
  return b;
}

Vector pdBand::cholesky(const Vector& RHS) const {
  double temp = 0;
  int low, up;
  int nrows = this->getnRows();
  pdBand G(nrows, bW);

  for(int i = 0; i < nrows; i++){
    if(pA[i][0] == 0){
      cout << "Singular system!" << endl;
      exit(1);
    }
    temp = pA[i][0];
    low = std::max(0, i - bW);

    for(int k = low; k < i; k++)
      temp -= G[k][i - k]*G[k][i - k];

    G.pA[i][0] = sqrt(temp);

    up = std::min(i + bW, nrows - 1);

    for(int j = (i + 1); j <= up; j++){
      temp = pA[i][j - i];
      low = std::max(0, j - bW);
      for(int k = low; k < i; k++)
        temp -= G[k][i - k]*G[k][j - k];
      G.pA[i][j - i] = temp/G[i][0];
    }
  }
  cout << "Cholesky factor in band storage" << endl;
  G.printMatrix();
  Vector h = G.forward(RHS);
  Vector b = G.backward(h);
  std::cout << "Solution vector" << endl;
  return b;
}

Matrix pdBand::bandToFull() const {
  double** pTemp = new(std::nothrow) double*[getnRows()];
  if(pTemp == 0){
    cout << "Memory allocation for pTemp failed" << endl;
    exit(1);
```

```
  }
  for(int i = 0; i < getnRows(); i++){
    pTemp[i] = new(std::nothrow) double[getnRows()];
    if(pTemp[i] == 0){
      cout << "Memory allocation for pTemp [" << i << "] failed" << endl;
      exit(1);
    }
    for(int j = i; j < getnRows(); j++){
      if(j <= std::min(getnRows(), i + bW))
        pTemp[i][j] = pA[i][j - i];
      else pTemp[i][j] = 0.;
    }
  }
  for(int i = 1; i < getnRows(); i ++)
    for(int j = 0; j < i; j++)
      pTemp[i][j] = pTemp[j][i];
  Matrix A(getnRows(), getnRows(), pTemp);
  for(int i = 0; i < getnRows(); i++)
    delete[] pTemp[i];
  delete[] pTemp;
  return A;
}
```

The class **Vector** header file is as follows.

Listing D.3 vector.h

```
//vector.h
#ifndef VECTOR_H
#define VECTOR_H

class Vector{
  // class members
  double* pA;
  int nRows;
  void pointerCheck() const;

 public:

  //constructors and destructor
  Vector(int nrows, double const* pa);//constructor
  Vector(int nrows = 0, double b = 0.);//default constructor
  Vector(const Vector& v);
  ~Vector();

  //overloaded operators
  Vector& operator=(const Vector& v);
  Vector operator+(const Vector& v) const;
  Vector& operator+=(const Vector& v);
  Vector operator-(const Vector& v) const;
  Vector& operator-=(const Vector& v);
  Vector operator*(double b) const;
  double operator[](int i) const {return pA[i];}

  //utilities and accessors
  double dotProd(const Vector& v) const;
  void printVec() const;
  int getnRows() const {return nRows;}

  //friends
  friend class Matrix;
  friend class pdBand;
};

Vector operator*(double b, const Vector& v);
#endif
```

The method definitions are then provided in the listing below.

Listing D.4 *vector.cpp*

```cpp
//vector.cpp
#include <iostream>
#include <cstdlib>
#include <new>
#include "vector.h"

using std::cout; using std::endl;

Vector::Vector(int nrows, double const* pa){
  nRows = nrows;
  pA = new(std::nothrow) double[nRows];
  pointerCheck();

  for(int i = 0; i < nRows; i++)
    pA[i] = pa[i];
}

Vector::Vector(int nrows, double b){
  nRows = nrows;
  if(nRows == 0){
    pA = 0;
    return;
  }
  pA = new(std::nothrow) double[nRows];
  pointerCheck();

  for(int i = 0; i < nRows; i++)
    pA[i] = b;
}

Vector::Vector(const Vector& v){
  nRows = v.nRows;
  pA = new(std::nothrow) double[nRows];
  pointerCheck();

  for(int i = 0; i < nRows; i++)
    pA[i] = v.pA[i];
}

Vector::~Vector(){
  delete[] pA;
}

double Vector::dotProd(const Vector& v) const {
  double dot = 0;
  for(int i = 0; i < nRows; i++)
    dot += v.pA[i]*pA[i];
  return dot;
}

Vector& Vector::operator=(const Vector& v){

  if(this == &v)//avoid self assignment
    return *this;

  //define/redefine object's members
  if(nRows != v.nRows){

    if(pA != 0)
      delete[] pA;

    nRows = v.nRows;
    pA = new(std::nothrow) double[nRows];
    pointerCheck();
```

```
  }
  for(int i = 0; i < nRows; i++)
    pA[i] = v.pA[i];

  return *this;
}

Vector Vector::operator+(const Vector& v) const {
  if(nRows != v.nRows){
    cout << "Bad row dimensions in +!" << endl;
    exit(1);
  }
  Vector temp(nRows);
  for(int i = 0; i < nRows; i++)
    temp.pA[i]=this->pA[i] + v.pA[i];

  return temp;
}

Vector& Vector::operator+=(const Vector& v){
  if(nRows != v.nRows){
    cout << "Bad row dimensions in +=!" << endl;
    exit(1);
  }
  for(int i = 0; i < nRows; i++)
    pA[i] += v.pA[i];

  return *this;
}

Vector Vector::operator-(const Vector& v) const {
  if(nRows != v.nRows){
    cout << "Bad row dimensions in -!" << endl;
    exit(1);
  }
  Vector temp(nRows);
  for(int i = 0; i < nRows; i++)
    temp.pA[i] = this->pA[i] - v.pA[i];

  return temp;
}

Vector& Vector::operator-=(const Vector& v){
  if(nRows != v.nRows){
    cout << "Bad row dimensions in -=!" << endl;
    exit(1);
  }
  for(int i = 0; i < nRows; i++)
    pA[i] -= v.pA[i];

  return *this;
}

Vector Vector::operator*(double b) const {
  Vector temp(nRows);
  for(int i = 0; i < nRows; i++)
    temp.pA[i] = b*pA[i];

  return temp;
}

void Vector::pointerCheck() const {
  if(pA == 0){
    cout << "Memory allocation for pA failed" << endl;
    exit(1);
  }
}
```

```
void Vector::printVec() const {
  for(int i = 0; i < nRows; i++)
    cout << pA[i] << " " << endl;
}

Vector operator*(double b, const Vector& v){
  Vector B = v;
  return B*b;
}
```

APPENDIX E

The ranGen class

The "complete" version of the class ranGen that was used in Chapter 8 and elsewhere is given in this appendix. We begin with the header file below.

Listing E.1 ranGen.h

```
//ranGen.h
#ifndef RANGEN_H
#define RANGEN_H

enum unifType {FM2, WH};
enum normType {BM, C};

class ranGen{

  unifType T;
  normType NT;
  unsigned long seed1, seed2, seed3;

  double gCauchy(double x) const;
  double GinvCauchy(double u) const;
  double fNorm(double x) const;

 public:

  ranGen(unifType t = WH, normType nt = BM) : T(t), NT(nt) {}
  ~ranGen(){};

  void ranUnif(int n, double* pu);
  void ranNorm(int n, double* pNorm, double mu = 0,
               double sig = 1);

  void setSeed(unsigned long s1);
};
#endif
```

There are two types of uniform random number generators: the Fishman-Moore II (FM2) generator and the Wichmann-Hill (WH) generator. These are packaged together in the method ranUnif. Both the Cauchy accept-reject (C) and the Box-Muller (BM) methods are provided as options for generating random normals in the method ranNorm. Selection between the different generation algorithms is accomplished by specification of one of the symbolic constants FM2, WH, C and BM that are defined using an enum statement. The default choices of WH and BM for the uniform and normal generator types are set in the class constructor. The ranNorm method allows for specification of the mean and standard deviation of the distribution with default values that produce standard normals.

The other class members are three unsigned long seeds for the generators. These values are set by the user with the public setSeed method that takes only one unsigned long argument. It then uses an integer form of the FM2 generator to set the two additional seeds for the WH generator.

The method definitions are given in the next listing.

Listing E.2 ranGen.cpp

```cpp
//ranGen.cpp
#include "ranGen.h"
#include <iostream>
#include <cmath>

//anonymous namespace
namespace {
  double pi = 2.*acos(0.);
  double sqrtwopi = sqrt(2.*pi);
  unsigned long a1 = 171, m1 = 30269, a2 = 172, m2 = 30307,
    a3 = 170, m3 = 30323, m4 = 2147483647, a4 = 630360016;
}

void ranGen::ranUnif(int n, double* pu) {
  if(T == WH){
    double temp = 0;
    for(int i=0; i< n; i++){
      seed1=(a1*seed1)%m1;
      seed2=(a2*seed2)%m2;
      seed3=(a3*seed3)%m3;
      temp = (((double)seed1)/((double)m1) + ((double)seed2)/((double)m2)
             + ((double)seed3)/((double)m3));
      if(temp > 2.) temp -= 2.;
      if(temp > 1.) temp -= 1.;
      pu[i] = temp;
    }
  }
  else{
    for(int i=0; i< n; i++){
      seed1=(a4*seed1)%m4;
      pu[i]=((double)seed1)/((double)m4);
    }
  }
}

void ranGen::ranNorm(int n, double* pNorm, double mu,
                     double sig) {
  double* pu = new double(2);
  int nFill = 0, iAccept = 0, nReject = 0;

  if(NT == BM){
    double spare = 0, mult = 0, Rsqr = 0;

    while(nFill < n){
      if(iAccept == 1){//if we have one left, use it now
        pNorm[nFill] = spare;
        iAccept = 0;//next time we need to generate another pair

        nFill += 1;
      }
      else{
        while(iAccept == 0){

          //generate two uniforms on [-1, 1]
          ranUnif(2,pu);
          pu[0] = 2*pu[0]- 1;
          pu[1] = 2*pu[1] - 1;

          Rsqr = pu[0]*pu[0] + pu[1]*pu[1];
          if(Rsqr < 1 && Rsqr != 0){//accept the result
            mult = sqrt(-2.*log(Rsqr)/Rsqr);
            pNorm[nFill] = mu + sig*pu[0]*mult;
            spare = mu + sig*pu[1]*mult;//save this one for next time
            nFill += 1;
```

```
                  iAccept = 1;//now we have one left over
                }
                else nReject += 2;
            }
          }
        }
      }
    }

    else{
      double y;
      double c = sqrt(2*pi/exp(1.));

      while(nFill < n){
        iAccept = 0;
        while(iAccept == 0){
          ranUnif(2, pu);
          y = GinvCauchy(pu[1]);
          if(c*gCauchy(y)*pu[0] <= fNorm(y)){//accept the result
            pNorm[nFill] = mu + sig*y;
            nFill += 1;//augment the number of samples
            iAccept = 1;//end the interior while loop
          }
          else nReject += 1;
        }
      }
    }
}

void ranGen::setSeed(unsigned long s1){
  if(T == WH){
    seed1 = s1;
    seed2=(a4*seed1)%m4;
    seed3=(a4*seed2)%m4;
  }
  else
    seed1 = s1;
}

double ranGen::gCauchy(double x) const {
  return 1./((1. + x*x)*pi);
}

double ranGen::GinvCauchy(double u) const {
  return tan(pi*(u - .5));
}

double ranGen::fNorm(double x) const {
  return exp(-.5*x*x)/sqrtwopi;
}
```

There are several global constants that are needed in various functions in the class. These could simply have been defined at the outset outside of any function allowing them to have global scope. However, name clashes can occur when the class is used in other programs. This is particularly true for the constant **pi**. To avoid this the constants can be wrapped in an anonymous namespace which is what has been done here. Then, the constants exist only inside this particular file (i.e., ranGen.cpp) and can only be accessed by the functions in that file.

References

C. Aggarwal, editor. *Data Streams Models and Algorithms*, volume 31 of *Advances in Database Systems*. Springer, New York, NY, 2007.

J. Ahrens and U. Dieter. Extensions of Forsythe's method for random sampling from the normal distribution. *Mathematics of Computation*, 27:927–937, 1973.

G. Andrews. *Number Theory*. Dover, New York, NY, 1971.

S. Arora and B. Barak. *Computational Complexity: A Modern Approach*. Cambridge University Press, 2009.

M. Austern. Technical report on C++ standard library extensions. Technical Report ISO/IEC TR 19768, June 2005.

P. Becker. *The C++ Standard Library Extensions: A Tutorial and Reference*. Addison-Wesley Professional, 2006.

C. Bélisle. Convergence theorems for a class of simulated annealing algorithms on $\mathcal{R}^{\lceil}$. *Journal of Applied Probability*, 29:885–895, 1992.

A. Ben-Israel and T. Greville. *Generalized Inverses Theory and Applications, Second Edition*. Springer, New York, NY, 2003.

Y. Benjamini and Y. Hochberg. Controlling the false discovery rate: a powerful and practical approach to multiple testing. *Journal of the Royal Statistical Society, Series B*, 57:289–300, 1995.

Y. Benjamini and D. Yekutieli. The control of the false discovery rate in multiple testing under dependency. *Annals of Statistics*, 29:1165–1188, 2001.

A. Björck. *Numerical Solutions for Least Squares Problems*. SIAM, Philadelphia, PA, 1996.

G. Booch, R. Maksimchuk, M. Engel, B. Young, J. Conallen, and K. Houston. *Object-Oriented Analysis and Design with Applications: 3rd Edition*. Addison-Wesley, New York, NY, 2007.

G. Box and M. Muller. A note on the generation of random normal deviates. *Annals of Mathematical Statistics*, 29:610–611, 1958.

A. Broder and M. Mitzenmacher. Network applications of Bloom filters: a survey. *Internet Mathematics*, 1:485–509, 2005.

T. Budd. *Classic Data Structures in C++*. Addison-Wesley, New York, NY, 1994.

R. Byrd, P. Lu, J. Nocedal, and C. Zhu. A limited memory algorithm for bound constrained optimization. *SIAM Journal of Scientific Computing*, 16:1190–1208, 1995.

D. Cameron, J. Elliott, M. Loy, E. Raymond, and B. Rosenblatt. *Learning GNU Emacs: Third Edition*. O'Reilly, Sebastopol, CA, 2004.

L. Carter and M. Wegman. Universal classes of hash functions. *Journal of Computer and System Sciences*, 18:143–154, 1979.

J. Chambers. *Programming with Data: A Guide to the S Language*. Springer, New York, NY, 1998.

T. Chan and J. Lewis. Computing standard deviations: accuracy. *Communications of the Association for Computing Machinery*, 22:526–531, 1979.

T. Chan, G. Golub, and R. LeVeque. Algorithms for computing the sample variance: analysis and recommendations. *American Statistician*, 37:242–247, 1983.

R. Chandra, L. Dagum, D. Kohr, D. Maydan, J. McDonald, and R. Menon. *Parallel Programming in OpenMP*. Morgan Kaufmann, San Francisco, CA, 2001.

R. Cheng and G. Feast. Some simple gamma variate generators. *Applied Statistics*, 28:290–295, 1979.

P. Coddington. Random number generators for parallel computers. *NHSE Review*, 1, 1996.

J. Conover. *Practical Nonparametric Statistics, Third Edition*. Wiley, New York, NY, 1998.

S. Conte and C. de Boor. *Elementary Numerical Analysis, Second Edition*. McGraw Hill, New York, NY, 1972.

T. Cormen, C. Leiserson, R. Rivest, and C. Stein. *Introduction to Algorithms, Second Edition*. McGraw Hill, New York, NY, 2003.

G. Cormode and S. Muthukrishnan. An improved data stream summary: the count-min sketch and its applications. *Journal of Algorithms*, 55:58–75, 2005.

A. Davidson and D. Hinkley. *Bootstrap Methods and Their Application*. Cambridge University Press, Cambridge, UK, 1997.

L. Devroye. *Non-Uniform Random Variate Generation*. Springer Verlag, New York, NY, 1986.

A. Drozdek. *Data Structures and Algorithms in C++:Third Edition*. Thomson Course Technology, Mountain View, CA, 2005.

B. Eckel. *Thinking in C++. Second Edition. Volume One: Introduction to Standard C++*. Prentice Hall, Upper Saddle River, NJ, 2000.

R. Eubank. *Nonparametric Regression and Spline Smoothing: Second Edition*. Dekker, New York, NY, 1999.

R. Eubank and J. Kim. A simple second order smoother. *Journal of Statistical Computation and Simulation*, 61:271–285, 1998.

M. Euster, J. Knaus, C. Porzelius, M. Schmidberger, and E. Vicedo. Hands-on tutorial for parallel computing with R. *Computational Statistics*, 26:219–239, 2011.

G. Fishman and R. Moore. A statistical evaluation of multiplicative congruential generators with period $2^{31} - 1$. *Journal of the American Statistical Association*, 77:129–136, 1982.

G. Fishman and R. Moore. An exhaustive analysis of multiplicative congruential random number generators with modulus $2^{31} - 1$. *SIAM Journal of Scientific and Statistical Computing*, 7:24–45, 1986.

B. Flaming. *Practical Algorithms in C++*. Wiley, New York, NY, 1995.

B. Flowers. *An Introduction to Numerical Methods in C++*. Oxford University Press, Oxford, UK, 2006.

W. Ford and W. Topp. *Data Structures with C++ Using STL: Second Edition*. Prentice Hall, Upper Saddle River, NJ, 2002.

J. Gentle. *Random Number Generation and Monte Carlo Methods*. Springer, New York, NY, 2003.

R. Gentleman. *R Programming for Bioinformatics*. CRC Press, 2009.

D. Goldberg. What every computer scientist should know about floating-point arithmetic. *ACM Computing Surveys*, 23:5–48, 1991.

G. Golub and C. Van Loan. *Matrix Computations, Third Edition*. Johns Hopkins, Baltimore, ME, 1996.

B. Gough. *An Introduction to GCC*. Network Theory Limited, Bristol, UK, 2005.

H. Joe. Tests of uniformity for sets of lotto numbers. *Statistics and Probability letters*, 16:181–188, 1993.

N. Josuttis. *The C++ Standard Library: A Tutorial and Reference*. Addison Wesley, Indianapolis, IN, 1999.

T. Kaneko and B. Liu. On local round-off error in floating-point arithmetic. *Journal of the Association for Computing Machinery*, 20:391–398, 1973.

A. Karl, R. Eubank, and D. Young. Parallel random number generation in C++ and R using RngStream. 2011.

J. Kiefer. Sequential minimax search for a maximum. *Proceedings of the American Mathematical Society*, 4:502–506, 1953.

A. Kinderman and J. Ramage. Computer generation of normal random variables. *Journal of the American Statistical Association*, 71:893–896, 1976.

D. Knuth. *The Art of Computer Programming, Volume 2: Seminumerical Algorithms*. Addison-Wesley, Upper Saddle River, NJ, 1998a.

D. Knuth. *The Art of Computer Programming, Volume 3: Sorting and Searching*. Addison-Wesley, Upper Saddle River, NJ, 1998b.

S. Kochan and P. Wood. *Unix Shell Programming, Third Edition*. SAMS, Indianapolis, IA, 2003.

I. Koren. *Computer Arithmetic Algorithms: Second Edition*. A. K. Peters, Natick, MA, 2002.

P. L'Ecuyer. Random numbers for simulation. *Communications of the ACM*, 33:85–98, 1990.

P. L'Ecuyer. Combined multiple recursive random number generators. *Operations Research*, 44:816–822, 1996.

P. L'Ecuyer. Good parameters and implementation for combined multiple recursive random number generators. *Operations Research*, 47:159–164, 1999.

P. L'Ecuyer and R. Simard. Testu01: A C library for empirical testing of random number generators. *ACM Transactions on Mathematical Software*, 33:Article 22, 2007.

P. L'Ecuyer and S. Tezuka. Structural properties for two classes of combined random number generators. *Mathematics of Computation*, 57:735–746, 1991.

P. L'Ecuyer, R. Simard, J. Chen, and W. D. Kelton. An object-oriented random-number package with many long streams and substreams. Technical report, University of Montreal, 2001.

P. L'Ecuyer, R. Simard, J. Chen, and W. D. Kelton. An object-oriented random-number package with many long streams and substreams. *Operations Research*, 50:1073–1075, 2002.

G. Leech, P. Rayson, and A. Wilson. *Word Frequencies in Written and Spoken English: based on the British National Corpus*. Longman, London, 2001.

D. Lehmer. Mathematical methods in large-scale computing units. In *Proceedings of the 2nd Symposium on Large-Scale Digital Calculating Machinery*, pages 141–146, Cambridge, MA, 1949.

D. Lehmer. Random number generation on the BRL high-speed computing machines by M. L. Juncosa. *Mathematical Reviews*, 15:559, 1954.

T. Lewis and W. Payne. Generalized feedback shift register pseudorandom number algorithm. *Journal of the Association for Computing Machinery*, 20:456–468, 1973.

M. Lu. *Arithmetic and Logic in Computer Systems*. Wiley, New York, NY, 2004.

D. Malik. *C++ Programming: Program Design Including Data Structures, Third Edition*. Thomson Course Technology, Mountain View, CA, 2007.

R. Mansharamani. An overview of discrete event simulation methodologies and implementation. *Sādhanā*, 22:611–627, 1997.

G. Marsaglia. Random numbers fall mainly in the planes. *Proceedings of the National Academy of Sciences of the United States of America*, 61:25–28, 1968.

G. Marsaglia. The structure of linear congruential sequences. In *S. K. Zaremba, Editor, Applications of Number Theory to Numerical Analysis*, pages 248–285. Academic Press, 1972.

G. Marsaglia. Random number generators. *Journal of Modern Applied Statistical Methods*, 2:2–13, 2003.

G. Marsaglia and W. Tsang. Some difficult-to-pass tests for randomness. *Journal of Statistical Software*, 7, 2002.

G. Marsaglia and A. Zaman. Monkey tests for random number generators. *Computers and Mathematics with Applications*, 26:1–10, 1993.

M. Mascagni and A. Srinivasan. Algorithm 806: Sprng: A scalable library for pseudorandom number generation. *ACM Transactions on Mathematical Software*, 26:436–461, 2000.

M. Matsumoto and Y. Kurita. Twisted gfsr generators. *ACM Transactions on Modeling and Computer Simulation*, 2:179–194, 1992.

M. Matsumoto and N. Takuji. Mersenne twister: a 623-dimensionality equidistribution uniform pseudo-random number generator. *ACM Transactions on Modeling and Computer Simulation*, 8:3–30, 1998.

J. Monahan. *Numerical Methods of Statistics*. Cambridge University Press, Cambridge, UK, 2001.

P. Murrell. *R Graphics*. Chapman and Hall, New York, NY, 2006.

S. Muthukrishnan. Data streams: algorithms and applications. *Foundations and Trends in Theoretical Computer Science*, 1:1–129, 2005.

H. Niederreiter. *Random Number Generation and Quasi-Monte Carlo Methods*. SIAM, Philadelphia, PA, 1992.

H. Niederreiter and R. Lidl. *Introduction to Finite Fields and Their Applications*. Cambridge University Press, New York, NY, 1986.

R. Pagh and F. Rodler. Cuckoo hashing. *Journal of Algorithms*, 51:122–144, 2004.

C. Papadimitriou. *Computational Complexity*. Addison-Wesley, 1994.

S. Park and K. Miller. Random number generators: good ones are hard to find. *Communications of the Association of Computing Machinery*, 31:1192–1201, 1988.

W. Payne, J. Rabung, and T. Bogyo. Coding the Lehmer pseudo random number generator. *Communications of the Association of Computing Machinery*, 12:85–86, 1969.

S. Prata. *C++ Primer Plus, Fifth Edition*. SAMS, Indianapolis, IA, 2005.

W. Press, S. Teukolsky, W. Vetterling, and B. Flannery. *Numerical Recipes in C++: The Art of Scientific Computing, Second Edition.* Cambridge University Press, Cambridge, UK, 2005.

W. Pugh. Skip lists: a probabilistic alternative to balanced trees. *Communications of the ACM,* 33:668–676, 1990.

M. Quinn. *Parallel Programming in C with MPI and OpenMP.* McGraw Hill, New York, NY, 2003.

A. Robbins, L. Lamb, and E. Hannah. *Learning the vi and Vim Editors: Seventh Edition.* O'Reilly Media, Inc., Cambridge, UK, 2008.

C. Robert and G. Casella. *Monte Carlo Statistical Methods, Second Edition.* Springer, New York, NY, 2004.

S. Ross. *Simulations: Fourth Edition.* Academic Press, New York, NY, 2006.

A. Rukhin. Testing randomness: a suite of statistical procedures. *Theory of Probability and Its Applications,* 45:111–132, 2001.

M. Saito and M. Matsumoto. SIMD-oriented fast Mersenne Twister: a 128-bit pseudorandom number generator. In *Monte Carlo and Quasi-Monte Carlo Methods,* pages 607 – 622. Springer, 2006.

M. Schmidberger, M. Morgan, D. Eddelbuettel, H. Yu, and Tierney. L. State of the art in parallel computing with R. *Journal of Statistical Software,* 31, 2009.

R. Sedgewick. *Algorithms in C++, Parts 1–4.* Addison Wesley, Upper Saddle River, NJ, 3rd edition, 1998.

R. Shumway and D. Stoffer. *Time Series Analysis and Its Applications: With R Examples.* Springer, New York, NY, 2010.

B. Silverman. *Density Estimation.* Chapman Hall, London, UK, 1986.

D. Sleator and R. Tarjan. Self-adjusting binary search trees. *Journal of the ACM,* 32:652–686, 1985.

J. Spall. *Introduction to Random Search and Optimization.* Wiley, New York, NY, 2003.

P. Spector. *Data Manipulation with R.* Springer, New York, NY, 2008.

R. Stallman, R. McGrath, and P. Smith. *GNU Make: A Program for Directed Compilation.* GNU Press, Boston, MA, 2004.

P. Sterbenz. *Floating Point Computation.* Prentice-Hall, Englewood Cliffs, NJ, 1974.

D. Stevenson. A proposed standard for binary floating-point arithmetic. *IEEE Computer,* 14:51–62, 1981.

B. Stroustrup. *The C++ Programming Language, Third Edition.* Addison-Wesley, Upper Saddle River, NJ, 1997.

R. Tausworthe. Random numbers generated by linear recurrence modulo two. *Mathematics of Computation,* 19:201–209, 1965.

D. Vandervoorde and N. Josuttis. *C++ Templates: The Complete Guide.* Addison Wesley, New York, NY, 2003.

W. Venerables and B. Ripley. *S Programming.* Springer, New York, NY, 2000.

J. Verzani. *Using R for Introductory Statistics.* Chapman & Hall, New York, NY, 2005.

G. Wahba. *Spline Models for Observational Data.* SIAM, Philadelphia, PA, 1990.

M. Wegman and L. Carter. New hash functions and their use in authentication and set equality. *Journal of Computer and System Sciences,* 22:265–279, 1981.

D. West. Updating mean and variance estimates: an improved method. *Communications of the Association for Computing Machinery,* 22:532–535, 1979.

A. Wichmann and J. Hill. Algorithm AS183: an efficient and portable pseudo-random number generator. *Applied Statistics,* 31:188–190, 1982.

J. Wilkinson. *rounding Error in Algebraic Processes.* Her Majesty's Stationery Office, London, UK, 1063.

E. Youngs and E. Cramer. Some results relevant to choice of sum-of-product algorithms. *Technometrics,* 13:657–665, 1971.

C. Zhu, R. Byrd, P. Lu, and J. Nocedal. Algorithm 778: L-BFGS-B: Fortran subroutines for large-scale bound-constrained optimization. *ACM Transactions on Mathematical Software,* 23:550–560, 1997.

N. Zierler. Primitive trinomials whose degree is a Mersenne prime. *Information and Control,* 15:67–69, 1969.

N. Zierler and J. Brillhart. On primitive trinomials (mod 2). *Information and Control*, 13:541–554, 1968.

N. Zierler and J. Brillhart. On primitive trinomials (mod 2), ii. *Information and Control*, 14:566–569, 1969.

Index

Printed and bound by CPI Group (UK) Ltd, Croydon, CR0 4YY

17/10/2024

01775672-0007